Classroom Manual for

Automatic Transmissions and Transaxles

Second Edition

Classroom Manual for Automatic Transmissions and Transaxles

Second Edition

Jack Erjavec

Columbus State Community College
Columbus, Ohio

Delmar Publishers

an International Thomson Publishing company 

Albany • Bonn • Boston • Cincinnati • Detroit • London • Madrid
Melbourne • Mexico City • New York • Pacific Grove • Paris • San Francisco
Singapore • Tokyo • Toronto • Washington

NOTICE TO THE READER

Publisher does not warrant or guarantee any of the products described herein or perform any independent analysis in connection with any of the product information contained herein. Publisher does not assume, and expressly disclaims, any obligation to obtain and include information other than that provided to it by the manufacturer.

The reader is expressly warned to consider and adopt all safety precautions that might be indicated by the activities described herein and to avoid all potential hazards. By following the instructions contained herein, the reader willingly assumes all risks in connection with such instructions.

The publisher makes no representations or warranties of any kind, including but not limited to, the warranties of fitness for particular purpose or merchantability, nor are any such representations implied with respect to the material set forth herein, and the publisher takes no responsibility with respect to such material. The publisher shall not be liable for any special, consequential, or exemplary damages resulting, in whole or in part, from the readers' use of, or reliance upon, this material.

Illustration by: David Kimble
Cover Design: Michele Canfield

DELMAR STAFF
Publisher: Alar Elken
Acquisitions Editor: Vernon R. Anthony
Developmental Editor: Catherine A. Wein
Project Editor: Megeen Mulholland
Production Coordinator: Karen Smith
Art and Design Coordinator: Michele Canfield
Editorial Assistant: Betsy Hough

Printed in the United States of America

For information, contact:

Delmar Publishers
3 Columbia Circle, Box 15015
Albany, New York 12212-5015

International Thomson Publishing Europe
Berkshire House 168-173
High Holborn
London, WC1V7AA
England

Thomas Nelson Australia
102 Dodds Street
South Melbourne, 3205
Victoria, Australia

Nelson Canada
1120 Birchmont Road
Scarborough, Ontario
Canada M1K 5G4

International Thomson Editores
Campos Eliseos 385, Piso 7
Col Polanco
11560 Mexico DF Mexico

International Thomson Publishing GmbH
Königswinterer Strasse 418
53227 Bonn
Germany

International Thomson Publishing Asia
221 Henderson Road
#05-10 Henderson Building
Singapore 0315

International Thomson Publishing Japan
Hirakawacho Kyowa Building, 3F
2-2-1 Hirakawacho
Chiyoda-ku, Tokyo 102
Japan

1 2 3 4 5 6 7 8 9 10 XXX 04 03 02 01 00 99 98

Library of Congress Cataloging-in-Publication Data
Erjavec, Jack.
Classroom manual and shop manual for automatic transmissions and transaxles / Jack Erjavec. -- 2nd ed.
p. cm. -- (Today's technician)
Includes bibliographical references and index.
Contents: 1. Classroom manual -- 2. Shop manual.
ISBN 0-8273-8637-0 (alk. paper)
1. Automobiles -- Transmission devices, Automatic -- Maintenance and repair. 2. Automobiles -- Transaxles -- Maintenance and repair.
I. Title. II. Title: Automatic transmissions and tranbsaxles. III. Series.
TL736.E73 1998
629.2'446'0288 -- DC21 98-7315
CIP

CONTENTS

PREFACE

Thanks to the support that the *Today's Technician* series has received from those who teach automotive technology, Delmar Publishers is able to live up to its promise to provide new editions every three years. We have listened to our critics and our fans and present this new revised edition. By revising our series every three years, we can and will respond to changes in the industry, changes in the certification process, and to the ever-changing needs of those who teach automotive technology.

The *Today's Technician series,* by Delmar Publishers, features textbooks that cover all mechanical and electrical systems of automobiles and light trucks. Principal titles correspond with the eight major areas of ASE (National Institute for Automotive Service Excellence) certification. Additional titles include remedial skills and theories common to all of the certification areas and advanced or specialized subject areas that reflect the latest technological trends.

Each title is divided into two manuals: a Classroom Manual and a Shop Manual. Dividing the material into two manuals provides the reader with the information needed to begin a successful career as an automotive technician without interrupting the learning process by mixing cognitive and performance-based learning objectives.

Each Classroom Manual contains the principles of operation for each system and subsystem. It also discusses the design variations used by different manufacturers. The Classroom Manual is organized to build upon basic facts and theories. The primary objective of this manual is to allow the reader to gain an understanding of how each system and subsystem operates. This understanding is necessary to diagnose the complex automobile systems.

The understanding acquired by using the Classroom Manual is required for competence in the skill areas covered in the Shop Manual. All of the high priority skills, as identified by ASE, are explained in the Shop Manual. The Shop Manual also includes step-by-step instructions for diagnostic and repair procedures. Photo Sequences are used to illustrate many of the common service procedures. Other common procedures are listed and are accompanied with fine-line drawings and photographs that allow the reader to visualize and conceptualize the finest details of the procedure. The Shop Manual also contains the reasons for performing the procedures, as well as when that particular service is appropriate.

The two manuals are designed to be used together and are arranged in corresponding chapters. Not only are the chapters in the manuals linked together, the contents of the chapters are also linked. Both manuals contain clear and thoughtfully selected illustrations. Many of the illustrations are original drawings or photos prepared for inclusion in this series. This means that the art is a vital part of each manual.

Highlights of this Edition—Classroom Manual

The text was updated throughout, to include the latest developments. Some of these new topics include OBD-II systems, lab scopes, elecronic transmission control systems, and continuously variable transmissions. As well as the latest electronically controlled automatic transmissions and transaxles.

Highlights of this Edition—Shop Manual

The text was updated throughout to include the latest developments. Some of these new topics include the use of lab scopes to test various components, the testing of computerized transmission systems, and the latest torque converter control circuits.

Located at the end of each chapter are two new features: Job Sheets and ASE Challenge Questions. The Job Sheets provide a format for students to perform some of the tasks covered in the chapter. In addition to walking a student through a procedure, step-by-step, these Job Sheets challenge the student by asking why or how something should be done. Thereby making the students think about what they are doing.

Speaking of challenging questions, each chapter ends with a group of questions that reflect the content of an ASE exam. These questions are not merely end-of-chapter questions, they represent the contents of an actual ASE test. These questions, of course, are in addition to the ASE style end-of-chapter questions that were in the first edition.

Jack Erjavec, Series Advisor

Classroom Manual

To stress the importance of safe work habits, the Classroom Manual dedicates one full chapter to safety. Included in this chapter are common safety practices, safety equipment, and safe handling of hazardous materials and wastes. This includes information on MSDS sheets and OSHA regulations. Other features of this manual include:

A Bit of History

This feature gives the student a sense of the evolution of the automobile. It not only contains nice-to-know information, but also should spark some interest in the subject matter.

Marginal Notes

New terms are pulled out and defined. Common trade jargon also appears in the margin and gives some of the common terms used for components. This allows the reader to speak and understand the language of the trade, especially when conversing with an experienced technician.

References to the Shop Manual

Reference to the appropriate page in the Shop Manual is given whenever necessary. Although the chapters of the two manuals are synchronized, material covered in other chapters of the Shop Manual may be fundamental to the topic discussed in the Classroom Manual.

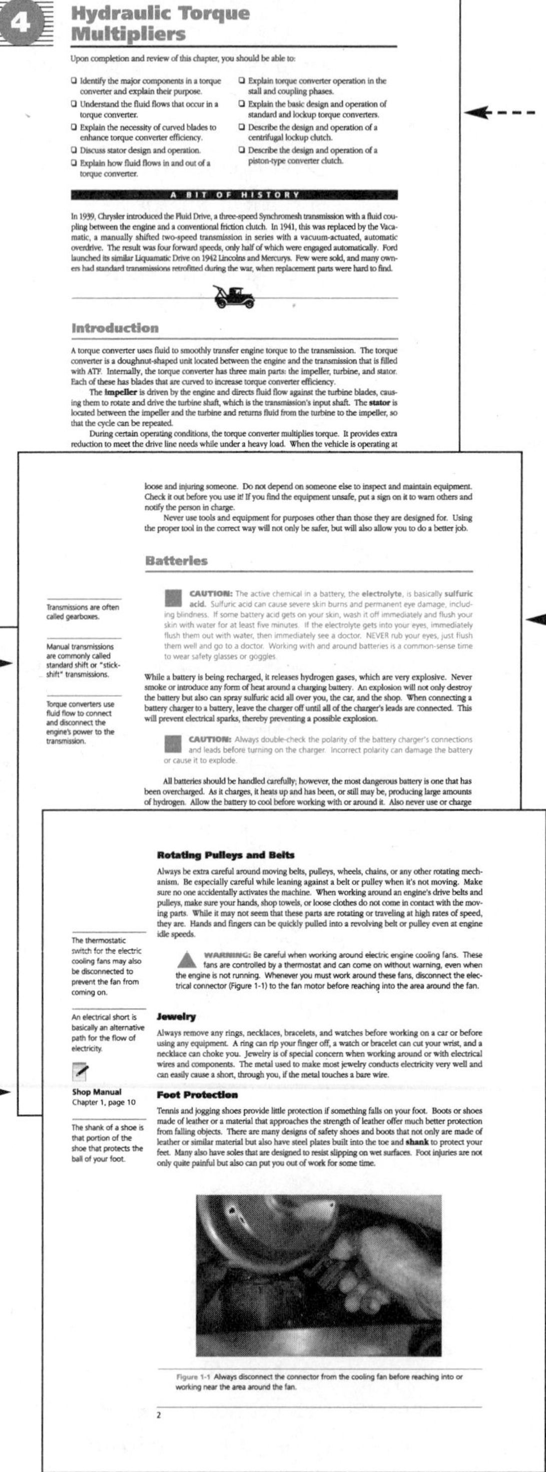

CHAPTER 4

Hydraulic Torque Multipliers

Upon completion and review of this chapter, you should be able to:

- Identify the major components in a torque converter and explain their purpose.
- Understand the fluid flows that occur in a torque converter.
- Explain the necessity of curved blades to enhance torque converter efficiency.
- Discuss stator design and operation.
- Explain how fluid flows in and out of a torque converter.
- Explain torque converter operation in the stall and coupling phases.
- Explain the basic design and operation of standard and lockup torque converters.
- Describe the design and operation of a centrifugal lockup clutch.
- Describe the design and operation of a piston-type converter clutch.

A BIT OF HISTORY

In 1939, Chrysler introduced the Fluid Drive, a three-speed Synchromesh transmission with a fluid coupling between the engine and a conventional friction clutch. In 1941, this was replaced by the Vacamatic, a manually shifted two-speed transmission in series with a vacuum-actuated, automatic overdrive. The result was four forward speeds, only half of which were engaged automatically. Ford launched its similar Liquamatic Drive on 1942 Lincolns and Mercurys. Few were sold, and many owners had standard transmissions retrofitted during the war, when replacement parts were hard to find.

Introduction

A torque converter uses fluid to smoothly transfer engine torque to the transmission. The torque converter is a doughnut-shaped unit located between the engine and the transmission that is filled with ATF. Internally, the torque converter has three main parts: the impeller, turbine, and stator. Each of these has blades that are curved to increase torque converter efficiency.

The **impeller** is driven by the engine and directs fluid flow against the turbine blades, causing them to rotate and drive the turbine shaft, which is the transmission's input shaft. The **stator** is located between the impeller and the turbine and returns fluid from the turbine to the impeller, so that the cycle can be repeated.

During certain operating conditions, the torque converter multiplies torque. It provides extra reduction to meet the drive line needs while under a heavy load. When the vehicle is operating at

---

loose and injuring someone. Do not depend on someone else to inspect and maintain equipment. Check it out before you use it! If you find the equipment unsafe, put a sign on it to warn others and notify the person in charge.

Never use tools and equipment for purposes other than those they are designed for. Using the proper tool in the correct way will not only be safer, but will also allow you to do a better job.

Batteries

Transmissions are often called gearboxes.

Manual transmissions are commonly called standard shift or "stick-shift" transmissions.

Torque converters use fluid flow to connect and disconnect the engine's power to the transmission.

CAUTION: The active chemical in a battery, the **electrolyte**, is basically **sulfuric acid.** Sulfuric acid can cause severe skin burns and permanent eye damage, including blindness. If some battery acid gets on your skin, wash it off immediately and flush your skin with water for at least five minutes. If the electrolyte gets into your eyes, immediately flush them out with water, then immediately see a doctor. NEVER rub your eyes, just flush them well and go to a doctor. Working with and around batteries is a common-sense time to wear safety glasses or goggles.

While a battery is being recharged, it releases hydrogen gases, which are very explosive. Never smoke or introduce any form of heat around a charging battery. An explosion will not only destroy the battery but also can spray sulfuric acid all over you, the car, and the shop. When connecting a battery charger to a battery, leave the charger off until all of the charger's leads are connected. This will prevent electrical sparks, thereby preventing a possible explosion.

CAUTION: Always double-check the polarity of the battery charger's connections and leads before turning on the charger. Incorrect polarity can damage the battery or cause it to explode.

All batteries should be handled carefully; however, the most dangerous battery is one that has been overcharged. As it charges, it heats up and has been, or still may be, producing large amounts of hydrogen. Allow the battery to cool before working with or around it. Also never use or charge

---

Rotating Pulleys and Belts

Always be extra careful around moving belts, pulleys, wheels, chains, or any other rotating mechanism. Be especially careful while leaning against a belt or pulley when it's not moving. Make sure no one accidentally activates the machine. When working around an engine's drive belts and pulleys, make sure your hands, shop towels, or loose clothes do not come in contact with the moving parts. While it may not seem that these parts are rotating or traveling at high rates of speed, they are. Hands and fingers can be quickly pulled into a revolving belt or pulley even at engine idle speeds.

The thermostatic switch for the electric cooling fans may also be disconnected to prevent the fan from coming on.

WARNING: Be careful when working around electric engine cooling fans. These fans are controlled by a thermostat and can come on without warning, even when the engine is not running. Whenever you must work around these fans, disconnect the electrical connector (Figure 1-1) to the fan motor before reaching into the area around the fan.

An electrical short is basically an alternative path for the flow of electricity.

Jewelry

Always remove any rings, necklaces, bracelets, and watches before working on a car or before using any equipment. A ring can rip your finger off, a watch or bracelet can cut your wrist, and a necklace can choke you. Jewelry is of special concern when working around or with electrical wires and components. The metal used to make most jewelry conducts electricity very well and can easily cause a short, through you, if the metal touches a bare wire.

Shop Manual
Chapter 1, page 10

Foot Protection

The shank of a shoe is that portion of the shoe that protects the ball of your foot.

Tennis and jogging shoes provide little protection if something falls on your foot. Boots or shoes made of leather or a material that approaches the strength of leather offer much better protection from falling objects. There are many designs of safety shoes and boots that not only are made of leather or similar material but also have steel plates built into the toe and **shank** to protect your feet. Many also have soles that are designed to resist slipping on wet surfaces. Foot injuries are not only quite painful but also can put you out of work for some time.

Figure 1-1 Always disconnect the connector from the cooling fan before reaching into or working near the area around the fan.

2

Cognitive Objectives

These objectives define the contents of the chapter and define what the student should have learned upon completion of the chapter.
Each topic is divided into small units to promote easier understanding and learning.

Cautions and Warnings

Throughout the text, cautions are given to alert the reader to potentially hazardous materials or unsafe conditions. Warnings are also given to advise the student of things that can go wrong if instructions are not followed or if a nonacceptable part or tool is used.

Terms to Know

A list of new terms appears next to the Summary. Definitions for these terms can be found in the Glossary at the end of the manual.

Summary

Terms to Know

Caustic
Corrosivity
Electrolyte
Ignitability
Reactivity
Shank
Sulfuric acid
Toxicity
Volatile

- You have an important role in creating an accident-free work environment. Your appearance and work habits will go a long way toward preventing accidents. Common sense and concern for others will result in fewer accidents.
- Accidents can be prevented by not having anything dangle near rotating equipment and parts.
- Jewelry should never be worn when working on a car. Any type of jewelry can get caught in rotating parts or can short out an electrical circuit.
- Safety glasses should be worn whenever you are working under a vehicle or when you are using machining equipment, grinding wheels, chemicals, compressed air, or fuels.
- Gasoline spills should be cleaned up immediately and gasoline should be stored only in approved containers.
- Dirty and oily rags should be stored in approved containers.
- All equipment should be inspected for safety hazards before being used.
- Tools and equipment should be used only for the purposes for which they were designed.
- Fire, heat, and sparks should be kept away from a battery. These could cause the battery to explode.
- Accidents can happen. When they do, you should respond immediately to prevent further injury or damage.
- All employees have the right to know what hazardous materials they are using to perform their jobs. Most of the needed information is contained on an MSDS.
- Hazardous wastes must be disposed of properly.

Review Questions

Short Answer Essays

1. Why are safe work habits considered to be an indication of a professional attitude?
2. How does the way you dress affect your personal safety?
3. When should you wear eye protection?
4. How should gasoline-soaked rags be stored?
5. List five basic rules for using tools and equipment safely.
6. What should you do if an accident occurs in the shop?
7. If a chemical gets into someone's eye, what two things should you do?
8. What sort of information is given on an MSDS?
9. How should hazardous wastes be disposed of?

Summaries

Each chapter concludes with summary statements that contain the important topics of the chapter. These are designed to help the reader review the contents.

Review Questions

Short answer essay, fill-in-the-blank, and multiple-choice type questions follow each chapter. These questions are designed to accurately assess the student's competence in the stated objectives at the beginning of the chapter.

Review Questions

Short Answer Essays

1. What are the primary purposes of a vehicle's drive train?
2. Why does torque increase when a smaller gear drives a larger gear?
3. How are gear ratios calculated?
4. Why are transmissions equipped with many different forward gear ratios?
5. What is the primary difference between a transaxle and a transmission?
6. Why are U-joints and CV joints used in the drive line?
7. What does a differential unit do to the torque it receives?
8. What is the purpose of the torque converter assembly? How does it work?
9. What kinds of gears are commonly used in today's automotive drive trains?
10. When are ball or roller bearings used?

Fill in the Blanks

1. The main components of the drive train are the ______________, ______________, and ______________ ______________.
2. The rotating or turning effort of the engine's crankshaft is called ______________.
3. Gears are used to apply torque to other rotating parts of the drive train and to ______________ torque.
4. Torque is calculated by multiplying the applied force by the ______________ from the center of the ______________ to the point where the force is exerted.
5. Torque is measured in ______________-______________ and ______________-______________.
6. Gear ratios are determined by dividing the number of teeth on the ______________ gear by the number of teeth on the ______________ gear.
7. Reverse gear is accomplished by adding a ______________ to a two-gear set.
8. The torque converter assembly is composed of a ______________, an ______________, and a ______________.
9. In FWD cars, the transmission and drive axle are located in a single assembly called a ______________.
10. In RWD cars, the drive axle is connected to the transmission through a ______________ ______________.

The J1930 List of Terminology

Located in the Appendix, this list serves as a reference to the acceptable industry terms as defined by SAE.

SAE J1930 Revised SEP95

TABLE 1—CROSS REFERENCE AND LOOK UP

Existing Usage	Acceptable Usage	Acceptable Acronized Usage
A/C (Air Conditioning)	Air Conditioning	A/C
A/C Cycling Switch	Air Conditioning Cycling Switch	A/C Cycling Switch
A/T (Automatic Transaxle)	Automatic Transaxle[1]	A/T[1]
A/T (Automatic Transmission)	Automatic Transmission[1]	A/T[1]
AAT (Ambient Air Temperature)	Ambient Air Temperature	AAT
AC (Air Conditioning)	Air Conditioning	A/C
ACC (Air Conditioning Clutch)	Air Conditioning Clutch	A/C Clutch
Accelerator	Accelerator Pedal	AP
Accelerator Pedal Position	Accelerator Pedal Position[1]	APP[1]
ACCS (Air Conditioning Cyclic Switch)	Air Conditioning Cycling Switch	A/C Cycling Switch
ACH (Air Cleaner Housing)	Air Cleaner Housing[1]	ACL Housing1
ACL (Air Cleaner)	Air Cleaner[1]	ACL[1]
ACL (Air Cleaner) Element	Air Cleaner Element[1]	ACL Element[1]
ACL (Air Cleaner) Housing	Air Cleaner Housing[1]	ACL Housing[1]
ACL (Air Cleaner) Housing Cover	Air Cleaner Housing Cover[1]	ACL Housing Cover[1]
ACS (Air Conditioning System)	Air Conditioning System	A/C System
ACT (Air Charge Temperature)	Intake Air Temperature[1]	IAT[1]
Adaptive Fuel Strategy	Fuel Trim[1]	FT[1]
AFC (Air Flow Control)	Mass Air Flow	MAF
AFC (Air Flow Control(	Volume Air Flow	VAF
AFS (Air Flow Sensor)	Mass Air Flow Sensor	MAF Sensor
AFS (Air Flow Sensor)	Volume Air Flow Sensor	VAF Sensor
After Cooler	Charge Air Cooler[1]	CAC[1]
AI (Air Injection)	Secondary Air Injection[1]	AIR[1]
AIP (Air Injection Pump)	Secondary Air Injection Pump[1]	AIR Pump[1]
AIR (Air Injection Reactor)	Pulsed Secondary Air Injection[1]	PAIR[1]
AIR (Air Injection Reactor)	Secondary Air Injection[1]	AIR[1]
AIRB (Secondary Air Injection Bypass)	Secondary Air Injection Bypass[1]	AIR Bypass[1]
AIRD (Secondary Air Injection Diverter)	Secondary Air Injection Diverter[1]	AIR Diverter[1]
Air Cleaner	Air Cleaner[1]	ACL[1]
Air Cleaner Element	Air Cleaner Element[1]	ACL Element[1]
Air Cleaner Housing	Air Cleaner Housing[1]	ACL Housing[1]
Air Cleaner Housing Cover	Air Cleaner Housing Cover[1]	ACL Housing Cover[1]
Air Conditioning	Air Conditioning	A/C
Air Conditioning Sensor	Air Conditioning Sensor	A/C Sensor
Air Control Valve	Secondary Air Injection Control Valve[1]	AIR Control Valve[1]
Air Flow Meter	Mass Air Flow Sensor[1]	MAF Sensor[1]
Air Flow Meter	Volume Air Flow Sensor[1]	VAF Sensor[1]
Air Intake System	Intake Air System[1]	IA System[1]
Air Flow Sensor	Mass Air Flow Sensor[1]	MAF Sensor[1]
Air Management 1	Secondary Air Injection Bypass[1]	AIR Bypass[1]
Air Management 2	Secondary Air Injection Diverter[1]	AIR Diverter[1]
Air Temperature Sensor	Intake Air Temperature Sensor[1]	IAT Sensor[1]
Air Valve	Idle Air Control Valve[1]	IAC Valve[1]
AIV (Air Injection Valve)	Pulsed Secondary Air Injection[1]	PAIR[1]
ALCL (Assembly Line Communication Link)	Data Link Connector[1]	DLC[1]
Alcohol Concentration Sensor	Flexible Fuel Sensor[1]	FF Sensor[1]
ALDL (Assembly Line Diagnostic Link)	Data Link Connector[1]	DLC[1]

339

Shop Manual

To stress the importance of safe work habits, the Shop Manual also dedicates one full chapter to safety. Other important features of this manual include:

Tools Lists

Each chapter begins with a list of the Basic Tools needed to perform the tasks included in the chapter. Whenever a Special Tool is required to complete a task, it is listed in the margin next to the procedure.

Marginal Notes

Page numbers for cross-referencing appear in the margin. Some of the common terms used for components, and other bits of information, also appear in the margin. This provides an understanding of the language of the trade and helps when conversing with an experienced technician.

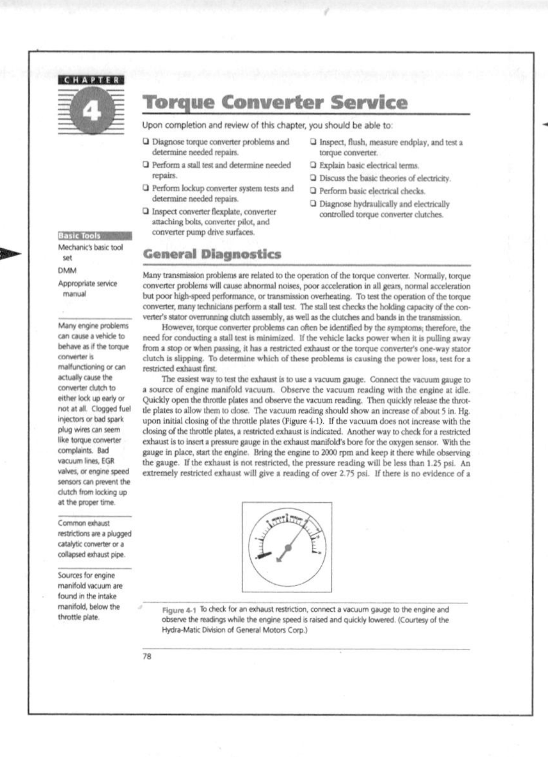

CHAPTER 4

Torque Converter Service

Upon completion and review of this chapter, you should be able to:

- Diagnose torque converter problems and determine needed repairs.
- Perform a stall test and determine needed repairs.
- Perform lockup converter system tests and determine needed repairs.
- Inspect converter flexplate, converter attaching bolts, converter pilot, and converter pump drive surfaces.
- Inspect, flush, measure endplay, and test a torque converter.
- Explain basic electrical terms.
- Discuss the basic theories of electricity.
- Perform basic electrical checks.
- Diagnose hydraulically and electrically controlled torque converter clutches.

Basic Tools

Mechanic's basic tool set

DMM

Appropriate service manual

General Diagnostics

Many transmission problems are related to the operation of the torque converter. Normally, torque converter problems will cause abnormal noises, poor acceleration in all gears, normal acceleration but poor high-speed performance, or transmission overheating. To test the operation of the torque converter, many technicians perform a stall test. The stall test checks the holding capacity of the converter's stator overrunning clutch assembly, as well as the clutches and bands in the transmission.

However, torque converter problems can often be identified by the symptoms; therefore, the need for conducting a stall test is minimized. If the vehicle lacks power when it is pulling away from a stop or when passing, it has a restricted exhaust or the torque converter's one-way stator clutch is slipping. To determine which of these problems is causing the power loss, test for a restricted exhaust first.

The easiest way to test the exhaust is to use a vacuum gauge. Connect the vacuum gauge to a source of engine manifold vacuum. Observe the vacuum reading with the engine at idle. Quickly open the throttle plates and observe the vacuum reading. Then quickly release the throttle plates to allow them to close. The vacuum reading should show an increase of about 5 in. Hg. upon initial closing of the throttle plates (Figure 4-1). If the vacuum does not increase with the closing of the throttle plates, a restricted exhaust is indicated. Another way to check for a restricted exhaust is to insert a pressure gauge in the exhaust manifold's bore for the oxygen sensor. With the gauge in place, start the engine. Bring the engine to 2000 rpm and keep it there while observing the gauge. If the exhaust is not restricted, the pressure reading will be less than 1.25 psi. An extremely restricted exhaust will give a reading of over 2.75 psi. If there is no evidence of a

Many engine problems can cause a vehicle to behave as if the torque converter is malfunctioning or can actually cause the converter clutch to either lock up early or not at all. Clogged fuel injectors or bad spark plug wires can seem like torque converter complaints. Bad vacuum lines, EGR valves, or engine speed sensors can prevent the clutch from locking up at the proper time.

Common exhaust restrictions are a plugged catalytic converter or a collapsed exhaust pipe.

Sources for engine manifold vacuum are found in the intake manifold, below the throttle plate.

Figure 4-1 To check for an exhaust restriction, connect a vacuum gauge to the engine and observe the readings while the engine speed is raised and quickly lowered. (Courtesy of the Hydra-Matic Division of General Motors Corp.)

78

Performance Objectives

These objectives define the contents of the chapter and identify what the student should have learned upon completion of the chapter. These objectives also correspond to the list of required tasks for ASE certification. *Each ASE task is addressed.*

Although this textbook is not designed to simply prepare someone for the certification exams, it is organized around the ASE task list. These tasks are defined generically when the procedure is commonly followed and specifically when the procedure is unique for specific vehicle models. Imported and domestic model automobiles and light trucks are included in the procedures.

Photo Sequences

Many procedures are illustrated in detailed Photo Sequences. These detailed photographs show the students what to expect when they perform particular procedures. They also can provide a student a familiarity with a system or type of equipment, that the school may not have.

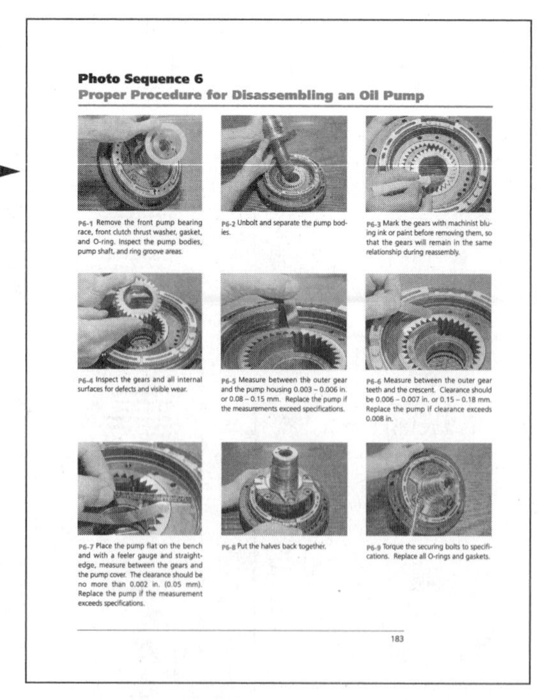

Photo Sequence 6

Proper Procedure for Disassembling an Oil Pump

P6-1 Remove the front pump bearing race, front clutch thrust washer, gasket, and O-ring. Inspect the pump bodies, pump shaft, and ring groove areas.

P6-2 Unbolt and separate the pump bodies.

P6-3 Mark the gears with machinist bluing ink or paint before removing them, so that the gears will remain in the same relationship during reassembly.

P6-4 Inspect the gears and all internal surfaces for defects and visible wear.

P6-5 Measure between the outer gear and the pump housing 0.003 – 0.006 in. or 0.08 – 0.15 mm. Replace the pump if the measurements exceed specifications.

P6-6 Measure between the outer gear teeth and the crescent. Clearance should be 0.006 – 0.007 in. or 0.15 – 0.18 mm. Replace the pump if clearance exceeds 0.008 in.

P6-7 Place the pump flat on the bench and with a feeler gauge and straightedge, measure between the gears and the pump cover. The clearance should be no more than 0.002 in. (0.05 mm). Replace the pump if the measurement exceeds specifications.

P6-8 Put the halves back together.

P6-9 Torque the securing bolts to specifications. Replace all O-rings and gaskets.

183

Service Tips

Whenever a shortcut or special procedure is appropriate, it is described in the text. These tips are generally those things commonly done by experienced technicians

Cautions and Warnings

Throughout the text, cautions are given to alert the reader to potentially hazardous materials or unsafe conditions. Warnings are also given to advise the student of things that can go wrong if instructions are not followed or if a nonacceptable part or tool is used.

SERVICE TIP: R-12 is being phased out and will not be used as the refrigerant in automotive air conditioning systems. A new refrigerant, R-134A, is being used. R-134A is less harmful to the atmosphere, but it requires the same safe-handling practices as R-12.

CAUTION: Phosgene gas is poisonous and will make you sick or fatally ill.

WARNING: R-12 should never be released into the atmosphere; it should always be captured and reclaimed by special equipment whenever an A/C system is opened. It has been determined that R-12 has an adverse effect on the earth's ozone layer.

Lifting Heavy Objects

When lifting a heavy object like a transmission, use a hoist or have someone else help you. If you must work alone, *always* lift heavy objects with your legs, not your back. Bend down with your legs, not your back, and securely hold the object you are lifting. Then stand up, keeping the object close to you (Figure 1-2). Trying to "muscle" something with your arms or back can result in severe damage to your back and can end your career and limit what you do for the rest of your life!

Before lifting a heavy object, position yourself at the angle the object should be lifted at and with your feet close to it. Then bend your knees, with your back straight, and lower yourself enough to firmly grab the object. Begin to lift by straightening your knees and keeping your elbows as straight as possible and the object close to your body. Continue lifting by straightening your legs and keeping your back straight. To place the object onto a workbench, turn your entire body. Never twist at the hips. Lower the object to the bench by bending your knees and place one end on the bench. While keeping your back straight, slide the rest of the object onto the bench.

Safe Work Areas

Classroom Manual
Chapter 1, page 3

Familiarize yourself with the layout of the shop. Find out where fire extinguishers, first-aid kits, eyewash stations (Figure 1-3), and other safety items are located. Take time to read the operating

Figure 1-2 Begin to lift a heavy object by keeping it close to you and by straightening your knees and keeping your arms as straight as possible.

Figure 1-3 A typical emergency eyewash station. (Courtesy of Brodhead-Garrett)

References to the Classroom Manual

Reference to the appropriate page in the Classroom Manual is given whenever necessary. Although the chapters of the two manuals are synchronized, material covered in other chapters of the Classroom Manual may be fundamental to the topic discussed in the Shop Manual.

Customer Care

This feature highlights those little things a technician can do or say to enhance customer relations.

Battery Safety

When possible, you should disconnect the negative cable of the battery of a car before you disconnect any electrical wire or component. This prevents the possibility of a fire or electrical shock. It also eliminates the possibility of an accidental short, which can ruin the car's electrical system. This is especially true of newer cars that are equipped with many electronic and computerized controls. Any electrical arcing can cause damage to the components.

Classroom Manual
Chapter 1, page 5

To properly disconnect the battery, you should disconnect the negative or ground cable first, then disconnect the positive cable. Since electrical circuits require a ground to be complete, by removing the ground cable you eliminate the possibility of a circuit accidentally becoming completed. When reconnecting the battery, connect the positive cable first, then the negative.

CAUTION: Never smoke or cause sparks around a battery. The slightest increase in heat can cause the battery to explode.

CUSTOMER CARE: To keep your customer happy, always record the stations set on the radio before disconnecting the battery. The absence of power will remove the stations from the radio's memory. Also be sure to reset those stations and the clock after the battery has been reconnected. To eliminate the need to reset everything, you can connect a 9-volt battery to the cigarette lighter socket. Adapters to do this are available from tool suppliers or one can be made easily. The 9-volt battery will also keep other memories alive and will retain any codes in the computer.

Jump Starting

To avoid hurting yourself or the car's electrical system, carefully follow these steps when using jumper cables (Figure 1-5) or a booster battery to help start an engine. If the battery is a sealed battery equipped with a charge indicator lamp, do not follow this procedure if the battery lamp indicates a need for replacing the battery.

1. Remove the vent caps of both batteries. (If the battery is a sealed type, do not attempt to pry off the top.
2. Cover the vent holes with a wet rag or cloth.
3. Make sure that the two cars are not touching each other.
4. Turn off all electrical accessories on both cars.
5. Connect the positive sides of both batteries with the positive booster cable.
6. Connect the negative booster cable to the negative post of the booster battery.
7. Connect the other end of the booster cable to a known good ground on the car being jumped.
8. Start the engine of the booster car.
9. Start the engine of the dead car.

Job Sheets

Located at the end of each chapter, the Job Sheets provide a format for students to perform procedures covered in the chapter. A reference to the ASE Task addressed by the procedure is referenced on the Job Sheet.

Job Sheet 3 (3)

Name ______________________ Date __________

Identification of Special Tools

Upon completion of this job sheet, you will be able to use the service manual to determine what special tools are required to properly service a particular transmission.

Tools and Materials

Appropriate service manual

Procedure

Your instructor will assign you a transmission type. You will use the service manual to identify and describe the special tools recommended for servicing this transmission. You will also identify any tools your shop has that can be used in place of the factory specified tools.

1. The transmission assigned to you is:
   Transmission Model ______________________
   from: year _____ make ____________ model ____________
   The vehicle's VIN is: ______________________
2. The manual you will use to find the information:
   ____
3. List the special tools referenced in the overhaul section of the service manual for this transmission (include in your list what part or service each of the tools is used for:
   ____
4. List the special tools that are available for your use in the shop:
   ____
5. List the available tools that can be used in place of the specified tools:
   ____

Instructor Check ______________________

37

Case Studies

Case Studies concentrate on the ability to properly diagnose the systems. Each chapter ends with a case study in which a vehicle has a problem, and the logic used by a technician to solve the problem is explained.

CASE STUDY

A customer with a late-model Ford pickup equipped with an E4OD transmission had complained that every time he hit a bump, the transmission would downshift. Then, when driven on smooth surfaces, the transmission would begin to cycle on its own between third and fourth gears.

The technician originally thought the problem was caused by a bad fourth gear clutch that couldn't hold the planetary gear set in overdrive or by a problem in the lockup torque converter circuit. Beginning the diagnosis with a visual inspection of the transmission and the many sensors that provide information to the E4OD's control module, the technician found faulty signals from the **manual lever position (MLP) sensor.** These signals explained the erratic shifting, because the computer uses the sensor to determine what gear the transmission should be in and how much modified line pressure to supply.

The technician used the required special tool to realign the sensor with the gear selector, but still found that the resistance readings were out of specifications. The sensor was replaced and the problem of erratic shifting solved.

Terms to Know

Terms in this list can be found in the Glossary at the end of the manual.

Terms to Know

Aeration	Kickdown switch	PM generator
Default mode	Limp-in	T.V. linkage
DMM	MAP sensor	Throttle position sensor
Grounding switches	MLP	Transmission identification number
Inhibitor switch	Neutral safety switch	Vehicle speed sensor
Kick down	Oxidation	

ASE Style Review Questions

Each chapter contains ASE style review questions that reflect the performance objectives listed at the beginning of the chapter. These questions can be used to review the chapter as well as to prepare for the ASE certification exam.

ASE-Style Review Questions

1. While diagnosing noises apparently coming from a transaxle assembly:
Technician A says a knocking sound at low speeds is probably caused by worn CV joints.
Technician B says a clicking noise heard when the vehicle is turning is probably caused by a worn or damaged outboard CV joint.
Who is correct?
A. A only **C.** Both A and B
B. B only **D.** Neither A nor B

2. *Technician A* says if the shift for all forward gears is delayed, a slipping front or forward clutch is normally indicated.
Technician B says a slipping rear clutch is indicated when there is a delay or slip when the transmission shifts into any forward gear.
Who is correct?
A. A only **C.** Both A and B
B. B only **D.** Neither A nor B

Diagnostic Chart

Chapters include detailed diagnostic charts linked with the appropriate ASE task. These charts list common problems and most probable causes. They also list a page reference in the Classroom Manual for better understanding of the system's operation and a page reference in the Shop Manual for details on the procedure necessary for correcting the problem.

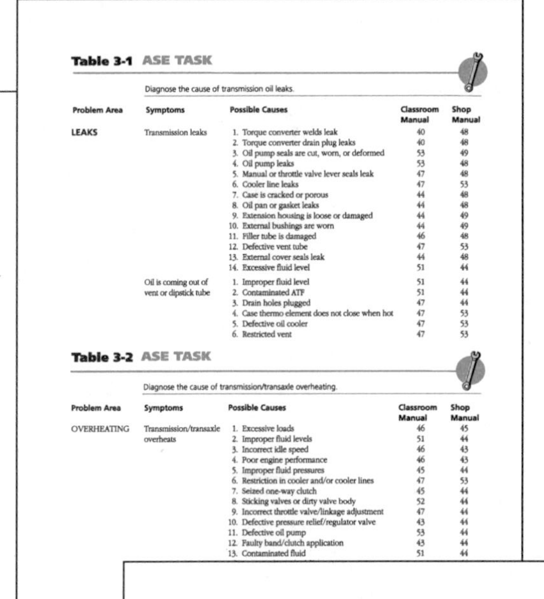

Table 3-1 ASE TASK

Diagnose the cause of transmission oil leaks.

Problem Area	Symptoms	Possible Causes	Classroom Manual	Shop Manual
LEAKS	Transmission leaks	1. Torque converter welds leak	40	48
		2. Torque converter drain plug leaks	40	48
		3. Oil pump seals are cut, worn, or deformed	53	49
		4. Oil pump leaks	53	48
		5. Manual or throttle valve lever seals leak	47	48
		6. Cooler line leaks	47	53
		7. Case is cracked or porous	44	48
		8. Oil pan or gasket leaks	44	48
		9. Extension housing is loose or damaged	44	49
		10. External bushings are worn	44	49
		11. Filler tube is damaged	46	48
		12. Defective vent tube	47	53
		13. External cover seals leak	44	48
		14. Excessive fluid level	51	44
	Oil is coming out of vent or dipstick tube	1. Improper fluid level	51	44
		2. Contaminated ATF	51	44
		3. Drain holes plugged	47	44
		4. Case thermo element does not close when hot	47	53
		5. Defective oil cooler	47	53
		6. Restricted vent	47	53

Table 3-2 ASE TASK

Diagnose the cause of transmission/transaxle overheating.

Problem Area	Symptoms	Possible Causes	Classroom Manual	Shop Manual
OVERHEATING	Transmission/transaxle overheats	1. Excessive loads	46	45
		2. Improper fluid levels	51	44
		3. Incorrect idle speed	46	43
		4. Poor engine performance	46	43
		5. Improper fluid pressures	45	44
		6. Restriction in cooler and/or cooler lines	47	53
		7. Seized one-way clutch	45	44
		8. Sticking valves or dirty valve body	52	44
		9. Incorrect throttle valve/linkage adjustment	47	44
		10. Defective pressure relief/regulator valve	43	44
		11. Defective oil pump	53	44
		12. Faulty band/clutch application	43	44
		13. Contaminated fluid	51	44

ASE Practice Examination

A 50-question ASE practice exam, located in the Appendix, is included to test students on the content of the complete Shop Manual.

Appendix A

ASE Practice Examination

Final Exam Automatic Transmission/Transaxle E2

1. Assembly of the final drive gearing is being discussed.
Technician A says a torque wrench is used to tighten bearing adjusting nuts.
Technician B says a torque wrench should be used to install the drive axle retainer.
Who is correct?
A. A only **C.** Both A and B
B. B only **D.** Neither A nor B

2. A vehicle experiences engine flare in low gear only.
Technician A says the torque converter lockup clutch is slipping.
Technician B says the transmission oil pump is not providing the required pressure.
Who is correct?
A. A only **C.** Both A and B
B. B only **D.** Neither A nor B

3. The results of a pressure test are being discussed.
Technician A says low idle pressure may be caused by a defective exhaust gas recirculation system.
Technician B says low Neutral and Park pressures may indicate a fluid leakage past the clutch and servo seals.
Who is correct?
A. A only **C.** Both A and B
B. B only **D.** Neither A nor B

4. The vehicle creeps in neutral.
Technician A says a too high engine idle speed could be the cause.
Technician B says a too tight clutch pack may be the problem.
Who is correct?
A. A only **C.** Both A and B
B. B only **D.** Neither A nor B

5. *Technician A* says over-torque valve body fasteners may cause a lack of engine braking in manual low.
Technician B says a lack of engine braking in manual third may be caused by a bad overrunning clutch.
Who is correct?
A. A only **C.** Both A and B
B. B only **D.** Neither A nor B

6. The transmission's output shaft and its sealing components are being discussed.
Technician A says the shaft and all of its sealing components must be replaced if nicks and scratches are found in the shaft's sealing area.
Technician B says all of the shaft's seals and rings must be replaced during a rebuild.
Who is correct?
A. A only **C.** Both A and B
B. B only **D.** Neither A nor B

7. The vehicle will only upshift to second at full throttle. This could be caused by any of the following EXCEPT:
A. Clogged oil passages
B. Low fluid level
C. Bad clutch pack
D. Open upshift switch

8. The vehicle will not move in any gear.
Technician A says a misadjusted T.V. cable could be the cause.
Technician B says leakage at the oil pump and/or valve body could cause this condition.
Who is correct?
A. A only **C.** Both A and B
B. B only **D.** Neither A nor B

375

Instructor's Guide

The Instructor's Guide is provided free of charge as part of the *Today's Technician Series* of automotive technology textbooks. It contains Lecture Outlines, Answers to Review Questions, Pretest and Test Bank including ASE style questions.

Classroom Manager

The complete ancillary package is designed to aid the instructor with classroom preparation and provide tools to measure student performance. For an affordable price, this comprehensive package contains:

Instructor's Guide
200 Transparency Masters
Answers to Review Questions
Lecture Outlines and Lecture Notes
Printed and Computerized Test Bank
Laboratory Worksheets and Practicals

Reviewers

I would like to extend a special thanks to those who saw things I overlooked and for their contributions:

Henry Barber
Guilford Technical Community College
Jamestown, NC

Leonard L. Flores, Jr.
St. Phillips College
San Antonio, TX

Robert L. Haines
Delta College
University Center, MI

Lawrence LeGree
Montcalm Community College
Sidney, MI

Norris Martin
Texas State Technical College—Waco
Waco, TX

Terry Ristig
Yauapai College
Prescott, AZ

John Thorp
Illinois Central College
E. Peoria, IL

Robert L. Van Dyke
Denver Automotive and Deisel College
Denver, CO

Contributing Companies

I would also like to thank these companies who provided technical information and art for this edition:

American Honda Motor Co., Inc.
Blackhawk Automotive, Inc.
Brodhead-Garrett
Chrysler Corporation
CRC Chemicals
Ford Motor Company
Goodson Shop Supplies
Lincoln Automotive, St. Louis, MO
Matco Tools
Mazda
Nissan North America Inc.
Snap-on Tools Corporation
Society of Automotive Engineers, Inc.
Portions of materials contained herein have been reprinted with permission of General Motors Corporation, Service Technology Group.

CHAPTER 1

Safety

Upon completion and review of this chapter, you should be able to:

- ❑ Explain how safety practices are part of professional behavior.
- ❑ Dress safely and professionally.
- ❑ Recognize fire hazards.
- ❑ Inspect equipment and tools for unsafe conditions.
- ❑ Work properly around batteries.
- ❑ Explain the procedures for responding to an accident.
- ❑ Identify substances that could be regarded as hazardous materials.

Introduction

Safety is the condition of being safe from undergoing or causing hurt, injury, or loss. Everyone in a shop has this job. You should work safely to protect yourself and the people around you. Perhaps the best single safety rule is "Think before you act." Too often people working in shops gamble by working in an unsafe way. Often they win and no one gets hurt, but all it takes is one accident and all of their past winnings are lost. By gambling and perhaps saving five minutes, you can lose an eye or a hand. By acting first and thinking last, you can ruin your back and lose your career. Accidents in a shop can be prevented by others and by YOU. Safe work habits also prevent damage to the vehicles and equipment in the shop.

It has been said that 50% of all shop accidents could have been prevented by a single individual, the technician.

Personal Safety

Neat work habits are also safe habits. Cleaning up spills and keeping equipment and tools out of the path of others prevent accidents. This is only common sense, but too often time is not taken to remove these safety hazards. It is the rush to complete a job that usually results in unsafe conditions.

An accident is something that happens unintentionally and is a consequence of doing something else.

Neat work habits are also the sign of a professional and an individual who takes pride in his or her profession and work. A true professional takes time to clean his or her tools and work area. A professional does a better job, makes more money, and has fewer accidents than others who don't take the time to be neat and safe.

With a professional attitude, you do not clown around in the shop, you do not throw items in the shop, and you do not create an unsafe condition for the sake of saving time. Instead of ignoring basic safety rules to save time, a professional saves time by effective diagnostics and proper procedures.

Shop Manual
Chapter 1, page 2

Dress and Appearance

A professional is concerned with the way he or she looks. How you dress and appear to others says something not only about your personality but also about your attitude, including your attitude about safety. Clothing that hangs out freely, such as shirttails, can create a safety hazard and cause serious injury. Nothing you wear should be allowed to dangle in the engine compartment or around equipment. If you have long hair, it should be tied up or tucked under a hat. Shirts should be tucked in and buttoned and long sleeves buttoned or carefully rolled up.

Keep your clothing clean. If you spill gasoline or oil on yourself, change that item of clothing immediately. Oil against your skin for a prolonged period of time can produce rashes or other allergic reactions. Gasoline can irritate cuts and sores. Besides being bad for your health, greasy uniforms certainly do not look professional.

Rotating Pulleys and Belts

Always be extra careful around moving belts, pulleys, wheels, chains, or any other rotating mechanism. Be especially careful while leaning against a belt or pulley when it's not moving. Make sure no one accidentally activates the machine. When working around an engine's drive belts and pulleys, make sure your hands, shop towels, or loose clothes do not come in contact with the moving parts. While it may not seem that these parts are rotating or traveling at high rates of speed, they are. Hands and fingers can be quickly pulled into a revolving belt or pulley even at engine idle speeds.

The thermostatic switch for the electric cooling fans may also be disconnected to prevent the fan from coming on.

WARNING: Be careful when working around electric engine cooling fans. These fans are controlled by a thermostat and can come on without warning, even when the engine is not running. Whenever you must work around these fans, disconnect the electrical connector (Figure 1-1) to the fan motor before reaching into the area around the fan.

Jewelry

An electrical short is basically an alternative path for the flow of electricity.

Always remove any rings, necklaces, bracelets, and watches before working on a car or before using any equipment. A ring can rip your finger off, a watch or bracelet can cut your wrist, and a necklace can choke you. Jewelry is of special concern when working around or with electrical wires and components. The metal used to make most jewelry conducts electricity very well and can easily cause a short, through you, if the metal touches a bare wire.

Shop Manual
Chapter 1, page 10

Foot Protection

The shank of a shoe is that portion of the shoe that protects the ball of your foot.

Tennis and jogging shoes provide little protection if something falls on your foot. Boots or shoes made of leather or a material that approaches the strength of leather offer much better protection from falling objects. There are many designs of safety shoes and boots that not only are made of leather or similar material but also have steel plates built into the toe and **shank** to protect your feet. Many also have soles that are designed to resist slipping on wet surfaces. Foot injuries are not only quite painful but also can put you out of work for some time.

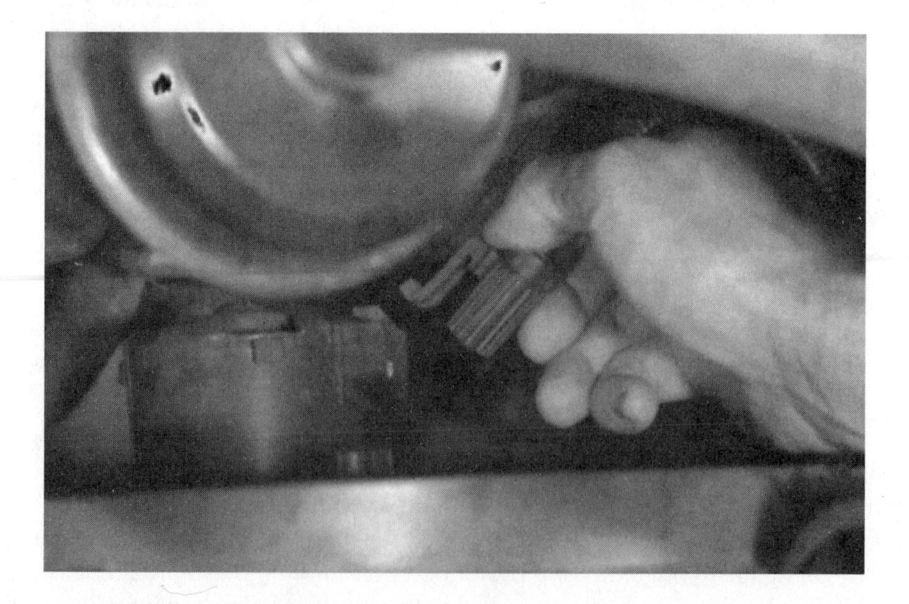

Figure 1-1 Always disconnect the connector from the cooling fan before reaching into or working near the area around the fan.

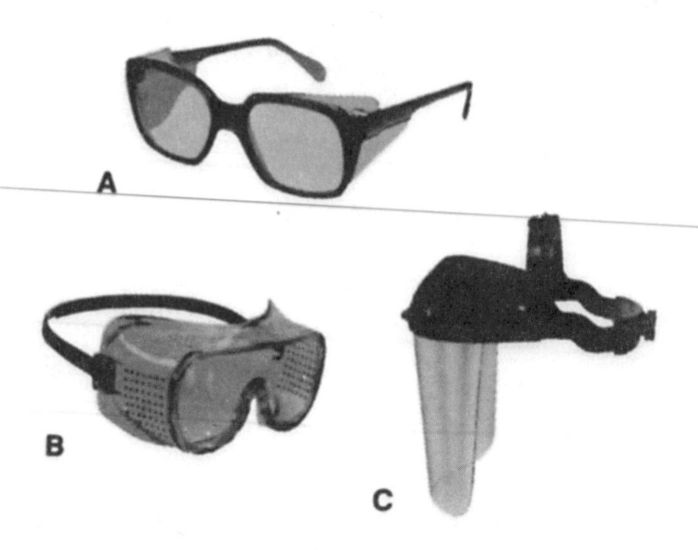

Figure 1-2 Different types of eye protection worn by automotive technicians: (A) safety glasses; (B) goggles; (C) face shield. (Courtesy of Goodson Shop Supplies)

Safety Glasses

Whenever and wherever there is a possibility of chemicals, dirt, or metal particles flowing in the air, you should wear some sort of eye protection. Dirt, grease, or rust can get into your eyes and cause serious damage to your eye. Safety glasses or goggles (Figure 1-2) should be worn when you are working under a vehicle and when you are using any machining equipment, wire buffing wheels, grinding wheels, chemicals, compressed air, or fuels. Chemicals, such as automatic transmission fluid, can cause serious eye irritations, which can lead to blindness. The lenses of safety glasses should be made of tempered glass or safety plastic. Common sense should tell you to wear safety glasses nearly all of the time you are working in the shop. Some schools and repair shops require that you do.

Shop Manual
Chapter 1, page 10

Hand Protection

Good hand protection is often overlooked. A scrape, cut, or burn can seriously impair your ability to work for many days. A well-fitted pair of heavy work gloves should be worn while hand grinding, welding, or when handling chemicals or high-temperature components. Special rubber gloves are recommended for handling **caustic** chemicals.

A caustic material has the ability to destroy or eat through something. Caustic materials are considered extremely corrosive.

Fire Hazards and Prevention

Many items around a typical shop are potential fire hazards. These include gasoline, diesel fuel, cleaning solvents, and dirty rags. Each of these should be treated as potential fire bombs.

Shop Manual
Chapter 1, page 3

Gasoline

Gasoline is present in shops so often that its dangers are often forgotten. A slight spark or an increase in heat can cause a fire or explosion. Gasoline fumes are heavier than air; therefore, when an open container of gasoline is sitting about, the fumes spill out over the sides of the container onto the floor. These fumes are more easily ignited than liquid gasoline and can easily explode.

CAUTION: Never siphon gasoline or diesel fuel with your mouth. These liquids are poisonous and can make you sick or fatally ill.

Figure 1-3 Safety can for gasoline storage. (Courtesy of Brodhead-Garrett)

Never smoke around gasoline or in a shop filled with gasoline fumes; even the droppings of hot ashes can ignite the gasoline. If the engine has a gasoline leak or if you have caused a leak by disconnecting a fuel line, wipe it up immediately and stop the leak. While stopping the leak, be extra careful not to cause any sparks. Make sure that any grinding or welding that may be taking place in the area is stopped until the spill is totally cleaned up and the floor has been flushed with water.

Always immediately wipe up gasoline that has spilled on the floor. If vapors are present in the shop, have the doors open and the ventilating system turned on to get rid of those fumes. Remember, it takes only a small amount of fuel mixed with air to cause combustion. The rags used to wipe up the gasoline should be taken outside to dry, then kept in a safety container until they are disposed of or cleaned.

Gasoline should always be stored in approved containers (Figure 1-3) and never in a glass bottle or jar. If the glass jar is knocked over or dropped, a terrible explosion could take place. Never use gasoline to clean parts. Never pour gasoline into a carburetor air horn to start the car.

The volatility of a substance is a statement of how easily the substance vaporizes or explodes.

Diesel Fuel

Diesel fuel is not as **volatile** as gasoline, but it should be stored and handled in the same manner as gasoline. It is also not as refined as gasoline and tends to be a very dirty fuel. It normally contains many impurities, including active microscopic organisms that can be highly infectious. If diesel fuel happens to get on an open cut or sore, immediately wash thoroughly.

The flammability of a substance is a statement of how well the substance supports combustion.

Solvents

Cleaning solvents are not as volatile as gasoline, but they are still **flammable**. These should be stored and treated in the same way as gasoline.

Shop Manual
Chapter 1, page 3

Rags

Oily or greasy rags can also be a source for fires. These rags should be stored in an approved container and never thrown out with normal trash. Like gasoline, oil is a **hydrocarbon** and can ignite with or without a spark or flame.

A hydrocarbon is a substance composed of hydrogen and carbon atoms.

Safe Tools and Equipment

Shop Manual
Chapter 1, page 3

Whenever you are using any equipment, make sure you use it properly and that it is set up according to the manufacturer's instructions. All equipment should be properly maintained and periodically inspected for unsafe conditions. Frayed electrical cords or loose mountings can cause serious injuries. All electrical outlets should be equipped to allow for the use of three-pronged electrical cords. The third prong allows for a safety ground connection. All equipment with rotating parts should be equipped with safety guards (Figure 1-4) that reduce the possibility of the parts coming

Figure 1-4 Equipment such as grinding wheels should be equipped with safety shields and tool guards. (Courtesy of the Snap-On Tools Corp.)

loose and injuring someone. Do not depend on someone else to inspect and maintain equipment. Check it out before you use it! If you find the equipment unsafe, put a sign on it to warn others and notify the person in charge.

Never use tools and equipment for purposes other than those they are designed for. Using the proper tool in the correct way will not only be safer, but will also allow you to do a better job.

Shop Manual
Chapter 1, page 5

Batteries

CAUTION: The active chemical in a battery, the **electrolyte**, is basically **sulfuric acid.** Sulfuric acid can cause severe skin burns and permanent eye damage, including blindness. If some battery acid gets on your skin, wash it off immediately and flush your skin with water for at least five minutes. If the electrolyte gets into your eyes, immediately flush them out with water, then immediately see a doctor. NEVER rub your eyes, just flush them well and go to a doctor. Working with and around batteries is a common-sense time to wear safety glasses or goggles.

Shop Manual
Chapter 1, page 7

While a battery is being recharged, it releases hydrogen gases, which are very explosive. Never smoke or introduce any form of heat around a charging battery. An explosion will not only destroy the battery but also can spray sulfuric acid all over you, the car, and the shop. When connecting a battery charger to a battery, leave the charger off until all of the charger's leads are connected. This will prevent electrical sparks, thereby preventing a possible explosion.

CAUTION: Always double-check the polarity of the battery charger's connections and leads before turning on the charger. Incorrect polarity can damage the battery or cause it to explode.

All batteries should be handled carefully; however, the most dangerous battery is one that has been overcharged. As it charges, it heats up and has been, or still may be, producing large amounts of hydrogen. Allow the battery to cool before working with or around it. Also never use or charge a battery that has frozen electrolyte. Extreme amounts of hydrogen can be released from the electrolyte as it thaws. If a battery does explode and battery acid gets on you, immediately flood the acid off, then get medical help.

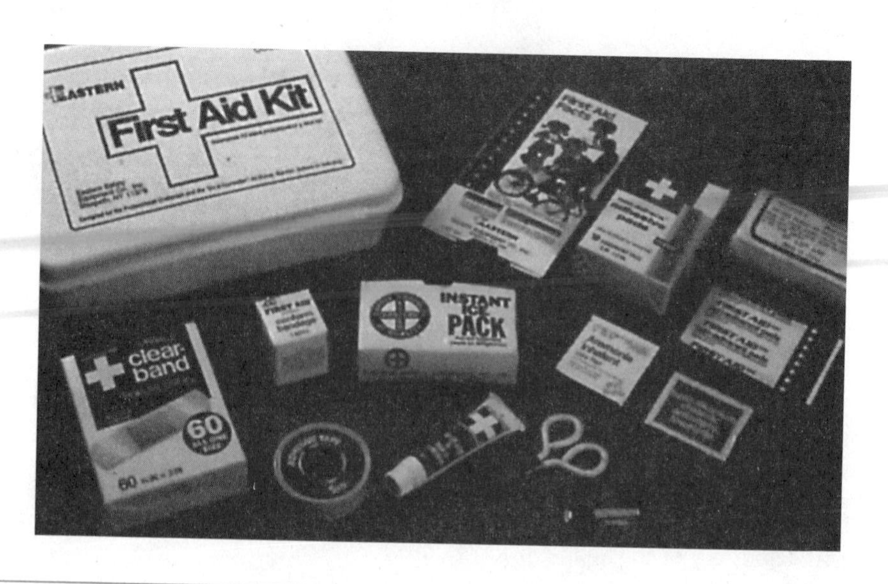

Figure 1-5 A typical first-aid kit.

Accidents

Accidents can and do happen. If there is an accident, the quicker you respond to it, the less damage there will be. For example, if battery acid gets in your eye, the sooner your eye is flushed and examined by a doctor, the better chance you have of not being blinded. Never pass off an eye injury as something unimportant. Often, when a chemical gets in your eye, it takes a few hours to cause great discomfort and by that time the damage may already be done.

Your instructor or supervisor should be informed immediately of all accidents that occur in the shop. It's a good idea to post an up-to-date list of emergency telephone numbers next to the telephone. The numbers should include a doctor, hospital, and the fire and police departments. The work area should also have a first-aid kit for treating minor injuries (Figure 1-5). Facilities for flushing eyes should also be near or in the shop area. Know where they are.

First Aid

The American Red Cross offers many low-cost but thorough courses on first aid. You will realize the importance of these classes the first time you have to give first aid to someone or when someone must give it to you. The knowledge of how to treat certain injuries can save someone's life.

Hazardous Materials

Many solvents and other chemicals used in an auto shop have warning and caution labels that should be read and understood by everyone that uses them. These are typically considered hazardous materials. Also many service procedures generate what are known as hazardous wastes. Dirty solvents and liquid cleaners are good examples of these.

Right-to-Know Law

While at work, every employee is protected by "right-to-know" laws concerning hazardous materials and wastes. The general intent of these laws is for employers to provide a safe working place, as it relates to hazardous materials. All employees must be trained about their rights under the legislation, the nature of the hazardous chemicals in their workplace, the labeling of chemicals, and the information about each chemical listed and described on Material Safety Data Sheets (MSDS). These sheets (Figure 1-6) are available from the manufacturers and suppliers of the chemicals.

CRC MATERIAL SAFETY DATA SHEET SILOO

CRC Industries, Inc. • 885 Louis Drive • Warminster, PA 18974 • (215) 674-4300

PRODUCT NAME CLEAN-R-CARB (AEROSOL) #-MSDS05079
PRODUCT- 5079,5079T,5081,5081T
(Page 1 of 2)

1. INGREDIENTS	CAS #	ACGIH TLV	OSHA PEL	OTHER LIMITS	%
Acetone	67-64-1	750 ppm	750 ppm		2-5
Xylene	1330-20-7	100 ppm	100 ppm		68-75
2-Butoxy Ethanol	111-76-2	25 ppm	25 ppm	(skin)	3-5
Methanol	67-56-1	200 ppm	200 ppm		3-5
Detergent	-	NA	NA		0-1
Propane	74-98-6	NA	1000 ppm		10-20
Isobutane	75-28-5	NA	NA	1000ppm	10-20

2. PHYSICAL DATA : (without propellent)
Specific Gravity : 0.865 Vapor Pressure : ND
% Volatile : > 99
Boiling Point : 176 F initial Evaporation Rate : Moderately fast
Freezing Point : ND Vapor Density : ND
Appearance and Odor: pH: NA
A clear colorless liquid, aromatic odor.

Solubility : Partially soluble in water.

3. FIRE AND EXPLOSION DATA
Flashpoint : -40 F Method : TCC
Flammable Limits : propellent LEL:1.8 UEL:9.5
Extinguishing Media : CO2, dry chemical, foam
Unusual Hazards : Aerosol cans may explode when heated above 120 F.

4. REACTIVITY AND STABILITY
Stability : Stable
Hazardous decomposition products
: CO2, carbon monoxide (thermal)

Materials to avoid : Strong oxidizing agents and sources of ignition.

5. PROTECTION INFORMATION
Ventilation : Use mechanical means to insure vapor conc. is below TLV.

Respiratory : Use self-contained breathing apparatus above TLV.

Gloves : Solvent resistant Eye & Face : Safety glasses
Other Protective Equipment: Not normally required for aerosol product usage.

Figure 1-6 A sample of a Material Safety Data Sheet (MSDS). (Courtesy of CRC Industries, Inc.)

They detail the chemical composition and precautionary information for all products that can present health or safety hazards.

You should become familiar with the intended purposes of the various materials, the recommended protective equipment, accident and spill procedures, and any other information regarding the safe handling of hazardous materials. This training must be given annually to employees and provided to new employees as part of their job orientation. The Canadian equivalents to the MSDS are called Workplace Hazardous Materials Information Systems (WHMIS).

CAUTION: When handling any hazardous material, always wear the appropriate safety protection. Always follow the correct procedures while using the material and be familiar with the information given on the MSDS for that material.

All hazardous materials should be properly labeled, indicating what health, fire, or reactivity hazard each chemical poses, and what protective equipment is necessary when handling each chemical. The manufacturer of the hazardous material must provide all warnings and precautionary information, which must be read and understood by all users before they use the material. You should pay great attention to the label information. By doing so, you will use the substance in the proper and safe way, thereby preventing hazardous conditions.

All hazardous materials used in the shop should be listed and posted for the employees to see. Shops must maintain documentation on the hazardous chemicals in the workplace, proof of training programs, records of accidents or spill incidents, satisfaction regarding employee requests for specific chemical information via the MSDS, and a general right-to-know compliance procedure manual utilized within the shop.

Hazardous Wastes

Waste materials are considered hazardous if they are on the EPA list of known harmful materials or if they have one or more of the following characteristics: **ignitability, corrosivity, reactivity,** or EP **toxicity.** A complete list of EPA hazardous wastes can be found in the Code of Federal Regulations. No material is considered hazardous waste until the shop is finished using it and is ready to dispose of it. All hazardous wastes must be disposed of properly.

Reactivity is a statement of how easily a substance can cause or be part of a chemical reaction.

A substance that has high ignitability is one that can catch fire quickly.

Corrosivity is a statement that defines how likely it is that a substance will destroy or eat away at other substances.

A substance with high toxicity is very poisonous.

Summary

Terms to Know

Caustic

Corrosivity

Electrolyte

Ignitability

Reactivity

Shank

Sulfuric acid

Toxicity

Volatile

- ❑ You have an important role in creating an accident-free work environment. Your appearance and work habits will go a long way toward preventing accidents. Common sense and concern for others will result in fewer accidents.
- ❑ Accidents can be prevented by not having anything dangle near rotating equipment and parts.
- ❑ Jewelry should never be worn when working on a car. Any type of jewelry can get caught in rotating parts or can short out an electrical circuit.
- ❑ Safety glasses should be worn whenever you are working under a vehicle or when you are using machining equipment, grinding wheels, chemicals, compressed air, or fuels.
- ❑ Gasoline spills should be cleaned up immediately and gasoline should be stored only in approved containers.
- ❑ Dirty and oily rags should be stored in approved containers.
- ❑ All equipment should be inspected for safety hazards before being used.
- ❑ Tools and equipment should be used only for the purposes for which they were designed.
- ❑ Fire, heat, and sparks should be kept away from a battery. These could cause the battery to explode.
- ❑ Accidents can happen. When they do, you should respond immediately to prevent further injury or damage.
- ❑ All employees have the right to know what hazardous materials they are using to perform their jobs. Most of the needed information is contained on an MSDS.
- ❑ Hazardous wastes must be disposed of properly.

Review Questions

Short Answer Essays

1. Why are safe work habits considered to be an indication of a professional attitude?
2. How does the way you dress affect your personal safety?
3. When should you wear eye protection?
4. How should gasoline-soaked rags be stored?
5. List five basic rules for using tools and equipment safely.
6. What should you do if an accident occurs in the shop?
7. If a chemical gets into someone's eye, what two things should you do?
8. What sort of information is given on an MSDS?
9. How should hazardous wastes be disposed of?
10. What do all employees of a shop have the right to know?

Fill in the Blanks

1. Basic safety rules center around two general behaviors: ____________________ ____________________ and ____________________ for others.

2. Safety glasses should be worn whenever you are working under a vehicle or when you are using ____________________ ____________________, ____________________ ____________________, ____________________, ____________________ ____________________, or ____________________.

3. ____________________, ____________________, and ____________________ should be kept away from a battery. These could cause the battery to explode.

4. Most of the information needed about the hazardousness of a chemical is found on a ____________________.

5. Safe ____________________ habits can prevent ____________________.

6. The ____________________ of a substance is a statement of how easily the substance vaporizes or explodes.

7. There are many items around a typical shop that pose a fire hazard. These include: ____________________, ____________________ ____________________, ____________________ ____________________, and ____________________ ____________________.

8. The ____________________ of a substance is a statement of how well the substance supports combustion.

9. ____________________ and ____________________ shoes provide little protection if something falls on your foot. Boots or shoes made of ____________________ or a material that approaches the strength of ____________________ offer much better protection from falling objects.

10. The general intent of ____________________ ____________________ ____________________ laws is for employers to provide a safe working place, as it relates to hazardous materials.

ASE Style Review Questions

1. While discussing ways to prevent fires:
Technician A says never work around gasoline spills; wait until the gasoline vaporizes before continuing to work.
Technician B says all dirty and oily rags should be kept in a pile until the end of the day, then they should be moved to a suitable container.
Who is correct?
A. A only
B. B only
C. Both A and B
D. Neither A nor B

2. While discussing accidents:
Technician A says accidents can be prevented by not having anything dangle near rotating equipment and parts.
Technician B says most accidents can be prevented by using common sense.
Who is correct?
A. A only
B. B only
C. Both A and B
D. Neither A nor B

3. While discussing what to do when an accident does occur:
Technician A says technicians should immediately determine the cause of the accident.
Technician B says technicians should respond immediately to prevent further injury or damage.
Who is correct?
A. A only **C.** Both A and B
B. B only **D.** Neither A nor B

4. While discussing safe storage and use of gasoline:
Technician A says gasoline spills should be cleaned up immediately.
Technician B says gasoline should only be stored in approved containers.
Who is correct?
A. A only **C.** Both A and B
B. B only **D.** Neither A nor B

5. While discussing the use of equipment in the shop:
Technician A says unsafe equipment should have its power disconnected and be marked with a sign to warn others not to use it.
Technician B says that all equipment should be inspected for safety hazards before being used.
Who is correct?
A. A only **C.** Both A and B
B. B only **D.** Neither A nor B

6. While discussing hazardous wastes:
Technician B says hazardous wastes must be properly disposed of.
Technician B says a material is not considered a hazardous waste until the shop is finished using it.
Who is correct?
A. A only **C.** Both A and B
B. B only **D.** Neither A nor B

7. While discussing ways to create an accident-free work environment:
Technician A says everyone in the shop should take full responsibility for ensuring safe work areas.
Technician B says the appearance and work habits of technicians can help prevent accidents.
Who is correct?
A. A only **C.** Both A and B
B. B only **D.** Neither A nor B

8. While discussing tools and equipment:
Technician A says that safe hand tools are clean tools.
Technician B says all tools and equipment should be used only for the purposes for which they were designed.
Who is correct?
A. A only **C.** Both A and B
B. B only **D.** Neither A nor B

9. While discussing hazardous materials:
Technician A says the employer must inform terminated or retired employees of the hazardous materials they used on a particular job.
Technician B says nearly all of the information about the hazardousness of a substance can be found in a service manual.
Who is correct?
A. A only **C.** Both A and B
B. B only **D.** Neither A nor B

10. While discussing jewelry:
Technician A says jewelry should never be worn when working on a car.
Technician B says jewelry can conduct electricity.
Who is correct?
A. A only **C.** Both A and B
B. B only **D.** Neither A nor B

Drive Train Theory

Upon completion and review of this chapter, you should be able to:

- ❑ Identify the major components of a vehicle's drive train.
- ❑ Explain how a set of gears can increase torque.
- ❑ Define and determine the ratio between two meshed gears.
- ❑ State the purpose of a transmission.
- ❑ Describe the basic operation of a planetary gear set.
- ❑ Describe the major differences between a transmission and a transaxle.
- ❑ State the purpose of a torque converter assembly.
- ❑ Describe the differences between a typical FWD and RWD car.
- ❑ State the purpose of a differential.
- ❑ Identify and describe the various gears used in modern drive trains.
- ❑ Identify and describe the various bearings used in modern drive trains.

Introduction

An automobile can be divided into four major systems or basic components: (1) the engine, which serves as a source of power; (2) the drive train, which transmits the engine's power to the car's wheels; (3) the chassis, which supports the engine and body and includes the brake, steering, and suspension systems; and (4) the car's body, interior, and accessories, which include the seats, heater and air conditioner, lights, windshield wipers, and other comfort and safety features.

Basically, a drive train has four main purposes: to connect and disconnect the engine's power to the wheels, to select different speed ratios, to provide a way to move the car in reverse, and to control the power to the drive wheels for safe turning of the vehicle. The main components of the drive train are the transmission, differential, and drive axles.

Nearly all of the questions on the ASE Automatic Transmission and Transaxles Certification Test are based on the transmission itself; few questions relate to other drive train components.

The combination of the engine and the drive train is sometimes referred to as the vehicle's power train.

Engine

Although the engine (Figure 2-1) is a major system by itself, its output should be considered a component of the drive train. An engine develops a rotary motion or **torque** that, when multiplied by the transmission gears, will move the car under a variety of conditions. The engine produces power by burning a mixture of fuel and air in its combustion chambers. Combustion causes a high pressure in the cylinders, which forces the pistons downward. Connecting rods transfer the downward movement of the pistons to the crankshaft, which rotates by the force on the pistons.

Most automotive engines are four-stroke cycle engines. The opening and closing of the intake and exhaust valves are timed to the movement of the piston. As a result, the engine passes through four different events or strokes during one combustion cycle. The four strokes are called the intake, compression, power, and exhaust strokes. As long as the engine is running, this cycle of events repeats itself, resulting in the production of engine torque.

During the intake stroke, the volume of the cylinder increases (Figure 2-2), which creates a low pressure in the cylinder. Due to a law of nature in which a higher pressure will always move to a lower pressure, outside air moves into the cylinder in an attempt to equalize the lower pressure within the cylinder. This movement of air further attempts to fill the cylinder with a fresh mixture of air and fuel for combustion.

The amount of **vacuum** formed on the intake stroke depends on the speed of the engine and the amount of air that is able to enter the cylinders. It also depends on the cylinder's ability to seal when the piston is on its intake stroke. The throttle plates control both of these. Under normal conditions, the plates control engine speed by controlling the amount of air that enters the cylinders.

A rear-wheel-drive car is best described as a car that moves by the power exerted by its rear wheels.

A stroke is one complete up or down movement of the piston in its cylinder.

The low pressure formed on the intake stroke is commonly referred to as engine vacuum.

Vacuum can be best defined as any pressure lower than atmospheric pressure.

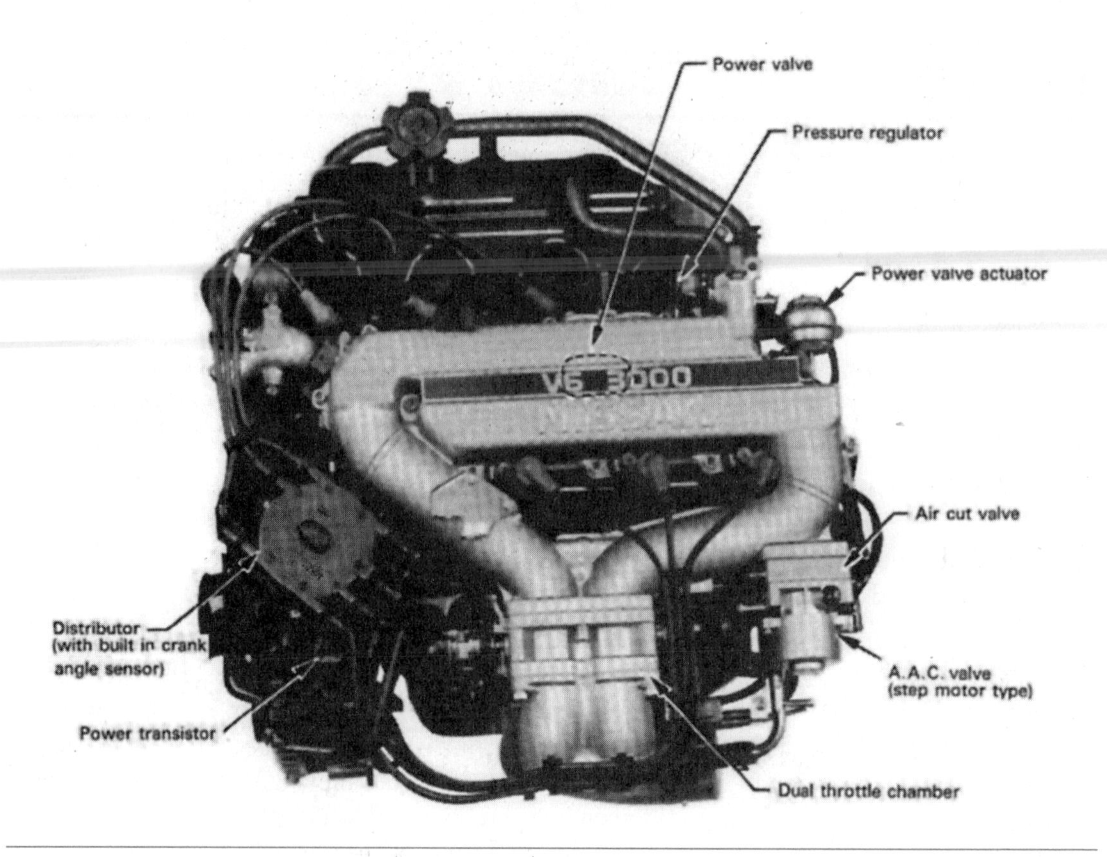

Figure 2-1 A typical late-model V-6 engine. (Courtesy of Nissan Motor Co., Ltd.)

The amount of load on the engine determines how much the plates must be opened to maintain a particular engine speed. When there is a light load, such as while the vehicle is maintaining a cruising speed on a highway, the throttle plates need to be opened only slightly to maintain the desired speed. Therefore, large amounts of vacuum are formed in the cylinders during the intake stroke.

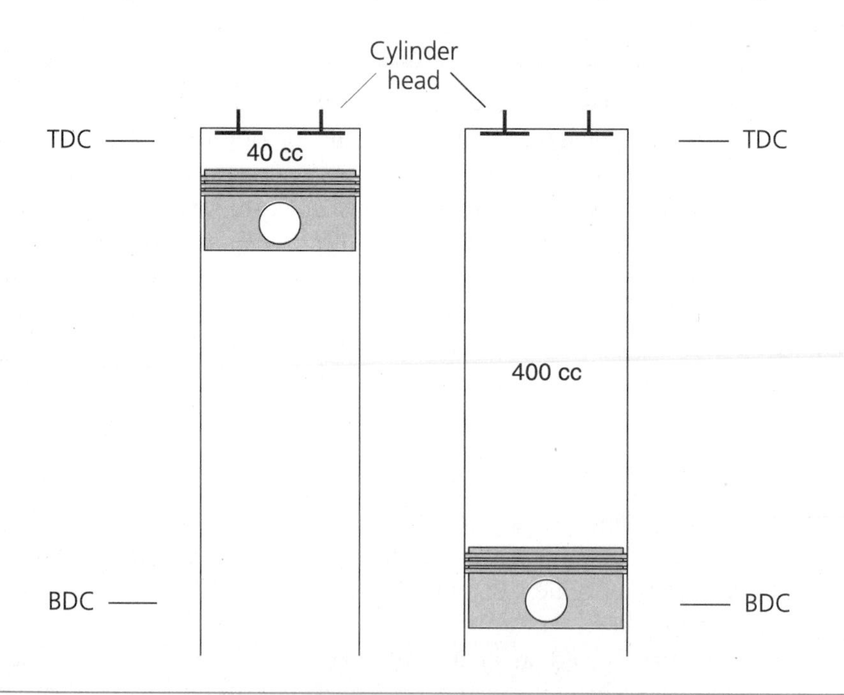

Figure 2-2 On the intake stroke, the piston moves down and increases the volume of the cylinder, which creates low pressure in the cylinder.

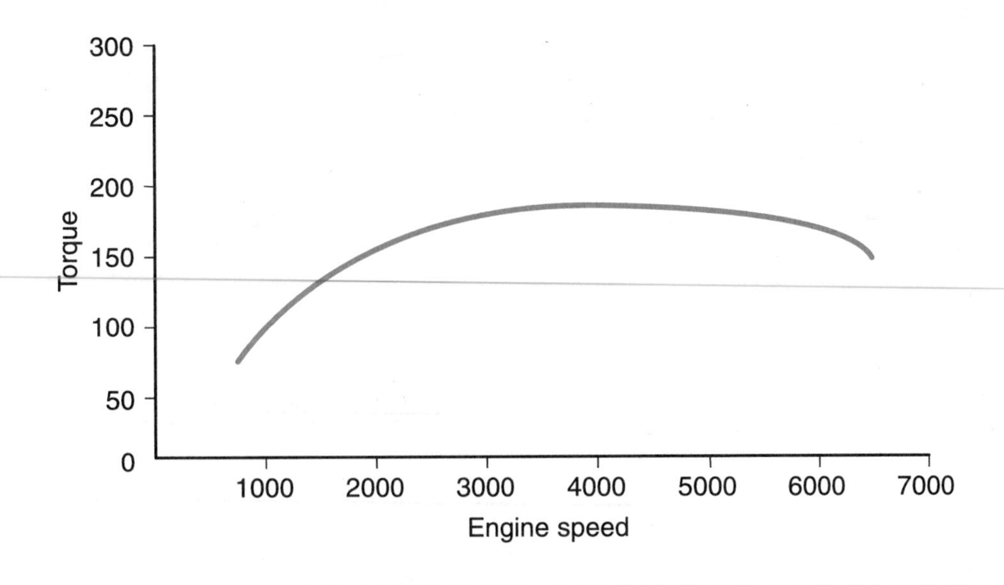

Figure 2-3 The amount of torque produced by an engine varies with the speed of the engine.

When the engine is under a heavy load, the throttle plates must be opened further to maintain the same speed. This allows more air to enter the cylinders and decreases the amount of vacuum formed during the intake stroke. Therefore, the amount of engine vacuum formed during the intake stroke is primarily controlled by **engine load.** As engine load increases, vacuum decreases.

Engine Torque

The rotating or turning effort of the engine's crankshaft is called **engine torque.** Engine torque is measured in **foot-pounds (ft.-lbs.)** or in the metric measurement **Newton-meters (Nm).** Most engines produce a maximum amount of torque while operating within a range of engine speeds and loads. When an engine reaches the maximum speed of that range, torque is no longer increased. This range of engine speeds is normally referred to as the engine's torque curve (Figure 2-3). Ideally, the engine should operate within its torque curve at all times.

To convert foot-pounds to Newton-meters, multiply the number of foot-pounds by 1.355.

As a car equipped with a manual transmission is climbing a steep hill, its driving wheels slow down, which causes engine speed to decrease, which reduces the engine's output. The driver must **downshift** the transmission, which increases engine speed and allows the engine to produce more torque. When the car reaches the top of the hill and begins to go down, its speed and the speed of the engine rapidly increase. The driver can now upshift, which allows the engine's speed to decrease and places it back in the torque curve.

Downshifting is shifting to a lower gear.

As a car equipped with an automatic transmission is climbing up a steep hill, the speed of its driving wheels and engine decreases. However, the driver does not need to downshift the transmission. The transmission will sense the increased load and automatically downshift. This downshifting increases engine speed and allows the engine to produce more torque. When the car reaches the top of the hill and begins to go down, the transmission will automatically upshift to decrease the engine's speed.

Torque Multiplication

Measurements of **horsepower** (Figure 2-4) indicate the amount of work being performed and the rate at which it is being done. The term "power" actually means a force at work; that is, doing work over a period of time. The drive line can transmit power and multiply torque, but it cannot multiply power. When power flows through one gear to another, the torque is multiplied in proportion to the different gear sizes. Torque is multiplied, but the power remains the same, since the torque is multiplied at the expense of rotational speed.

One horsepower is the equivalent of moving 33,000 pounds over a distance of one foot in one minute.

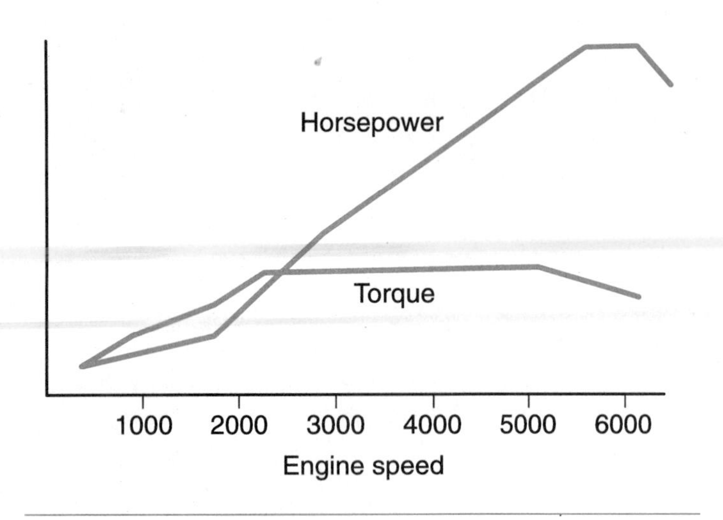

Figure 2-4 After an engine has reached its peak torque, its horsepower output increases with an increase in engine speed.

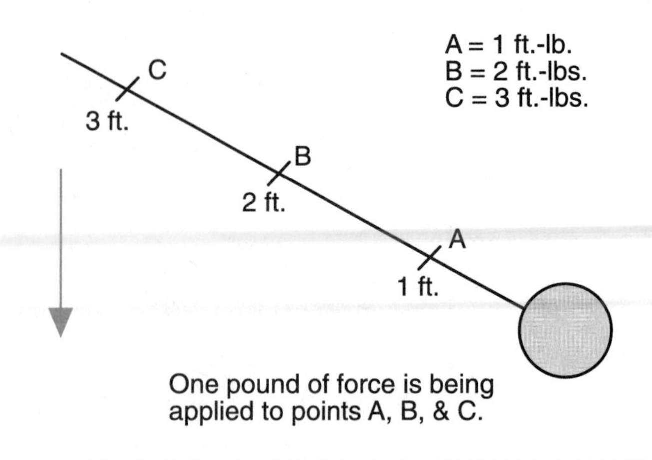

Figure 2-5 Torque is calculated by multiplying the force (1 pound) by the distance from the center of the shaft to the point (Points A, B, and C) where force is exerted.

Basic Gear Theory

The main components of all drive trains are gears. Gears apply torque to other rotating parts of the drive train and are used to multiply torque. As gears with different numbers of teeth mesh, each rotates at a different speed and torque. Torque is calculated by multiplying the force by the distance from the center of the shaft to the point where the force is exerted (Figure 2-5).

Shop Manual
Chapter 2, page 27

For example, if you tighten a bolt with a wrench that is 1 ft. long and apply a force of 10 lbs. to the wrench, you are applying 10 ft.-lbs. of torque to the bolt. Likewise, if you apply a force of 20 lbs. to the wrench, you are applying 20 ft.-lbs. of torque. You could also apply 20 ft.-lbs. of torque by applying only 10 lbs. of force if the wrench is 2 ft. long (Figure 2-6).

The distance from the center of a circle to its outside edge is called the radius. On a gear, the radius is the distance from the center of the gear to the point on its teeth where force is applied.

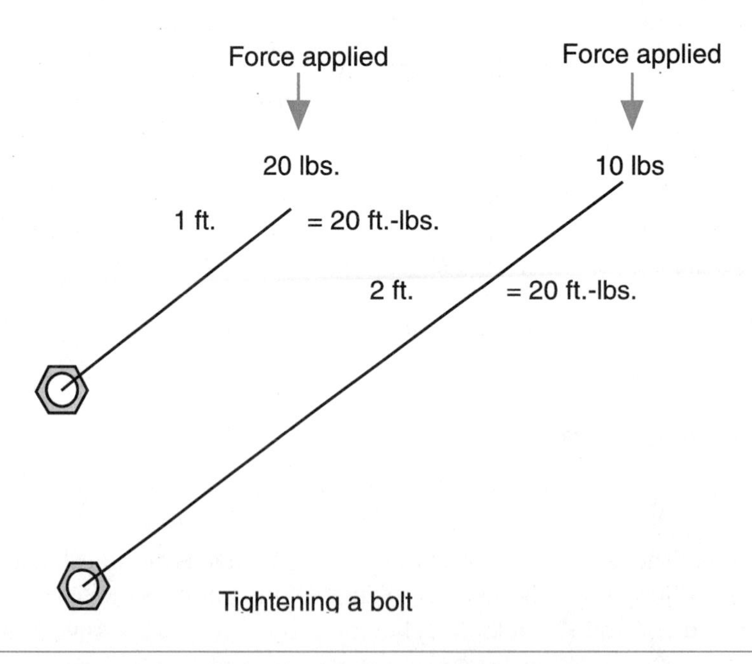

Figure 2-6 The torque applied to both bolts is 20 ft.-lbs.

If a tooth on the driving gear is pushing against a tooth on the driven gear with a force of 25 lbs. and the force is applied at a distance of 1 ft., which is the radius of the driving gear, a torque of 25 ft.-lbs. is applied to the driven gear. The 25 lbs. of force from the teeth of the smaller (driving) gear is applied to the teeth of the larger (driven) gear. If that same force is applied at a distance of 2 ft. from the center, the torque on the shaft at the center of the driven gear will be 50 ft.-lbs. The same force is acting at twice the distance from the shaft center (Figure 2-7).

The amount of torque that can be applied from a power source is proportional to the distance from the center at which it is applied. If a fulcrum or pivot point is placed closer to the object being moved, more torque is available to move the object; but the lever must move farther than when the fulcrum is farther away from the object. The same principle is used for gears in mesh: a small gear will drive a large gear more slowly, but with greater torque.

The **meshing** of gears describes the fit of one tooth of one gear between two teeth of the other gear.

A drive train consisting of a driving gear with 24 teeth and a radius of 1 inch and a driven gear with 48 teeth and a radius of 2 inches will have a torque multiplication factor of 2 and a speed reduction of 1/2. Thus, it doubles the amount of torque applied to it at half the speed (Figure 2-8). The radii between the teeth of a gear act as levers; therefore, a gear that is twice the size of another has twice the lever arm length of the other.

Gear ratios are normally expressed in terms of some number to one (1) and use a colon (:) to show the numerical comparison; e.g., 3.5:1, 1:1, 0.85:1.

Gear ratios express the mathematical relationship of one gear to another. Gear ratios can be varied by changing the diameter and number of teeth of the gears in mesh. A gear ratio also expresses the amount of torque multiplication between two gears. The ratio is obtained by dividing the diameter or number of teeth of the driven gear by the diameter or teeth of the driving gear. If the smaller driving gear had 11 teeth and the larger gear had 44 teeth, the ratio would be 4:1 (Figure 2-9). The gear ratio tells you how many times the driving gear has to turn to rotate the driven gear once. With a 4:1 ratio, the smaller gear must turn four times to rotate the larger gear once.

The larger gear turns at one-fourth the speed of the smaller gear, but has four times the torque of the smaller gear. In gear systems, **speed reduction** means torque increase. For example, when a typical four-speed transmission is in first gear, there is a speed reduction of 12:1 from the engine to the drive wheels, which means that the crankshaft turns 12 times to turn the wheels once. The resulting torque is 12 times the engine's output: therefore, if the engine produces 100 ft.-lbs. of torque, a torque of 1200 ft.-lbs. is applied to the drive wheels.

Pulleys can also be used to change speed and torque. Because they are typically connected by a drive belt, the direction of the driven pulley is the same as the direction of the drive pulley.

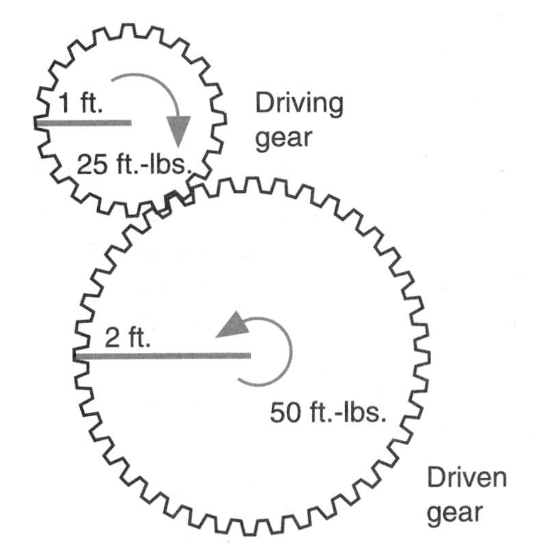

Figure 2-7 The driven gear will turn at half the speed but twice the torque because it is two times larger than the driving gear.

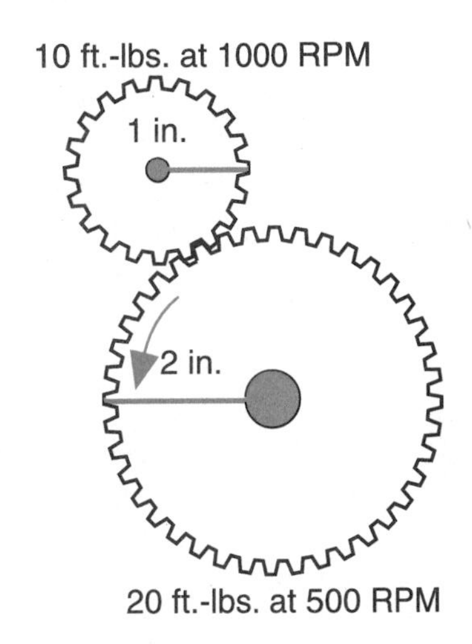

Figure 2-8 The one-inch gear will turn the two-inch gear at half its speed but twice the torque.

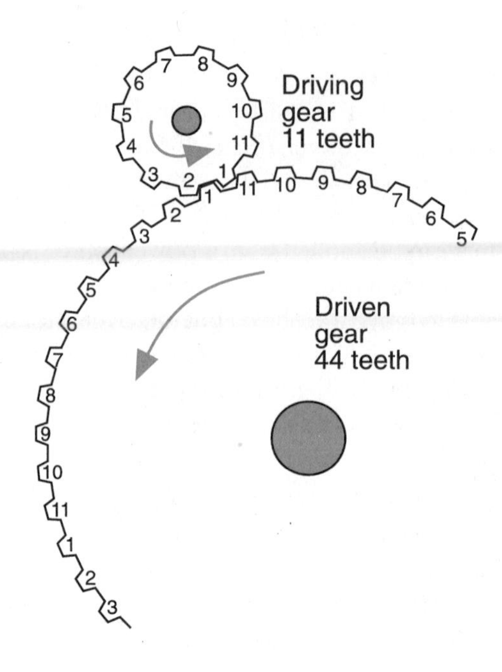

Figure 2-9 The driving gear must rotate four times to rotate the driven gear once. The ratio of the gear set is 4:1.

However, the relationship of size has the same effect as the size of gears. When the drive pulley has the same diameter as the driven pulley, the two will rotate at the same speed and with the same torque. When the drive pulley is smaller than the driven pulley, the driven pulley will turn at a lower rotational speed but with greater torque. Likewise, when the drive pulley is larger than the driven pulley, the driven pulley will rotate faster but with less torque. Pulleys are used with drive belts to operate some engine components such as generators, power steering pumps, and air conditioning compressors. Pulleys are also the basis for the operation of constant variable ratio transmissions.

Constant variable ratio transmissions are found on a few cars (for example, the Subaru Justy and Honda Civic). These transmissions automatically change torque and speed ranges without requiring a change in engine speed. The premise behind this transmission design is to keep the engine operating within a fixed speed range. This allows for improved fuel economy and decreased emission levels.

Transmission

Transmissions are often called gearboxes.

The transmission is mounted to the rear of the engine and is designed to allow the car to move forward and in reverse. It also has a **neutral** position. In this position, the engine can run without applying power to the drive wheels. Therefore, although there is input to the transmission when the vehicle is in neutral, there is no output from the transmission because the driving gears are not engaged to the output shaft.

Manual transmissions are commonly called standard shift or "stick-shift" transmissions.

There are two basic types of transmissions: automatic and manual transmissions. Nearly all automatic transmissions use a combination of a torque converter and a **planetary gear** system to change gear ratios automatically. A manual transmission (Figure 2-10) is an assembly of gears and shafts that transmits power from the engine to the drive axle, and changes in gear ratios are controlled by the driver.

Torque converters use fluid flow to connect and disconnect the engine's power to the transmission.

By moving the shift lever on a manual transmission and depressing the clutch pedal, various gear and speed ratios can be selected. The gears in a transmission are selected to give the driver a choice of both speed and torque. An automatic transmission or transaxle selects gear ratios according to engine speed, power train load, vehicle speed, and other operating factors. Little effort is

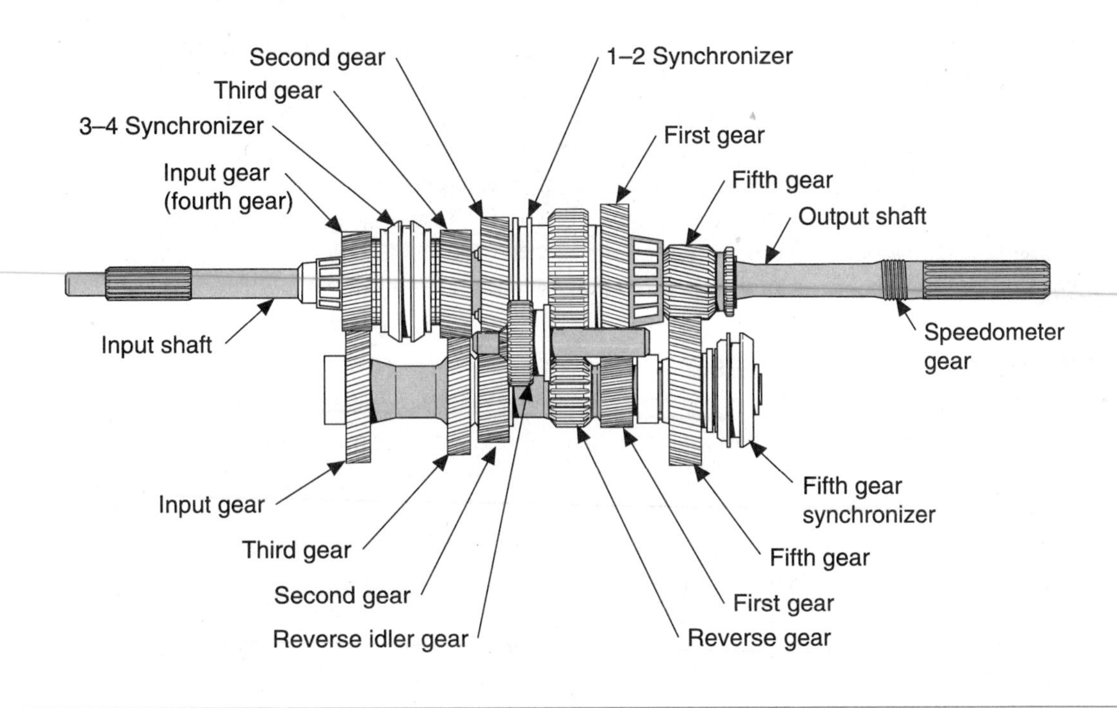

Figure 2-10 The arrangement of the gears and shafts in a typical 5-speed manual transmission. (Reprinted with the permission of Ford Motor Co.)

needed on the part of the driver, because both upshifts and downshifts occur automatically. A driver-operated clutch is not needed to change gears. The driver can also manually select a lower forward gear, reverse, neutral, or park. Depending on the forward range selected, the transmission can also provide engine braking during deceleration.

Lower gears allow for lower vehicle speeds but more torque. Higher gears provide less torque but higher vehicle speeds. Gear ratios state the ratio of the number of teeth on the driven gear to the number of teeth on the drive gear. There is often much confusion about the terms high and low gear ratios. A gear ratio of 4:1 is lower than a ratio of 2:1. Although numerically the 4 is higher than the 2, the 4:1 gear ratio allows for lower speeds and hence is termed a low gear ratio.

Like manual transmissions, automatic transmissions provide various gear ratios that match engine speed to the vehicle's speed (Figure 2-11). However, an automatic transmission is able to shift between gear ratios by itself. There is no need for a manually operated clutch to assist in the

First and second gears are the low gears in a typical transmission, whereas third, fourth, and fifth are the high gears.

Shift levers are typically called "shifters."

Gear	Engine output torque	Engine speed	Gear ratio	Transmission output torque	Transmission output speed	
1	200 ft.-lbs.	2000 rpm	4:1	800 ft.-lbs.	500 rpm	Underdrive
2	200 ft.-lbs.	2000 rpm	2:1	400 ft.-lbs.	1000 rpm	Underdrive
3	200 ft.-lbs.	2000 rpm	1:1	200 ft.-lbs.	2000 rpm	Direct drive
4	200 ft.-lbs.	2000 rpm	0.5:1	100 ft.-lbs.	4000 rpm	Overdrive

Figure 2-11 The torque output and output speed of a gear set are inversely proportional to each other.

change of gears, as is the case for a manual transmission. Also, an automatic transmission can remain engaged in a gear without stalling the engine while the vehicle is stopped.

High gear is a common term used for the top gear in each type of transmission.

Today, most automatic transmissions have four forward speeds. These speeds or gears are identified as first, second, third, and fourth. Different gear ratios are necessary because an engine develops relatively little power at low engine speeds. Without the aid of gears, the engine must be turning at a fairly high speed before it can deliver enough power to get the car moving. Through selection of the proper gear ratio, torque applied to the drive wheels can be multiplied, preventing the need to run the engine at high speeds to initiate and maintain movement.

Transmission Gears

Transmissions contain several combinations of large and small gears. In low or first gear, a small gear drives a large gear on another shaft. This reduces the speed of the larger gear but increases its turning force or torque and offers the proper gear ratio for starting movement or pulling heavy loads. First gear is primarily used to initiate movement. It has the lowest gear ratio of any gear in a transmission. It also allows for the most torque multiplication.

The ratio of second gear does not offer the same amount of torque multiplication as does first gear; however, it does offer a substantial amount. Second gear is used when the need for torque multiplication is less than the need for vehicle speed and acceleration. Since the car is already in motion, less torque is needed to move the car.

Direct drive is characterized by the transmission's output shaft rotating at the same speed as its input shaft.

Third gear allows for a further decrease in torque multiplication, while increasing vehicle speed and encouraging fuel economy. This gear typically provides a **direct drive** (1:1) ratio, so that the amount of torque that enters the transmission is also the amount of torque that passes through and out of the transmission output shaft. This gear is used at cruising speeds and promotes fuel economy. While the car is in third gear, it lacks the performance characteristics of the lower gears.

Overdrive causes the output shaft of the transmission to rotate faster than the input shaft.

Many of today's transmissions have a fourth gear, called an **overdrive** gear. Overdrive gears have ratios of less than 1:1 (typically 0.75:1). These ratios are achieved by using a small driving gear meshed with a smaller driven gear. Output speed is increased and torque is reduced. The purpose of overdrive is to promote fuel economy and reduce operating noise while maintaining highway cruising speed.

Through the use of an additional gear in mesh with two other speed gears, the direction of the incoming torque is reversed and the transmission output shaft rotates in the opposite direction of the forward gears. Because reverse gear ratios are typically based on the drive and driven gears used for first gear, only low speeds can be obtained in reverse.

Final drive ratio is also called the overall gear ratio.

The transmission's gear ratios are further increased by the gear ratio of the ring and pinion gears in the drive axle assembly (Figure 2-12). Typical axle ratios are between 2.5:1 and 4.5:1. The final drive gear ratio is calculated by multiplying the transmission gear ratio by the axle ratio. If a transmission is in first gear with a ratio of 3.63:1 and has an axle drive ratio of 3.52:1, the **overall gear ratio** is 12.87:1. If third gear has a ratio of 1:1, using the same axle ratio, the overall gear ratio is 3.52:1.

The ring gear in a planetary gear set is called the internal gear by some manufacturers.

While a manual transmission must be disconnected from the engine briefly by disengaging the clutch each time the gears are shifted, an automatic transmission does its gear shifting while it is engaged to the engine. This is accomplished through the use of constantly meshing planetary gears.

A planetary gear set consists of a **ring gear**, a **sun gear**, and several planet gears all mounted in the same plane (Figure 2-13). The ring gear has its teeth on its inner surface and the sun gear, concentric with the ring gear, has its teeth on its outer surface. The planet gears are spaced evenly around the sun gear and mesh with both the ring and sun gears. The ring, sun, and planet gears each have their own shaft or carrier to rotate on.

Planet gears are often referred to as pinion gears.

By applying the engine's torque to one of the gears in a planetary gear set and preventing another member of the set from moving, torque multiplication is available on the third set of gears.

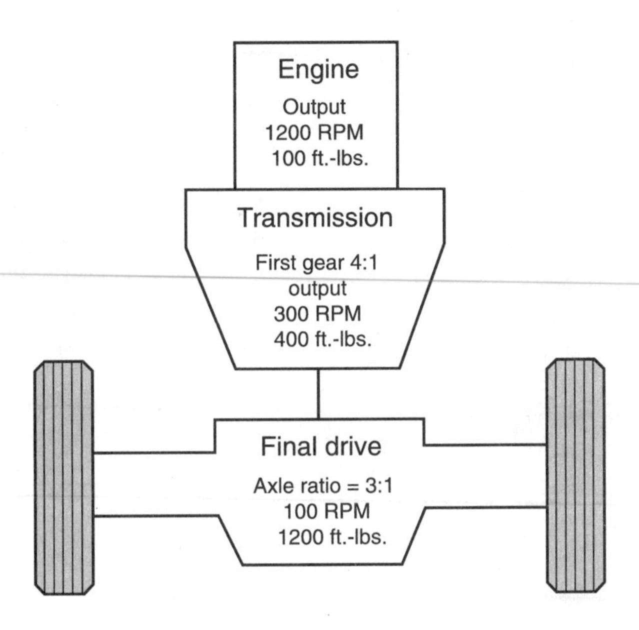

Figure 2-12 The torque increase and subsequent speed reduction of a typical vehicle's drive train.

Brake bands (Figure 2-14) or clutch packs (Figure 2-15) attached to the individual gear carriers and shafts are hydraulically activated to direct power flow from the engine to any of the gears and to hold any of the gears from rotating. This allows gear ratio changes and the reversing of power flow while the engine is running.

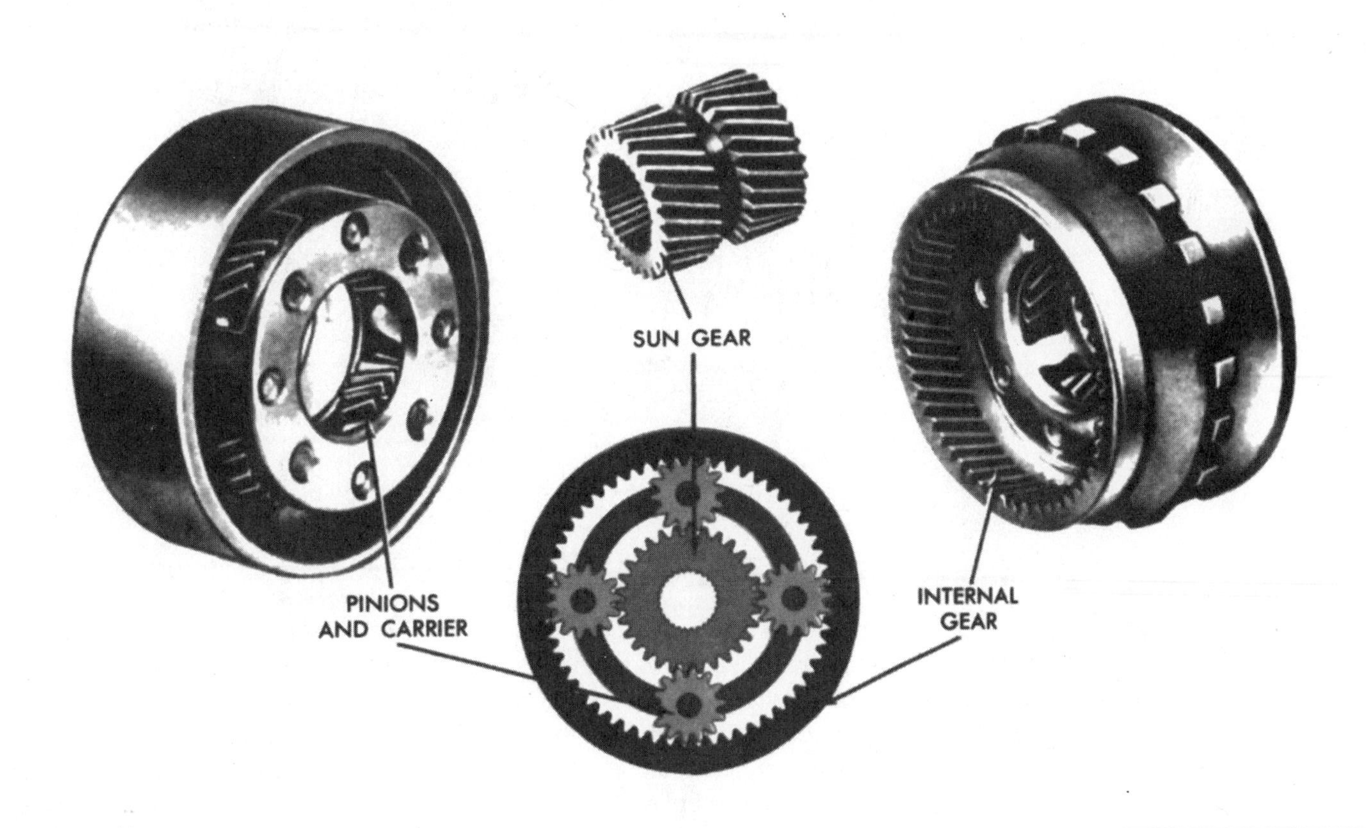

Figure 2-13 A planetary gear set. (Courtesy of the Hydra-Matic Division of General Motors Corp.)

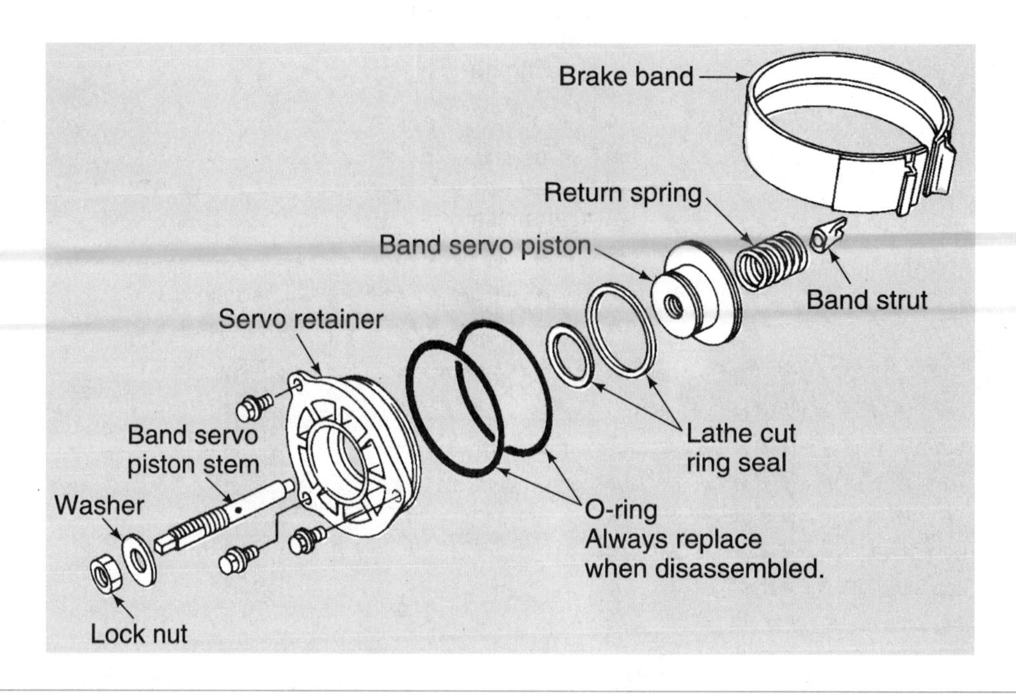

Figure 2-14 A typical band assembly. (Courtesy of Nissan Motor Co., Ltd.)

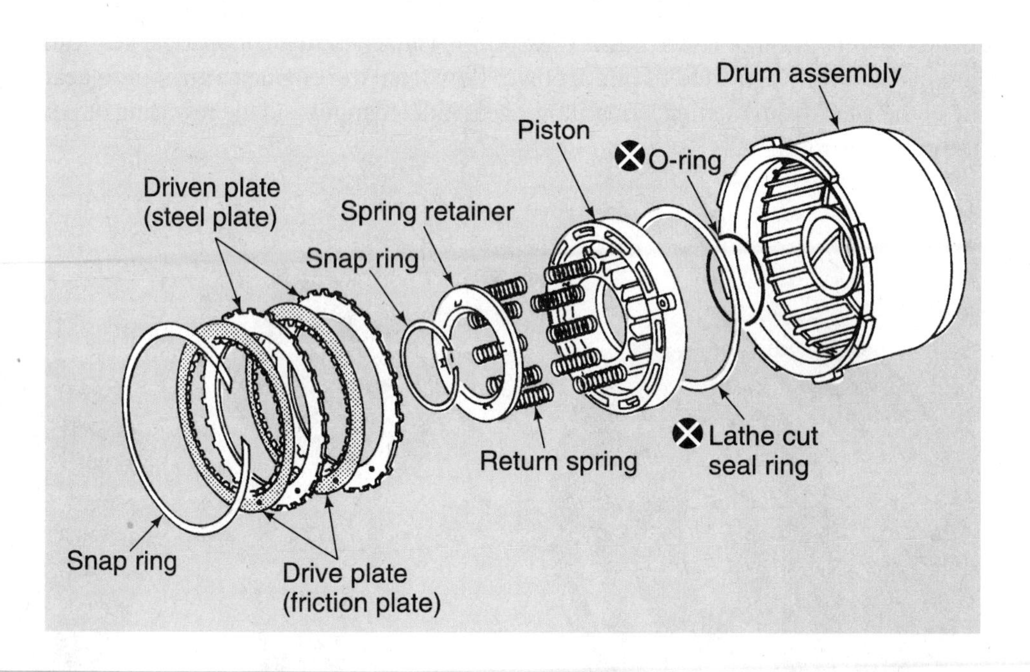

Figure 2-15 A typical multiple-disc clutch assembly. (Courtesy of Nissan Motor Co., Ltd.)

Planetary Gear Ratios

Calculating the gear ratios of a planetary gear set is much the same as calculating ratios of a **spur gear,** except different formulas are used. These formulas are necessary because there are three sets of gears in mesh and some of the gears have internal teeth. However, the ratios are still based on the number of teeth on the drive and driven gears. In most cases, the carrier walks around a held gear. To make one complete rotation, the pinions of the carrier must mesh with all the teeth of the held gear. Because the carrier links the drive and the driven gears, the total number of teeth on the drive and driven gears must be used to calculate the gear ratio. To do this, the number of teeth on the drive gear is added to the number of teeth on the driven gear. This sum is then divided by the number of teeth on the drive gear. Therefore, the formula for calculating the gear ratio of a planetary gear set is:

$$\text{gear ratio} = \frac{\text{drive gear} + \text{driven gear}}{\text{drive gear}}$$

For example, let's look at a gear set with a sun gear that has 25 teeth and a ring gear with 75 teeth. In low or first gear, the sun gear is the drive gear. The ratio of this combination is calculated as follows:

$$\text{gear ratio} = \frac{25 + 75}{25}$$

$$\text{gear ratio} = \frac{100}{25} = 4{:}1$$

When the ring gear is in drive gear, such as second gear operation, the ratio changes.

$$\text{gear ratio} = \frac{25 + 75}{75}$$

$$\text{gear ratio} = \frac{100}{75} = 1.33{:}1$$

When either the ring gear or sun gear is held (this happens in overdrive), the ratio is determined by adding the number of teeth on the gears and using this formula, when the driven gear is the ring gear:

$$\text{gear ratio} = \frac{\text{driven gear}}{\text{drive gear} + \text{driven gear}}$$

$$\text{gear ratio} = \frac{75}{100} = 0.75{:}1$$

When the gear set is in direct drive, the members of the gear set are locked together and rotate as a single member. Because of this action, no gear reduction takes place and the ratio is simply 1:1.

When the transmission is in reverse, a simple driven to drive relationship exists. Using the sun and ring gears from the previous examples, the ratio of reverse gear would be 3:1 when the sun gear is the drive gear:

$$\text{gear ratio} = \frac{\text{driven gear}}{\text{drive gear}}$$

$$\text{gear ratio} = \frac{75}{25} = 3{:}1$$

The engine's power is transmitted to the transmission through a torque converter (Figure 2-16). The torque converter drives an oil pump, which transmits fluid to a control-valve assembly. This valve assembly provides the hydraulic fluid needed to activate the various brake bands and clutch packs. The **valve body** controls the flow of the fluid throughout the transmission in response to the inputs it receives about engine and vehicle speeds and loads.

A valve body is a complex assembly of valves and is normally mounted in the transmission's oil pan

Most newer automatic transmissions rely on data received from electronic sensors and use an electronic control unit to operate solenoids in the valve body to shift gears. Older automatic transmissions rely on mechanical and vacuum signals that determine when the transmission should shift.

Electronically controlled automatic transmissions have many advantages over the older designs, including more precise shifting of gears, greater fuel economy, and increased reliability.

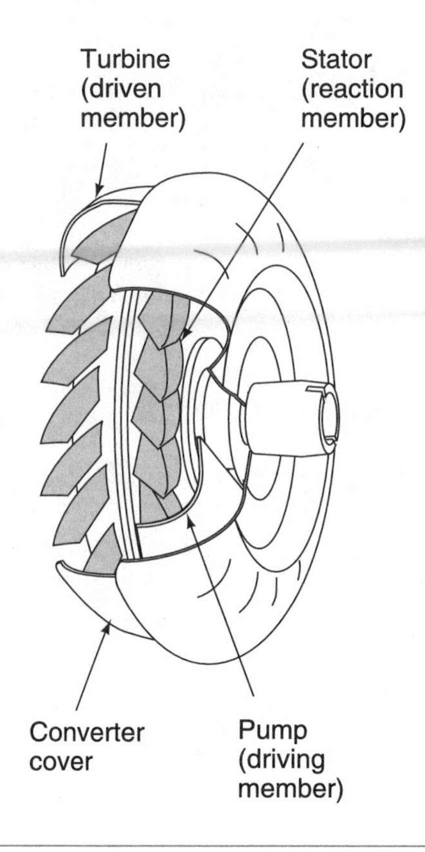

Figure 2-16 A torque converter. (Courtesy of the Hydra-Matic Division of General Motors Corp.)

When shifting gears, older designs rely on the action of a cable-operated or vacuum-controlled modulator to adjust fluid pressure in the transmission's lines to activate spring-loaded valves. Because this action takes a small amount of time, shifting delays and slippage result. The shifting of electronic transmissions is precisely controlled by a computer, which gathers information from many sensors, including those for throttle position, temperature, engine load, and vehicle speed. The computer processes this information every few milliseconds and sends electrical signals to the shift solenoids, which control the shift valves of the valve body. Electronically controlled solenoids also match transmission line-pressure to engine torque for better shift feel than mechanically controlled transmissions.

An automatic transmission is connected to the engine by a fluid-filled torque converter. The rotary motion of the engine's crankshaft is transferred from the **flywheel,** through the torque converter, to the transmission. This rotary motion is then delivered by the transmission to the **differential** and transferred by axle shafts to the tires, which push against the ground to move the car.

Automatic transmission fluid is often referred to as ATF and is at times simply called transmission fluid.

The torque converter consists of an impeller, which is attached to the engine's crankshaft, a mating turbine, which is attached to the transmission's input shaft, and a torque-multiplying stator, which is mounted between the turbine and the impeller.

The torque converter operates by hydraulic force generated by automatic transmission fluid. The torque converter changes or multiplies torque transmitted by the engine's crankshaft and directs it through the transmission. The torque converter also automatically engages and disengages engine power to the transmission in response to engine speed.

As the engine rotates, the impeller directs the transmission fluid at the blades of the turbine (Figure 2-17). The turbine is driven by the force exerted by the moving fluid. Since the turbine is connected to the transmission's input shaft, engine output is transferred to the transmission.

While operating at normal idle speeds, the engine does not rotate fast enough to allow the impeller to direct fluid against the turbine with enough force to cause it to turn. This lack of

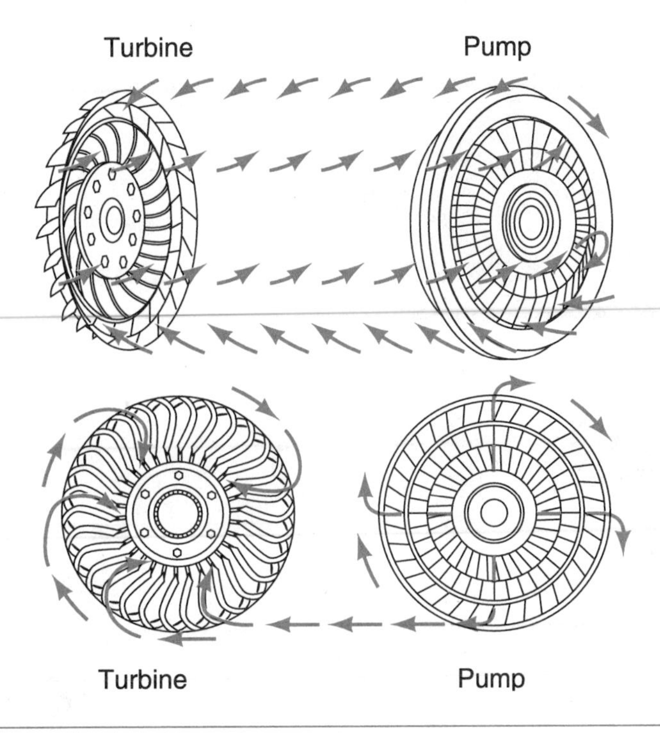

Figure 2-17 Basic oil flow through a torque converter. (Courtesy of the Hydra-Matic Division of General Motors Corp.)

hydraulic coupling enables the vehicle to stand still without stalling the engine when the gears are engaged. The coupling and uncoupling action of the impeller to the turbine is similar to the operation of the clutch in a manual transmission-equipped vehicle.

The clutch assembly of a vehicle equipped with a manual transmission is mounted to a flywheel, which is a large, heavy disc attached to the rear end of the crankshaft (Figure 2-18). In addition to providing a friction surface and mounting for the clutch, the flywheel also dampens crankshaft vibrations, adds inertia to the rotation of the crankshaft, and serves as a large gear for the starter motor. Automatic transmissions do not require the use of a heavy flywheel, rather the weight of the fluid-filled torque converter mounted to a lightweight flywheel is used for the same purposes.

The flywheel for automatic transmissions is called a **flexplate.**

Today's cars are designed to transfer the engine's power to either the front or rear wheels. In an FWD car, the transmission and driving axle are both located in one cast aluminum housing called a transaxle assembly (Figure 2-19). All of the driving components are located compactly at the front of the vehicle. One of the major advantages of front-wheel drive is that the weight of the power train components is placed over the driving wheels, which provides for improved traction on slippery road surfaces.

FWD is the standard abbreviation for front-wheel drive.

Shop Manual
Chapter 2, page 25

Figure 2-18 A typical flywheel mounted to the rear of an engine's crankshaft.

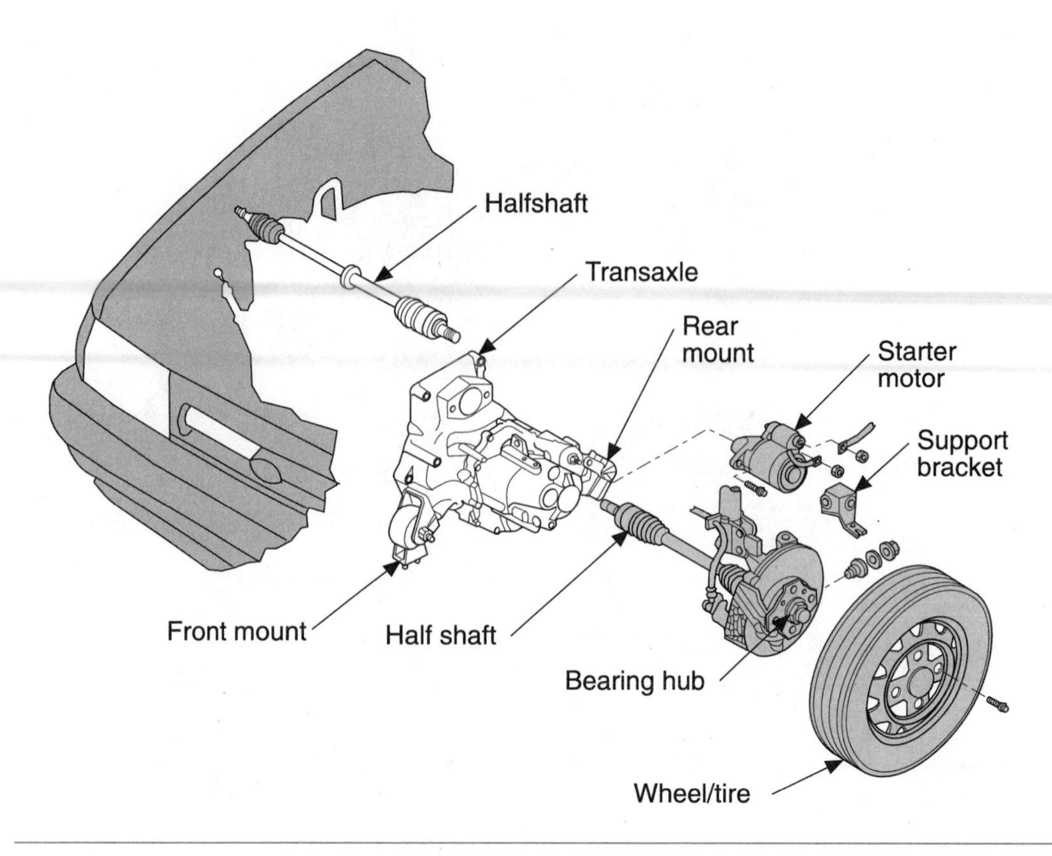

Figure 2-19 A typical transaxle assembly. (Reprinted with the permission of Ford Motor Co.)

RWD is the standard abbreviation for rear-wheel drive.

RWD cars have the power train, with the exception of the engine, located beneath the body. The engine is mounted at the front of the chassis and the related power train components extend to the rear driving wheels. The transmission's internal parts are located within an aluminum or cast iron housing called the transmission case assembly. The driving axle is located at the rear of the vehicle in a separate housing called the rear axle assembly. A drive shaft connects the output of the transmission to the rear axle.

A BIT OF HISTORY

The French-built Panhard, 1892, was the first vehicle to have its power generated by a front-mounted, liquid-fueled, internal combustion engine and transmitted to the rear driving wheels by a clutch, transmission, differential, and drive shaft.

Drive Line

The car's **drive shaft** (Figure 2-20) and its joints are often called the drive line. The drive line transmits torque from the transmission to the driving wheels. RWD cars use a long drive shaft that connects the transmission to the rear axle. The engine and drive line of FWD cars are located between the front driving wheels.

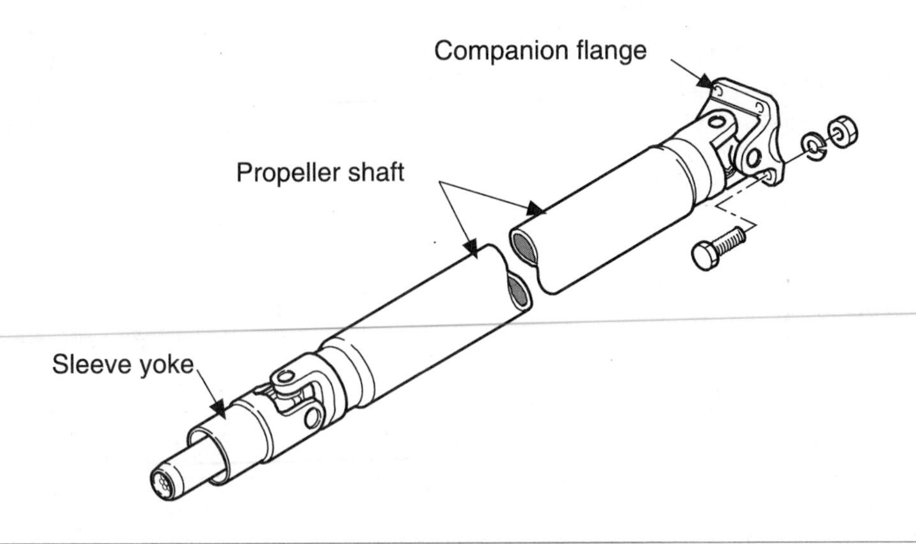

Figure 2-20 A typical RWD drive shaft. (Courtesy of Nissan Motor Co., Ltd.)

Drive Line for RWD Cars

A drive shaft is a steel tube normally consisting of two universal joints and a slip joint. The drive shaft transfers power from the transmission output shaft to the rear drive axle. A differential in the axle housing transmits the power to the rear wheels, which then move the car forward or backward.

Drive shafts differ in construction, lining, length, diameter, and type of slip joint. Typically, the drive shaft is connected at one end to the transmission and at the other end to the rear axle, which moves up and down with wheel and spring movement.

Drive shafts are typically made of hollow, thin-walled steel, aluminum, or carbon fiber tubing with the universal joint yokes welded or glued at either end. **Universal joints** (Figure 2-21) allow

Universal joints are most often called U-joints.

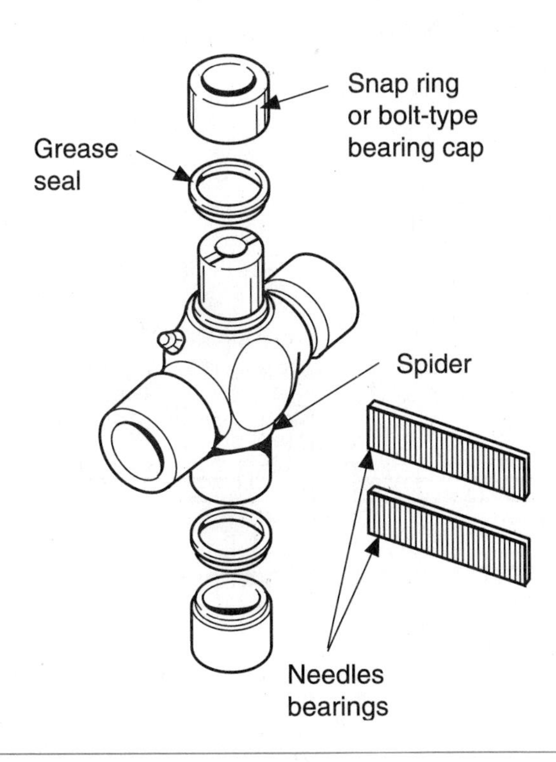

Figure 2-21 An exploded view of a universal joint. (Reprinted with the permission of Ford Motor Co.)

the drive shaft to change angles in response to the movements of the rear axle assembly. As the angle of the drive shaft changes, its length must also change. The **slip yoke** normally fitted to the front universal joint allows the shaft to remain in place as its length requirements change.

Rear Axle Assembly

The rear axle housing encloses the complete rear-wheel driving axle assembly. In addition to housing the parts, the axle housing also serves as a place to mount the vehicle's rear suspension and braking system. The rear axle assembly serves two other major functions: it changes the direction of the power flow 90 degrees and acts as the final gear reduction unit (Figure 2-22).

The rear axle consists of two sets of gears: the ring and pinion gears and the differential gears. When torque leaves the transmission, it flows through the drive shaft to the ring and pinion gears, where it is further multiplied. By considering the engine's torque curve, the car's weight, and the tire size, manufacturers are able to determine the best rear axle gear ratios for proper acceleration, hill-climbing ability, fuel economy, and noise level limits.

The primary purpose of the differential gear set is to allow a difference in driving wheel speed when the vehicle is rounding a corner or curve. The differential also transfers torque equally to both driving wheels when the vehicle is traveling in a straight line and both wheels have the same traction.

The torque on the ring gear is transmitted to the differential, where it is split and sent to the two driving wheels. When the car is traveling in a straight line, both driving wheels travel the same distance at the same speed. However, when the car is making a turn, the outer wheel must travel farther and faster than the inner wheel. When the car is steered into a 90-degree turn to the right and the inner wheel turns on a 30-foot radius, the inner wheel travels about 46 feet. The outer wheel, being nearly 5 feet from the inner wheel, travels nearly 58 feet (Figure 2-23).

Without some means for allowing the drive wheels to rotate at different speeds, the wheels would skid when the car turns. This would result in little control during turns and in excessive tire wear. The differential eliminates these troubles by allowing the outer wheel to rotate faster as turns are made.

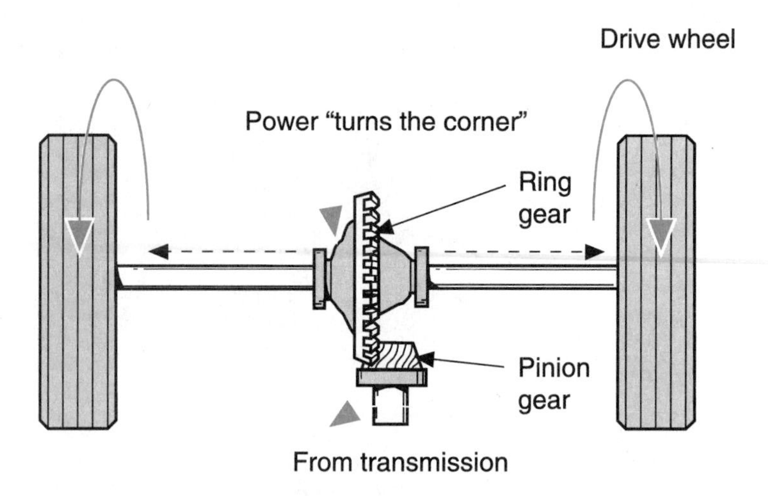

Figure 2-22 The gears in a drive axle assembly not only multiply the torque output but also transmit power "around the corner" to the drive wheels.

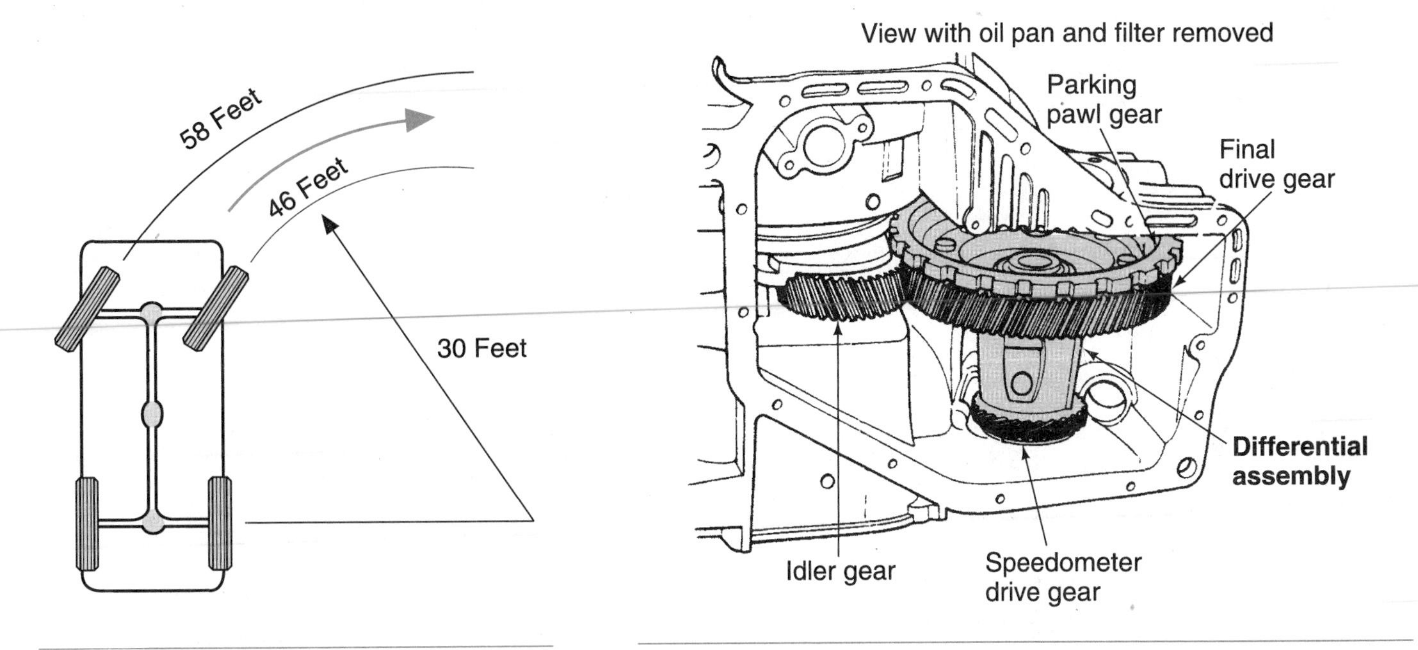

Figure 2-23 Travel of a vehicle's wheels when it is turning a corner.

Figure 2-24 A differential and final drive assembly in a transaxle housing. (Reprinted with the permission of Ford Motor Co.)

Differential Design

On FWD cars, the differential unit is normally part of the transaxle assembly (Figure 2-24). On RWD cars, it is part of the rear axle assembly (Figure 2-25). A differential unit is located in a cast iron casting, the differential case, and is attached to the center of the ring gear. Located inside the case are the differential pinion shafts and gears and the axle side gears.

Pinion gears are typically mounted or attached to a shaft and supply the input to a gear set.

The differential assembly revolves with the ring gear. Axle side gears are splined to the rear axle or front axle drive shafts.

When an automobile is moving straight ahead, both wheels are free to rotate. Engine power is applied to the pinion gear, which rotates the ring gear. Beveled pinion gears are carried around by the ring gear and rotate as one unit. Each axle receives the same power, so each wheel turns at the same speed.

When the car turns a sharp corner, only one wheel rotates freely. Torque still comes in on the pinion gear and rotates the ring gear, carrying the beveled pinions around with it. However, one axle is held stationary and the beveled pinions are forced to rotate on their own axis and "walk around" their gear. The other side is forced to rotate because it is subjected to the turning force of the ring gear, which is transmitted through the pinions.

During one revolution of the ring gear, one gear makes two revolutions, one with the ring gear and another as the pinions "walk around" the other gear. As a result, when the drive wheels have unequal resistance applied to them, the wheel with the least resistance turns more revolutions. As one wheel turns faster, the other turns proportionally slower.

To prevent a loss of power on slippery surfaces, a differential lock is often used to lock the two axles together until the slippery spot is passed, at which point they are released. These differentials are referred as to **limited-slip** or **traction-lock** differentials. When the car is proceeding in a straight line, the differential gears are locked against rotation due to gear reaction. When the vehicle turns a corner or a curve, the differential pinion gears rotate around the differential pinion

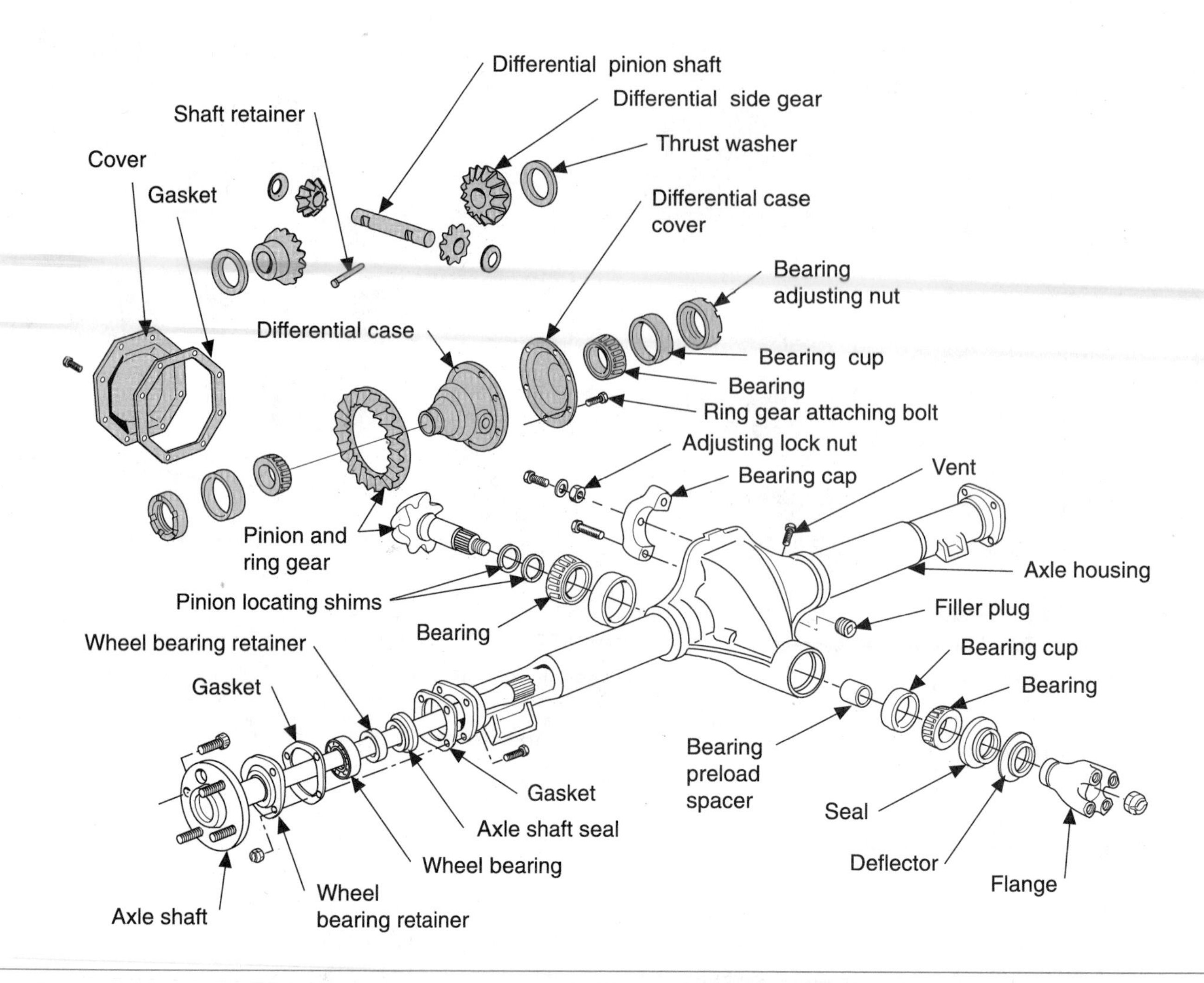

Figure 2-25 An exploded view of a rear drive axle and differential assembly. (Reprinted with the permission of Ford Motor Co.)

shaft. The differential pinion gears allow the inside axle shaft and driving wheels to slow down. On the opposite side, the pinion gears allow the outside wheels to accelerate. Both driving wheels resume equal speeds when the vehicle completes the corner or curve. This differential action improves vehicle handling and reduces driving wheel tire wear.

Driving Axles

CV joints allow the angle of the axle shafts to change with no loss in rotational speed.

4WD or 4 x 4 are the standard abbreviations for four-wheel drive.

Full-time 4WD is also called "all-wheel drive."

On RWD automobiles, the axle shafts or drive axles are located within the hollow horizontal tubes of the axle housing. The purpose of an axle shaft is to transmit the torque from the differential's side gears to the driving wheels. Axle shafts are heavy steel bars splined at the inner end to mesh with the axle side gear in the differential. The driving wheel is bolted to the wheel flange at the outer end of the axle shaft. The car's wheels rotate with the axles, which allow the car to move.

The drive axles of a FWD car extend from the sides of the transaxle to the drive wheels. **Constant velocity (CV) joints** are fitted to the axles to allow the axles to move with the car's suspension and steering systems (Figure 2-26).

4WD vehicles, especially those that are used off the road, can deliver power to all four wheels. The driver can select two-wheel or four-wheel drive. Some four-wheel-drive vehicles are always engaged in four-wheel drive; these vehicles are commonly said to have "full-time" four-

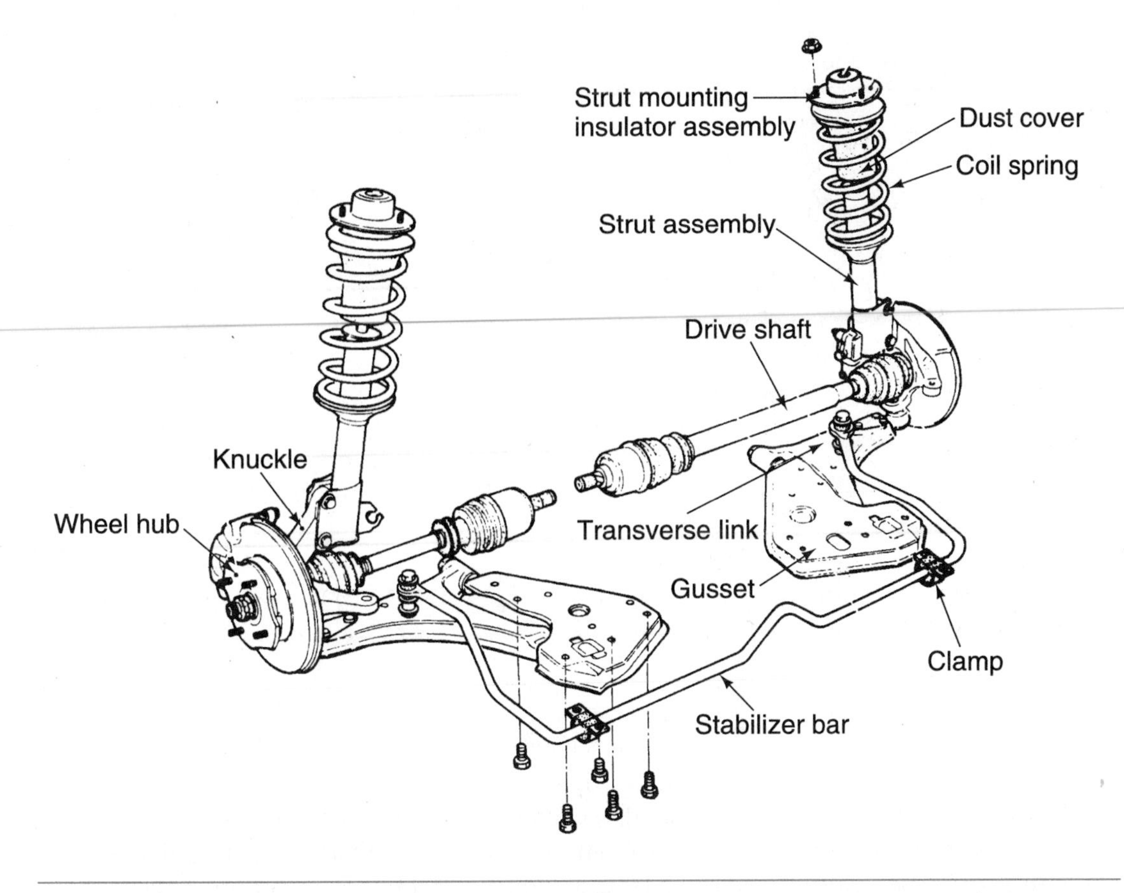

Figure 2-26 Typical location of FWD axle shafts and their CV joints. (Courtesy of Nissan Motor Co., Ltd.)

wheel drive. 4WD vehicles are equipped with two drive shafts, one to the front drive axle and the other to the rear drive axle. Each shaft is equipped with universal joints and a slip joint. Torque from the transmission enters a transfer case. The **transfer case** has gears that can be engaged to send power to only the rear wheels or to both the front and the rear wheels.

Types of Gears

Gears are normally used to transmit torque from one shaft to another. These shafts may operate in line, parallel to each other, or at an angle to each other. These different applications require a variety of gear designs, which vary primarily in the size and shape of the teeth.

In order for gears to mesh, they must have teeth of the same size and design. Meshed gears have at least one pair of teeth engaged at all times. Some gear designs allow for contact between more than one pair of teeth. Gears are normally classified by the type of teeth they have and by the surface on which the teeth are cut.

Automobiles use a variety of gear types to meet the demands of speed and torque. The most basic type of gear is the spur gear, which has its teeth parallel to and in alignment with the center of the gear. Early manual transmissions used straight-cut spur gears (Figure 2-27), which were easier to machine but were noisy and difficult to shift. Today these gears are used mainly for slow speeds, such as reverse, to avoid excessive noise and vibration. They are commonly used in simple devices such as hand or powered winches.

Helical gears (Figure 2-28) are like spur gears except that their teeth have been twisted at an angle from the gear center line. These gears get their name from being cut in a helix, which is a

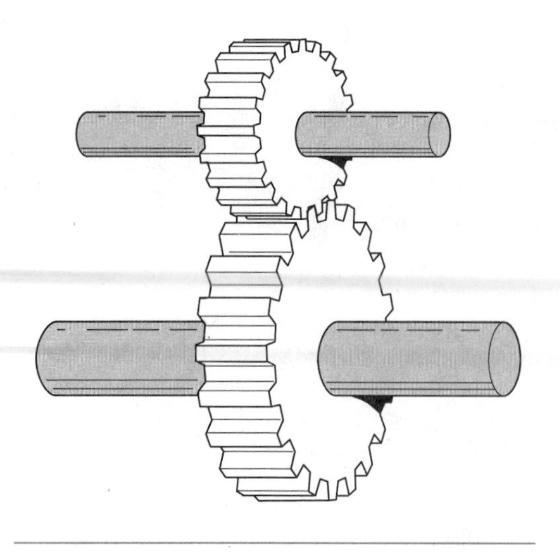

Figure 2-27 Spur gears have teeth cut straight across the gear's edge and parallel to the shaft.

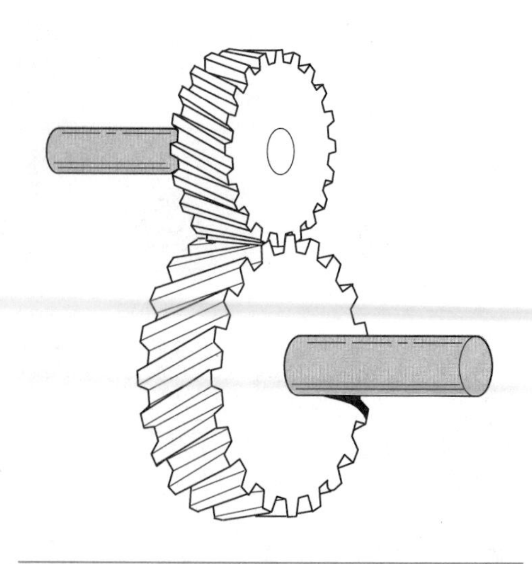

Figure 2-28 Helical gears have teeth cut at an angle to the gear's axis of rotation.

form of curve. This curve is more difficult to machine, but it is used because it reduces gear noise. Engagement of these gears begins at the tooth tip of one gear and rolls down the trailing edge of the teeth. This angular contact tends to cause side thrusts, which a bearing must absorb. However, helical spur gears are quieter in operation and have greater strength and durability than straight spur gears, simply because the contacting teeth are longer. Helical spur gears are widely used in transmissions today because they are quieter at high speeds and are durable.

Herringbone gears are actually double helical gears with teeth angles reversed on opposite sides. This causes the thrust produced by one side to be counterbalanced by the thrust produced by the other side. The two sets of teeth are often separated at the center by a narrow gap for better alignment and to prevent oil from being trapped at the apex. Herringbone gears are best suited for quiet, high-speed, low-thrust applications where heavy loads are applied. Large turbines and generators frequently use herringbone gears because of their durability.

Bevel gears are shaped like a cone with its top cut off. The teeth point inward toward the peak of the cone. These gears permit the power flow to "turn a corner." Spiral bevel gears (Figure 2-29) have their teeth cut obliquely on the angular faces of the gears. The most commonly used spiral beveled gear set is the ring and pinion gears used in heavy truck differentials.

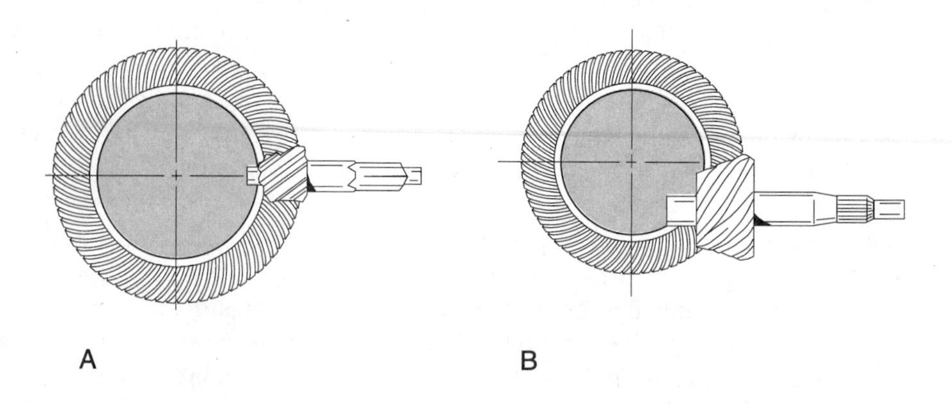

Figure 2-29 (A) Spiral bevel drive axle gears and (B) hypoid gears. (Reprinted with the permission of Ford Motor Co. of Canada, Ltd.)

The hypoid gear resembles the spiral bevel gear, but the pinion drive gear is located below the center of the ring gear. Its teeth and general construction are the same as the spiral bevel gear. The most common use for hypoid gears is in modern differentials. Here, they allow for lower body styles by lowering the transmission drive shaft.

The worm gear is actually a screw that is capable of high-speed reductions in a compact space. Its mating gear has teeth that are curved at the tips to permit a greater contact area. Power is supplied to the worm gear, which drives the mating gear. Worm gears usually provide right-angle power flows.

Some tractors and off-road vehicles use worm gears as final drive gears.

Rack and pinion gears convert straight-line motion into rotary motion, and vice versa. Rack and pinion gears also change the angle of power flow with some degree of speed change. The teeth on the rack are cut straight across the shaft, whereas those on the pinion are cut like a spur gear. These gear sets can provide control of arbor presses and other devices where slow speed is involved. Rack and pinion gears also are commonly used in automotive steering boxes.

Internal gears have their teeth pointing inward and are commonly used in the planetary gear set used in automatic transmissions and transfer cases. These are gear sets in which an outer ring gear has internal teeth that mate with teeth on smaller planetary gears. These gears, in turn, mesh with a center or sun gear. Many changes in speed and torque are possible, depending on which parts are held stationary and which are driven. In a planetary gear set, one gear is normally the input, another is prevented from moving, or is held, and the third gear is the output gear. Planetary gears are widely used because each set is capable of more than one speed change. The gear load is spread over several gears, reducing stress and wear on any one gear.

Bearings

Gears are either securely attached to a shaft or designed to move freely on the shaft. The ease with which the gears rotate on the shaft or the shaft rotates with the gears partially determines the amount of power needed to rotate them. If they rotate with great difficulty due to high friction, much power is lost. High friction will also cause excessive wear to the gears and shaft. To reduce the friction, bearings are fitted to the shaft and/or gears.

The simplest type of bearing is a cylindrical hole formed in a piece of material, into which the shaft fits freely. The hole usually has a brass or bronze lining, or **bushing,** which not only reduces the friction but also allows for easy replacement when wear occurs. Bushings usually fit tightly in the hole.

Bushings are often referred to as plain bearings.

Ball or **roller bearings** are used wherever friction must be minimized. With these types of bearings, rolling friction replaces the sliding friction that occurs with plain bearings. Typically, two bearings are used to support a shaft instead of a single long bushing. Bearings have three purposes: they support a load, maintain alignment of a shaft, and reduce rotating friction.

Most bearings are capable of withstanding only loads that are perpendicular to the axis of the shaft. Such loads are called radial loads and bearings that carry them are called radial or journal bearings.

A journal is the area on a shaft that rides on the bearing.

To prevent the shaft from moving in the axial direction, shoulders or collars may be formed on it or secured to it. If both collars are made integral with the shaft, the bearings must be split or made into halves, and the top half or cap bolted in place after the shaft has been put in place. The collars or shoulders withstand any end thrusts, and bearings designed this way are termed **thrust bearings.**

A needle bearing is a small roller bearing.

Some thrust bearings look similar to a thick washer fitted with needle bearings on its flat surface. These are typically called thrust needle bearings and are commonly used in automatic transmissions (Figure 2-30). Automatic transmissions also use items that look like thrust needle bearings without the needle bearings. These are thrust washers and are used to control endplay.

A single row journal or radial ball bearing has an inner race made of a ring of case-hardened steel with a groove or track formed on its outer circumference for a number of hardened steel balls to run upon. The outer race is another ring that has a track on its inner circumference. The balls fit

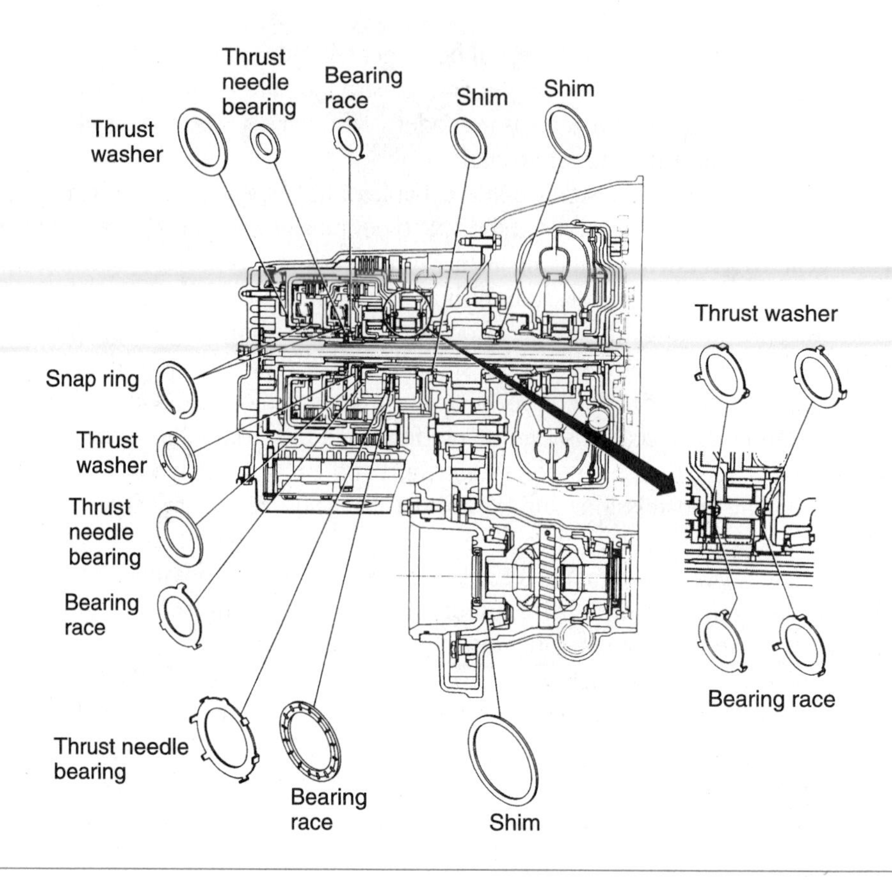

Figure 2-30 Typical location of thrust needle bearings and their races. (Courtesy of Nissan Motor Co., Ltd.)

between the two tracks and roll around in the tracks as either race turns. The balls are kept from rubbing against each other by some form of cage. These bearings can withstand radial loads and can also withstand a considerable amount of axial thrust. Therefore, they are often used as combined journal and thrust bearings.

A bearing designed to take only radial loads has only one of its races machined with a track for the balls. Other bearings are designed to take thrust loads in only one direction. If this type of bearing is installed incorrectly, the slightest amount of thrust will cause the bearing to come apart.

Another type of ball bearing uses two rows of balls. These are designed to withstand considerable amounts of radial and axial loads. Constructed like two single-row ball bearings joined together, these bearings are often used in rear axle assemblies.

Roller bearings (Figure 2-31) are used wherever it is desirable to have a large bearing surface and low amounts of friction. Large bearing surfaces are needed in areas of extremely heavy loads. The

Figure 2-31 A roller-type bearing. (Reprinted with the permission of Ford Motor Co.)

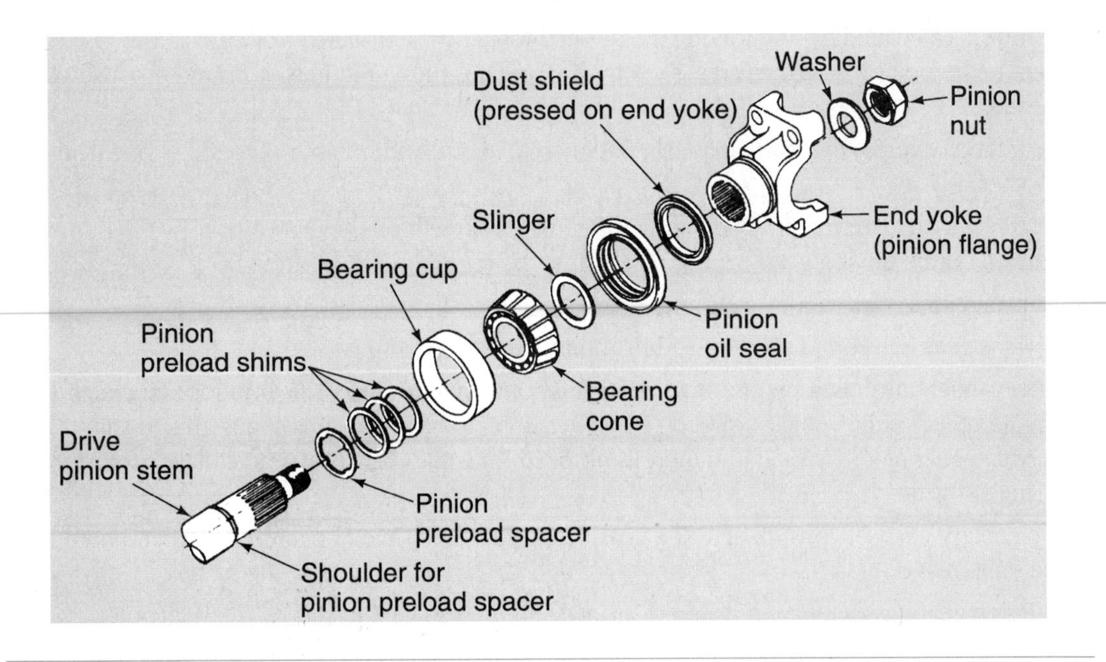

Figure 2-32 A tapered roller-type bearing is used to support the pinion shaft in this assembly. (Reprinted with the permission of Ford Motor Co.)

rollers are usually fitted between a journal of a shaft and an outer race. As the shaft rotates, the rollers turn and rotate in the race. Tapered roller bearings (Figure 2-32) are commonly used in drive axle assemblies. To control the unwanted movements of the shaft, a pair of tapered roller bearings is used.

General Maintenance

Most of the drive line needs little maintenance other than periodic oil changes. However, the fluid and filter of an automatic transmission should be replaced at specific time and mileage intervals. Always follow the recommended service intervals and use the fluid recommended by the manufacturer.

Excessive heat is a primary concern for automatic transmissions, especially late-model designs. Transmission fluid, which transmits power from the engine, also cools and lubricates the internal parts of the transmission. If the transmission is abused or is malfunctioning, the fluid may overheat, oxidize, and break down. This would cause damage to the internal seals, clutches, bands, and other important parts of the transmission. The condition and level of the fluid should be checked on a frequent and regular basis.

Summary

- The drive train has four primary purposes: to connect the engine's power to the drive wheels, to select different speed ratios, to provide a way to move the vehicle in reverse, and to control the power to the drive wheels for safe turning of the vehicle.
- The main components of the drive train are the transmission, differential, and drive axles.
- The rotating or turning effort of the engine's crankshaft is called engine torque.
- The amount of engine vacuum formed during the intake stroke is determined largely by the amount of load on the engine.
- Gears are used to apply torque to other rotating parts of the drive train and to multiply torque.

Terms to Know

Ball bearing
Bevel gear
Bushing
Constant velocity joint
Differential
Direct drive
Downshift
Drive shaft
Engine load
Engine torque
flexplate
Flywheel
Foot-pounds
Gear ratios
Helical gear
Herringbone gear
Horsepower
Hypoid gear
Limited-slip
Meshing
Neutral
Newton-meters
Overall gear ratio
Overdrive
Planetary gear
Pulley
Ring gear
Roller bearing
Slip yoke
Speed reduction
Spur gear
Sun gear
Thrust bearing
Torque
Traction-lock
Transfer case
Universal joint
Vacuum
Valve body

- ❑ Torque is calculated by multiplying the applied force by the distance from the center of the shaft to the point where the force is exerted. Torque is measured in foot-pounds and Newton-meters.
- ❑ Gear ratios express the mathematical relationship, in size and number of teeth, of one gear to another.
- ❑ Gear ratios are determined by dividing the number of teeth on the driven gear by the number of teeth on the driving gear.
- ❑ Transmissions offer various gear ratios through the meshing of various sized gears.
- ❑ Reverse gear is accomplished by adding a third gear to a two-gear set.
- ❑ Like manual transmissions, automatic transmissions provide various gear ratios that match engine speed to the vehicle's speed. However, an automatic transmission is able to shift between gear ratios by itself and there is no need for a manually operated clutch to assist in the change of gears.
- ❑ A planetary gear set consists of a ring gear, a sun gear, and several planet gears all mounted in the same plane.
- ❑ The ring gear has its teeth on its inner surface and the sun gear has its teeth on its outer surface. The planet gears are spaced evenly around the sun gear and mesh with both the ring and sun gears.
- ❑ By applying the engine's torque to one of the gears in a planetary gear set and preventing another member of the set from moving, torque multiplication is available on the third set of gears.
- ❑ Brake bands or clutch packs attached to the individual gear carriers and shafts are hydraulically activated to direct engine power to any of the gears and to hold any of the gears from rotating. This allows gear ratio changes and the reversing of power flow while the engine is running.
- ❑ An oil pump in the transmission provides the hydraulic fluid needed to activate the various brake bands and clutch packs.
- ❑ The valve body controls the flow of the fluid throughout the transmission and acts on the vacuum and mechanical signals it receives about engine and vehicle speeds and loads.
- ❑ Most newer automatic transmissions rely on data received from electronic sensors and use an electronic control unit to operate solenoids in the valve body to shift gears.
- ❑ In FWD cars, the transmission and drive axle are located in a single assembly called a transaxle. In RWD cars, the drive axle is connected to the transmission through a drive shaft.
- ❑ The drive shaft and its joints are called the drive line of the car.
- ❑ Universal joints allow the drive shaft to change angles in response to movements of the car's suspension and rear axle assembly.
- ❑ The rear axle housing encloses the entire rear-wheel driving axle assembly.
- ❑ The primary purpose of the differential is to allow a difference in wheel speeds when the vehicle is rounding a corner or curve. The ring and pinion gears in the drive axle also multiply the torque they receive from the transmission.
- ❑ On FWD cars, the differential is part of the transaxle assembly.
- ❑ The drive axles on FWD cars extend from the sides of the transaxle to the drive wheels. CV joints are fitted to the axles to allow the axles to move with the car's suspension and steering.
- ❑ 4WD vehicles typically use a transfer case that relays engine torque to both a front and rear driving axle.
- ❑ An understanding of gears and bearings is the key to effective troubleshooting and repair of drive line components.

Review Questions

Short Answer Essays

1. What are the primary purposes of a vehicle's drive train?
2. Why does torque increase when a smaller gear drives a larger gear?
3. How are gear ratios calculated?
4. Why are transmissions equipped with many different forward gear ratios?
5. What is the primary difference between a transaxle and a transmission?
6. Why are U-joints and CV joints used in the drive line?
7. What does a differential unit do to the torque it receives?
8. What is the purpose of the torque converter assembly? How does it work?
9. What kinds of gears are commonly used in today's automotive drive trains?
10. When are ball or roller bearings used?

Fill in the Blanks

1. The main components of the drive train are the ________________ , ________________ , and ________________ ________________ .
2. The rotating or turning effort of the engine's crankshaft is called ________________ ________________ .
3. Gears are used to apply torque to other rotating parts of the drive train and to ________________ torque.
4. Torque is calculated by multiplying the applied force by the ________________ from the center of the ________________ to the point where the force is exerted.
5. Torque is measured in ________________ - ________________ and ________________ - ________________ .
6. Gear ratios are determined by dividing the number of teeth on the ________________ gear by the number of teeth on the ________________ gear.
7. Reverse gear is accomplished by adding a ________________ ________________ to a two-gear set.
8. The torque converter assembly is composed of a ________________ , an ________________, and a ________________ .
9. In FWD cars, the transmission and drive axle are located in a single assembly called a ________________ .
10. In RWD cars, the drive axle is connected to the transmission through a ________________ ________________ .

ASE Style Review Questions

1. While discussing the purposes of a drive train:
   Technician A says that it connects the engine's power to the drive wheels.
   Technician B says that it controls the power to the drive wheels for safe turning of the vehicle.
   Who is correct?
   A. A only **C.** Both A and B
   B. B only **D.** Neither A nor B

2. While discussing the uses of gears:
   Technician A says gears are used to apply torque to other rotating parts of the drive train.
   Technician B says gears are used to multiply torque.
   Who is correct?
   A. A only **C.** Both A and B
   B. B only **D.** Neither A nor B

3. While discussing gear ratios:
   Technician A says they express the mathematical relationship, according to the number of teeth, of one gear to another.
   Technician B says they express the size difference of two gears by stating the ratio of the smaller gear to the larger gear.
   Who is correct?
   A. A only **C.** Both A and B
   B. B only **D.** Neither A nor B

4. While discussing torque converters:
   Technician A says they are used to transfer engine torque to the transmission.
   Technician B says the turbine rotates at engine speed at all times.
   Who is correct?
   A. A only **C.** Both A and B
   B. B only **D.** Neither A nor B

5. While discussing engine vacuum:
   Technician A says it always increases as engine load decreases.
   Technician B says it always increases with an increase of engine speed.
   Who is correct?
   A. A only **C.** Both A and B
   B. B only **D.** Neither A nor B

6. While discussing valve bodies:
   Technician A says the valve body relies on vacuum signals to determine the best time to cause a change in gear ratios.
   Technician B says the valve body's action may be controlled by electrical solenoids.
   Who is correct?
   A. A only **C.** Both A and B
   B. B only **D.** Neither A nor B

7. While discussing universal joints:
   Technician A says they minimize vibrations caused by the power pulses of the engine.
   Technician B says they allow the drive shaft to change angles in response to movements of the car's suspension and rear axle assembly.
   Who is correct?
   A. A only **C.** Both A and B
   B. B only **D.** Neither A nor B

8. While discussing the purpose of a differential:
   Technician A says it allows for equal wheel speed while the vehicle is rounding a corner or curve.
   Technician B says the ring and pinion in the drive axle multiply the torque they receive from the transmission.
   Who is correct?
   A. A only **C.** Both A and B
   B. B only **D.** Neither A nor B

9. While discussing FWD vehicles:
   Technician A says the differential is normally part of the transaxle assembly.
   Technician B says the drive axles extend from the sides of the transaxle to the drive wheels.
   Who is correct?
   A. A only **C.** Both A and B
   B. B only **D.** Neither A nor B

10. While discussing 4WD vehicles:
    Technician A says 4WD vehicles typically use a transfer case to transfer engine torque to both a front and a rear driving axle.
    Technician B says 4WD vehicles normally have two transmissions, two drive shafts, and two differentials.
    Who is correct?
    A. A only **C.** Both A and B
    B. B only **D.** Neither A nor B

General Theories of Operation

Upon completion and review of this chapter, you should be able to:

- ❑ Describe the differences between automatic transmissions and automatic transaxles.
- ❑ List the factors that determine when an automatic transmission will automatically shift.
- ❑ Describe the four basic systems of all automatic transmissions.
- ❑ Describe the operation and purpose of a torque converter.
- ❑ Describe the basic construction of automatic transmission housings, including the purpose of the various mechanical and electrical connections.
- ❑ List the various types of reaction members commonly used in automatic transmissions.
- ❑ Explain Pascal's Law and how it applies to the operation of an automatic transmission.
- ❑ Describe the purpose of a transmission's valve body.
- ❑ List and describe the various load-sensing devices used in an automatic transmission's hydraulic system.

Introduction

Automatic transmissions are used in many rear-wheel-drive and four-wheel-drive vehicles. Automatic transaxles are used in most front-wheel-drive vehicles. The major components of a transaxle (Figure 3-1) are the same as those of a transmission (Figure 3-2), except the transaxle assembly includes the final drive and differential gears in addition to the transmission. Automatic transmissions and transaxles shift automatically without the driver moving a gearshift lever or depressing a clutch pedal.

A BIT OF HISTORY

The first fully automatic "gear boxes" were introduced by Mercedes of Germany in 1914. These were very limited production cars built for high-ranking officials.

Basic Operation

An automatic transmission receives engine power through a torque converter, which is driven by the engine's crankshaft. Hydraulic pressure in the converter allows power to flow from the torque converter to the transmission's input shaft. The only time a torque converter is mechanically connected to the transmission is during the application of the torque converter clutch. This can only occur if the torque converter is fitted with a locking mechanism.

Automatic transmission fluid is typically referred to as ATF.

The input shaft drives a planetary gear set that provides the various forward gears, a neutral position, and one reverse gear. Power flow through the gears is controlled by multiple-disc clutches, one-way clutches, and friction bands. These hold a member of the gear set when force is applied to them. By holding different members of the planetary gear set, different gear ratios are possible. In all gear positions, one member is driven, one is held, and the other becomes the output.

Hydraulic pressure is routed to the correct driving or holding element by the transmission's valve body, which contains many hydraulic valves and controls the pressure of the hydraulic fluid and its direction.

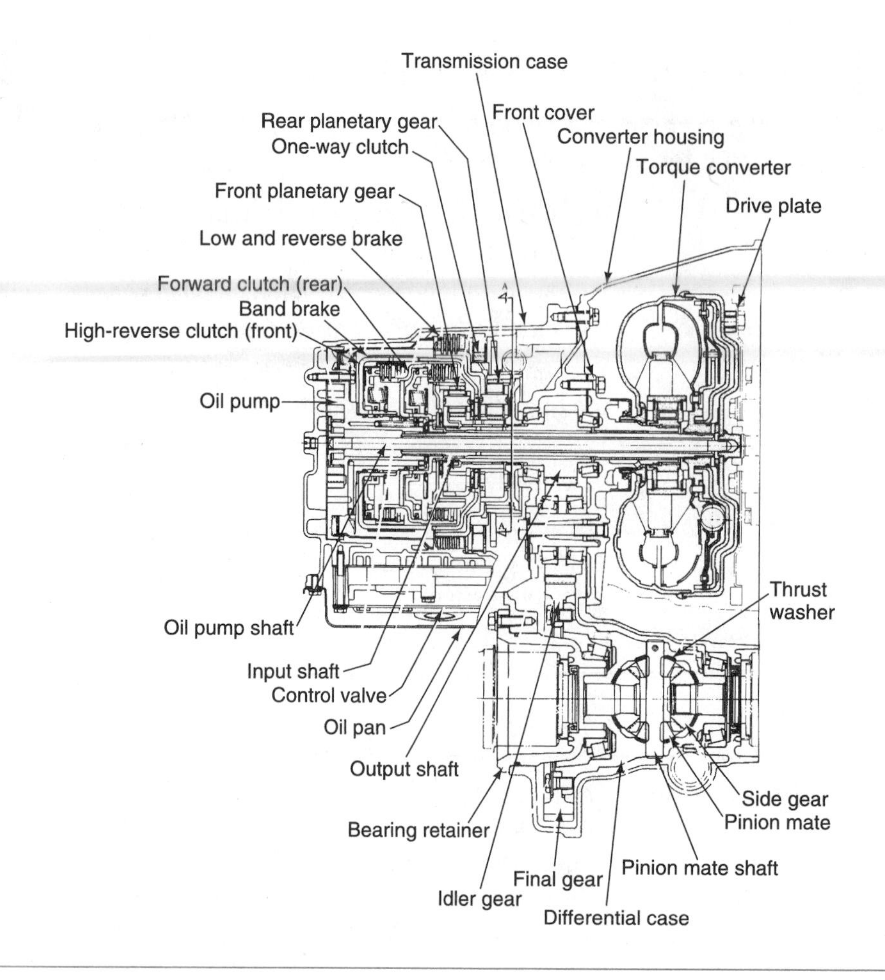

Figure 3-1 A typical automatic transaxle. (Courtesy of Nissan Motor Co., Ltd.)

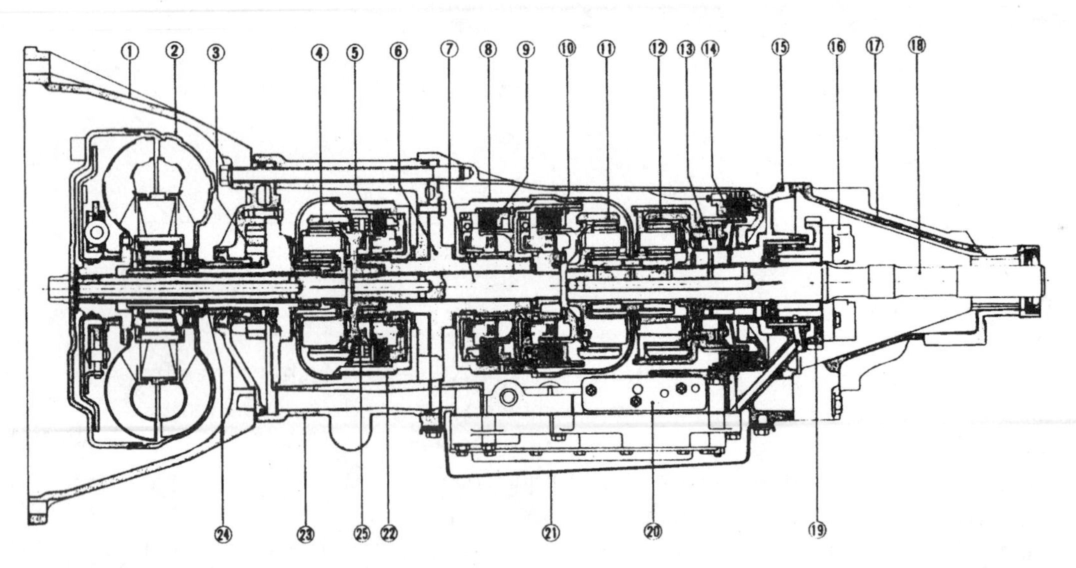

Figure 3-2 A typical automatic transmission. (Courtesy of Nissan Motor Co., Ltd.)

An automatic transmission or transaxle selects gear ratios according to engine speed, engine load, vehicle speed (Figure 3-3), and other operating conditions. Both upshifts and downshifts occur automatically. The transmission can also be manually selected into a lower forward gear, reverse, neutral, or park. Depending on the forward gear range selected, the transmission may provide engine braking during deceleration.

Upshifting is the process of moving up to a higher gear ratio.

For many years, the most widely used automatic transmissions and transaxles had three forward speeds and neutral, reverse, and park positions. Four-speed units, with an overdrive gear, are becoming the most common automatic transmission. An overdrive condition occurs when the transmission's input shaft rotates less than one full revolution for every one *full revolution of the output* shaft. Overdrive improves fuel economy and reduces engine noise. Overdrive also allows the engine to run at slower speeds at high vehicle speeds, which increases engine life. Most automatic transmissions also feature a lockup torque converter, which reduces the power lost through the operation of a conventional torque converter.

Downshifting is the process of moving down to a lower gear ratio.

Gear ratio is an expression of the number of revolutions made by a driving gear compared to the number of revolutions made by a driven gear.

A transmission can have a seven-position shift selector (Figure 3-4). The selector lever is linked to the manual lever on the transmission to select the desired gear range. These ranges are:

"P" (PARK)—The transmission is in neutral with its output shaft locked to the case by engaging a parking pawl within the parking gear. The engine can be started in this selector position and this is the only position in which the ignition key can be removed.

"R" (REVERSE)—The transmission is in reverse at a reduced or lower gear ratio.

"N" (NEUTRAL)—Like PARK, the transmission is in neutral and the engine can be started; however, the output shaft is not locked to the case and the parking brake should be applied.

"OD" (OVERDRIVE)—This is the normal driving gear range. Selection of this position provides for automatic shifting into all forward gears and allows for application and release of the converter clutch.

Shop Manual
Chapter 3, page 55

"D" (DRIVE)—Selection of this gear range provides automatic shifting into all forward gears except overdrive gear, and allows for the application and release of the converter clutch. This position is selected when overdrive is not desired, such as when traveling on hilly or mountainous roads or when towing a trailer. Operating in overdrive during those conditions places an extraordinary amount of load on the engine and can result in severe damage.

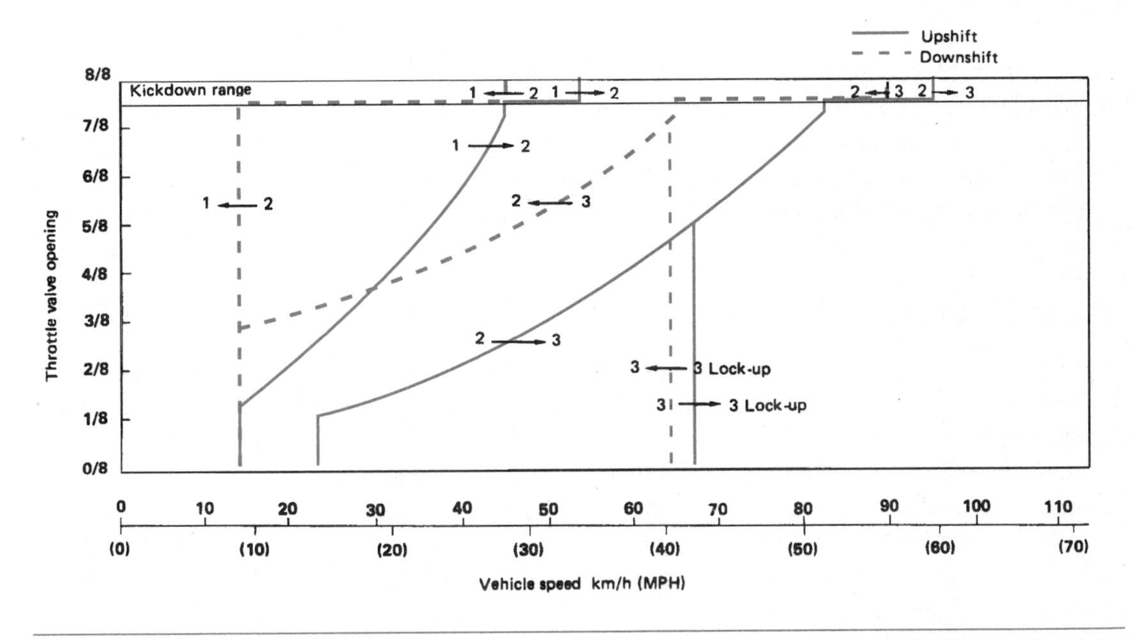

Figure 3-3 The shift schedule for a three-speed automatic transmission. (Courtesy of Nissan Motor Co., Ltd.)

Figure 3-4 Gear selection pattern for a seven-position shift selector.

"2" (MANUAL SECOND)—The selection of this gear position provides only second gear operation regardless of vehicle speed. Some transmissions, when MANUAL SECOND is selected, will start off in first gear then upshift to second. Others will start and remain in second gear. The selection of this gear range is wise for acceleration on slippery surfaces or for engine braking while going down steep hills.

"1" (MANUAL LOW)—In most transmissions, this shift selection allows only first gear operation. Some transmissions are designed to shift into second gear when engine speed is high and MANUAL LOW is selected. Selection of this gear at high speed results in a downshift into second gear. An automatic downshift to first will occur only when the vehicle's speed decreases to a predetermined level, normally about 30 mph. Once the transmission is in first gear, it will stay in low until the selector lever is moved to another position.

An automatic transmission automatically selects the gear ratio and torque output best suited for the existing operating conditions, vehicle load, and engine speed. A typical automatic transmission consists of four basic systems: the torque converter, planetary gear sets, hydraulic systeM, and **reaction members.**

Autostick

Some late-model cars, such as the Dodge Intrepid, feature a manual shift option that allows the driver to control the shifting of the transmission. Chrysler calls this option "autostick." The benefit of this option is simply driver control of gear changes without the use of a clutch pedal. When the shifter is moved into the autostick position, the transaxle remains in whatever gear it was using before autostick was activated. Moving the shifter toward the driver causes the transmission to upshift. Moving the shifter to the passenger side causes the transaxle to downshift. The selected gear is illuminated on the instrument panel.

Automatic transmission fluid can be referred to as transmission fluid.

The car can be launched in first, second, or third gear while in the autostick mode. Shifting into overdrive cancels the autostick option and the transmission resumes its normal overdrive operation. Although the driver has virtually full control of transmission shifting, there are limitations as to when shifts can be made. A shift into third or fourth gear cannot be made if the vehicle is going below 15 mph. Nor can the transmission be downshifted into first if the car is traveling at speeds greater than 41 mph. These safeguards prevent transmission and engine damage possible if those gears were selected.

Shop Manual
Chapter 3, page 48

Torque Converters

The torque multiplication action of a torque converter is similar to adding another reduction gear to the transmission.

Automatic transmissions use a **torque converter** to transfer engine torque from the engine to the transmission. The torque converter operates through hydraulic force provided by automatic transmission fluid.

The torque converter automatically engages and disengages power from the engine to the transmission in relation to engine rpm. When the engine is running at the correct idle speed, there is an insufficient amount of fluid flow in the converter to allow for power transfer through the torque converter. As engine speed increases, fluid flow increases and creates a sufficient amount of force to transmit engine power through the torque converter to the input shaft of the transmission. This action connects the engine to the transmission's planetary gear set and provides some torque multiplication during some driving conditions.

A BIT OF HISTORY

In 1941, Chrysler introduced a four-speed, semi-automatic transmission with a hydraulic coupling.

The torque converter also absorbs the shock of the transmission while it is changing gears. The main components of a torque converter are the cover, turbine, impeller, and stator (Figure 3-5). The **converter cover** is connected to the flywheel and transmits power from the engine into the torque converter. The **impeller** is driven by the converter cover. Its rear hub usually drives the transmission's oil pump. The **turbine** is splined to the **input shaft** and is driven by the force of the fluid from the impeller. The **stator** aids in redirecting the flow of fluid from the turbine to the impeller (Figure 3-6) and is equipped with a one-way clutch that allows it to remain stationary during maximum torque development.

The input shaft provides power flow from the torque converter into the planetary gear train.

The stator of a torque converter is called the reactor by some manufacturers.

The input shaft is also called the turbine shaft.

Gear Sets

A BIT OF HISTORY

The transmission in a Model "T" Ford was based on a planetary gear set that was manually controlled.

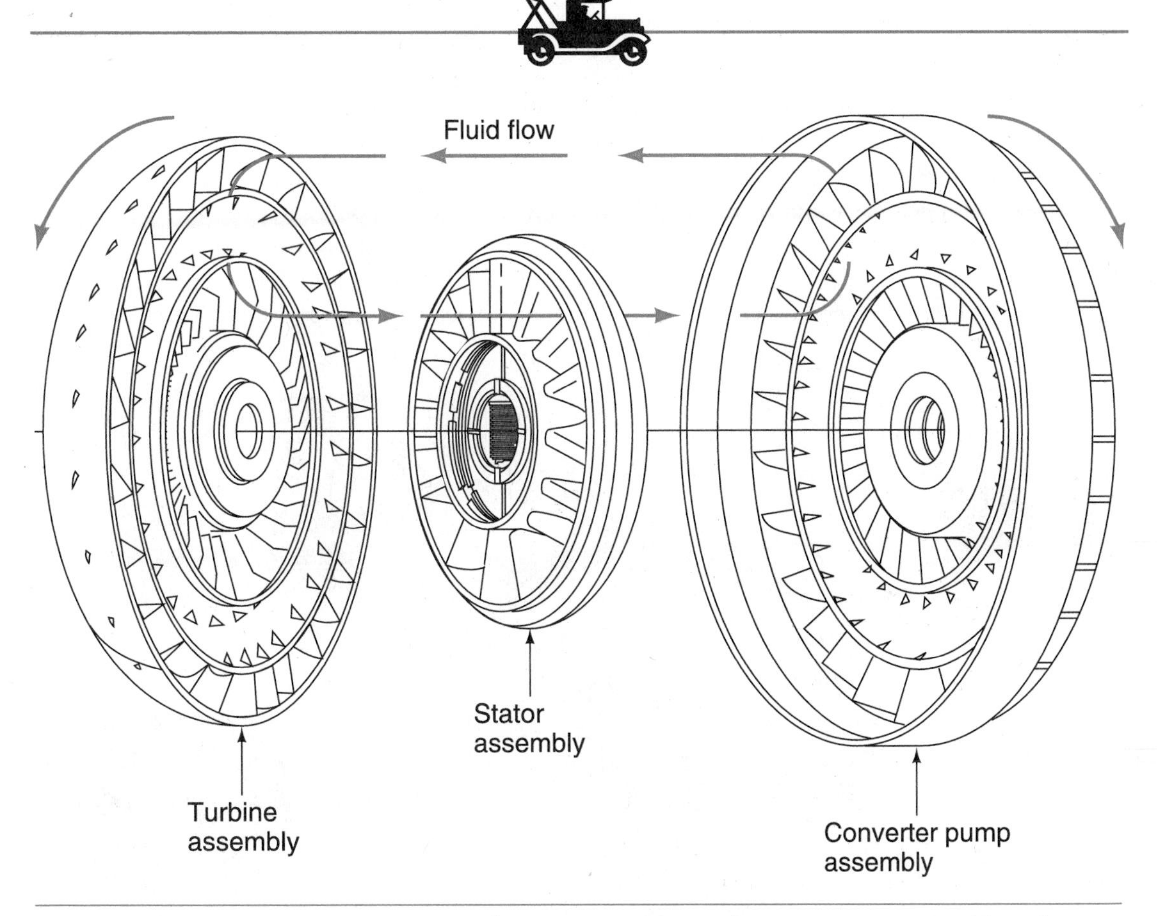

Figure 3-5 The main components of a torque converter. The converter cover and impeller (pump assembly) are shown as a solid unit. (Courtesy of the Hydra-Matic Division of General Motors Corp.)

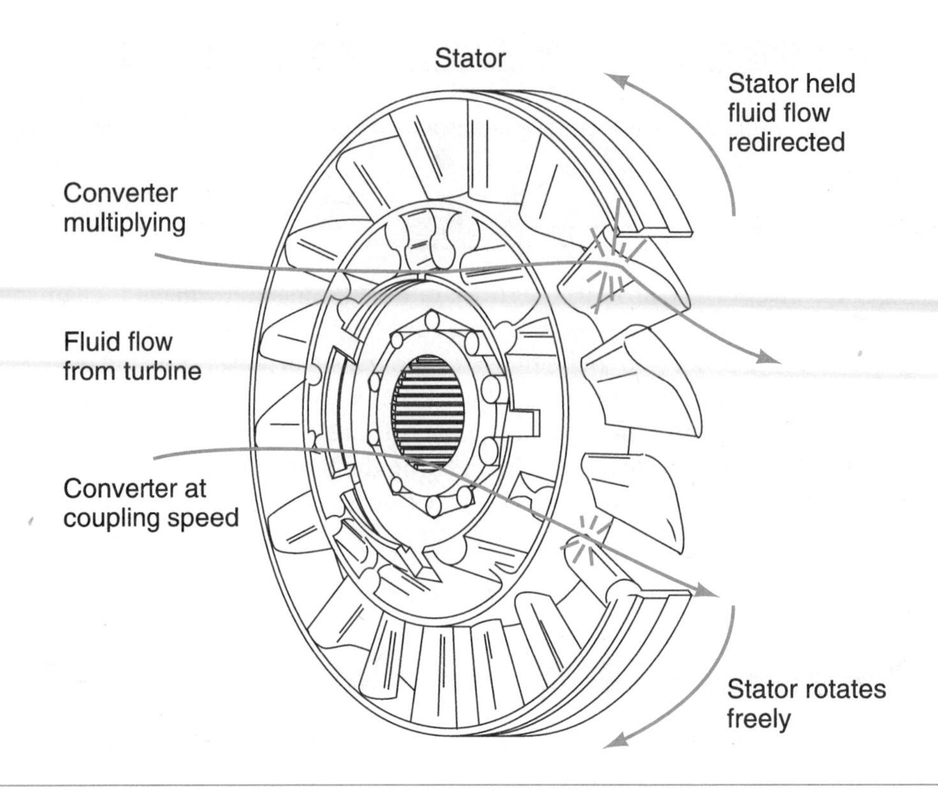

Figure 3-6 The stator aids in directing the fluid flow from the turbine to the impeller. (Courtesy of the Hydra-Matic Division of General Motors Corp.)

A reduction ratio results in increased torque with a reduced speed.

Planetary gear sets provide for the different gear ratios in an automatic transmission. The gear ratios are selected manually by the driver or automatically by the hydraulic control system, which engages and disengages the clutches and bands used to shift gears. A single planetary gear set consists of a sun gear, a planet carrier with two or more planet pinion gears, and an interval ring gear (Figure 3-7). From a single planetary gear set, two reverses and five forward speeds are possible. The five forward speeds are two reduction ratios, a direct drive, and overdrive ratios. These different speeds are obtained by holding one or more gear set members and driving another.

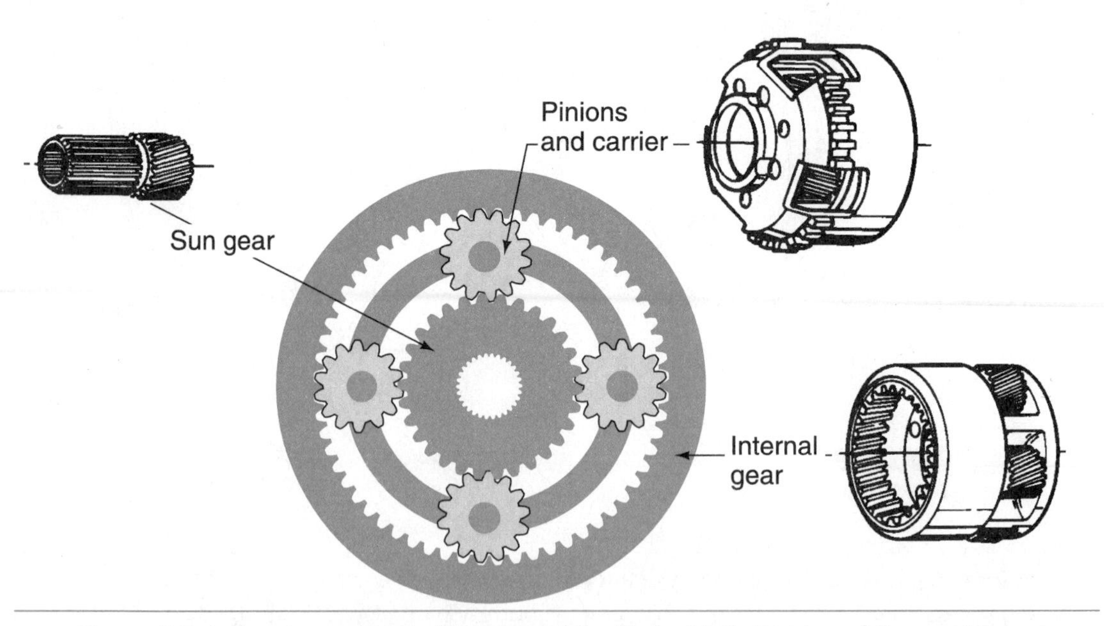

Figure 3-7 A planetary gear set. (Courtesy of the Hydra-Matic Division of General Motors Corp.)

Hydraulic Controls

Automatic transmission fluid (ATF) is a special oil designed to allow for proper transmission operation. Transmissions are equipped with a fluid cooler that prevents the overheating of the fluid, which can result in damage to the transmission. The transmission's oil pump is the source of all pressures with the hydraulic system. It provides a constant supply of oil under pressure to operate, lubricate, and cool the transmission. **Pressure-regulating valves** change the oil's pressure to control the shift points and shift quality of the transmission. **Flow-directing valves** direct the pressurized oil to the appropriate apply device to cause a change in gear ratios. The hydraulic system also keeps the torque converter filled with fluid.

Shop Manual
Chapter 3, page 43

Reaction members are those parts of an automatic transmission that hold or drive members of the planetary gear set in order to change gears. One-way overrunning clutches, bands, and multiple-disc clutch packs are examples of reaction members.

Electronic Controls

Electrical and electronic circuits are used to perform work or control the operation of a device. An electrical switch is a simple electrical control that turns a circuit on or off, and consequently turns the electrical device connected to the circuit on or off. Switches can be opened or closed by the driver, mechanical linkages, or a predetermined condition. The latter typically is a low or high hydraulic pressure.

Electrical sensors are also a type of control; however, rather than merely opening and closing a circuit, sensors vary the flow of electricity through the circuit. Sensors are typically potentiometers or variable resistors that respond to changes in conditions. The sensor's resistance reflects the condition it is monitoring.

Solenoids and motors are commonly used in automobiles. These devices convert electrical energy into mechanical energy. A motor provides for rotating power, whereas a solenoid provides for linear movement. Solenoids in automatic transmissions are used to control the direction and flow of hydraulic pressure.

Shop Manual
Chapter 3, page 50

The front of the housing surrounding the torque converter is shaped like a bell and is often called a bell housing.

Until recently, all automatic transmissions were controlled by hydraulic circuits. However, many transmissions now control the operation of the torque converter and transmission through a computer. Based on information received from various electronic sensors and switches, a computer can control the operation of the torque converter and transmission's shift points. Computer-controlled electrical solenoids are typically used to control shifting and pressures (Figure 3-8).

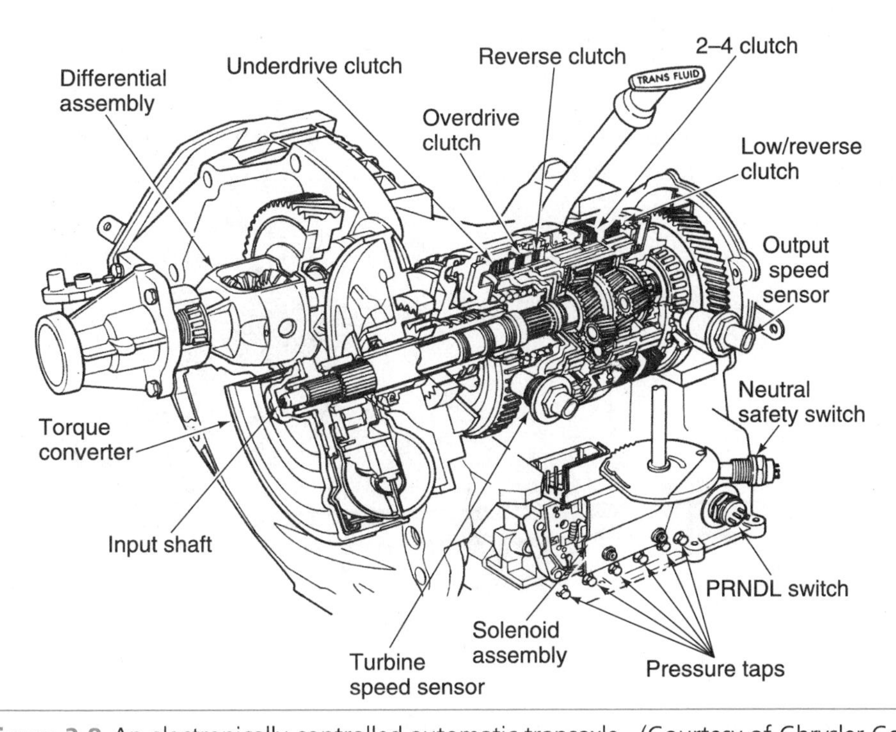

Figure 3-8 An electronically controlled automatic transaxle. (Courtesy of Chrysler Corp.)

Transmission and Transaxle Housings

Shop Manual
Chapter 3, page 49

Extension housings conceal the transmission's output shaft.

The wheelbase of a vehicle is best defined as the distance between the front and rear axles.

Automatic transmission housings are typically aluminum castings. Aluminum housings are lightweight and are designed to allow for quick heat dissipation. The torque converter and transmission housings are normally cast as a single unit; however, in some designs the torque converter housing is a separate casting (Figure 3-9). By bolting the torque converter housing to the transmission housing, engineers can use the same transmission with different-sized torque converters to meet the specific needs of particular types of vehicles. Also, at the rear of some transmission housings is a bolted-on extension housing (Figure 3-10). This allows for the use of housings of different lengths; therefore, the transmission can be used on vehicles with different wheelbases.

Transaxle housings (Figure 3-11) are typically composed of two or three separate castings bolted together. One of these castings is the torque converter housing. Since FWD vehicles do not use a drive shaft to convey power to a drive axle, most transaxles are not fitted with an extension housing. Although they serve many of the same purposes as a transmission, transaxle housings are considerably different in appearance from transmission housings.

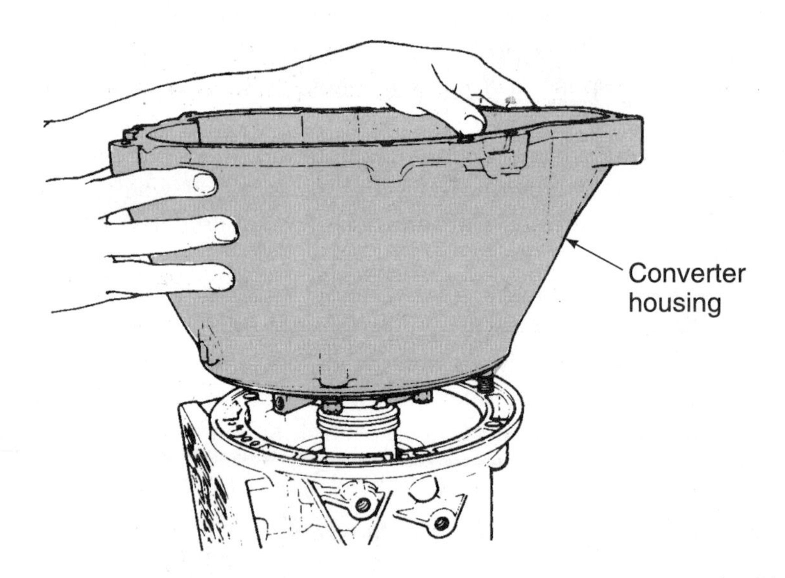

Figure 3-9 A torque converter housing cast separately from the transmission housing. (Reprinted with the permission of Ford Motor Co.)

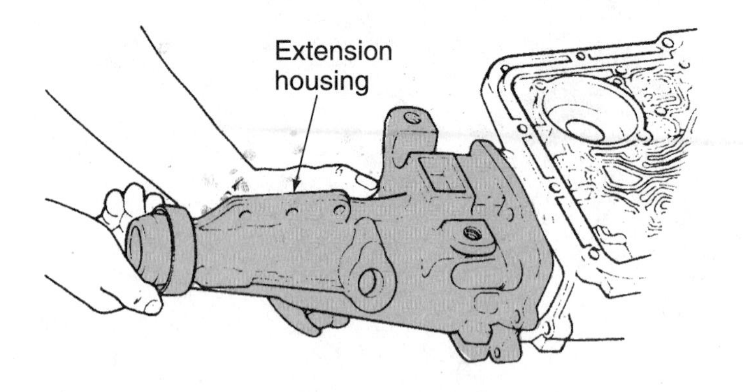

Figure 3-10 A transmission housing with a bolted-on extension housing. (Reprinted with the permission of Ford Motor Co.)

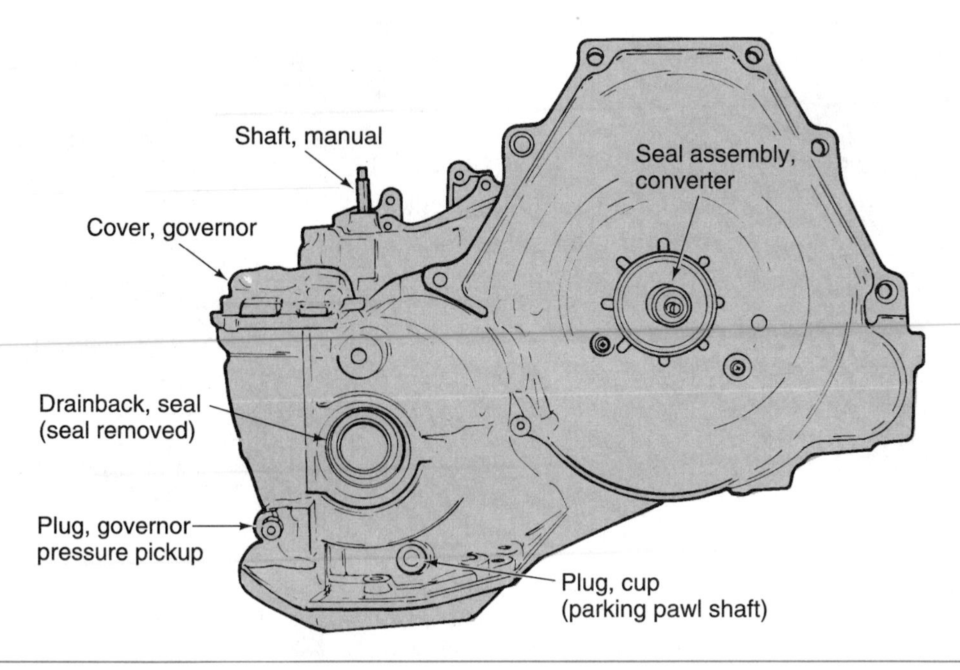

Figure 3-11 Side view of a transaxle housing. Note that the converter housing is an integral part of the housing. (Courtesy of Chevrolet Motor Division of General Motors Corp.)

A BIT OF HISTORY

The 1936 Imperia, which was built in Belgium, had a fully automatic transmission with a V-8 engine and front-wheel drive.

Transmission housings are cast to secure and accommodate key components of the transmission. For example, linear keyways are cut into the inside diameter of the housing. These keyways are designed to hold a multiple-disc brake assembly in position in the housing (Figure 3-12). Some transmission housings have round structures projecting from the side. These are the cylinders of the **servo** assemblies in which pistons, springs, and seals are fitted to operate the bands of the transmission. On some transaxles, these projections house the **governor** assembly (Figure 3-13) and are sealed with a removable cover.

Shop Manual
Chapter 3, page 55

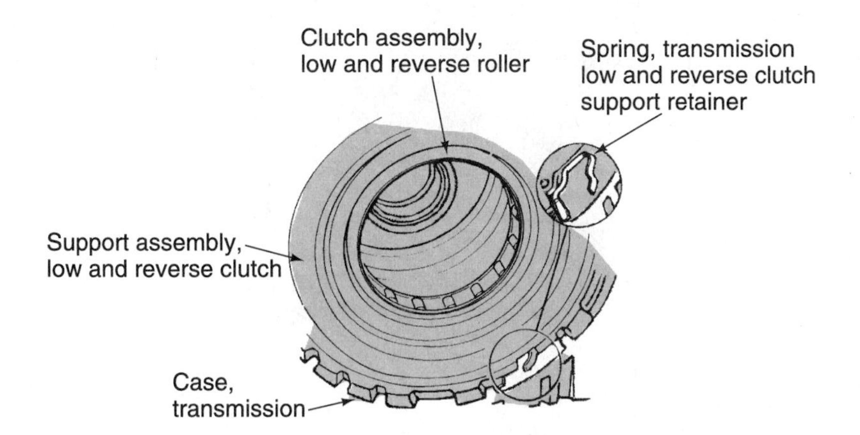

Figure 3-12 A multiple-disc clutch pack fitted into the keyways cast into the transmission housing. (Courtesy of the Hydra-Matic Division of General Motors Corp.)

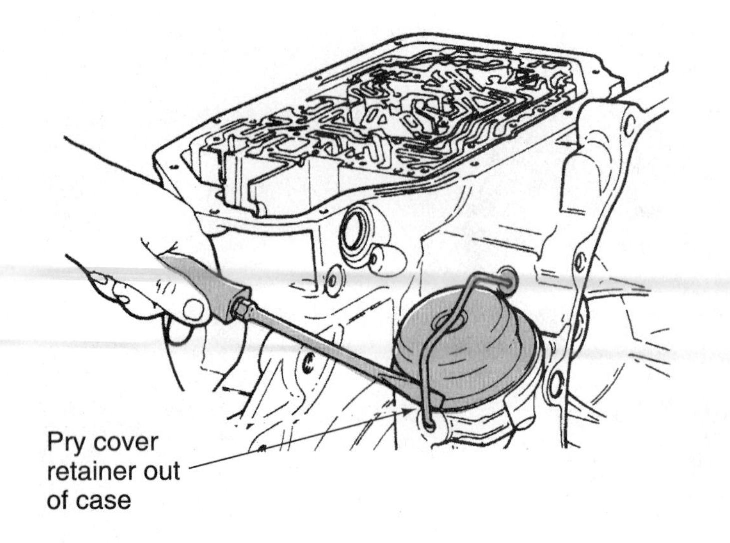

Figure 3-13 The governor housing of a transaxle. (Reprinted with the permission of Ford Motor Co.)

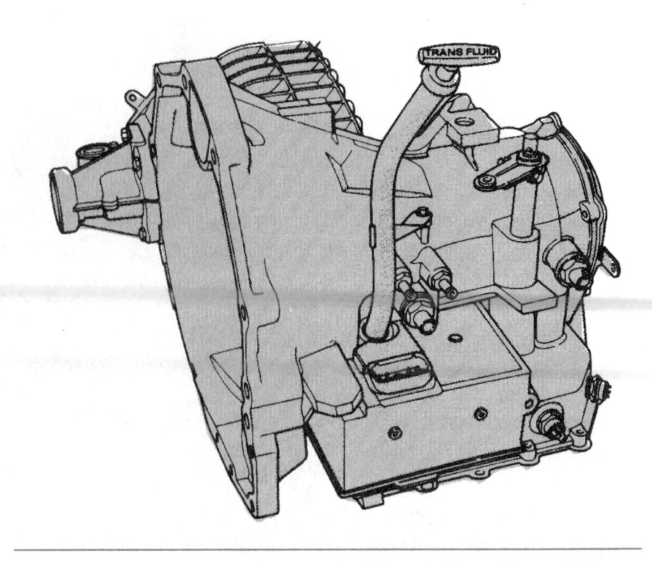

Figure 3-14 Location of oil-level dipstick and filler tube on a transaxle. (Courtesy of Chrysler Corp.)

The transmission's fluid dipstick and filler tube are used to check the level of ATF and add ATF to the transmission. This tube is normally located at the front of a transaxle housing (Figure 3-14) and at the right front of a transmission housing.

Some housings have band-adjusting screws located on the opposite side of the housing from the servo assemblies. These screws are adjusted to compensate for band wear. The band-adjusting screws can have square or hex heads and are locked in place with a jam nut.

Mechanical Connections

Shop Manual
Chapter 3, page 50

The speedometer drive assembly is geared to the **output shaft.** A flexible drive cable normally connects the drive assembly to the speedometer in the instrument panel. Transmissions without a speedometer drive have a wiring harness connected to a vehicle speed sensor, also located at the output shaft.

The common acronym for a vehicle speed sensor is VSS.

At the rear or side of some transmission housings is a round metal canister with a tubular opening; this is the **vacuum modulator** (Figure 3-15). A vacuum line from the engine's intake manifold is connected to the opening. This supplies vacuum to operate the modulator. The

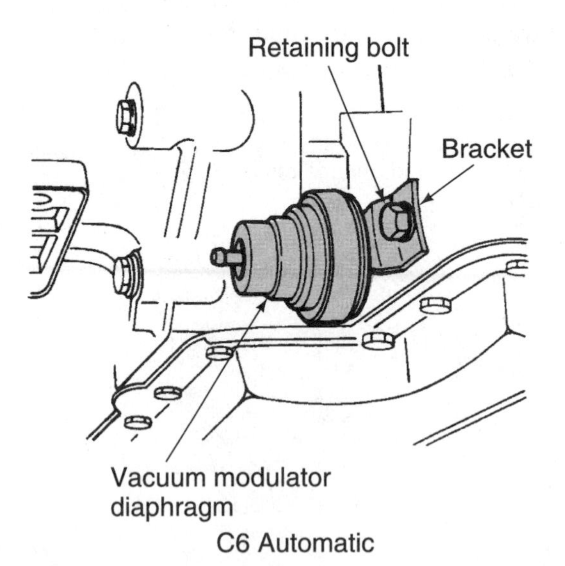

Figure 3-15 Typical location and mounting of a vacuum modulator. (Reprinted with the permission of Ford Motor Co.)

vacuum modulator is a **load-sensing device** that increases or decreases fluid pressure in response to the vacuum signal. Transmissions not equipped with a vacuum modulator use a throttle cable or electronics to sense engine load and change fluid pressures. Increasing the fluid pressure holds the planetary control units tightly to reduce the chance of slipping while under heavy load.

Linkages

The throttle cable is normally connected to a lever located at the left front of the transmission housing. The **throttle valve** is moved by the throttle lever, which is moved by the throttle linkage at the carburetor or throttle body of the fuel injection system (Figure 3-16). The throttle lever moves according to the position of the throttle pedal. Some automatic transmissions use a rod-type linkage assembly, not a cable, to move the throttle lever. The other lever is the manual shift lever. As the gearshift selector is moved into its various positions, the manual lever moves the **manual shift valve** in the transmission's valve body (Figure 3-17). The movement of the gear selector is directed to the manual shift lever via a cable or shift rod assembly. Each gear position of the lever or shift valve is held in place by the internal linkage seating itself into the various seats or detents for the gear range selected.

Shop Manual
Chapter 3, page 60

A detent is a half-round seat that a ball or roller fits into to hold a lever in position.

Cooler Lines

The housings are also fitted with an ATF cooler outlet and return line fittings (Figure 3-18). ATF leaves the torque converter and is directed to the transmission cooler in the vehicle's radiator. Some vehicles have an auxiliary cooler separate from the radiator. The cooler return line sends the cooled fluid back to the transmission housing.

Most transmissions are also equipped with plugs that can be removed to connect hydraulic pressure gauges for testing the operation of the transmission (Figure 3-19).

Shop Manual
Chapter 3, page 53

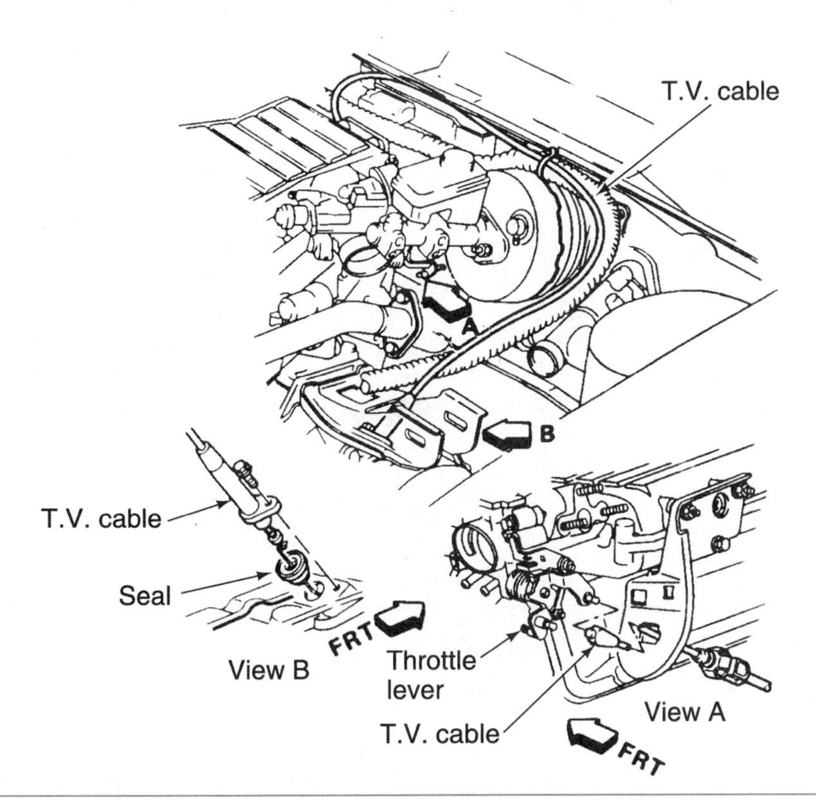

Figure 3-16 Typical routing of throttle valve linkage. (Courtesy of Chevrolet Motor Division of General Motors Corp.)

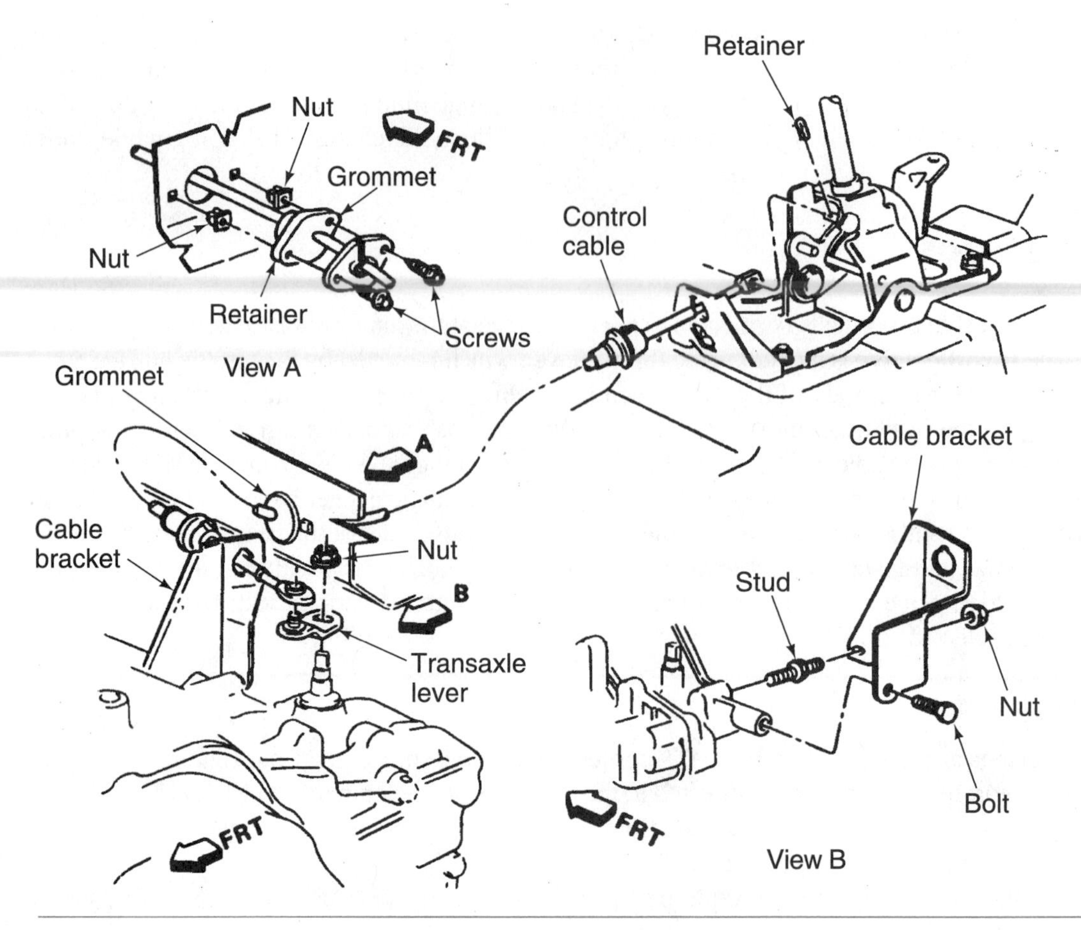

Figure 3-17 Typical shift cable routing for a floor-mounted shifter. (Courtesy of Chevrolet Motor Division of General Motors Corp.)

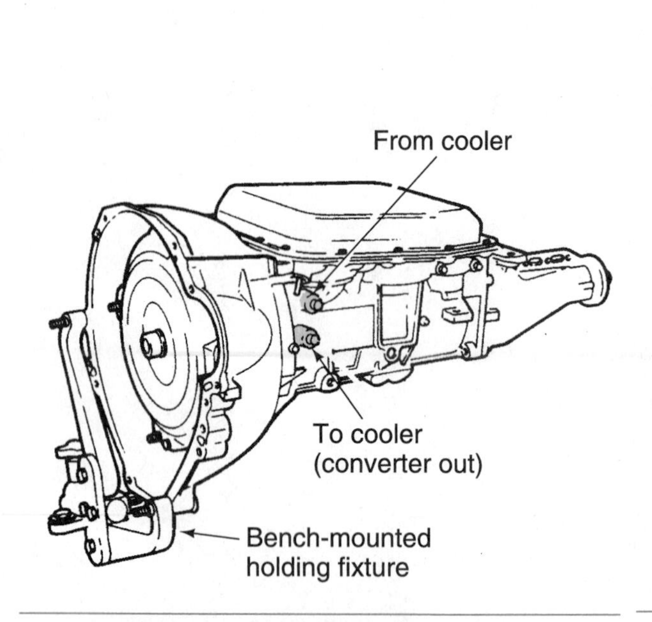

Figure 3-18 Typical location of fluid cooler inlet and outlet ports. (Reprinted with the permission of Ford Motor Co.)

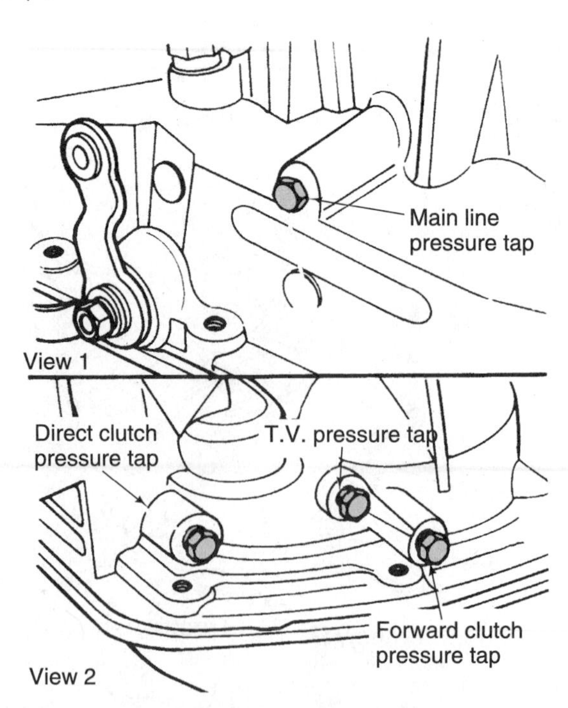

Figure 3-19 Typical location of pressure taps for connecting a pressure gauge to a transmission for diagnostic purposes. (Reprinted with the permission of Ford Motor Co.)

Electrical Connections

Shop Manual
Chapter 3, page 50

Many different electrical switches, sensors, and connectors may be connected to the housing (Figure 3-20). As more electronics are used to control and monitor the functioning of the transmission, more connectors and electrical devices will be found at the housing. One of the more basic switches is the neutral safety switch, which allows the engine to start only when the transmission is in neutral or park. The backup light switch is sometimes incorporated with the neutral safety switch, as are other shift position switches. These switches may be individual switch assemblies. Connectors for shift solenoids may also be present on the outside of the case.

Mounts

Shop Manual
Chapter 3, page 51

The weight of the transmission or transaxle is supported by the engine and its mounts and by a transmission mount (Figure 3-21). These mounts are not only critical for proper operation of the transmission, but they also isolate transmission noise and vibrations from the passenger compartment. The mountings for the engine and transmission keep the power train in proper alignment with the rest of the drive train and help maintain proper adjustment of the various linkages attached to the housing.

Planetary Gearing

Planetary gears provide for the different gear ratios needed to move a vehicle in the desired direction at the correct speed. A planetary gear set consists of a sun gear, planet gears, and a ring gear. These gears are typically helically cut gears, which offer quiet operation.

The sun gear is at the center of the planetary gear set; just s the earth's sun is at the center of our solar system. Planet gears surround the sun gear, just as the earth and other planets surround the sun in our solar system. These gears are mounted and supported by the planet carrier and each gear spins on its own separate shaft. The planet gears are in constant mesh with the sun and ring gears. The ring gear is the outer gear of the gear set. It has internal teeth and surrounds the rest of the gear set. Its gear teeth are in constant mesh with the planet gears. The number of planet gears used in a planetary gear set varies according to the loads the transmission is designed to face. For heavy loads, the number of planet gears is increased to spread the work load over more gear teeth.

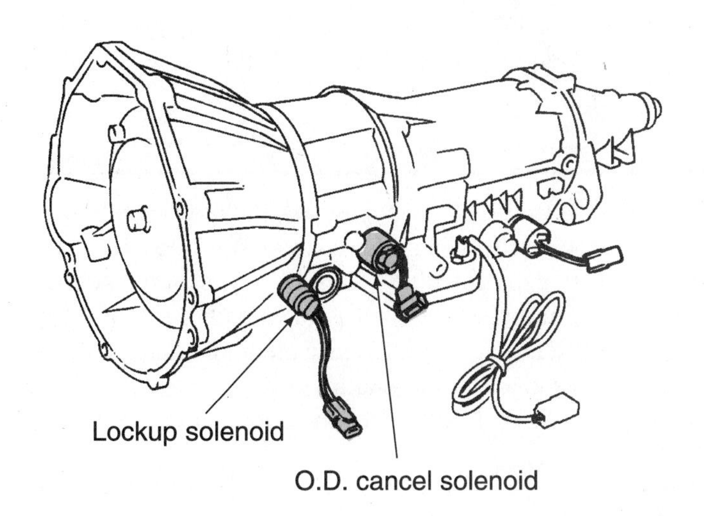

Figure 3-20 Various electrical sensor, solenoid, and switch connections to the transmission housing. (Courtesy of Nissan Motor Co., Ltd.)

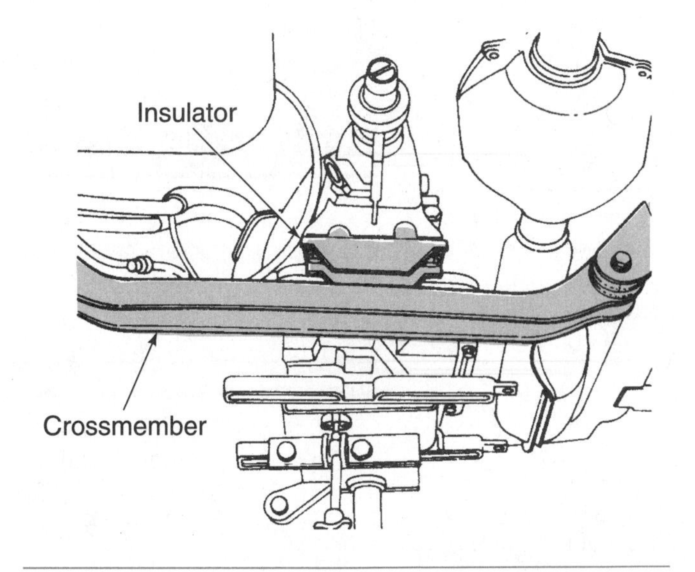

Figure 3-21 Typical rear transmission-to-frame mount. (Reprinted with the permission of Ford Motor Co.)

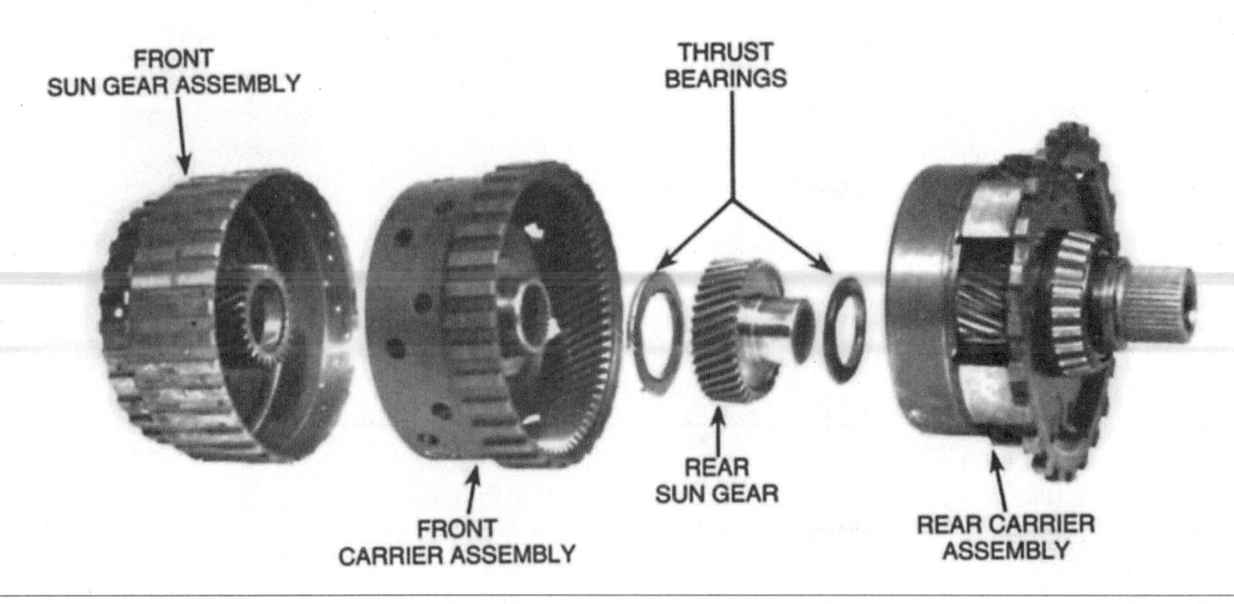

Figure 3-22 A compound planetary gear set. (Courtesy of Chrysler Corp.)

The planetary gear set can provide a gear reduction or overdrive, direct drive or reverse, or a neutral position. Because the gears are in constant mesh, gear changes are made without engaging or disengaging gears, as is required in a manual transmission. Rather, clutches and brake bands are used to either hold or release different members of the gear set to get the proper direction of rotation and gear ratio.

Compound Planetary Gear Sets

When two or more planetary gear sets are used in a transmission, they are collectively called a **compound planetary gear system.**

A limited number of gear ratios is available from a single planetary gear set. To increase the number of available gear ratios, gear sets can be added. The typical automatic transmission with three or four forward speeds has at least two planetary gear sets.

In automatic transmissions, two or more planetary gear sets (Figure 3-22) are connected together to provide the various gear ratios needed to efficiently move a vehicle. There are two common designs of compound gear sets: the Simpson gear set in which two planetary gear sets share a common sun gear; and the Ravingeaux gear set, which has two sun gears, two sets of planet gears, and a common ring gear. Some transmissions are fitted with an additional single planetary gear set, which is used to provide an "add-on" overdrive gear.

Hydraulic Basics

Hydraulics is the study of liquids and fluids in motion or under pressure.

An automatic transmission is a complex hydraulic circuit. To better understand how an automatic transmission works, a good understanding of how basic hydraulic circuits work is needed. A simple hydraulic system has liquid, a pump, lines to carry the liquid, control valves, and an output device. The liquid must be available from a continuous source such as an oil pan or sump. An oil pump is used to move the liquid through the system. The lines to carry the liquid may be pipes, hoses, or a network of internal bores or passages in a housing (Figure 3-23). Control valves are used to regulate hydraulic pressure and direct the flow of the liquid. The output device is the unit that uses the pressurized liquid to do work.

A fluid is a liquid or gas that is capable of flowing.

Hydraulics involves the use of a liquid or fluid. All matter in the universe exists in at least one of three basic forms: solid, liquid, and gas. A fluid is something that does not have a definite shape; therefore, liquids and gases are fluids. A characteristic of all fluids is that they conform to the shape of their container. A major difference between a gas and a liquid is that a gas will always fill a sealed container whereas a liquid may not. A gas will readily expand or compress

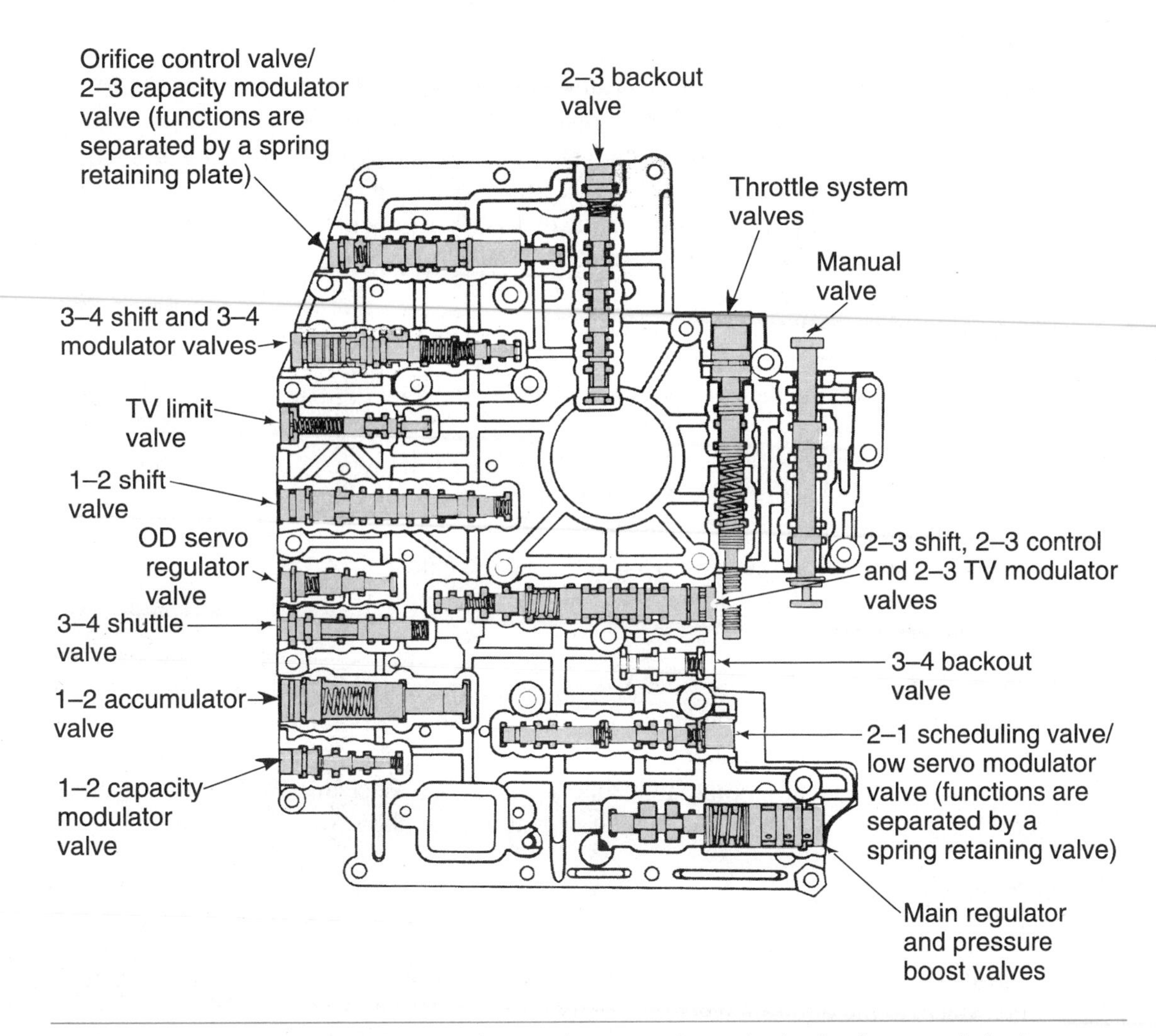

Figure 3-23 The various bores, valves, and passages in a valve body of a transmission's hydraulic system. (Reprinted with the permission of Ford Motor Co.)

according to the pressure exerted on it. A liquid will typically not compress, regardless of the pressure on it. Therefore, liquids are considered noncompressible fluids. Liquids will, however, predictably respond to pressures exerted on them. Their reaction to pressure is the basis of al l hydraulic applications.

Pascal's Law states that pressure exerted on a confined liquid or fluid is transmitted undiminished and equally in all directions and acts with equal force on all areas.

Over 300 years ago a French scientist, Blaise Pascal, determined that if you had a liquid-filled container with only one opening and you applied force to the liquid through that opening, the force would be evenly distributed throughout the liquid. This explains how pressurized liquid can operate the hydraulic devices within an automatic transmission. The oil pump pressurizes the transmission fluid and the fluid is delivered to various reaction members. When the pressure is great enough to activate the reaction member, it will activate and stay that way until the pressure decreases. The transmission's **valve body** distributes the pressurized fluid to the appropriate hydraulic gears according to the operating conditions of the vehicle.

ATF

The fluid used in an automatic transmission's hydraulic system is ATF. ATF is important to the operation of the transmission. The operation of the transmission and torque converter generates much heat. The ATF circulating through the transmission and torque converter and over the parts of the transmission cools the transmission. It does this by absorbing the heat. The heated fluid moves to the transmission fluid cooler where the heat is removed. The cooled ATF then returns to the transmission to be heated again.

Shop Manual
Chapter 3, page 43

The ATF also lubricates the moving parts in the transmission. This too reduces the heat produced by the transmission. As the fluid lubricates and cools the transmission, it also cleans the parts. The dirt is carried by the fluid to a filter where the dirt is removed.

Another critical job of ATF is its role in shifting gears. ATF moves under pressure throughout the transmission and causes various valves to move. The amount the valves move depends on the pressure of the fluid. This is how automatic shifting occurs. The pressure of the ATF changes with changes in engine speed and load. Hydraulic pressure is developed by the transmission's oil pump and is controlled by various regulating valves that sense engine load and vehicle speed.

ATF is also used to operate the various apply devices (clutches and bands) in the transmission. At the appropriate time, a switching valve opens and sends pressurized fluid to the apply device, which engages or disengages a gear. The valves and hydraulic circuits are contained in the valve body.

ATF is a lubricant designed specifically for automatic transmissions. There are several types of ATF available, each of which is refined for a specific application. Automatic transmissions are equipped with an oil filter or screen designed to filter particles out of the circulating oil. These filters are normally located below the valve body and in the oil pan.

Reaction Members

Shop Manual
Chapter 3, page 45

Reaction members are those parts of an automatic transmission that hold members of the planetary gear set in order to change gears. One-way overrunning clutches, bands, and multiple-disc clutch packs are examples of reaction members. One-way overrunning clutches are purely mechanical devices, whereas clutches and bands are hydraulically controlled mechanical devices. Most automatic transmissions use more than one type of these reaction members; some use all three.

Valve Body

For efficient operation of the transmission, the bands and clutches must be released and applied at the proper time. It is the responsibility of the hydraulic control system to control the hydraulic pressure being sent to the different reaction members. Central to the hydraulic control system is the valve body assembly. This assembly is usually made of two or three main parts: a valve body, **separator plate**, and **transfer plate** (Figure 3-24). These parts are bolted as a single unit

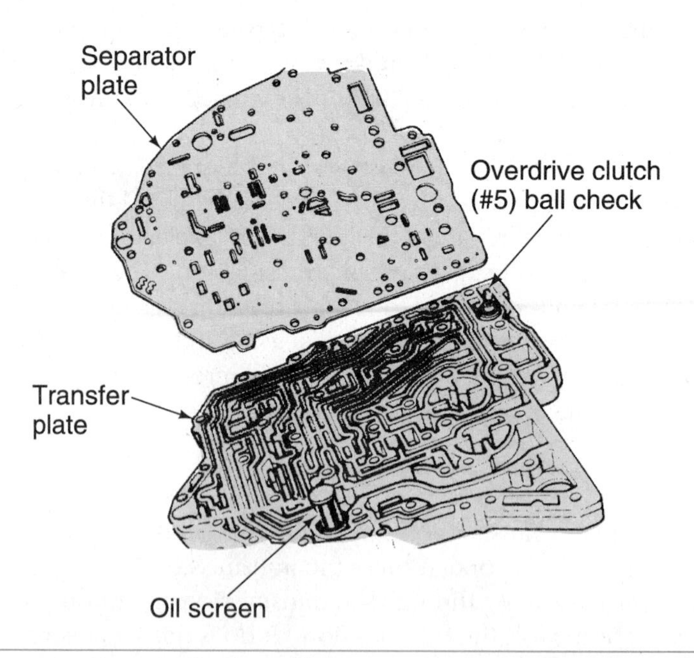

Figure 3-24 The separator and transfer plates of a valve body. (Courtesy of Chrysler Corp.)

to the transmission housing. The valve body is machined from aluminum or iron and has many precisely machined bores and fluid passages. Various valves are fitted into the bores, and the passages direct fluid to various valves and other parts of the transmission. The separator and transfer plates are designed to seal off some of these passages and allow fluid to flow through specific passages.

The purpose of a valve body is to sense and respond to engine and vehicle load, as well as to meet the needs of the driver. Valve bodies are normally fitted with three different types of valves: spool valves, **check ball valves,** and **poppet valves** (Figure 3-25). The purpose of these valves is to start, stop, or use movable parts to regulate and direct the flow of fluid throughout the transmission.

Oil Flow

The source of fluid flow through the transmission is the oil pump. The three most common types of oil pumps used in an automatic transmission are **gear-type** (Figure 3-26), **rotor-type,** and **vane-type** (Figure 3-27). Oil pumps are driven by the pump drive hub of the torque converter or oil pump shaft and converter cover on transaxles. Therefore, whenever the torque converter cover is rotating, the oil pump is driven. The oil pump creates fluid flow throughout the transmission. The valve body regulates and directs the fluid flow to meet the needs of the transmission.

Oil pumps are capable of creating excessive amounts of pressure, which can damage the transmission. Therefore, the transmission is equipped with a pressure regulator valve.

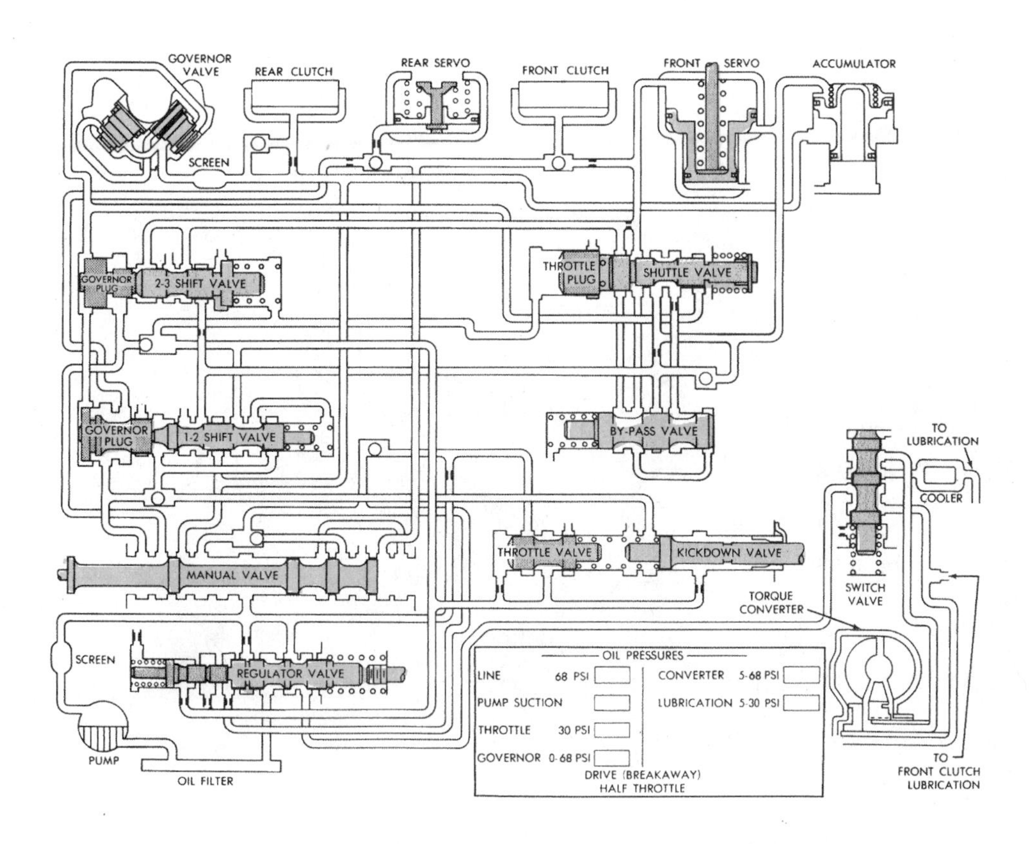

Figure 3-25 Identification of various valves in a typical valve body. (Courtesy of Chrysler Corp.)

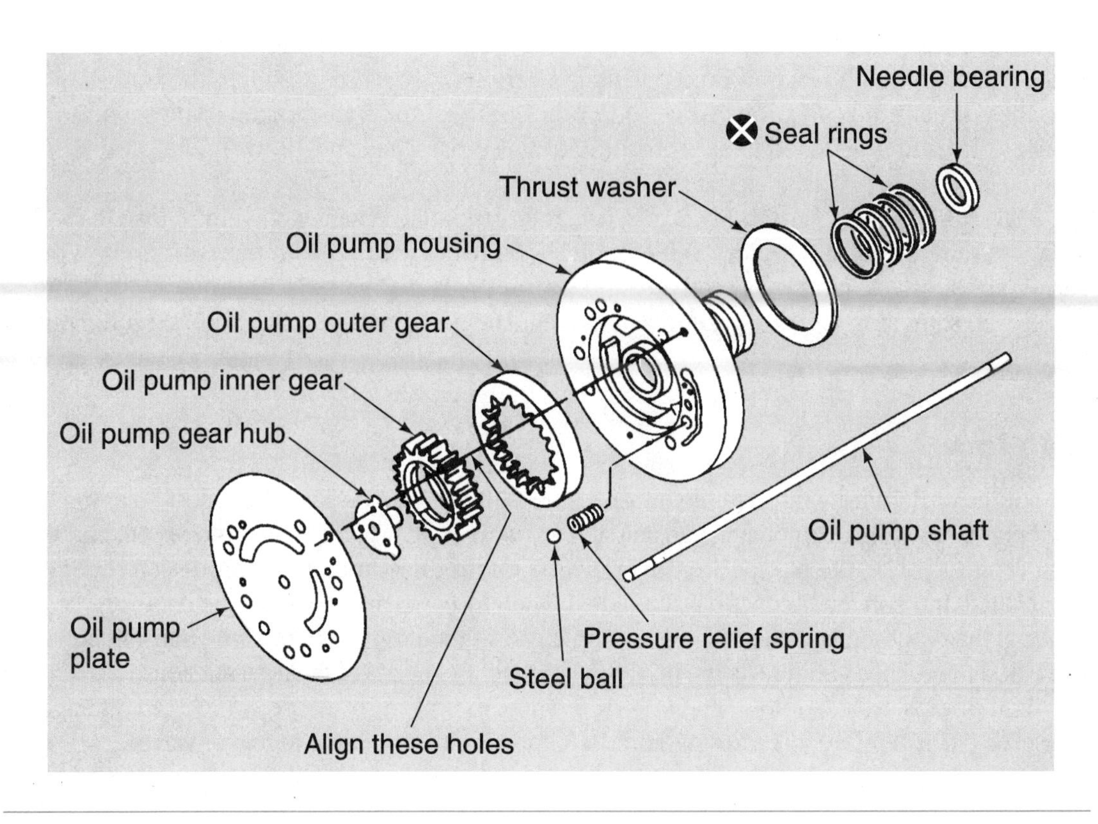

Figure 3-26 A typical gear-type oil pump. (Courtesy of Nissan Motor Co., Ltd.)

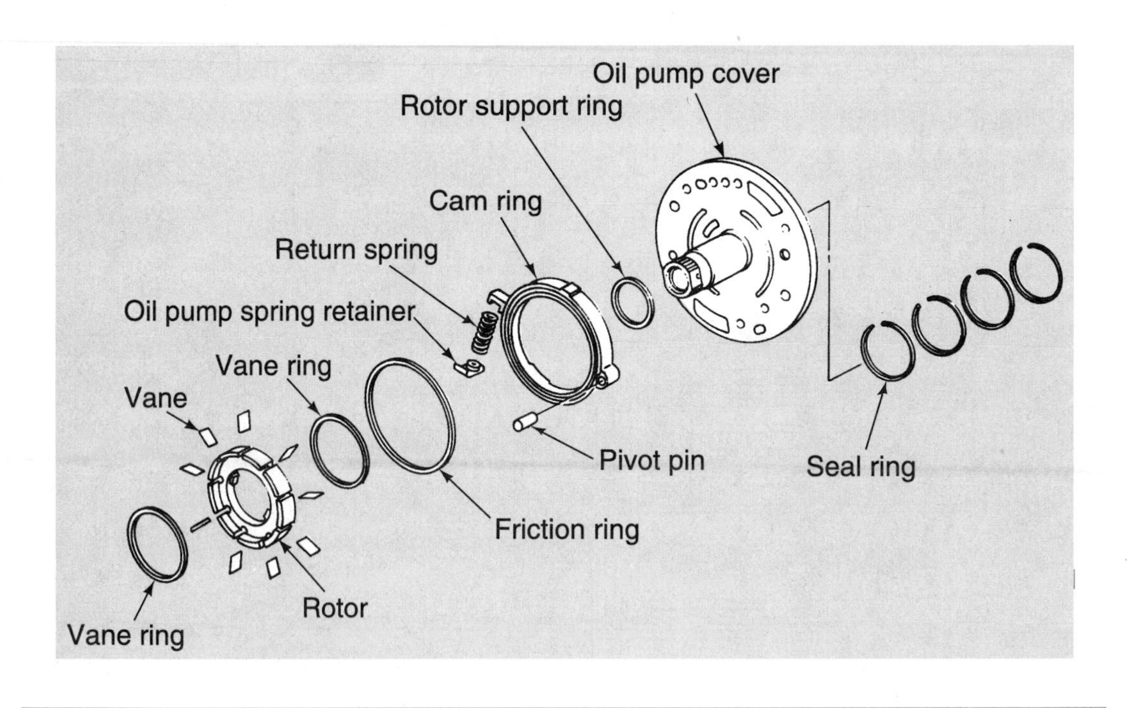

Figure 3-27 A typical vane-type oil pump. (Courtesy of Nissan Motor Co., Ltd.)

Pressure Regulator Valve

The **pressure regulator valve** is usually located in the oil pump or valve body. It maintains basic fluid pressure. Pressure regulating valves are typically spool valves that toggle back and forth in their bore to open and close an exhaust passage. By opening the exhaust passage, the valve decreases the pressure of the fluid. As soon as the pressure decreases to a predetermined amount, the spool valve moves to close off the exhaust port and pressure again begins to build. The action of the spool valve regulates the fluid pressure.

Pressure Boosts

When the engine is operating under heavy load conditions, fluid pressure must be increased to increase the holding capacity of a reaction member. This is accomplished by sending pressurized fluid to one side of the pressure regulator's spool valve. This pressure works against the spool valve's normal movement to open the exhaust port and allows pressure to build to a higher point than normal.

Engine load can be monitored electronically through the use of various electronic sensors that send information to an electronic control unit, which in turn controls the pressure at the valve body. Load can also be monitored by throttle position. Throttle pedal movement moves a **throttle valve** in the valve body via a throttle cable. When the throttle pedal is opened, the throttle valve opens and applies pressure to the pressure regulator. This delays the opening of the pressure regulator valve, which allows for an increase in pressure. When the driver lets off the throttle pedal, the pressure regulator valve is free to move and normal pressure is maintained.

Shop Manual
Chapter 3, page 61

Low engine vacuum is an indication that the engine is under heavy load.

Many transmissions have been equipped with a vacuum modulator, which uses **engine vacuum** to change transmission pressure (Figure 3-28). The vacuum modulator allows for an

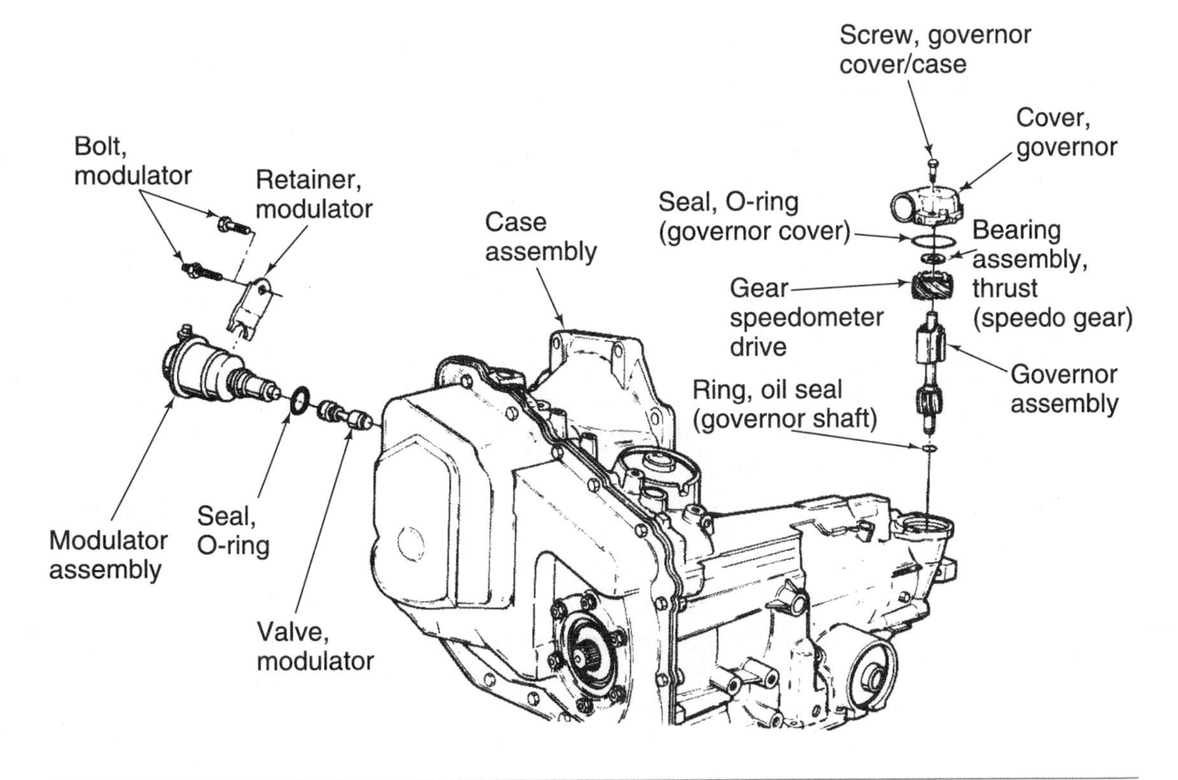

Figure 3-28 A vacuum modulator and governor assembly. (Courtesy of Buick Motor Division of General Motors Corp.)

The fluid pressure signal from the governor is called governor pressure.

increase in pressure when vacuum is low and decreases it when vacuum is high. When high vacuum is present at the modulator, the pressure regulator works normally and maintains normal pressure. However, when the vacuum is low, the modulator allows pressurized fluid to enter onto the side of the spool valve in the pressure regulator, which allows for an increase in pressure.

Governor Assembly

The governor assembly, usually located on or driven by the transmission's output shaft, senses road speed and sends a fluid pressure signal to the valve body to either upshift or downshift. When vehicle speed is increased, the pressure developed by the governor is directed to the shift valve. As the speed (and therefore the pressure) increases, the spring tension on the shift valve is overcome and the valve moves. This action causes an upshift. Likewise, a decrease in speed results in a decrease in pressure and a downshift.

Although the governor sends a signal that will force an upshift, engine load may cause a delay in the shift. This allows for operation in a lower gear when there is a heavy load and the vehicle needs the gear reduction. During heavy load operation, the governor pressure must be strong enough to overcome the high throttle pressure plus the spring tension on the shift valve before it can force an upshift. Because of this, the transmission will remain in a particular gear range until a higher-than-normal engine speed is reached. Electronically controlled transmissions use a vehicle speed sensor instead of a governor.

Kickdown Valve

Shop Manual
Chapter 3, page 63

The valve body is also fitted with a **kickdown** circuit, which provides a downshift when the driver requires additional power. When the throttle pedal is quickly opened wide, throttle pressure rapidly increases and directs a large amount of pressure onto the kickdown valve. This moves the kickdown valve, which opens a port and allows mainline pressure to flow against the shift valve. The spring tension on the shift valve, the kickdown pressure, and the throttle pressure will push on the end of the shift valve, causing it to move to the downshift position. This forces a quick downshift.

Shift Feel

An orifice is a small opening or restriction in a hydraulic circuit that causes a decrease in pressure on the downstream side of the fluid flow.

An accumulator is a hydraulic piston assembly that helps a servo apply a band or clutch smoothly. It does this by absorbing sudden pressure surges in the hydraulic circuit to the servo.

All transmissions are designed to change gears at the correct time, according to engine speed and load and driver intent. However, transmissions are also designed to provide for positive change of gear ratios without jarring the driver or passengers. If a band or clutch is applied too quickly, a harsh shift will occur. **Shift feel** is controlled by the pressure at which each reaction member is applied or released, the rate at which each is pressurized or exhausted, and the relative timing of the apply and release of the members.

To improve shift feel during gear changes, a band is often released while a multiple-disc clutch is being applied. The timing of these two actions must be just right or both components will be released or applied at the same time, which would cause engine flare-up or driveline shudder. Several other methods are used to smooth gear changes and improve shift feel.

Clutch packs sometimes contain a wavy spring-steel separator plate that helps smooth the application of the clutch. Shift feel can also be smoothed out by using a restricting **orifice** or an **accumulator** piston (Figure 3-29) in the band or clutch apply circuit. A restricting orifice in the passage to the apply piston restricts fluid flow and slows the pressure increase at the piston by limiting the quantity of fluid that can pass in a given time. An accumulator piston slows pressure buildup at the apply piston by diverting a portion of the pressure to a second piston in the same hydraulic circuit. This delays and smooths the application of a clutch or band.

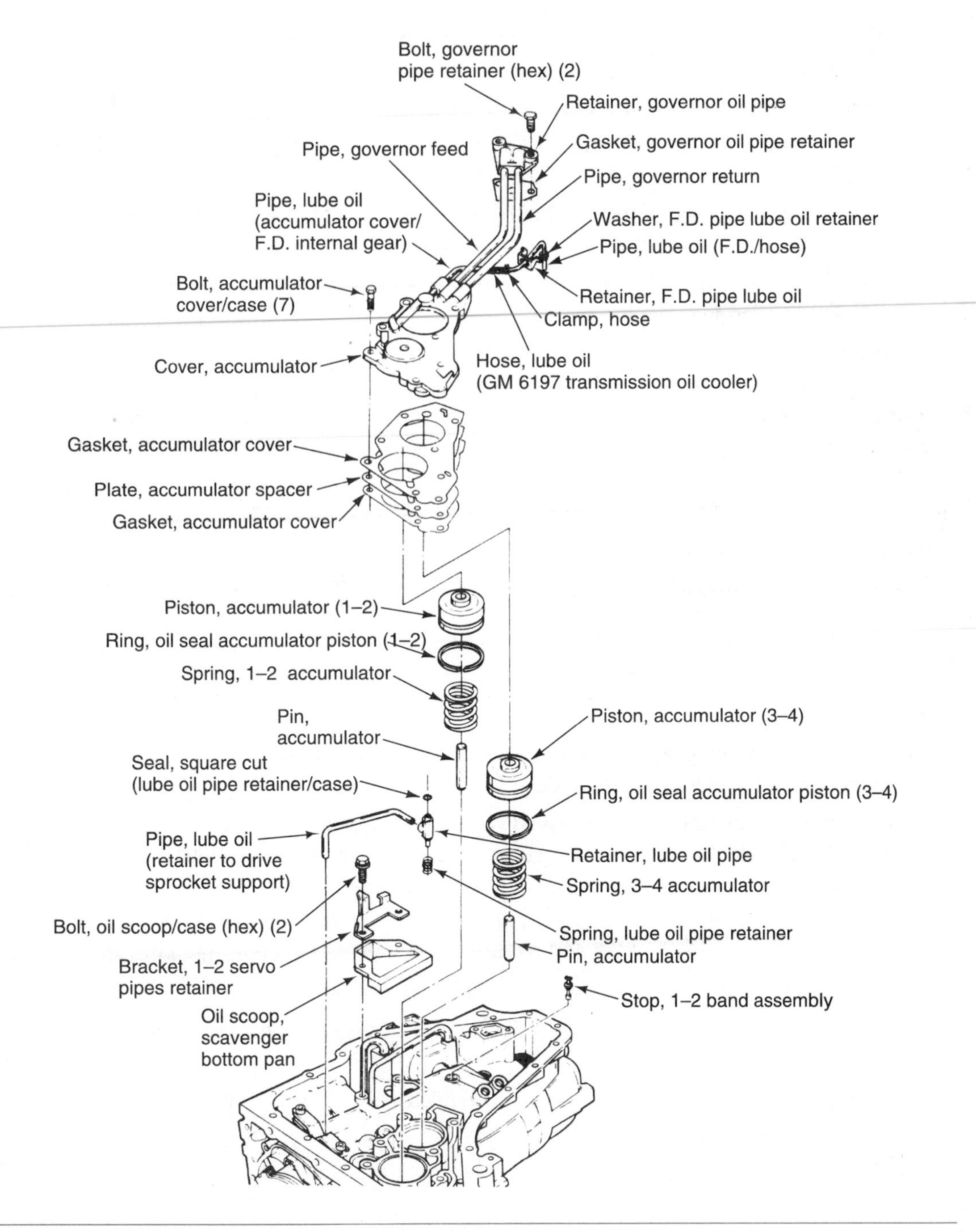

Figure 3-29 An accumulator assembly. (Courtesy of Buick Motor Division of General Motors Corp.)

Summary

- Automatic transmissions are used in many rear-wheel-drive and four-wheel-drive vehicles. Automatic transaxles are used in most front-wheel-drive vehicles.
- The major components of a transaxle are the same as those of a transmission, except the transaxle assembly includes the final drive and differential gears in addition to the transmission.
- The input shaft drives a planetary gear set, which provides the various forward gears, a neutral position, and one reverse gear.

Terms to Know

Accumulator

ATF

Check ball valves

Compound planetary gear system

Engine load
Engine vacuum
Flow-directing valves
Gear-type pump
Governor
Impeller
Input Shaft
Kickdown
Load-sensing device
Manual shift valve
Orifice
Output shaft
Poppet valves
Pressure-regulating valves
Pressure regulator valve
Reaction members
Rotor-type pump
Separator plate
Servo
Shift feel
Stator
Throttle valve
Torque converter
Torque converter cover
Transfer plate
Turbine
Vacuum modulator
Valve body
Vane-type pump

- ❑ Power flow through the gears is controlled by multiple-disc clutches, one-way clutches, and friction bands. These hold a member of the gear set when hydraulic pressure is applied to them, thereby making different gear ratios possible.
- ❑ An automatic transmission or transaxle selects gear ratios according to engine speed, engine load, vehicle speed, and other operating conditions.
- ❑ Gear ratio is an expression of the number of revolutions made by a driving gear compared to the number of revolutions made by a driven gear.
- ❑ The torque converter automatically engages or disengages power from the engine to the transmission in relation to engine rpm.
- ❑ The main components of a torque converter are the cover, turbine, impeller, and stator. The cover is connected to the flywheel and transmits power from the engine to the impeller, which drives the transmission's oil pump. The turbine is splined to the input shaft and is driven by the force of the fluid from the impeller. The stator redirects the flow of fluid from the turbine to the impeller.
- ❑ A single planetary gear set consists of a sun gear, a planet carrier with two or more planet pinion gears, and a ring gear.
- ❑ Transmissions are equipped with a fluid cooler that prevents overheating of the fluid, which could result in damage to the transmission.
- ❑ The transmission's oil pump provides a constant supply of oil under pressure to operate, lubricate, and cool the transmission.
- ❑ Pressure-regulating valves change the oil's pressure to control the shift points of the transmission. Flow-directing valves direct the pressurized oil to the appropriate reaction members to cause a change in gear ratios.
- ❑ Reaction members are those parts of an automatic transmission that hold or drive members of the planetary gear set in order to change gears. One-way overrunning clutches, bands, and multiple-disc clutch packs are examples of reaction members.
- ❑ Many transmissions rely on information sent from various electronic sensors and switches to a computer that controls the locking of the torque converter and the transmission's shift points.
- ❑ The speedometer drive assembly is geared to the output shaft.
- ❑ The vacuum modulator is a load-sensing device that increases or decreases fluid pressure in response to the vacuum signal.
- ❑ Transmissions not equipped with a vacuum modulator use a throttle cable or electronics to sense engine load and change fluid pressures.
- ❑ The throttle cable is connected to the throttle lever at the valve body and is moved according to the position of the throttle pedal.
- ❑ As the gearshift selector is moved into its various positions, the manual lever moves the manual shift valve in the transmission's valve body.
- ❑ When two or more planetary gear sets are used in a transmission, they are collectively called a compound planetary gear system.
- ❑ A simple hydraulic system has liquid, a pump, lines to carry the liquid, control valves, and an output device.
- ❑ Pascal's Law states that pressure exerted on a confined liquid or fluid is transmitted undiminished and equally in all directions and acts with equal force on all areas.
- ❑ ATF is important to the operation of the transmission. It serves as a fluid coupling between the engine and the transmission, it removes and dissipates heat from the transmission, it lubricates all of the moving parts of the transmission, and it conveys pressure to operate the bands and clutches.
- ❑ Typical reaction devices are bands, multiple-disc clutches, and overrunning clutches.

- ❑ Central to the hydraulic control system is the valve body assembly, which is typically made of two or three main parts: a valve body, a separator plate, and a transfer plate.
- ❑ The purpose of a valve body is to sense and respond to engine and vehicle load, as well as to meet the needs of the driver.
- ❑ Valve bodies are normally fitted with three different types of valves: spool valves, check ball valves, and poppet valves. The purpose of these valves is to start, stop, or use movable parts to regulate and direct the flow of fluid throughout the transmission.
- ❑ The three most common types of oil pumps used in automatic transmissions are: gear-type, rotor-type, and vane-type.
- ❑ The pressure regulator valve is located in the oil pump or valve body and maintains basic fluid pressure.
- ❑ When the engine is operating under heavy load conditions, fluid pressure is increased to increase the holding capacity of a clutch or band.
- ❑ Engine load can be monitored electronically through the use of various electronic sensors, which send information to an electronic control unit. Load can also be monitored by throttle position.
- ❑ Some transmissions are equipped with a vacuum modulator that allows for an increase in pressure when vacuum is low and a decrease in pressure when vacuum is high.
- ❑ The governor assembly senses road speed and sends a fluid pressure signal to the valve body to either upshift or downshift.
- ❑ The valve body is fitted with a kickdown circuit, which provides a downshift when the driver requires additional power.
- ❑ Shift feel is controlled by the pressure at which each reaction member is applied or released, the rate at which each is pressurized or exhausted, and the relative timing of the apply and release of the members.
- ❑ Clutch packs sometimes contain a wavy spring-steel separator plate that helps smooth the application of the clutch.
- ❑ An accumulator is a hydraulic piston assembly that helps a servo apply a band or clutch smoothly. It does this by absorbing sudden pressure surges in the hydraulic circuit to the servo.

Review Questions

Short Answer Essays

1. How can shift feel be controlled?
2. What are the major components of all automatic transmissions?
3. What is the purpose of the stator in a torque converter assembly?
4. List three functions of ATF.
5. How does engine load affect engine vacuum?
6. What is the purpose of a transmission fluid cooler?
7. Why must hydraulic line pressures increase when there is an increase of load on the engine?
8. Briefly explain how a torque converter works.
9. What are the primary purposes of the valves in the valve body?
10. What are purposes of the reaction members of a transmission?

Fill in the Blanks

1. Power flow through the gears is controlled by________________-________________ ________________, ________________-________________ ________________, and ________________, which hold a member of the gear set when force is applied to them.

2. The main components of a torque converter are the ________________, ________________, ________________, and ________________.

3. The three most common types of oil pumps used in automatic transmission are ________________-type, ________________-type, and ________________-type.

4. Oil pumps are driven by the ________________ of the torque converter; therefore, whenever the torque converter ________________ is rotating, the oil pump is driven.

5. The governor assembly senses ________________ speed and sends a fluid pressure signal to the ________________ ________________ to either upshift or downshift.

6. A transmission's oil pump provides a constant ________________ of oil under ________________ to ________________, ________________, and ________________ the transmission.

7. ________________ valves change the oil's pressure to control the shift points of the transmission, while ________________ valves direct the pressurized oil to the appropriate reaction members to cause a change in gear ratios.

8. Engine load can be monitored ________________, by ________________ ________________, or by ________________ ________________.

9. The purpose of a valve body is to ________________ and ________________ to engine and vehicle load, as well as to meet the needs or desires of the ________________.

10. An automatic transmission or transaxle selects gear ratios according to engine ________________, engine ________________, vehicle ________________, and other operating conditions.

ASE Style Review Questions

1. While discussing stators:
Technician A says a stator aids in directing the flow of fluid from the turbine to the impeller.
Technician B says a stator is equipped with a one-way clutch that allows it to remain stationary under certain conditions.
Who is correct?
A. A only
B. B only
C. Both A and B
D. Neither A nor B

2. While discussing manual shift valves:
Technician A says the gear selector lever and linkage moves the manual shift valve in the transmission.
Technician B says the manual shift valve, which is located in the valve body, moves indirectly to the movement of the throttle pedal.
Who is correct?
A. A only
B. B only
C. Both A and B
D. Neither A nor B

3. While discussing impellers:
Technician A says the torque converter's impeller is driven by the converter cover and its rear hub drives the transmission's oil pump.
Technician B says the impeller is splined to the input shaft and is driven by the force of the fluid from the turbine.
Who is correct?
A. A only **C.** Both A and B
B. B only **D.** Neither A nor B

4. While discussing oil pumps:
Technician A says oil pumps are driven by the pump drive hub of the torque converter.
Technician B says the oil pump is indirectly driven by the torque converter's stator.
Who is correct?
A. A only **C.** Both A and B
B. B only **D.** Neither A nor B

5. While discussing load-sensing devices:
Technician A says transmissions not equipped with a vacuum modulator use a throttle cable or electronics to sense engine load.
Technician B says increasing the fluid pressure holds the planetary control units tightly to reduce the chance of slipping while under heavy load.
Who is correct?
A. A only **C.** Both A and B
B. B only **D.** Neither A nor B

6. While discussing valve bodies:
Technician A says the purpose of a valve body is to increase oil pressures in response to increases in engine speed.
Technician B says valve bodies are normally fitted with three different types of valves that start, stop, or use movable parts to regulate and direct the flow of fluid throughout the transmission.
Who is correct?
A. A only **C.** Both A and B
B. B only **D.** Neither A nor B

7. While discussing engine load:
Technician A says changes in engine load cause changes in hydraulic pressure inside the transmission.
Technician B says engine load is monitored by throttle pedal movement or by engine vacuum.
Who is correct?
A. A only **C.** Both A and B
B. B only **D.** Neither A nor B

8. While discussing electricity:
Technician A says electrical switches are used to control the operation of other devices.
Technician B says electrical switches are used to open and close electrical circuits.
Who is correct?
A. A only **C.** Both A and B
B. B only **D.** Neither A nor B

9. While discussing the hydraulic system:
Technician A says the transmission's oil pump is the source of all pressures within the hydraulic system.
Technician B says the hydraulic system also keeps the torque converter filled with fluid.
Who is correct?
A. A only **C.** Both A and B
B. B only **D.** Neither A nor B

10. While discussing the levers that protrude from transmission housings:
Technician A says the throttle lever is moved by the engine vacuum sent to the modulator in response to engine load.
Technician B says that when the gear shift selector is moved, it moves the manual lever, which moves the manual shift valve in the transmission's valve body.
Who is correct?
A. A only **C.** Both A and B
B. B only **D.** Neither A nor B

Hydraulic Torque Multipliers

Upon completion and review of this chapter, you should be able to:

- ❑ Identify the major components in a torque converter and explain their purpose.
- ❑ Understand the fluid flows that occur in a torque converter.
- ❑ Explain the necessity of curved blades to enhance torque converter efficiency.
- ❑ Discuss stator design and operation.
- ❑ Explain how fluid flows in and out of a torque converter.
- ❑ Explain torque converter operation in the stall and coupling phases.
- ❑ Explain the basic design and operation of standard and lockup torque converters.
- ❑ Describe the design and operation of a centrifugal lockup clutch.
- ❑ Describe the design and operation of a piston-type converter clutch.

A BIT OF HISTORY

In 1939, Chrysler introduced the Fluid Drive, a three-speed Synchromesh transmission with a fluid coupling between the engine and a conventional friction clutch. In 1941, this was replaced by the Vacamatic, a manually shifted two-speed transmission in series with a vacuum-actuated, automatic overdrive. The result was four forward speeds, only half of which were engaged automatically. Ford launched its similar Liquamatic Drive on 1942 Lincolns and Mercurys. Few were sold, and many owners had standard transmissions retrofitted during the war, when replacement parts were hard to find.

Introduction

A torque converter uses fluid to smoothly transfer engine torque to the transmission. The torque converter is a doughnut-shaped unit located between the engine and the transmission that is filled with ATF. Internally, the torque converter has three main parts: the impeller, turbine, and stator. Each of these has blades that are curved to increase torque converter efficiency.

The **impeller** is driven by the engine and directs fluid flow against the turbine blades, causing them to rotate and drive the turbine shaft, which is the transmission's input shaft. The **stator** is located between the impeller and the turbine and returns fluid from the turbine to the impeller, so that the cycle can be repeated.

During certain operating conditions, the torque converter multiplies torque. It provides extra reduction to meet the drive line needs while under a heavy load. When the vehicle is operating at cruising speeds, the torque converter operates as a **fluid coupling** and transfers engine torque to the transmission. It also absorbs the shock from gear changing in the transmission. Not all of the engine's power is transferred through the fluid to the transmission; some is lost. To reduce the amount of power lost through the converter, especially at cruising speeds, manufacturers equip most of their current transmissions with a **lockup** converter.

The engagement of the converter clutch is based on both engine and vehicle speeds and the clutch is normally controlled by transmission hydraulics and on-board computer electronic controls. When the clutch engages, a mechanical connection exists between the engine and the drive wheels. This improves overall efficiency and fuel economy.

CVT Units

All automatic transmissions, except some designs of the constant variable transmission (CVT), use a torque converter to transfer power from the engine to the transmission. The **CVT** is a compact transmission that uses belts and pulleys to match the transmission's torque multiplication with the needs of the vehicle. These transmissions do not have fixed gear ratios. Rather, the size of the drive and driven pulleys changes in accordance to engine speed and load. This provides for constantly changing torque multiplication factors.

The same basic design of the CVT is used by different manufacturers. The primary design difference among the manufacturers relates to the method used to transmit engine torque to the transmission. Some manufacturers use a torque converter; some do not. Late-model Honda Civics are available with a CVT; these vehicles are not equipped with a torque converter. The transaxles have an internal start clutch that allows the vehicle to maintain an idle speed while it is in gear.

To transfer torque from the engine to the transmission, an electromagnetic clutch is used. Its operation is controlled by a clutch control unit. The clutch control unit switches current to energize and deenergize the electromagnetic clutch in response to inputs from various sensors. These sensors include a brake switch, accelerator pedal switches, and an inhibitor switch. The brake switch deenergizes the clutch when the vehicle is slowing or coming to a stop. The control unit uses inputs from the accelerator pedal switches to vary the amount of current to the electromagnetic clutch and to signal ratio changes to the adjustable pulleys inside the transmission. The inhibitor switch prevents clutch engagement when the gear selector is in the P or N position.

When the electromagnetic clutch is energized, engine torque is transferred to the transmission's input shaft and the drive pulley. At the end of the transmission's input shaft is the transmission's oil pump; therefore, the clutch control unit has indirect control of the oil pressure in the transmission. Pressure from the oil pump is used to control the **sheaves** of both the drive and driven pulleys. Line pressure is controlled by a three-way solenoid in the transmission valve body.

Moving the sheaves of the pulleys changes their operating diameter, which in effect changes the gear ratio.

Fluid Couplings

A BIT OF HISTORY

The use of fluid couplings actually began in the 1900s with steamship propulsion and later was used to help dampen the vibrations of large diesel engines. Just before WWII, Chrysler was the first American automobile manufacturer to use a fluid coupling. The fluid coupling was added to a manual transmission drive line, which allowed the engine to idle in gear, but a foot-operated clutch was still needed to shift gears.

The impeller and turbine are sometimes called the torus halves because of their convex shape.

A simple fluid coupling is composed of three basic members: a housing, an impeller, and a turbine (Figure 4-1). The impeller and the turbine are shaped like two halves of a doughnut. The impeller and turbine have internal vanes radiating from their centers. Both the impeller and the turbine are enclosed in the housing. ATF is forced into the housing by the transmission's oil pump. The housing is sealed and filled with fluid. The impeller is connected directly to the housing and is driven by the engine. The turbine, which is not attached to the engine, turns the transmission's input shaft as it rotates. The impeller acts like a pump and moves the fluid in the direction of its rotation. The moving fluid hits against the turbine vanes. This causes the turbine to rotate, which brings an input into the transmission.

The impeller is commonly referred to as the drive torus or the primary pump. Likewise, the turbine is referred to as the driven torus.

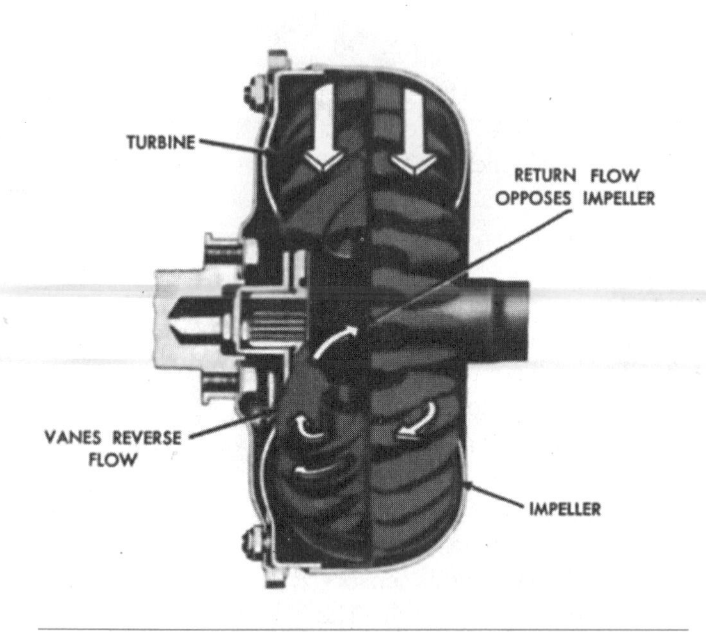

Figure 4-1 Coupling phase in a fluid coupler. (Reprinted with the permission of Ford Motor Co.)

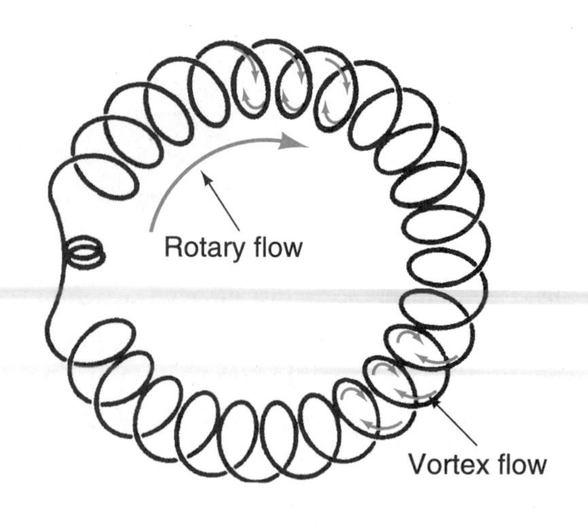

Figure 4-2 Difference between rotary and vortex flow. Note that vortex flow spirals its way around the converter.

Operation

The fluid coupling allows the vehicle to stop while in gear without stalling the engine because at low engine speeds the impeller doesn't rotate fast enough to drive the turbine. As engine speed increases, the force from the fluid flow off of the impeller increases and forces the turbine to rotate. The speed of the turbine increases as the force of the fluid flow increases and as the load on the transmission decreases or is overcome. Based on the fluid flow, a fluid coupling can transmit very little to very much force. Maximum efficiency is approximately 90 percent, but that figure depends on a number of factors: fluid type, impeller and turbine vane design, vehicle load, and engine torque.

Although fluid flow transfers the engine's power, some of the power is lost as the fluid flows. When the fluid flows from the impeller to the turbine, it forces the turbine to rotate in the direction of engine rotation. However, when the fluid leaves the turbine and moves back toward the impeller, its direction has changed and it is now moving in a direction opposite that of engine rotation. This puts an additional load on the engine and creates a turbulence in fluid flow, which also generates great amounts of heat.

Fluid Flow

Rotary flow occurs when the fluid path is in the same circular direction as the rotation of the impeller.

Vortex flow occurs when the fluid path is at a right angle to rotary oil flow and the direction of the impeller.

The operation of a fluid coupling is totally based on the flow of fluid inside the coupling. Two types of fluid flow take place inside the fluid coupling: rotary flow and vortex flow (Figure 4-2). These different flows complement each other, depending on the difference in speed between the impeller and the turbine. **Rotary flow** is the movement of the fluid in the direction of impeller rotation and results from the paddle action of the impeller vanes against the fluid. As this rotating fluid hits the blades of the slower turning or stationary turbine, it exerts a turning force onto the turbine.

Vortex oil **flow** is the fluid flow circulating between the impeller and turbine as the fluid moves from the impeller to the turbine and back to the impeller. This type of fluid flow is only present when there is a difference in rotational speeds between the impeller and the turbine.

Both types of flow can occur within a fluid coupling at the same time. When the impeller is rotating fast but turbine speed is restricted because of a load, most of the fluid within the housing moves with a vortex flow. However, as the load is overcome and turbine speed increases, more rotary flow occurs.

Within a fluid coupling, the fluid flows from the impeller to the turbine, then back to the impeller. As the fluid returns to the impeller, its direction has been changed by the turbine blades and it tends to work against impeller rotation and lowers the efficiency of the coupling. By adding stationary or reactionary vanes between the impeller and turbine, the fluid can be redirected so that it not only works with the impeller but can multiply the input torque during periods of heavy load. The unit that holds these reactionary vanes is the stator. When a stator is added to a fluid coupling, the assembly becomes a torque converter.

Torque Converters

A BIT OF HISTORY

The most important improvement to automatic transmissions was the result of Buick's development of a torque converter for tanks during the war. The converter was incorporated into the Dynaflow transmission in 1948, and eventually into other automatics.

The torque converter is a type of fluid coupling that connects the engine's crankshaft to the transmission's input shaft. The torque converter, working as a fluid coupling, smoothly transmits engine torque to the transmission. The torque converter allows some slippage between the engine and the transmission so that the engine will remain running when the vehicle is stopped while it is in gear. The torque converter also multiplies torque when the vehicle is under load to improve performance. Torque converters are more efficient than fluid couplings because they allow for increased fluid flow, which multiplies the torque from the engine when it is operating at low speeds.

Mountings

On all RWD and most FWD vehicles, the torque converter is mounted in line with the transmission's input shaft. Some transaxles use a drive chain to connect the converter's output shaft with the transmission's input shaft (Figure 4-3). Offsetting the input shaft from the centerline of the transmission gives the manufacturers more flexibility in the positioning of the transaxle. These transaxles typically have an extension housing that contains the torque converter and oil pump and is mounted to the rear of the engine.

Shop Manual
Chapter 4, page 83

The torque converter fits between the transmission and the engine and is normally supported by the transmission's oil pump housing and the engine's crankshaft. The rear hub of the converter's cover fits into the transmission's pump (Figure 4-4). The hub drives the oil pump whenever the engine is running.

Shop Manual
Chapter 4, page 84

The front of a torque converter is mounted to a flexplate, which is bolted to the end of the engine's crankshaft (Figure 4-5). The converter cover is normally fitted with studs that are used to tighten the cover to the flexplate. The flexplate is designed to be flexible enough to allow the front of the converter to move forward or backward if it expands or contracts because of heat or pressure. The centerlines of the converter and crankshaft are matched and this alignment is maintained by a pilot on the converter cover that fits into a recess in the crankshaft.

An externally toothed ring gear is normally pressed or welded to the outside diameter of the flexplate. The starter motor's drive gear meshes with the ring gear to crank the engine for starting.

The transmission's input shaft is supported by bushings in the stator support inside the torque converter. There is no mechanical link between the output of the engine and the input of the transmission. The fluid connects the power from the engine to the transmission. The combined weight of the fluid, torque converter, and flexplate serves as the flywheel for the engine.

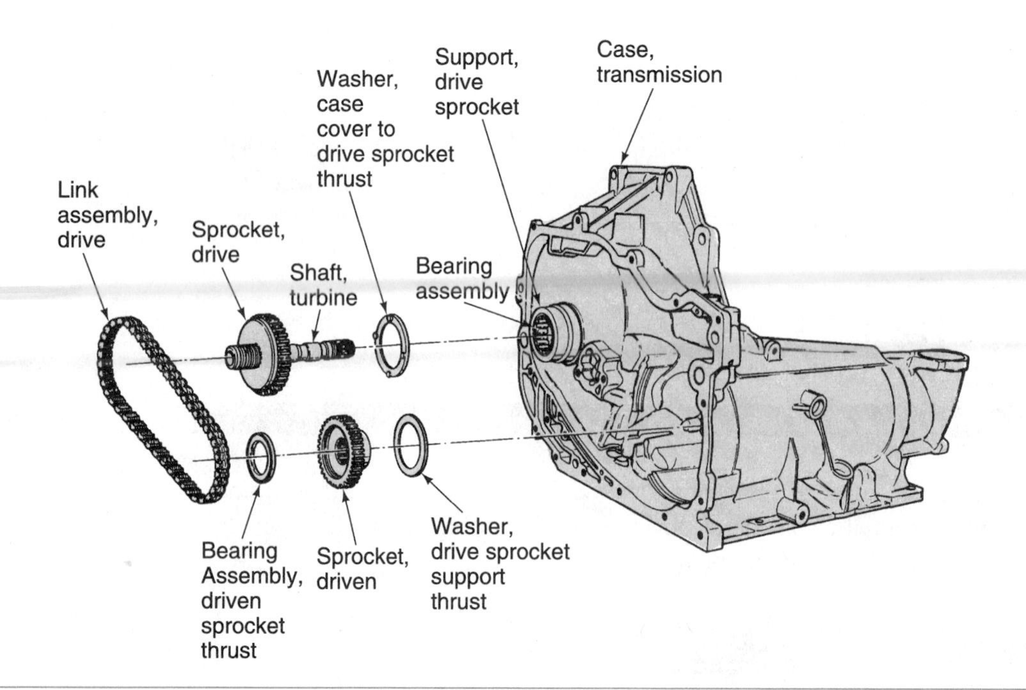

Figure 4-3 Typical drive chain setup for offsetting the torque converter and the input shaft. (Courtesy of the Hydra-Matic Division of General Motors Corp.)

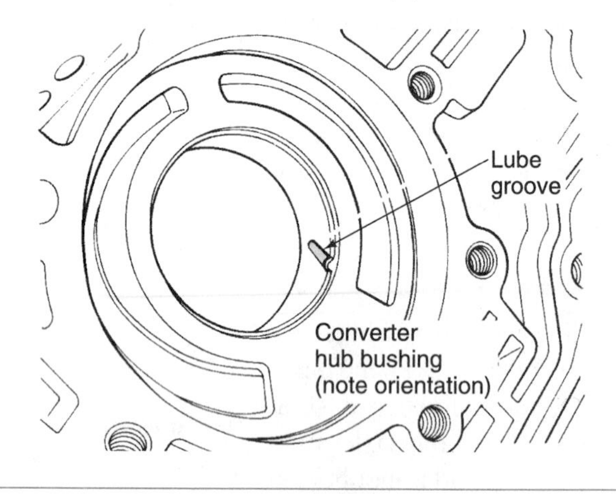

Figure 4-4 The converter's hub rides on a bushing in the transmission housing. (Reprinted with the permission of Ford Motor Co.)

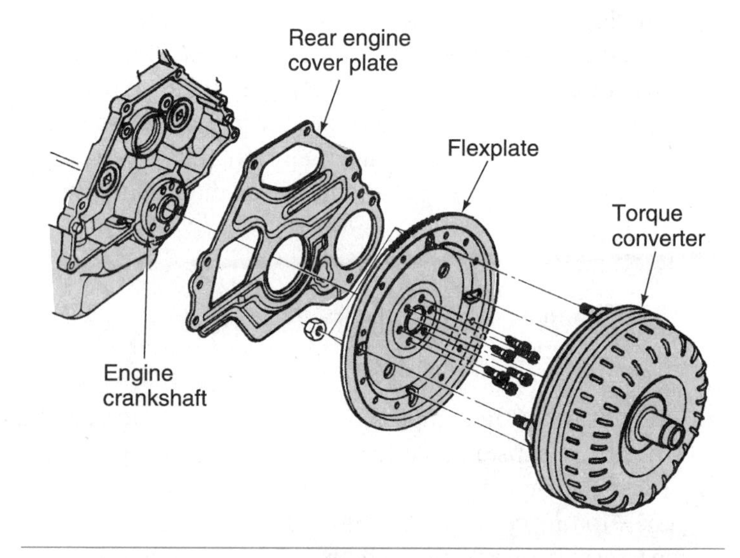

Figure 4-5 A torque converter is mounted to the flexplate, which is bolted to the engine's crankshaft. (Reprinted with the permission of Ford Motor Co.)

Construction

A BIT OF HISTORY

Variations on the basic three-element torque converter have been used. The Buick Dynaflow had two impellers, two stators, and one turbine in 1948. In 1953, the Twin-Turbine Dynaflow was released with two turbines, one impeller, and one stator. In 1956, Buick introduced a multiple-turbine torque converter that had a variable pitch stator. By the late 1960s, the industry, including Buick, had returned to the basic three-element converter.

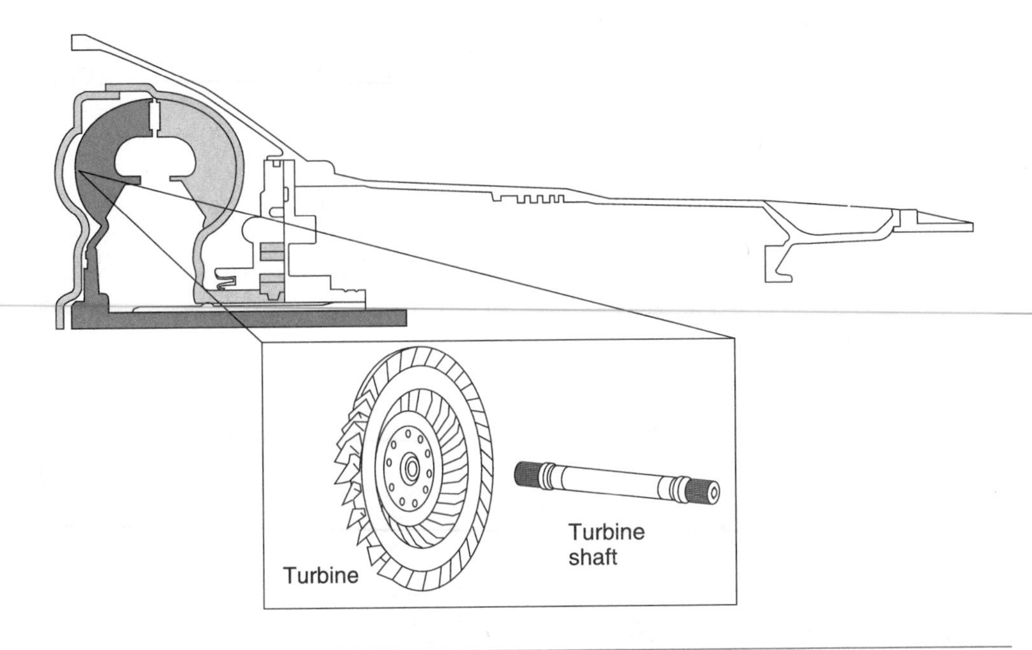

Figure 4-6 Location of turbine. Note that the turbine shaft extends from the turbine to the transmission and serves as the input shaft of the transmission. (Courtesy of the Hydra-Matic Division of General Motors Corp.)

An impeller is a finned wheel-like device that is turned by the engine to pump transmission fluid into the turbine, which drives the transmission's input shaft.

A turbine is a finned wheel-like device that receives fluid from the impeller and forces it back to the stator. The turbine transmits engine torque to the transmission's input shaft.

A stator is sometimes called a **reactor**.

A typical torque converter consists of three elements sealed in a single housing: the impeller, the turbine, and the stator. The impeller is the drive member of the unit and its fins are attached directly to the converter cover. Therefore, the impeller is the input device for the converter and always rotates at engine speed.

The turbine is the converter's output member and is coupled to the transmission's input shaft (Figure 4-6). The turbine is driven by the fluid flow from the impeller and always turns at its own speed. The fins of the turbine face toward the fins of the impeller. The impeller and the turbine have internal fins, but the fins point in opposite directions.

Shop Manual
Chapter 4, page 83

A BIT OF HISTORY

The Fordomatic had the first three-element torque converter of the design that almost all later transmissions have used.

The stator shaft is often called a ground sleeve or the stator support.

A stator is the centrally located finned wheel in a torque converter. It receives transmission fluid from the turbine and directs it back to the impeller. The stator also makes torque multiplication in a torque converter possible.

The stator is the reaction member of the converter (Figure 4-7). This assembly is about one-half the diameter of the impeller or turbine and is positioned between the impeller and turbine. The stator is not mechanically connected to either the impeller or turbine; rather, it fits between the turbine outlet and the inlet of the impeller. All of the fluid returning from the turbine to the impeller must pass through the stator. The stator redirects the fluid leaving the turbine back to the impeller. By redirecting the fluid so that it is flowing in the same direction as engine rotation, it allows the impeller to rotate more efficiently.

A stator is supported by a one-way clutch that is splined to the stator support shaft, which extends from the front of the transmission. The one-way clutch allows the stator to rotate only in the same direction as the impeller. The clutch allows the stator to freewheel when the impeller and turbine reach their coupling stage. The outer edge of the stator fins normally forms the inner edge of a three-piece fluid guide ring. This split guide ring directs fluid into a smooth, turbulence-free flow.

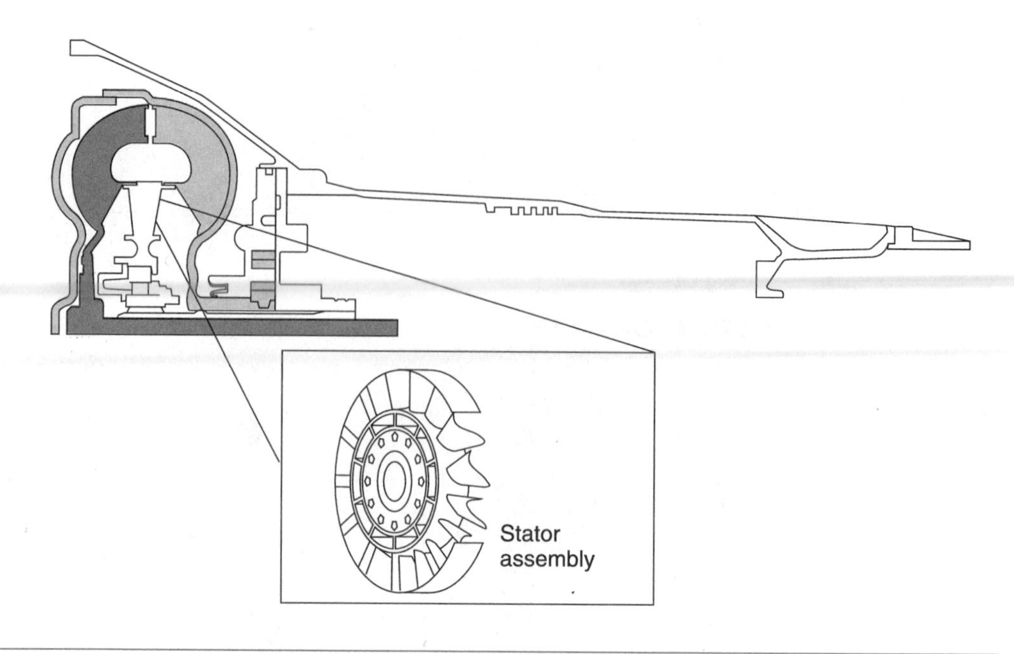

Figure 4-7 Location of stator assembly. Note that the stator is not attached to the turbine or the impeller. (Courtesy of the Hydra-Matic Division of General Motors Corp.)

Torque Converter Operation

The vanes of the impeller, turbine, and stator are sometimes referred to as the blades or fins.

The impeller rotates at engine speed whenever the engine is running. The impeller is composed of many curved vanes radiating out of an inner ring, which form passages for the fluid. As soon as the converter begins to rotate, the vanes of the impeller begin to circulate fluid.

The fluid in the torque converter is supplied by the transmission's oil pump. It enters through the converter's hub, then flows into the passages between the vanes. As the impeller rotates, the fluid is moved outward and upward through the vanes by centrifugal force because of the curved shape of the impeller. The faster the impeller rotates, the greater the centrifugal force becomes.

The fluid moves from the outer edge of the vanes into the turbine. As the fluid strikes the curved vanes of the turbine, it attempts to push the turbine into a rotation. Because the impeller is turning in a clockwise direction (as viewed from the front), the fluid also rotates in a clockwise direction as it leaves the vanes of the impeller. However, the turbine vanes are curved in the opposite direction of the impeller and the fluid turns the turbine in the same direction as the impeller.

The higher the engine speed, the faster the impeller turns, and the more force is transferred from the impeller to the turbine by the fluid. This explains why a torque converter allows the engine to idle in gear. When engine speed is low, the fluid does not have enough force to turn the turbine against the load on the drive train. The movement of the fluid in the converter is very weak and it is just circulating from the impeller to the turbine and back to the impeller. Therefore, little if any power is transmitted through the fluid coupling to the transmission.

As engine speed increases, the fluid is thrown at the turbine with a greater force, causing the turbine to rotate. Once the turbine begins to turn, engine power is transmitted to the transmission. However, the force from the fluid must be great enough to overcome the load of the vehicle before the turbine can rotate. Some of the energy in the moving fluid returns to the impeller as the torque converter responds to the torque requirements of the vehicle and to vortex flow. At low speeds, most of the energy in the fluid is lost as the fluid bounces back away from the turbine vanes.

Vortex flow is a continuous circulation of the fluid, outward in the impeller and inward in the turbine, around the split **guide rings** attached to the turbine and the impeller (Figure 4-8). The guide rings direct the vortex flow to provide for a smooth and turbulence-free fluid flow.

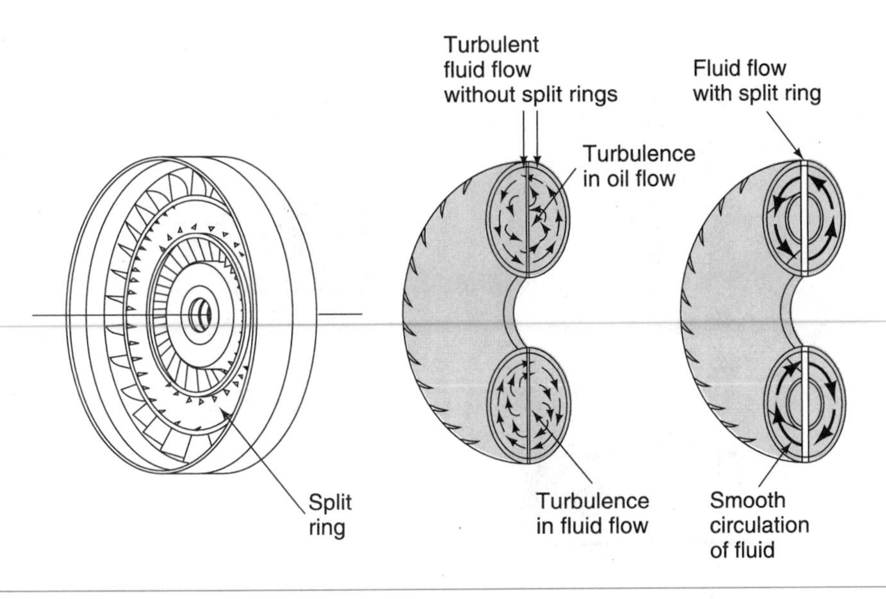

Figure 4-8 Location and action of the split guide rings. (Reprinted with the permission of Ford Motor Co.)

As the vortex flow continues, the fluid leaving the turbine to return to the impeller is moving in the opposite direction as crankshaft rotation. If the fluid were allowed to continue in this direction, it would enter the impeller as an opposing force and some of the engine's power would be used to redirect the flow of fluid. To prevent this loss of power, torque converters are fitted with a stator.

Torque Multiplication

Torque multiplication in a torque converter occurs when the vortex flow is redirected through the stator to increase flow to the turbine.

The stator receives the fluid thrown off by the turbine and redirects the fluid so that it reenters the impeller in the same direction as crankshaft rotation (Figure 4-9). The redirection of the fluid by the stator not only prevents a torque loss, but also provides for a multiplication of torque.

The stator is attached to a circular hub, which is mounted on a one-way clutch (Figure 4-10). This clutch assembly has an inner and outer race separated by spring-loaded roller bearings or

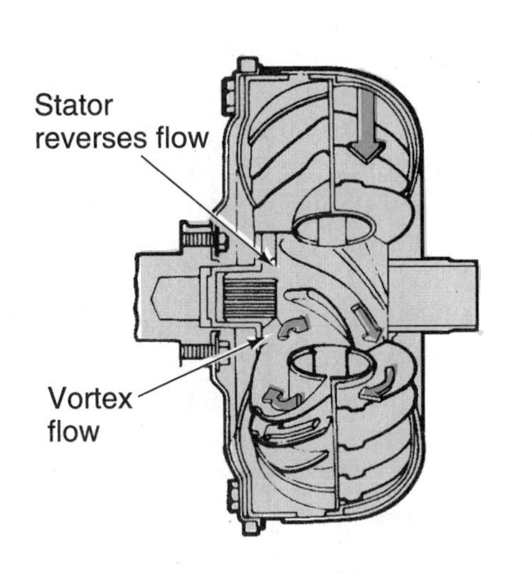

Figure 4-9 Action of the stator on the flow of fluid. (Reprinted with the permission of Ford Motor Co.)

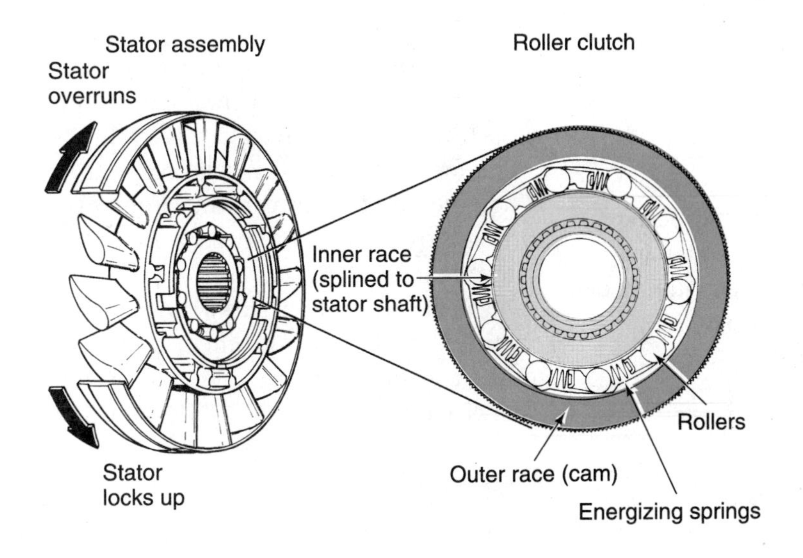

Figure 4-10 The one-way overrunning clutch in a stator assembly. (Courtesy of the Hydra-Matic Division of General Motors Corp.)

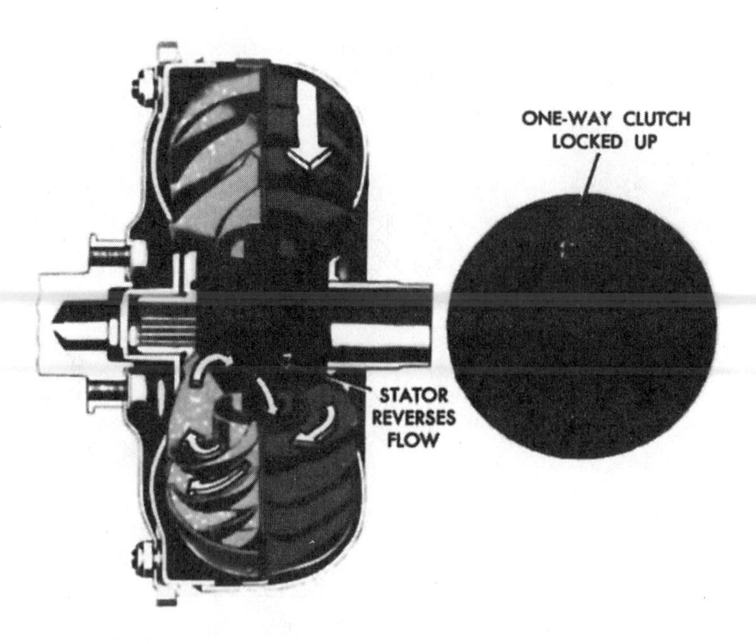

Figure 4-11 Fluid flow during torque multiplication. (Reprinted with the permission of Ford Motor Co.)

sprags. The inner race is splined to the stator support; therefore, it cannot turn. The outer race is fitted into the stator. The rollers or sprags are fitted between the two races and will allow the outer race to rotate in one direction only. The stator locks in the opposite direction of turbine rotation. When the stator is turned in the same direction as turbine rotation, the rollers are free between the races and the stator is able to turn (Figure 4-11).

The velocity of the vortex flow is the sum of the impeller-produced velocity plus the velocity of the fluid returning from the turbine and stator.

The fluid leaving the turbine has to pass through the stator blades before reaching the impeller. In passing through the stator, the direction of fluid flow is reversed by the curvature of the stator blade. The fluid now moves in a direction that aids in the rotation of the impeller. The impeller accelerates the movement of the fluid and it now leaves with nearly twice the energy and exerts a greater force on the turbine. This action results in a torque multiplication.

A torque converter has reached its coupling phase when turbine speed is about 90% of the speed of the impeller. At this point, there is no torque multiplication.

It is vortex flow that allows for torque multiplication. Torque multiplication occurs when there is high impeller speed and low turbine speed. Low turbine speed and the stator cause the returning fluid to have a high-velocity vortex flow. This allows the impeller to rotate more efficiently and increase the force of the fluid pushing the turbine in rotation. When the vehicle's torque requirements become greater than the output of the engine, the turbine slows down and causes an increase in vortex flow velocity. This causes an increase in torque multiplication.

As the vortex flow slows down, torque multiplication is reduced. Torque multiplication is obtained anytime the turbine is turning at less than 9/10 impeller speed. Most automotive torque converters are capable of maximum torque multiplication factors ranging from 1.7:1 to 2.8:1 (Figure 4-12).

The amount of torque multiplication a converter will produce is sometimes called the converter ratio.

Coupling Phase

Torque multiplication occurs because of the redirection of the fluid flow by the stator. This only takes place when the impeller is rotating faster than the turbine. As the speed of the turbine increases, the direction of the flow changes and there is less multiplication of torque. When the speed of the turbine nearly equals the speed of the impeller, fluid flows against the stator vanes in

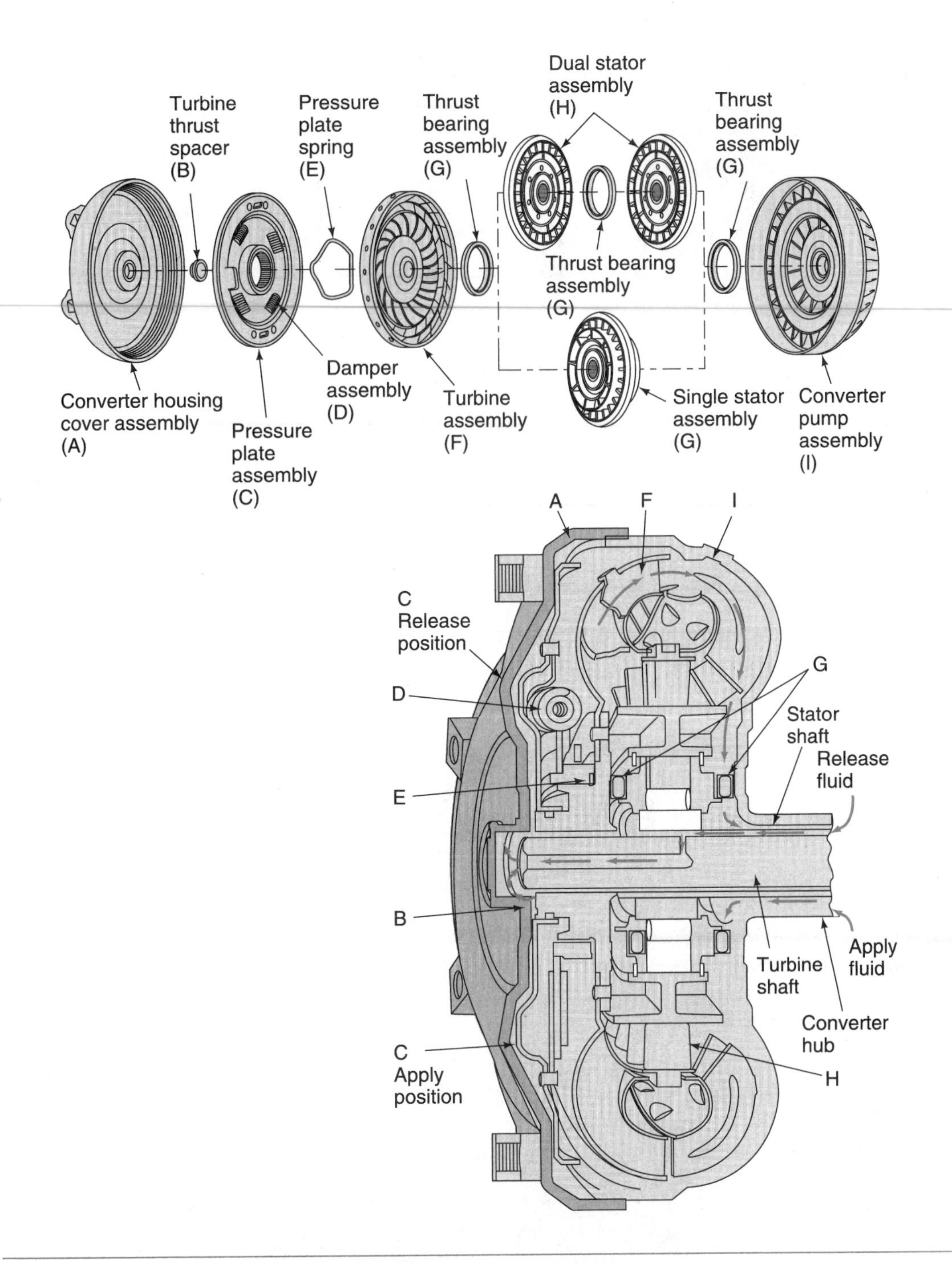

Figure 4-12 Typical construction of a torque converter. Note that this converter may be equipped with two stator assemblies for additional torque multiplication. (Courtesy of the Hydra-Matic Division of General Motors Corp.)

the same direction as the fluid from the impeller (Figure 4-13). This releases the one-way clutch and allows the stator to rotate freely (Figure 4-14). At this point, there is little vortex flow and the engine's torque is carried through the converter by the rotary flow of the fluid. This is the **coupling phase** of the torque converter and no torque multiplication takes place.

When the speed of the turbine reaches approximately 90 percent of the impeller's speed, coupling occurs. During coupling, the converter is acting like a fluid coupling transmitting engine torque to the transmission. The coupling phase does not occur at a specific speed or condition. It occurs whenever the speed of the turbine nearly equals the speed of the impeller.

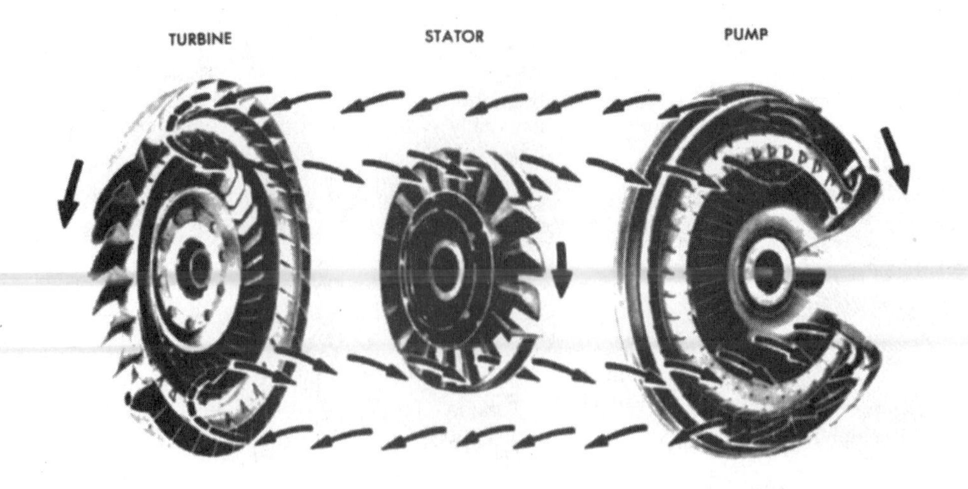

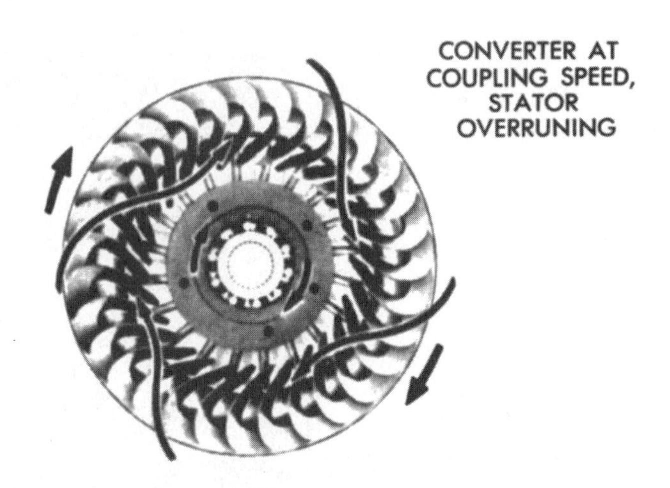

Figure 4-13 Fluid flow through the converter when the coupling phase has been achieved and the stator is overrunning. (Courtesy of the Hydra-Matic Division of General Motors Corp.)

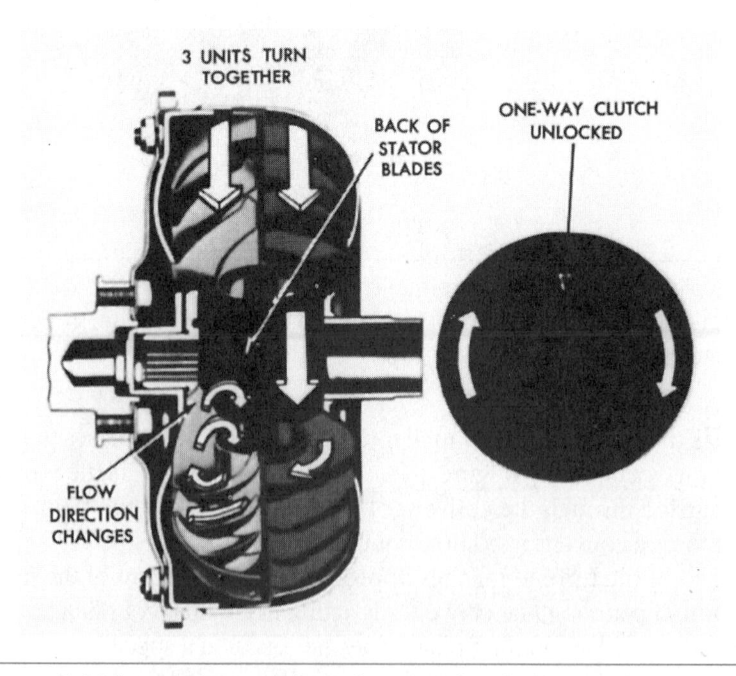

Figure 4-14 Fluid flow during coupling. (Reprinted with the permission of Ford Motor Co.)

Stall Speed

In spite of the advantages of using a torque converter with an automatic transmission, a torque converter is not a very efficient device. This is especially true at stall speeds when all of the engine torque that enters the torque converter is lost as heat and no torque is inputted into the transmission. From stall speed to the coupling speed, the efficiency of a converter increases to approximately 90 percent.

The condition called stall occurs when the turbine is held stationary and the impeller is spinning. **Stall speed** is the fastest speed an engine can reach while the turbine is at stall. Some stall occurs every time a vehicle starts moving forward or backward, as well as each time the vehicle is brought to a stop. Today, most torque converters have a stall speed of 1200 to 2800 rpm. Torque converters with a high stall speed are normally used with less powerful engines and those with a low speed are used with more powerful engines.

Shop Manual
Chapter 4, page 80

Stall speed represents the highest engine speed attained with the turbine stopped (at stall).

Converter Capacity

Although the construction of all torque converters is basically the same, the actual design of a converter is dictated by its application. The desired **converter capacity** and required stall speed are the two primary design considerations.

The capacity of a torque converter is related to stall speed; however, it is also an expression of the converter's ability to absorb and transmit engine torque with a limited amount of slippage. A low-capacity converter has a high stall speed, but it allows for a relatively large amount of slippage during the coupling phase. However, it also provides more torque multiplication and better acceleration.

A high-capacity converter allows for less slippage but has a low stall speed. This type of converter is very efficient at highway speeds, but it offers low torque multiplication. If a vehicle is equipped with a torque converter with the correct capacity, there will be a sort of balance between the amount of slippage, torque multiplication, and the stall speed.

The stall speed of a torque converter changes with a change in the diameter of the converter and the angles of the impeller, turbine, and stator vanes. Torque converter capacity or stall speed is matched to the size and power of an engine and to a vehicle's weight. The ideal stall speed will provide the best performance and fuel economy from the engine. The converter fitted to a transmission normally allows the stall speed to occur at the engine speed that produces the maximum amount of torque.

A converter with a high stall speed is sometimes referred to as a loose converter and one with a low stall speed is referred to as a tight converter.

The diameter of a torque converter directly affects the stall speed. Small diameter torque converters have high stall speeds and multiply torque at high engine speeds, but they do not couple until high engine speeds. Large diameter converters offer low stall speeds and torque multiplication at lower speeds. Typically, low output engines are fitted with small diameter converters to allow those engines to run at higher speeds and to allow the engine to operate when it is making most of its power.

High-performance cars are often fitted with a small diameter torque converter to raise the stall speed and to use the power available at high engine speeds.

Torque converter diameters vary from about 8 to 12 inches. Stall speed is actually determined by the circumference of the converter because this is the distance the fluid travels inside the converter. As the circumference increases, the speed of the fluid increases, at a given engine speed. The speed of the fluid in the impeller almost matches the speed of the converter's circumference.

The circumference of a 12-inch torque converter is nearly 50 percent larger than that of an 8-inch converter. Therefore, the 12-inch converter is capable of putting nearly 50 percent more speed and energy into the fluid. As a result, it will have a stall speed that is 50 percent lower than the 8-inch converter. To generate the increase in fluid speed and energy, more engine power is needed to utilize the capabilities of the converter.

The circumference of a converter is equal to the converter's diameter multiplied by pi or 3.1416.

The angle of the impeller and turbine vanes also determines the stall speed of a converter. The vanes can be angled forward or backward in relation to the direction of torque converter rotation. A forward angle produces higher fluid speeds; therefore, the stall speed tends to be lower. When the vanes have a backward angle, the stall speed increases. Changing the vane angle of an impeller or turbine is a common way to change torque converter capacity. This, of course, is not done by the technician. Rather, it is done during manufacturing to match the converter to a vehicle.

Torque converter capacity can also be changed by changing the angle of the stator vanes. As the angle of the vanes increases, more fluid will return to the impeller. Therefore, the stall speed will increase.

Variable Pitch Stator

In the late 1960s, some THM 300s and 400s were fitted with variable-pitch stator vanes. This was done in an attempt to improve vehicle performance during all modes of operation. By varying the pitch of the stator vanes, the converter behaved as a low or high stall converter, depending on the operating conditions. To change the angle of the vanes, the stator assembly was fitted with a hydraulic piston. Fluid flow to the piston was controlled by a valve and electric solenoid. A switch in the speedometer or on the throttle linkage activated the solenoid whenever vehicle speed was low and when the throttle was wide open. The activation of the solenoid allowed the piston to move the stator vanes to a greater angle, which provided a higher stall and maximum torque multiplication and acceleration. During all other operating conditions, the stator vanes were kept at less of an angle to allow for maximum fuel economy and quiet vehicle operation.

Cooling the Torque Converter

Slippage in a torque converter results from a loss of energy. Most of these losses are in the generation of frictional heat. Heat is produced as the fluid hits and pushes its internal members. To maintain the efficiency that it has, the fluid must be cooled. Excessive fluid heat would result in even more inefficiencies.

The transmission's oil pump continuously delivers pressurized ATF into the torque converter through a hollow shaft in the center of the torque converter assembly (Figure 4-15). A seal is used

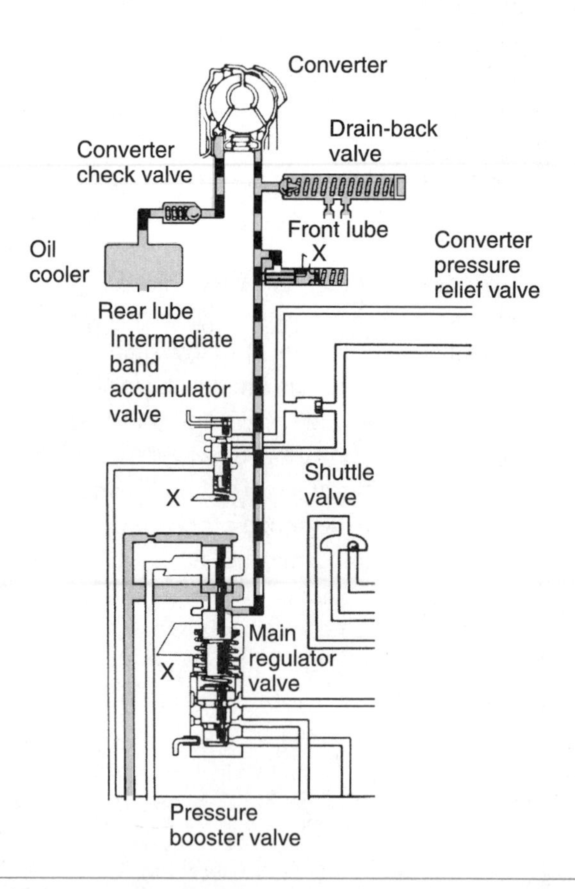

Figure 4-15 Typical oil circuit for converter and oil cooler. (Reprinted with the permission of Ford Motor Co.)

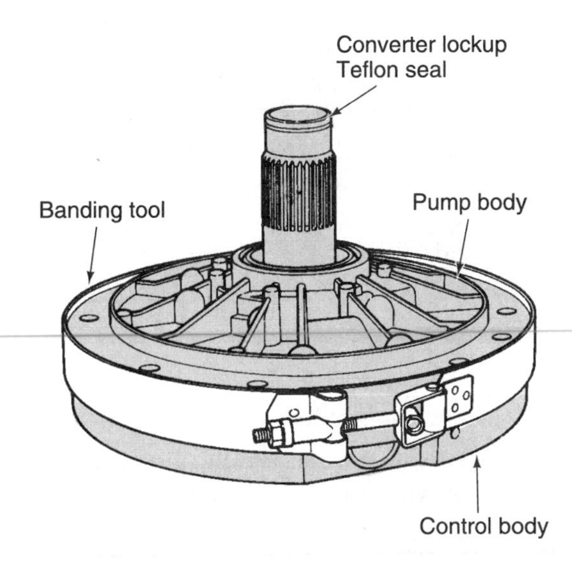

Figure 4-16 Location of seal on the hub of a converter. (Reprinted with the permission of Ford Motor Co.)

to prevent fluid from leaking at the point where the shaft enters the converter (Figure 4-16). The fluid is circulated through the converter and exits past the turbine through the turbine shaft, which is located within the hollow fluid feed shaft. From there, the fluid is directed to an external oil cooler and back to the transmission's oil pan. This cooling circuit ensures that the fluid flowing into the converter is cooled, thereby maintaining the converter's normal efficiency.

A few older transmissions were fitted with an air-cooled torque converter. The fluid in these converters does not flow to an external cooling device; they rely on the circulation of air to cool the fluid. Attached to the rear cover of these converters is a finned shroud. As the engine is running, air is drawn into the shroud to remove converter heat and the heated air is forced out.

Direct Drive

A BIT OF HISTORY

Packard's Ultramatic of 1949 featured a lockup torque converter, but it was dropped in 1957 due to its jerky action.

Up to 10 percent of the engine's energy is wasted in a pure fluid coupling. This wasted energy is in the form of heat. Torque converter slippage or the speed difference between the impeller and turbine during the coupling phase serves as evidence of the wasted energy. To eliminate this slippage, most late-model vehicles have a **lockup** torque converter that mechanically links the engine to the input of the transmission during some operating conditions. The results of using a lockup converter are improved fuel economy and reduced transmission fluid temperatures.

The lockup clutch assembly is commonly called a converter clutch.

Manufacturers have adapted automatic transmissions to provide a direct mechanical drive for input through the use of centrifugally applied clutches, hydraulically operated clutches, and planetary gear sets in the torque converter. The use of planetary gear sets is not as common as the use of clutches. Centrifugally applied clutches are also used; however, the most common lockup clutch systems rely on electronics to control a hydraulically operated clutch.

Shop Manual
Chapter 4, page XXX

When applied, the lockup clutch locks the turbine to the cover of the converter. Application of the clutch occurs at various operating speeds, depending on the model of the vehicle and the driving conditions. Most lockup converters consist of the three basic elements: impeller, turbine, and stator, plus a piston and clutch plate assembly, special thrust washers, and roller bearings.

The piston and clutch plate assembly has frictional material on the outer portion of the plate with a spring-cushioned **damper** assembly in the center. The clutch plate is splined to the turbine shaft and when the piston is applied, the plate locks the turbine to the converter. The thrust washers and roller bearings control the movements of and provide bearing surfaces for the components of the converter.

Lockup only occurs when the power train control module (or other control) determines that conditions are right for direct drive. The converter clutch should not connect the engine to the transmission when torque multiplication from the torque converter is needed, such as during acceleration. The clutch should be unlocked during braking; this prevents the engine from stalling. Likewise, the clutch should be unlocked when the vehicle is at a standstill while the transmission is in gear. Normally, during deceleration, the clutch should not be locked. If the clutch is locked during deceleration, fuel may be wasted and there will be high exhaust emissions. The clutch should be locked only when the engine is warm enough and is running at a great enough speed to prevent a shudder or stumble when it locks. The control systems use inputs from many different sensors to determine when the conditions are suitable for lockup.

A BIT OF HISTORY

The idea of using a converter clutch is not a new one. The Packard Ultramatic transmission, 1949, used a clutch in the converter to achieve third gear.

Centrifugal Lockup

In a **centrifugal lockup** converter, the turbine is driven directly by the converter cover through a centrifugal clutch that is splined to the turbine through a one-way clutch (Figure 4-17). The one-way clutch overruns and disconnects the turbine from the torque converter cover when the vehicle is decelerating or coasting. The turbine is allowed to overrun during coasting to provide maximum fuel economy. The one-way clutch is also considered a safety feature during hard braking. By allowing the lockup clutch to overrun, the engine will not stall as it attempts to overcome the load on the drive line.

The clutch assembly is located between the cover of the torque converter and the turbine. A centrifugal lockup converter can provide a mechanical link for the engine and the transmission in all gears. This lockup system functions without external controls as are necessary with hydraulically operated clutches.

Placed around the outside edge of the clutch assembly are several clutch shoes (Figure 4-18). As turbine speed increases, centrifugal force moves the shoes outward into contact with the converter cover, locking the turbine and the cover together. The amount of locking depends on the amount of force pushing the shoes onto the cover. The centrifugal force acting on the clutch shoes increases as the clutch assembly turns faster. With speed changes, the clutch shoes may totally lock the turbine to the cover or may allow some slippage.

The clutch shoes are arranged around the outside of the clutch disc. Each clutch shoe is connected by a spring and is positioned so that it can respond to centrifugal force. Each shoe is also faced with a frictional material pad.

As turbine speed and centrifugal force increase, the clutch shoes are thrown outward into contact with the inside diameter of the torque converter's cover. This provides for some mechanical

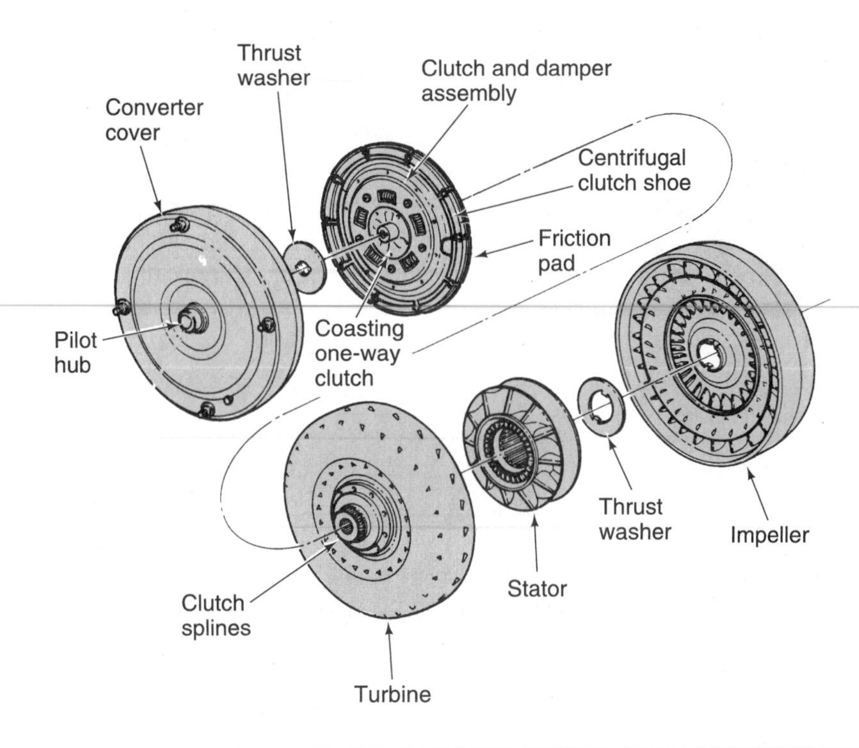

Figure 4-17 Centrifugal lockup clutch assembly. (Reprinted with the permission of Ford Motor Co.)

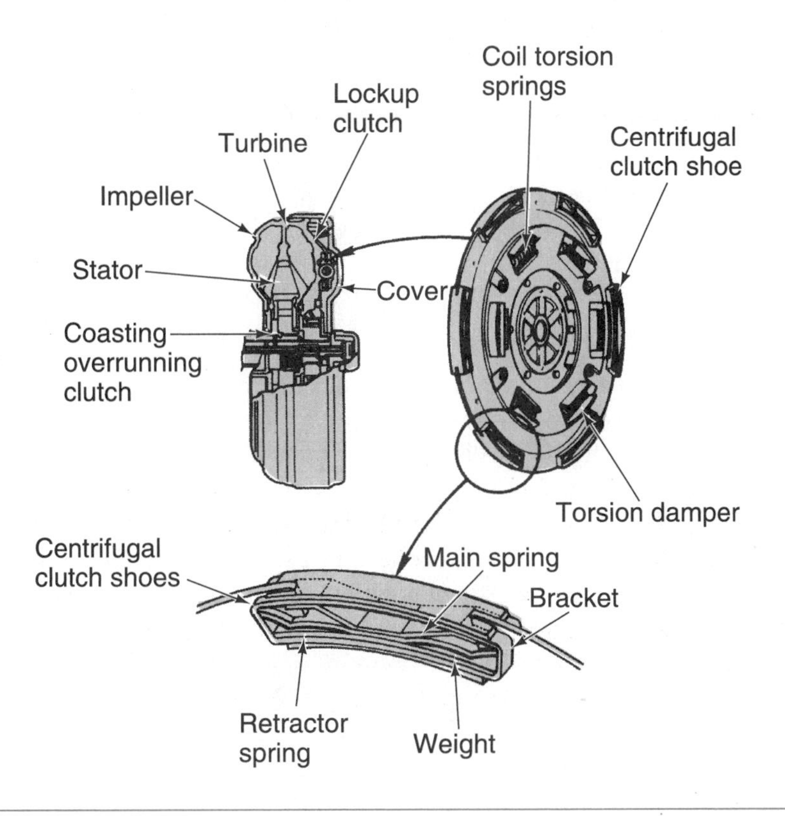

Figure 4-18 Clutch shoes on the clutch disc of a centrifugal lockup assembly. (Reprinted with the permission of Ford Motor Co.)

transferring of torque from the converter cover to the clutch. Power flows from the inside surface of the cover to the clutch assembly, which drives the turbine shaft. When the clutch is totally locked, there is no fluid coupling; rather, the power is transmitted mechanically.

Centrifugal force causes the shoes to pull against the tension of their spring. When the centrifugal force is decreased, the springs pull the shoes back toward the center of the disc. This action causes the clutch to disengage. Disengagement of the clutch happens gradually, just like engagement. The clutch shoes are designed to slip when the vehicle demands extra torque from the engine. This slippage allows for some torque multiplication. It is not unusual for the power flow to the transmission to be split between hydraulic and mechanical couplings. A friction-modified ATF allows the clutch shoes to slip without excessive friction when extra torque is needed.

Hydraulic Lockup

Shop Manual
Chapter 4, page 92

Another, and the most common, way to provide for a mechanical link between the engine and the transmission is through the use of a hydraulically operated lockup clutch. Introduced in the late 1970s, this design of lockup converter was controlled and operated totally by hydraulics and provided a mechanical link only when the transmission was in high gear. Today's transmissions lock up in more than one forward gear and rely on computer-controlled electrical solenoids to control the hydraulic pressure to the clutch assembly.

The operation of the hydraulic clutch is rather simple. However, the systems that control it can be fairly complex. The converter clutch is usually applied when the fluid flow through the torque converter is reversed by a valve. When the torque converter clutch control valve moves, the fluid begins to flow in a reversed direction. This forces the clutch disc or pressure plate against the front of the torque converter's cover. The position of the clutch now blocks the fluid flow through the converter and a mechanical link exists between the impeller and the turbine.

Normally, when the converter clutch is not engaged, fluid flows down the turbine shaft, past the clutch assembly, through the converter, and out past the outside of the turbine shaft. Normal torque converter pressures keep the clutch firmly against the turbine. There is no mechanical link and only a fluid coupling exists.

To engage the clutch, mainline pressure is directed between the plate and the turbine. This forces the plate into contact with the front inner surface of the cover and locks the turbine to the impeller. With the engagement of the clutch, the fluid in front of the clutch is squeezed out before the clutch is totally engaged. The presence of this fluid softens the engagement of the clutch and acts much like an accumulator.

Viscous Clutch

Shop Manual
Chapter 4, page 92

A viscous converter clutch (**VCC**) is used on some model automobiles to provide torque converter lockup. This design allows the clutch to lock in a very smooth manner with no engagement shock. It operates in the same way as a hydraulically applied clutch except that it uses a **viscous clutch** assembly to lock the impeller to the turbine. The viscous clutch assembly consists of a rotor, body, clutch cover, and silicone fluid (Figure 4-19). The **silicone fluid** is sealed between the cover and the body of the clutch assembly. It is the viscous silicone fluid that cushions the feel of clutch application.

The viscous converter clutch system from General Motors is referred to as the VCC system.

The viscous converter clutch is a self-contained fluid coupling with a built-in friction-faced pressure plate. When the clutch is engaged, the pressure plate is forced against the converter cover. Engine power is transmitted from the pressure plate through the fluid coupling to the transaxle's input shaft. The clutch's fluid coupling uses the viscous properties of thick silicone fluid between the closely spaced pressure plate and cover plate to transmit the power.

Late-model Cadillacs with a 440-T4/4T60 transmission are the most common vehicles equipped with a VCC.

When the clutch is applied, there is a constant but minor amount of slippage in the viscous unit, about 40 rpm at 60 mph. However, this slippage is nothing compared to a conventional torque converter without lockup. Engagement of the viscous clutch allows for improved fuel economy and reduced fluid operating temperatures. When the clutch is disengaged, the assembly operates in the same way as a conventional torque converter.

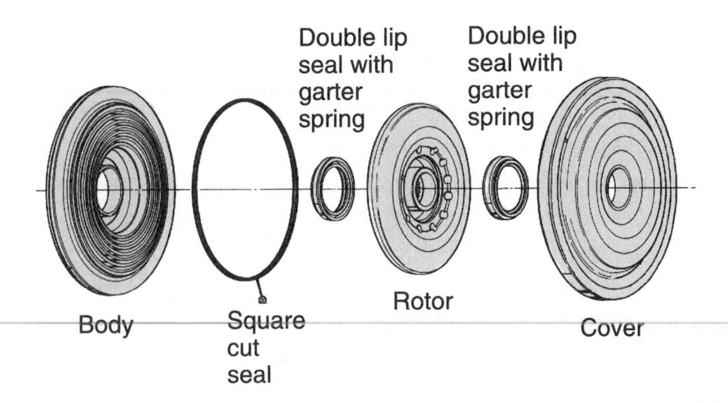

Figure 4-19 Viscous converter clutch assembly. (Courtesy of the Hydra-Matic Division of General Motors Corp.)

The viscous converter clutch is controlled by a solenoid, which is controlled by a control computer. The computer bases clutch engagement on conditions such as vehicle speed, throttle angle, transmission gear, transmission fluid temperature, engine coolant temperature, outside temperature, and barometric pressure, which it monitors. When the computer determines it is time to engage the clutch, it completes the ground circuit to the viscous clutch solenoid. Power is supplied to the solenoid by a fuse in the fuse panel.

The control computer will allow the clutch to engage (Figure 4-20) whenever internal transmission fluid pressures are correct, the engine is at its normal operating temperature, and one of the following sets of conditions exists: vehicle speed is approximately 25 mph, the transmission is in second gear, and the transmission oil temperature is below 200°F (93.3°C); or vehicle speed is at least 38 mph and the transmission oil temperature is between 200°F and 315°F (93.3°C to 157°C). If the temperature of the transmission oil exceeds 315°F (157°C), the viscous clutch is released to prevent transmission overheating.

Clutch Assembly

Shop Manual
Chapter 4, page 85

A typical hydraulic converter clutch has a plate that acts as a clutch piston and is splined to the front of the turbine. Friction material is bonded to either the forward face of the clutch plate or the

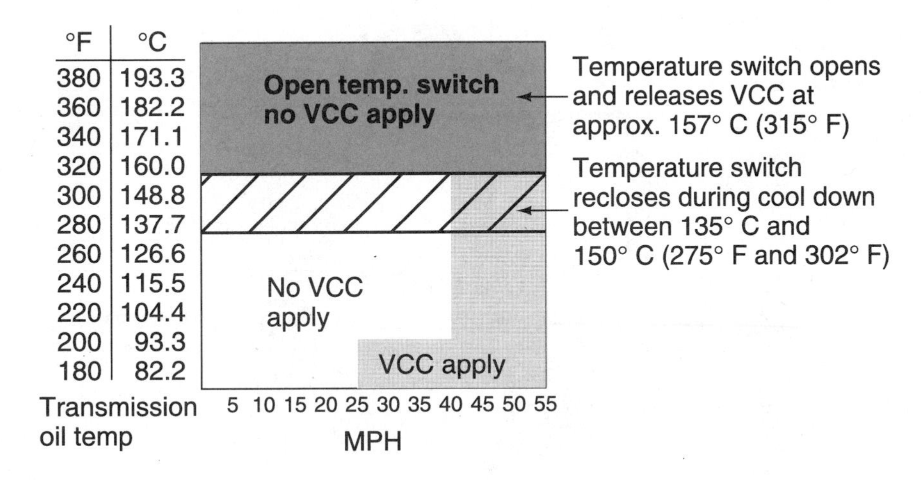

Figure 4-20 Chart showing the activity of a viscous clutch in response to changes in ATF temperature. (Courtesy of the Hydra-Matic Division of General Motors Corp.)

Torsional vibrations are slight speed fluctuations that occur during normal crankshaft revolutions. These vibrations can produce gear noise in a transmission, as well as a noticeable shaking of the vehicle.

inner front surface of the converter cover. A large ring of friction paper is the typical friction material used for a converter clutch.

A converter clutch disc is splined to the turbine so that it can drive the turbine when the frictional material is forced against the torque converter's cover. Most converter clutch assemblies are fitted with a damper assembly that directs the power flow through a group of coil springs. The damper springs are either located at the center of the disc or they are grouped around the outer edge of the disc. These springs are designed to absorb the normal torsional vibrations of an engine, which vary with changes in speed and load.

The **isolator springs** are placed evenly from the center of the disc and are sandwiched between two steel plates and riveted together. One of the plates is attached directly to the clutch assembly's hub and the other to the clutch disc. The two plates will move as a single unit after the plates have moved against the tension of the isolator springs. During lockup, the sudden application of torque to one plate is absorbed by the springs as they compress and start the other plate turning. The damper assembly acts as a shock absorber to the pulses of the engine's vibrations and softens the engagement of the clutch. When the converter clutch is not locked, the fluid inside the torque converter absorbs the torsional vibrations. Therefore, a damper assembly is not needed until the clutch mechanically connects the engine to the input of the transmission.

Converter Control Circuits

Many different systems have been used to control the application of the converter clutch. The systems vary from simple hydraulic controls at the valve body to complex computer-controlled solenoids (Figure 4-21). Simple systems limit the engagement of the lockup clutch in high gear only when the vehicle is traveling above a particular speed. More complex systems allow the clutch to engage at any time efficiency would be improved by doing so.

Most of the computer-controlled systems base clutch engagement on inputs from various sensors, which give information about the engine's fuel system, ignition system, vacuum, operating temperature, and vehicle speed. This information allows the computer to engage the clutch at exactly the right time according to operating conditions.

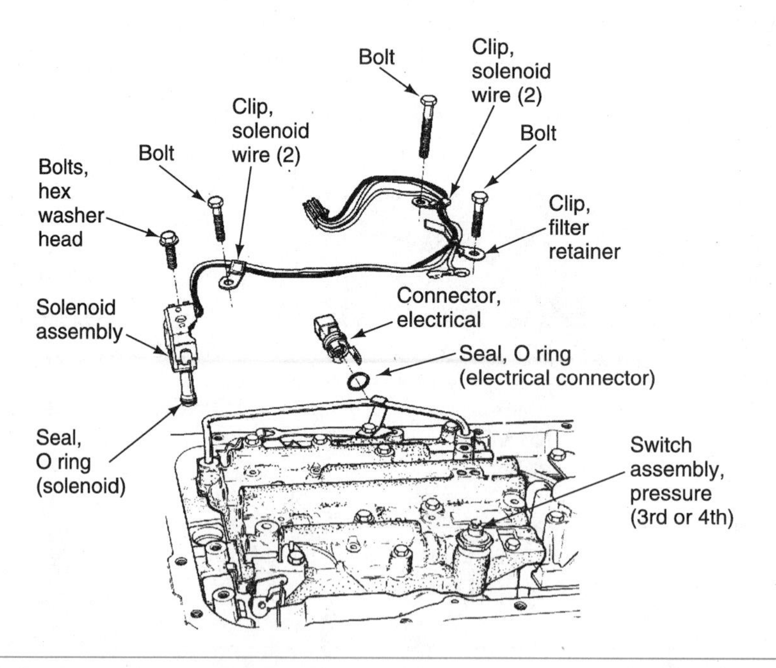

Figure 4-21 Typical location and mounting of a torque converter clutch solenoid. (Courtesy of the Hydra-Matic Division of General Motors Corp.)

The clutch is typically engaged by a clutch piston controlled by an electric solenoid and one or more spool valves. The solenoid controls pressurized fluid that moves the spool valve to move the clutch piston. The movement of the piston engages or disengages the converter clutch. When oil pressure moves the piston against the clutch disc, lockup occurs. As the piston moves away from the disc, the clutch unlocks.

A quick look at the systems used by different manufacturers shows the similarities in the systems. In spite of the different systems, they all function to provide a mechanical link between the engine and the transmission.

Shop Manual
Chapter 4, page 103

Chrysler Converter Controls

Early Chrysler electronically controlled lockup converter systems use an electric solenoid mounted on the valve body to engage and disengage the converter clutch. The operation of this solenoid is controlled by the engine control computer through a relay. The computer determines when to lock or unlock the torque converter based on information from the coolant temperature sensor, vacuum transducer, vehicle speed sensor, and carburetor or throttle ground switch. This information allows for precise converter clutch control, which leads to improved fuel economy, reduced transmission temperature, and reduced engine speeds.

A solenoid is an electric device capable of converting electrical energy into mechanical force.

Electronic control for the transmission was incorporated into the existing electronic combustion computer program for engine control. The system uses sensors already developed for use with the PCM to provide inputs for control of the torque converter lockup clutch. Prior to controlling the clutch electronically, Chrysler relied on hydraulic pressure to control the clutch. These hydraulic controls allowed clutch engagement when the transmission was in third gear only and operating above a preset speed, regardless of load.

Shop Manual
Chapter 4, page 81

With electronic control, the hydraulic lockup is still preset to the optimum lockup speed, but the solenoid-operated hydraulic valve is controlled by the computer; therefore, the actual time of lockup is controlled by the computer.

Prior to engaging the lockup clutch, the computer looks at engine coolant temperature. If the temperature is above 150°F, the computer scans other information from its sensors. If the temperature is below 150°F, the computer keeps the solenoid off until the temperature is reached. Once this temperature is reached, the computer looks at vehicle speed, engine vacuum, and throttle information to determine if the clutch should be engaged.

The vehicle speed sensor located on the speedometer cable sends an electronic signal to the computer, which reports that vehicle speed is above or below 40 mph. If the vehicle has reached the preset engagement speed, the computer will make sure the throttle is open. This prevents converter clutch engagement during coasting.

If the throttle is closed, regardless of other conditions, the clutch will remain off. When the throttle is open, the computer will wait a short time after it has been opened before it proceeds to engage the clutch.

The final factor that is looked at by the computer is engine vacuum. A **vacuum transducer** relays load information to the computer. Since engine vacuum relates to engine load, the computer will allow clutch engagement only during no-load or low-load conditions. The engine's vacuum must be above 4 inches and below 22 inches.

If all conditions are met, the computer energizes the clutch relay, which grounds the lockup solenoid (Figure 4-22). The activation of the solenoid moves the solenoid check ball against its seat. This prevents the solenoid valve from exhausting line pressure. Line pressure increases and forces the switch valve to move against coil spring tension. Pressure is then directed to the impeller drive hub and stator support to fill the torque converter with fluid. Line pressure flows from the impeller and turbine to fill the space behind the torque converter clutch piston (Figure 4-23). This high pressure forces the engagement of the clutch.

The acronym TCC stands for torque converter clutch.

To disengage the clutch, the solenoid is turned off. This moves the solenoid check ball away from its seat and fluid is exhausted. This action redirects the pressure inside the converter and relaxes the force on the clutch piston, thereby allowing the clutch to disengage.

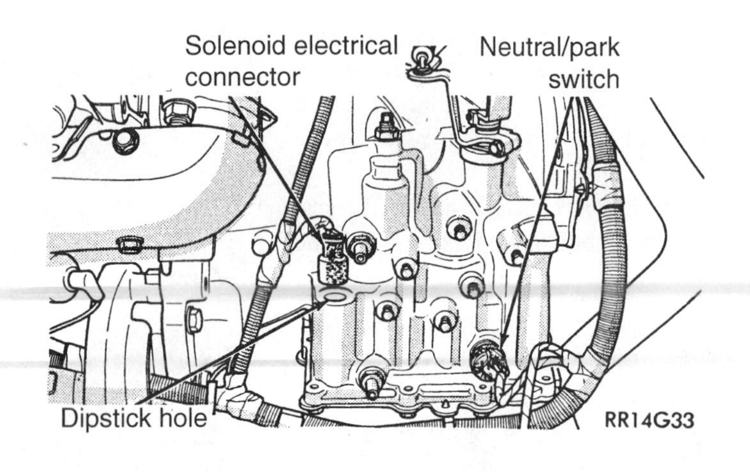

Figure 4-22 Location of the TCC solenoid wiring connector. (Courtesy of Chrysler Corp.)

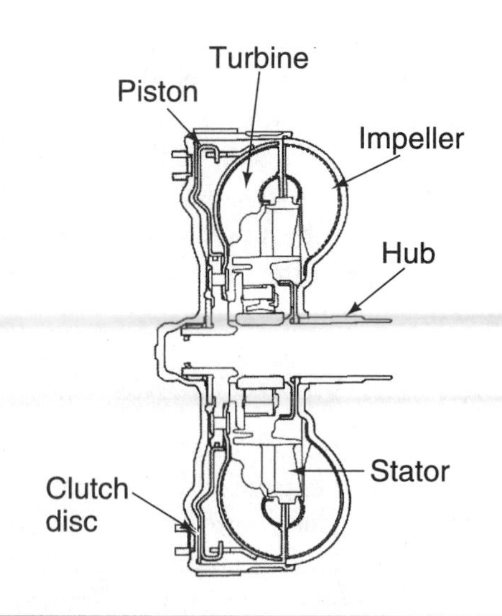

Figure 4-23 Hydraulic piston-type converter clutch assembly. (Courtesy of Chrysler Corp.)

The fail-safe valve prevents lockup clutch engagement until the transmission is in third gear.

Chrysler added a three-valve module to the transmission's valve body to provide for control of the lockup clutch's piston. This is controlled by governor pressure. When the vehicle's speed reaches a particular point, governor pressure forces the valve to move against the tension of its spring. The movement of the valve permits pressure to move to the fail-safe valve. When the transmission is in third gear, fluid flows to the **fail-safe valve** and moves it. Line pressure is now directed between the turbine shaft and the stator support to fill the torque converter and engage the clutch.

If the throttle is quickly opened during lockup, throttle pressure increases, causing the fail-safe valve to move and block the line pressure passage to the lockup valve. The clutch is disengaged as the fluid is redirected to the other side of the clutch's piston assembly.

Later-model Chrysler converter clutch systems operate in much the same way. However, as engine control computers became more complex with the addition of new sensors, so did converter clutch control. Slightly different conditions must be met before the computer will engage the clutch. The final determining factor, however, is the relationship between throttle position and vehicle speed.

The lockup solenoid is constantly receiving pressurized fluid. When it is deenergized, it exhausts the pressure, which prevents a buildup of pressure in the converter. When the computer completes the ground circuit for the solenoid, the exhaustion of fluid is stopped and the pressure builds. This buildup of pressure causes the engagement of the clutch.

Five conditions must exist before the computer will cause clutch engagement:

Shop Manual
Chapter 4, page 103

1. The coolant temperature must be at least 150°F.
2. The PARK/NEUTRAL switch must indicate that the transmission is in gear.
3. The brake switch must indicate that the brakes are not applied.
4. The vehicle must be traveling above 35 mph.
5. The signals from the throttle position sensor must be above a minimum voltage to indicate that the throttle is open.

A TP sensor is a sensor that monitors throttle position.

The computer also looks at the rate of change from the TP sensor. This rate indicates the driver's intent and helps the computer define the actual operating conditions.

If one of the following conditions exists, the solenoid will be activated and the converter clutch engaged. Lockup will occur under the following conditions:

36 mph and changes in voltage from the TP sensor are less than 0.16
37 mph and changes in voltage from the TP sensor are less than 1.44
55 mph and changes in voltage from the TP sensor are less than 1.74
65 mph and the voltage from the TP sensor is above a minimum voltage

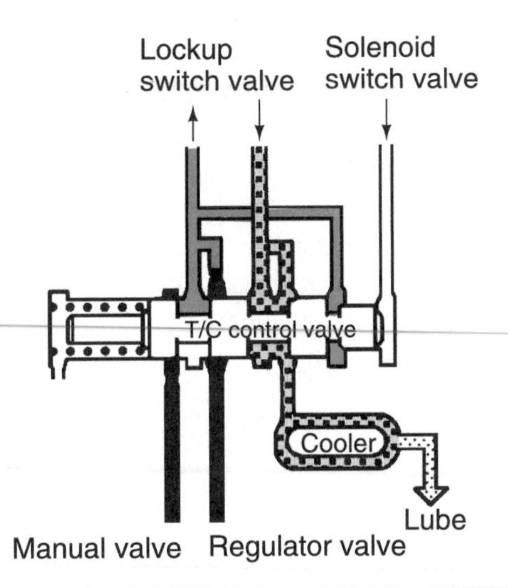

Figure 4-24 Controls for the fluid flow to and from the converter clutch assembly. (Courtesy of Chrysler Corp.)

Lockup will be interrupted or prevented during the following conditions:

38 mph and changes in voltage from the TP sensor are more than 0.30
39 mph and changes in voltage from the TP sensor are more than 0.96
60 mph and changes in voltage from the TP sensor are more than 1.54
70 mph and changes in voltage from the TP sensor are more than 2.24

If the vehicle is traveling at speeds less than 45 mph, the converter clutch will be disengaged any time the TP sensor rate of change increases or decreases by more than 0.040 volt per 11 milliseconds. When the vehicle is operating at speeds above 45 mph, the clutch will be disengaged whenever the TP sensor rate of change decreases by more than 0. 040 volt per 11 milliseconds.

Starting in 1989, Chryslers with the A-604/41TE transaxle have a torque converter that can lock up in second, third, and overdrive gear. The electronic modulated converter clutch **(EMCC)** system allows the converter clutch to partially engage between 23 and 47 mph. This feature limits the amount of converter slippage but serves to cushion the engine from the drive line. Full engagement occurs at about 50 mph and beyond. The converter clutch will lock according to engine speed and load in both third and overdrive gears. The clutch will also engage in second gear if the control unit receives a signal of high engine coolant temperature. This helps cool the engine. The transmission's hydraulic system includes a logic-controlled "solenoid torque converter clutch control valve" (Figure 4-24). This valve locks out the first gear so that the converter clutch is never engaged during first gear operation. The valve also redirects the fluid for first gear to the converter clutch valve. In order to downshift into first gear, the EMCC must receive information that the clutch is disengaged.

Most Chrysler FWD cars manufactured since 1994 are equipped with the A-604/41TE Ultradrive transaxle, except the larger models such as the New Yorker, LHS, Concorde, Intrepid, and Vision.

Shop Manual
Chapter 4, page 81

Ford Converter Clutch Controls

Electronically controlled hydraulic converter clutch circuits in Ford Motor Company products are controlled by the power train control module (PCM). This is the main processing device for the engine control system. The PCM receives information from various sensors and switches and generates output signals to control air/fuel mixture, emission controls, idle speed, ignition timing, and transmission operation. In addition, the system controls A/C compressor clutch operation, idle speed, transmission converter clutch operation, and third to fourth gear shifting on some models. Ford uses four slightly different system designs, each of which is based on particular transmission models, to control converter lockup. The main sensors for each of these designs are basically the same; however, the outputs are different.

Some Ford Motor Company transmissions and transaxles are equipped with either a centrifugal converter clutch or a split-torque-type torque converter.

Shop Manual
Chapter 4, page 103

CD4E Transaxle

Ford's CD4E transaxle is used in many of its FWD cars. The CD4E's torque converter clutch (Figure 4-25) is controlled by the power train control module. Under certain conditions, the PCM sends the appropriate signal to the TCC solenoid, which allows fluid pressure within the torque converter to force the piston plate and damper assembly against the cover, creating a mechanical link between the engine's crankshaft and the input shaft of the transaxle.

Pulse idth is the length of time something is energized.

The TCC solenoid is a **pulse width** modulated (PWM) style solenoid. The solenoid is used to control the apply and release of the bypass clutch in the torque converter. By modulating the pulse width of the solenoid, the pressure in the clutch circuit varies, thereby modulating the application and release of the bypass clutch in the torque converter. This allows for partial lockup during some operating conditions and for smooth engagement.

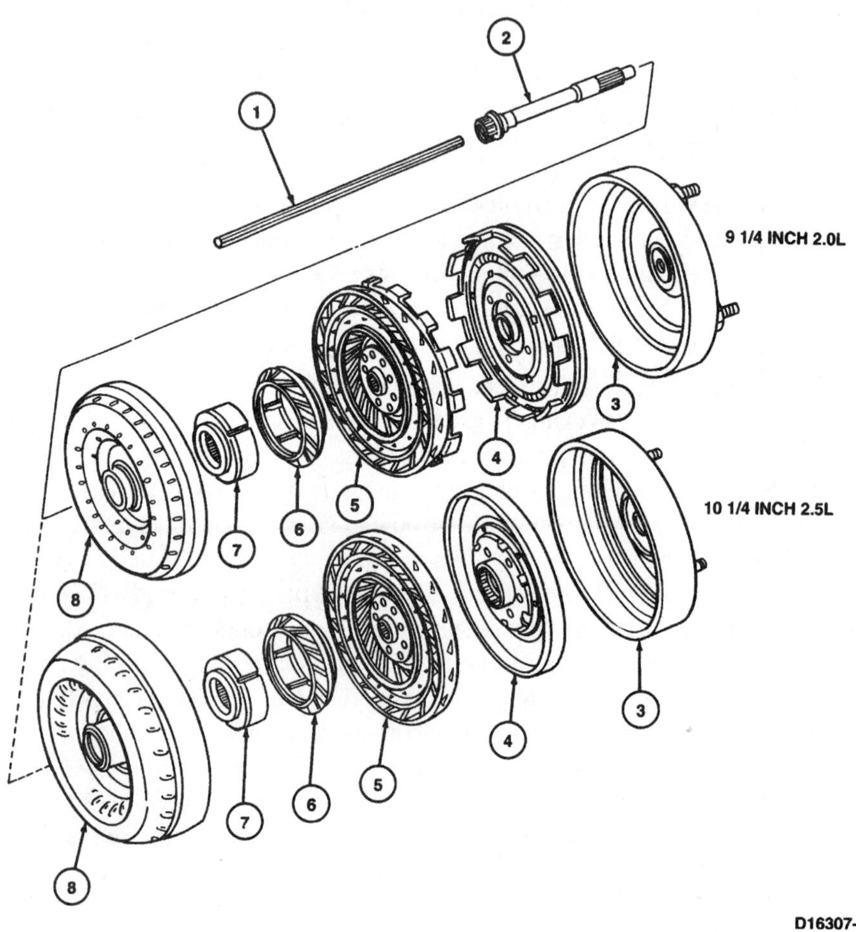

Item	Part Number	Description
2	7B328	Shaft—Pump Drive
2	7F213	Shaft Assy—Turbine
3	—	Cover (Part of 7902)
4	—	Plate—Torque Converter Clutch (Part of 7902)

Item	Part Number	Description
5	—	Turbine (Part of 7902)
6	—	Reactor (Part of 7902)
7	—	Thrust Bearing (Part of 7902)
8	—	Impeller (Part of 7902)

Figure 4-25 The two design variations for the lockup torque converter used in Ford's CD4E transaxle. (Reprinted with the permission of Ford Motor Co.)

AXOD, 4EAT, and E4OD Controls

The operations of the 4EAT and E4OD are very similar to the AXOD. The differences lie in the programming of the engine control computer. However, the 4EAT is less similar than the other two, in that it functions with only a lockup control valve and a lockup solenoid. The converter clutch of the AXOD and the E4OD is controlled by a converter clutch bypass solenoid, which vents line pressure to the oil pan when the converter clutch is released. The solenoid also redirects line pressure into an apply passage when the solenoid is energized to apply the clutch.

If the required operating conditions exist, the converter clutch will be applied at approximately 27 mph (43 km/h) when the transmission is operating in third gear, or approximately 35 mph (56 km/h) when it is in fourth gear. The converter clutch cannot be applied when the transmission is in first or second gear.

Engine coolant temperature and throttle position are monitored by the PCM and are used to determine if the converter clutch should be engaged. Before the converter clutch can be engaged, regardless of other conditions, the engine temperature must be at least 75°F. After the engine is warmed up and when the transmission is in third gear, the clutch will be engaged only if the throttle opening is greater than 8 percent and less than 59 percent. While in fourth gear, the throttle opening must be between 5 percent and 45 percent before the clutch is engaged. The PCM will also not allow clutch engagement in any gear if the throttle is closed, quickly opened, or opened wide.

When the converter clutch is disengaged, no current flows through the solenoid, preventing application of the converter clutch. When the solenoid is off, converter fluid moves from the converter regulator valve through the converter clutch control valve into a release passage. The release fluid then moves between the impeller and turbine shafts, pushing the converter clutch piston plate away from the converter cover. This releases the converter clutch (Figure 4-26).

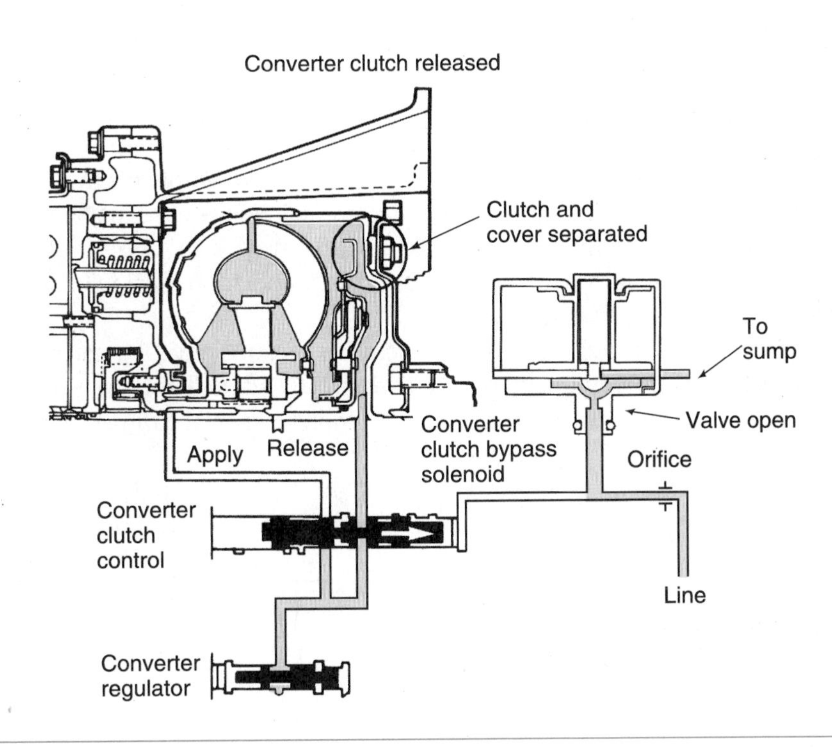

Figure 4-26 Fluid flow through the clutch circuit when the converter clutch is disengaged. (Reprinted with the permission of Ford Motor Co.)

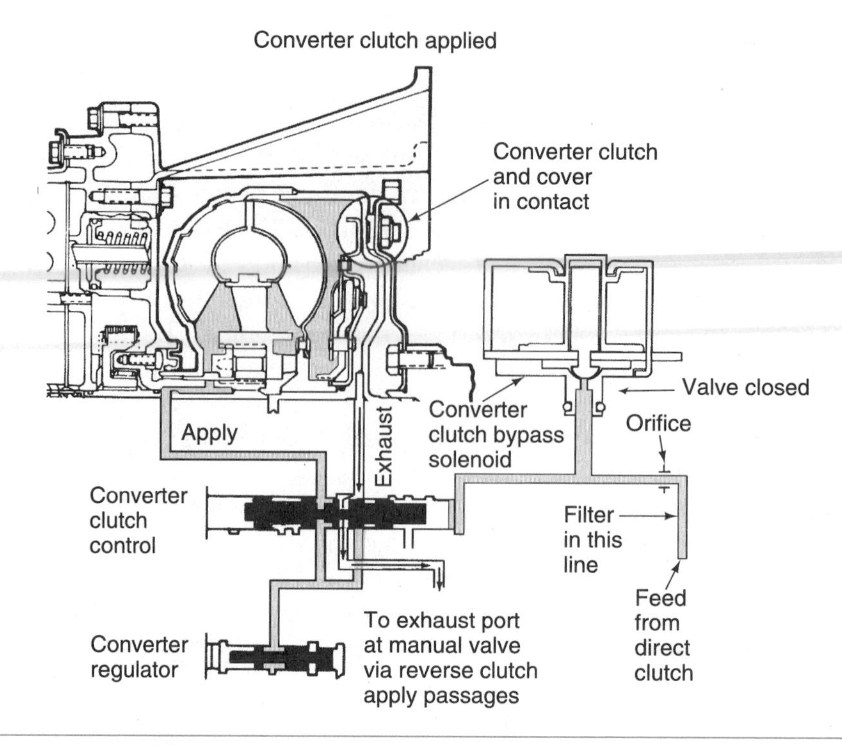

Figure 4-27 Fluid flow through the clutch circuit when the converter clutch is engaged. (Reprinted with the permission of Ford Motor Co.)

The converter clutch is applied when the converter clutch bypass solenoid is energized. This causes the release fluid between the converter cover and the clutch piston plate to be exhausted and allows fluid pressure to be applied to the back of the piston plate, causing it to contact the converter cover, thus locking the clutch to the cover (Figure 4-27).

AXOD-E/AX4S and AOD-E/4R70W Controls

Shop Manual
Chapter 4, page 103

A frequency is best defined as the number of cycles per second.

When it is installed in a Taurus or Sable model, the torque converter of an AXOD-E/AX4S functions in the same way as an AXOD. However, when the transmission is used in a Lincoln, a modulated lockup solenoid is used. The system operates in the same way as an AXOD, but the use of a pulse width modulated solenoid allows the PCM to vary clutch operation from full release to controlled slip to full apply.

The PCM turns the solenoid on and off at a constant **frequency.** If greater output pressure is needed, the pulse width is lengthened during each cycle. When less output pressure is needed, the pulse width is shortened. By varying the pulse width, the PCM can permit clutch slippage to obtain the best combination of performance and economy.

The converter control system in an AOD-E/4R70W also uses a pulse width modulated solenoid called the modulated converter clutch control (**MCCC**) solenoid. It operates in the same way as the AXOD-E/AX4S.

A4LD Controls

The converter clutch in an A4LD transmission is applied and released hydraulically, but it can be overridden electronically. The A4LD override solenoid (Figure 4-28) is normally energized to prevent fluid flow from unlocking the converter clutch.

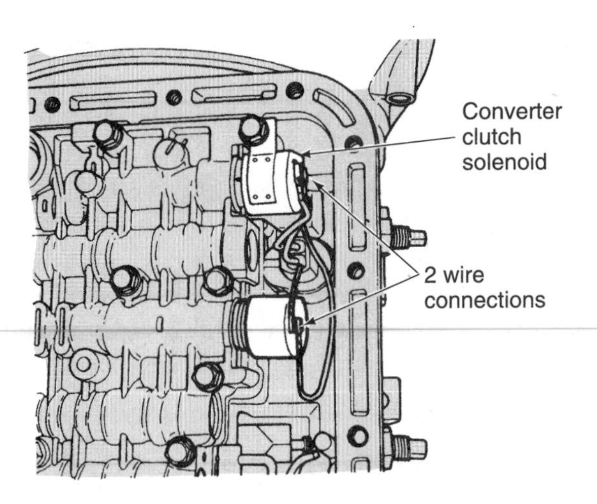

Figure 4-28 Location of converter clutch and override solenoids. (Reprinted with the permission of Ford Motor Co.)

A check valve is positioned in the circuit between the override solenoid and the converter clutch shift valve (Figure 4-29). This check valve operates according to line pressure and spring tension. When governor pressure is low, line pressure from the check valve keeps the converter clutch in the downshift position and the torque converter remains unlocked.

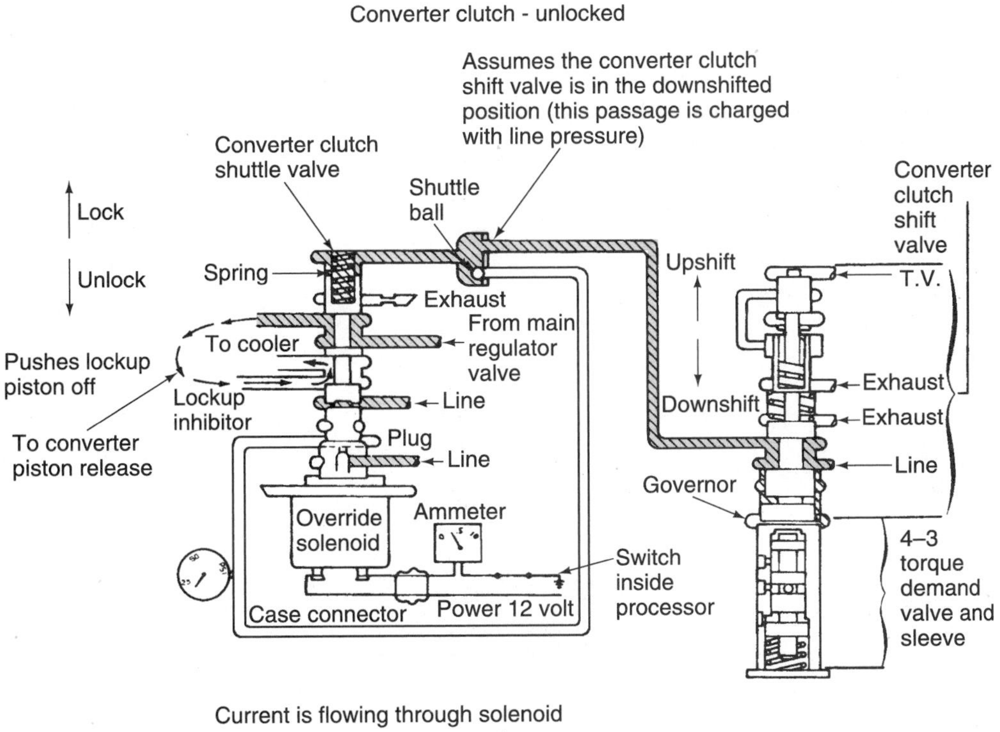

Figure 4-29 The flow of oil when the clutch is released or not locked to the converter's cover. (Reprinted with the permission of Ford Motor Co.)

As governor pressure increases, it moves the converter shift valve up, which exhausts fluid from the spring end of the check valve. This directs line pressure through the converter apply circuit. If governor pressure decreases enough, the converter clutch shift valve moves back down and the check valve returns to its normal position.

The override solenoid will be energized when the PCM detects a condition unfit for converter lockup. To unlock the converter, the PCM deenergizes the override solenoid. This allows line pressure to flow through the solenoid and into the lockup inhibition circuit. Pressure in the circuit moves the two-way check ball installed between the check valve and the converter shift valve to block the passageway. This allows the converter clutch shift valve to remain in the upshifted position, but it is ineffective. The check valve returns to its unlocked position and the converter apply fluid is exhausted. These actions release the clutch.

General Motors' Converter Clutch Controls

Shop Manual
Chapter 4, page 104

Computer command control is the common name given to GM's engine control computer.

The converter clutch shift valve is called the TCC shift valve and it controls the application and release of the clutch according to the position of the TCC regulator valve.

The torque converter clutch (**TCC**) assembly allows for a mechanical link between the impeller and the turbine during all gear ranges except PARK, REVERSE, NEUTRAL, and DRIVE range—first gear. On early vehicles equipped with computer command control, the converter clutch is engaged and disengaged by a solenoid (Figure 4-30). This solenoid is controlled by the brake switch, third gear clutch pressure switch, and the computer command control system. The computer controls the application of the clutch by providing a ground circuit for the TCC solenoid circuit. This solenoid assembly is energized by the PCM to redirect ATF to the clutch apply valve in the converter clutch control valve assembly. On vehicles that are not equipped with computer command control, converter clutch operation is controlled by the converter clutch shift valve and the solenoid, which are normally located together in the valve body assembly.

When the torque converter clutch solenoid's ground circuit is completed by the power train control module, fluid pressure is exhausted from between the converter's pressure plate and the converter cover, and converter clutch apply pressure from the converter clutch regulator valve pushes the converter pressure plate against the converter cover to apply the converter clutch. The apply feel of the clutch is controlled by the converter clutch regulator valve and the converter clutch accumulator (Figure 4-31).

When the TCC solenoid is deactivated, fluid pressure is applied between the converter cover and pressure plate. Converter feed pressure from the pressure regulator valve passes through the converter clutch apply valve into the release passage. This fluid is directed between the pump shaft and the turbine shaft and pushes the pressure plate away from the converter's cover to release the converter clutch (Figure 4-32).

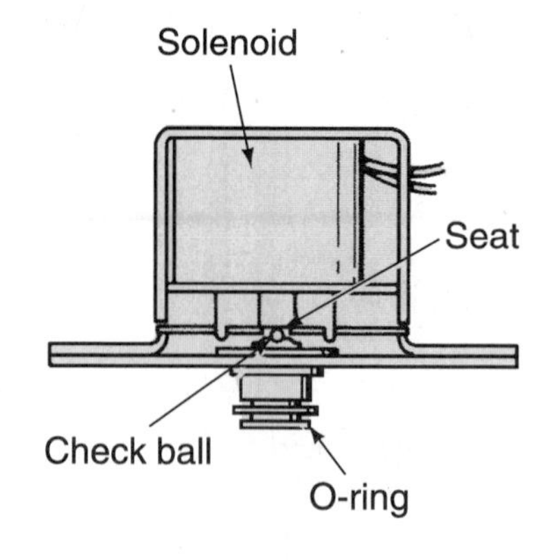

Figure 4-30 Typical clutch solenoid valve assembly. (Courtesy of the Hydra-Matic Division of General Motors Corp.)

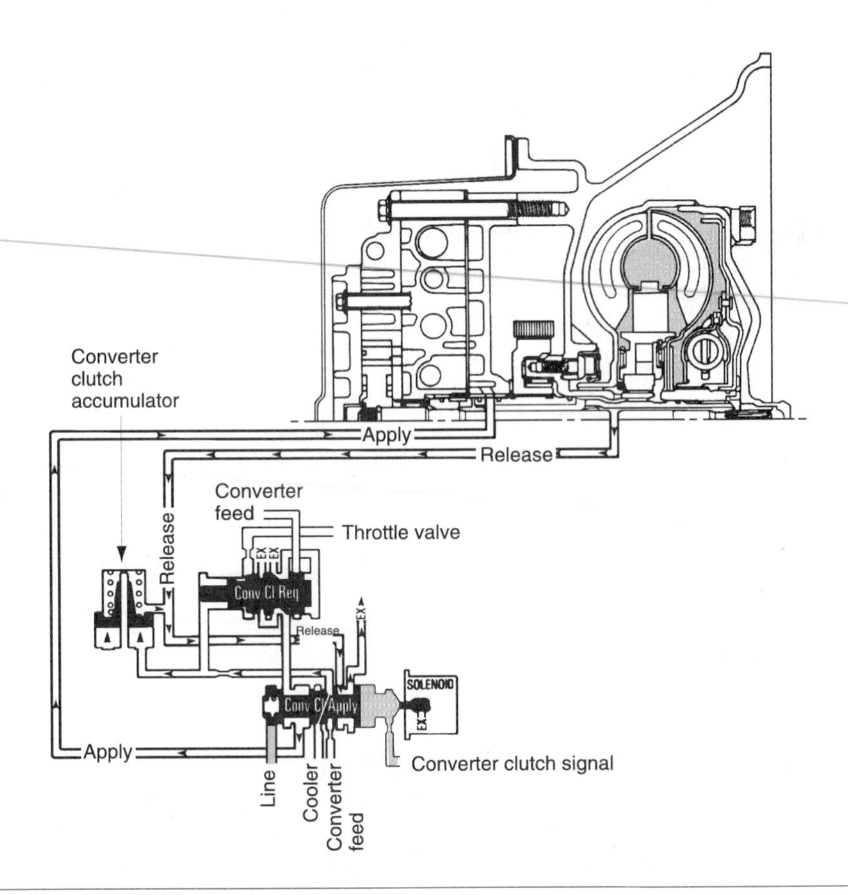

Figure 4-31 The converter clutch apply feel is controlled by the converter clutch regulator valve and the converter clutch accumulator. (Courtesy of the Hydra-Matic Division of General Motors Corp.)

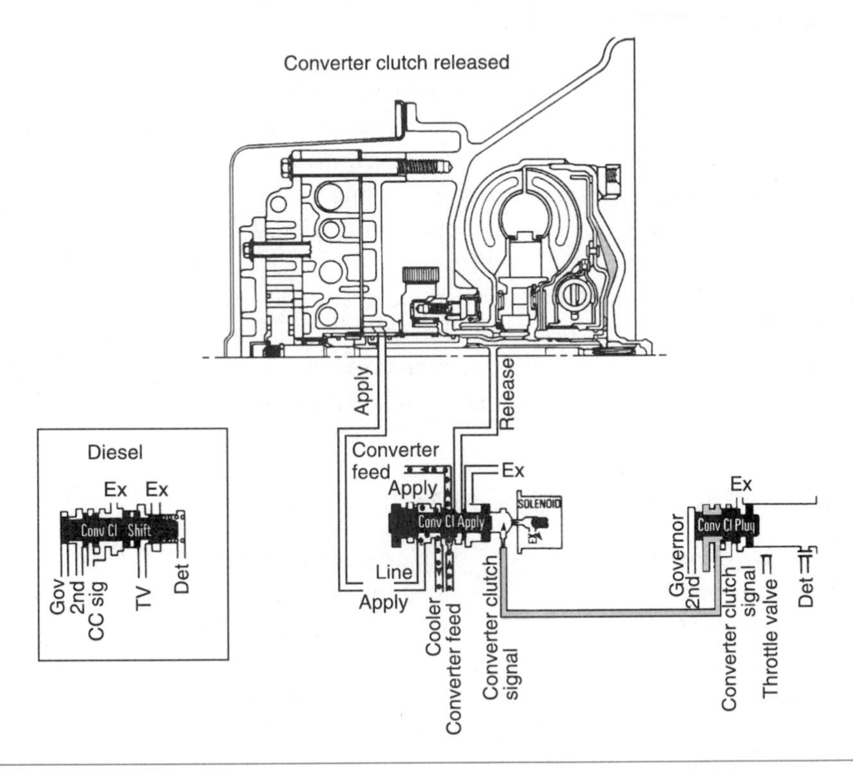

Figure 4-32 Oil circuit for a typical GM hydraulic converter clutch system. (Courtesy of the Hydra-Matic Division of General Motors Corp.)

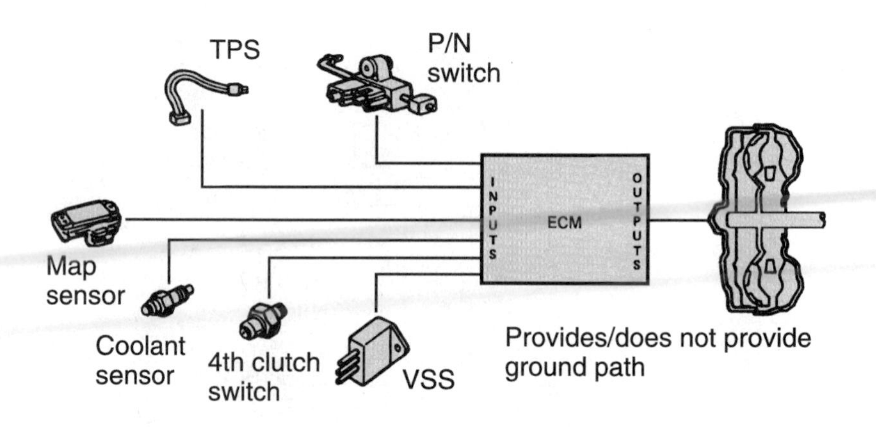

Figure 4-33 Typical layout for clutch control by a PCM in a GM transmission. (Courtesy of the Hydra-Matic Division of General Motors Corp.)

Various sensors provide input to the PCM for the control of the converter clutch (Figure 4-33). Not all models have the same sensors, but all systems operate in a similar way. Power from the ignition switch passes through the brake switch to the converter clutch solenoid. If the brake pedal is depressed with the clutch engaged, power to the TCC solenoid is interrupted. This releases the clutch to prevent the engine from stalling. The PCM receives information from the engine coolant temperature sensor and will not allow TCC operation until the coolant temperature is higher than 130°F to 150°F.

The computer also receives information from other sensors, such as the TP sensor, which provides the PCM with information on the position of the throttle. TCC operation is prevented when the throttle position signal is less than a specified value. Engine load is also looked at by the computer through inputs received from vacuum sensors, such as a MAP sensor.

Vehicle speed is a critical input, especially when the speed is compared to the other inputs. This combination of inputs can clearly define the exact operating conditions of the vehicle. One of two types of vehicle speed sensors is used to monitor vehicle speed for the system. A light emitting diode (LED) type is used in the instrument cluster on some models. Other models use a PM generator mounted in the transmission housing. Vehicle speed must be greater than a certain value before the TCC can be applied.

To prevent converter lockup in first and second gears, some systems have a third and fourth gear switch. This switch prevents TCC operation until the transmission is operating in third or fourth gear. The position of this switch may be monitored by the PCM or the switch may be wired in series with the power feed to the TCC solenoid. When the switch is open, no power will flow to the solenoid.

Honda Converter Clutch Controls

The torque converter of these transaxles can operate in various degrees of lockup: zero, partial, half, full, and cycling. Lockup is controlled by the transmission's hydraulic system and the transmission control module (**TCM**) (Figure 4-34). Three valves control the range of lockup according to two separate lockup control solenoid valves and a throttle valve. These lockup control solenoid valves are mounted on the torque converter housing and are controlled by the TCM. When the solenoid valves are activated, transmission modulator pressure changes.

When the gear selector is placed in DRIVE 4 and the transmission is operating in second, third, or fourth gear, the converter clutch may be applied. The clutch will only apply in Drive 3

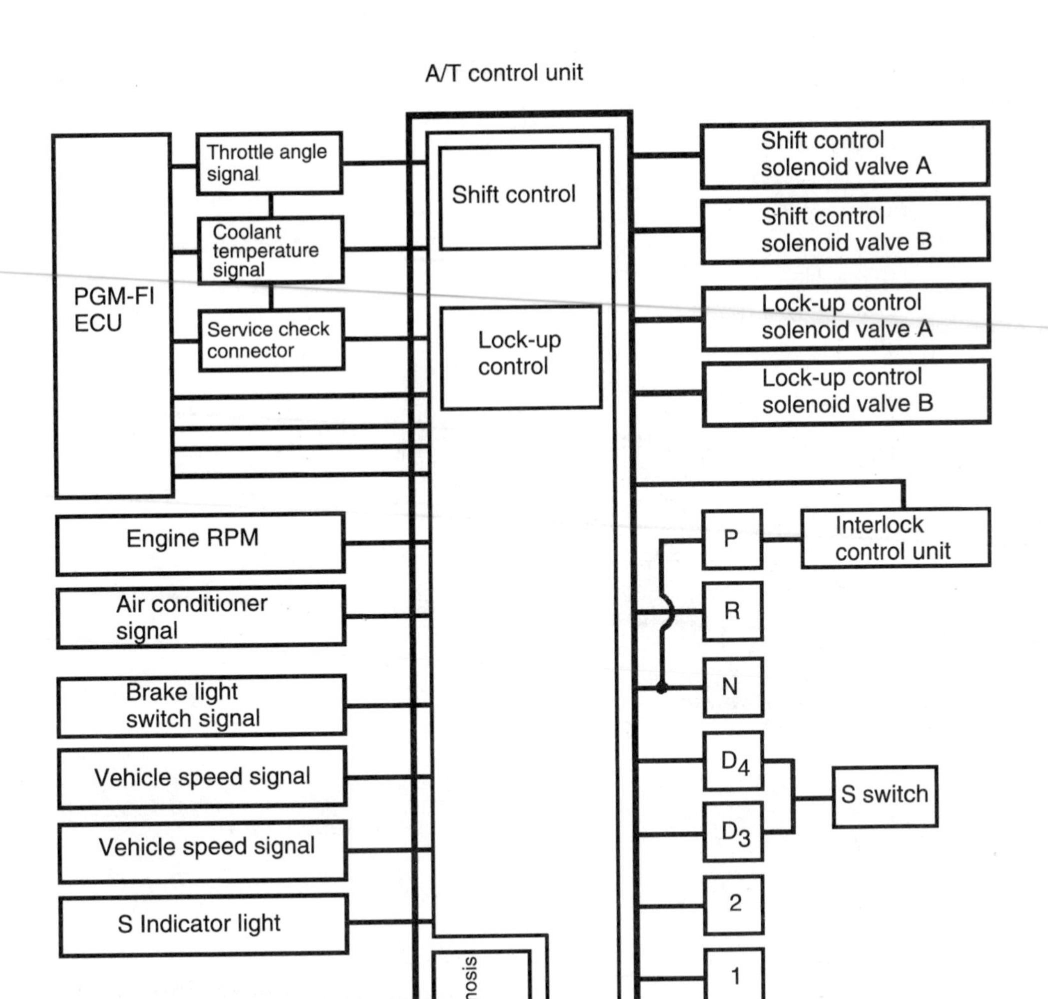

Figure 4-34 Typical layout for clutch control in a late-model Honda transaxle. (Courtesy of the Honda Motor Company, Ltd.)

when the transmission is operating in third gear. When the clutch is applied, the fluid between the converter cover and lockup piston is discharged and the converter oil exerts pressure through the piston against the converter cover. This firmly locks the converter's turbine to the converter's cover (Figure 4-35).

When the clutch is disengaged, the fluid in the converter flows in the opposite direction. As a result, the lockup piston is moved away from the converter cover. The pressurized fluid regulated by the modulator is present at both ends of the lockup shift valve and at one side of the lockup control valve. With equal pressure present at both ends of the lockup shift valve, the shift valve is moved to the right by the tension of the valve spring alone. This causes fluid to flow to the back side of the pressure plate and the clutch is pushed away from the converter's cover.

Partial lockup results from one solenoid valve being on and the other being off. One solenoid is on and releases the modulator pressure from one side of the shift lockup valve. Modulator pressure on the other side of the valve overcomes the tension of the spring, allowing the shift valve to move.

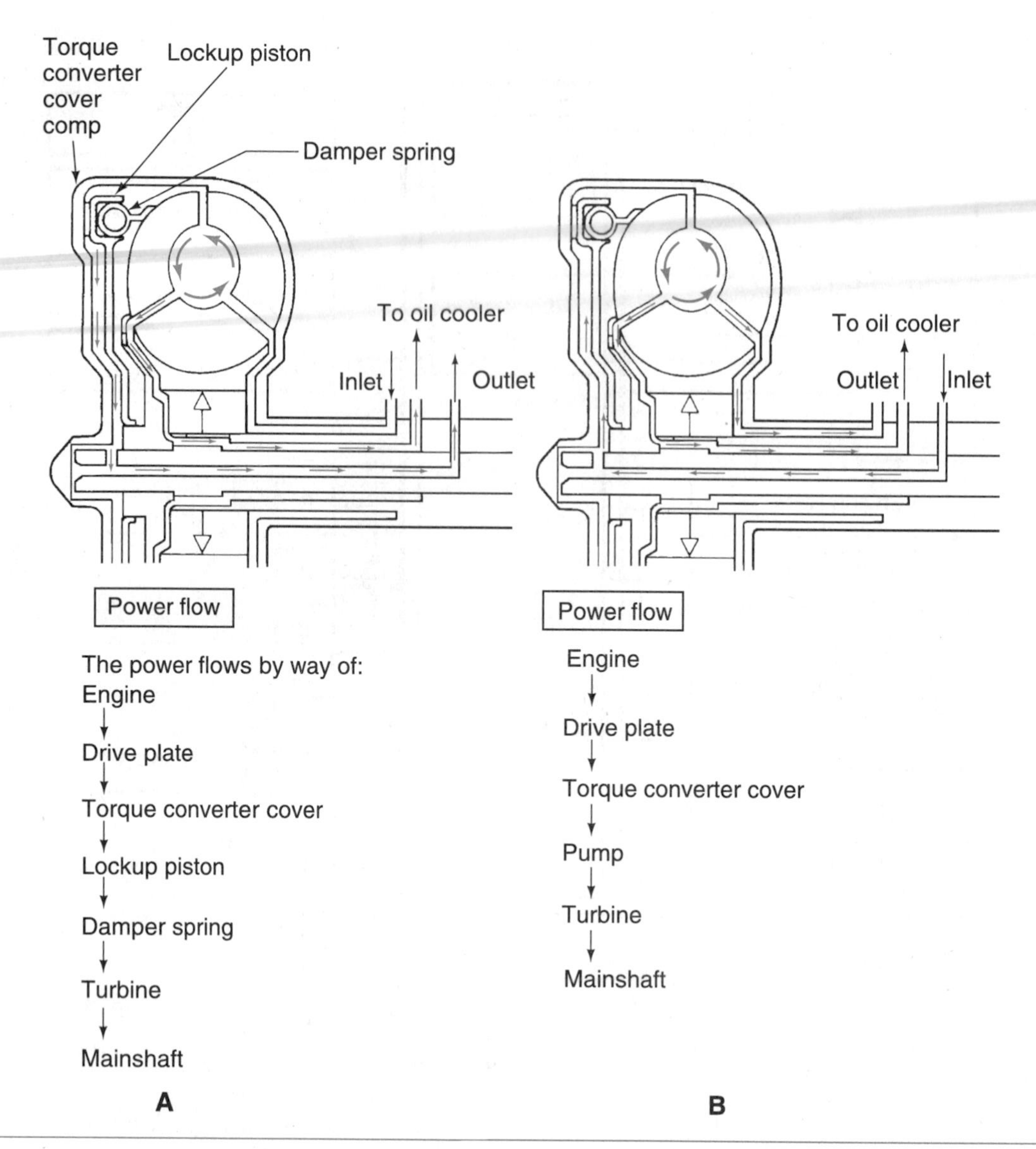

Figure 4-35 Power flow diagram and fluid flow in a Honda torque converter when there is lockup (A) and when the clutch is released (B). (Courtesy of the Honda Motor Co., Ltd.)

Modulator pressure is divided into two separate passages: one for the torque converter's inner pressure, which enters into the clutch assembly to engage the clutch, and one for torque converter back pressure, which enters between the pressure plate and the cover to disengage the clutch (Figure 4-36). Converter back pressure is regulated by the lockup control valve, whereas the position of the lockup timing valve is determined by the throttle pressure, tension of the valve spring, and the pressure regulated by the modulator. The position of the lockup control valve is determined by the back pressure of the lockup control valve and torque converter pressure, which is regulated by a check valve.

When the other solenoid is off, modulator pressure is maintained at one end of the lockup control valve, which moves it slightly to one side. This slight movement of the lockup control valve causes back pressure to decrease slightly. This results in a partial lockup of the clutch.

When both solenoids are turned on by the computer, the torque converter clutch is in the half-lock position. Modulator pressure is released, causing pressure to decrease on one side of

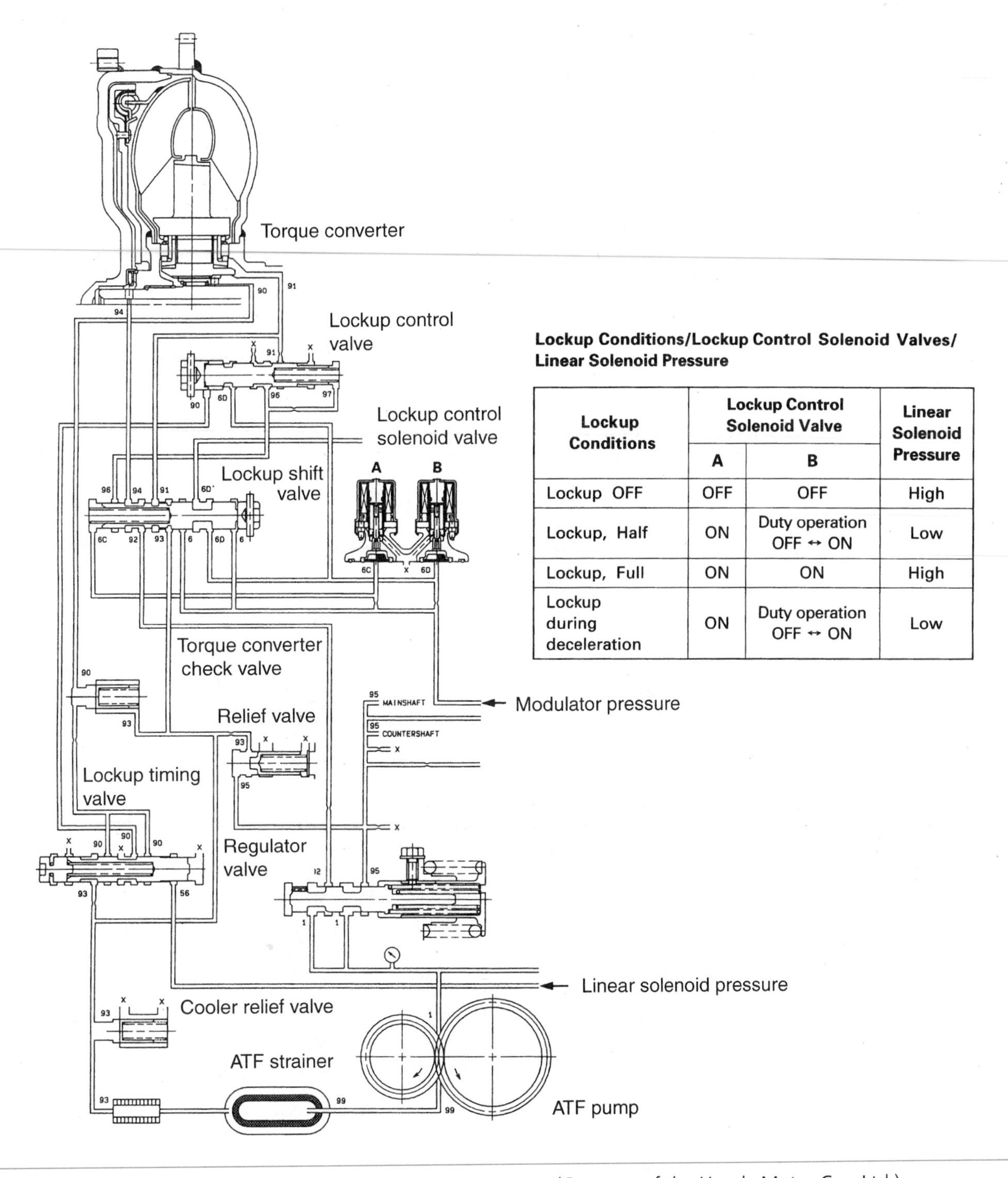

Lockup Conditions/Lockup Control Solenoid Valves/ Linear Solenoid Pressure

Lockup Conditions	Lockup Control Solenoid Valve		Linear Solenoid Pressure
	A	B	
Lockup OFF	OFF	OFF	High
Lockup, Half	ON	Duty operation OFF ↔ ON	Low
Lockup, Full	ON	ON	High
Lockup during deceleration	ON	Duty operation OFF ↔ ON	Low

Figure 4-36 Hydraulic system for the lockup torque converter. (Courtesy of the Honda Motor Co., Ltd.)

both the lockup control valve and the lockup timing valve. However, since throttle pressure is still low, the lockup timing valve remains to one side because of the tension of its spring. Because back pressure is also lower, a greater amount of pressure is present at the lockup clutch and it attempts to engage. However, some back pressure still exists and prevents the clutch from engaging fully.

Full lockup also occurs when both solenoid valves are on (Figure 4-37). But vehicle speed must increase, as must throttle pressure, which increases with throttle opening. The lockup timing

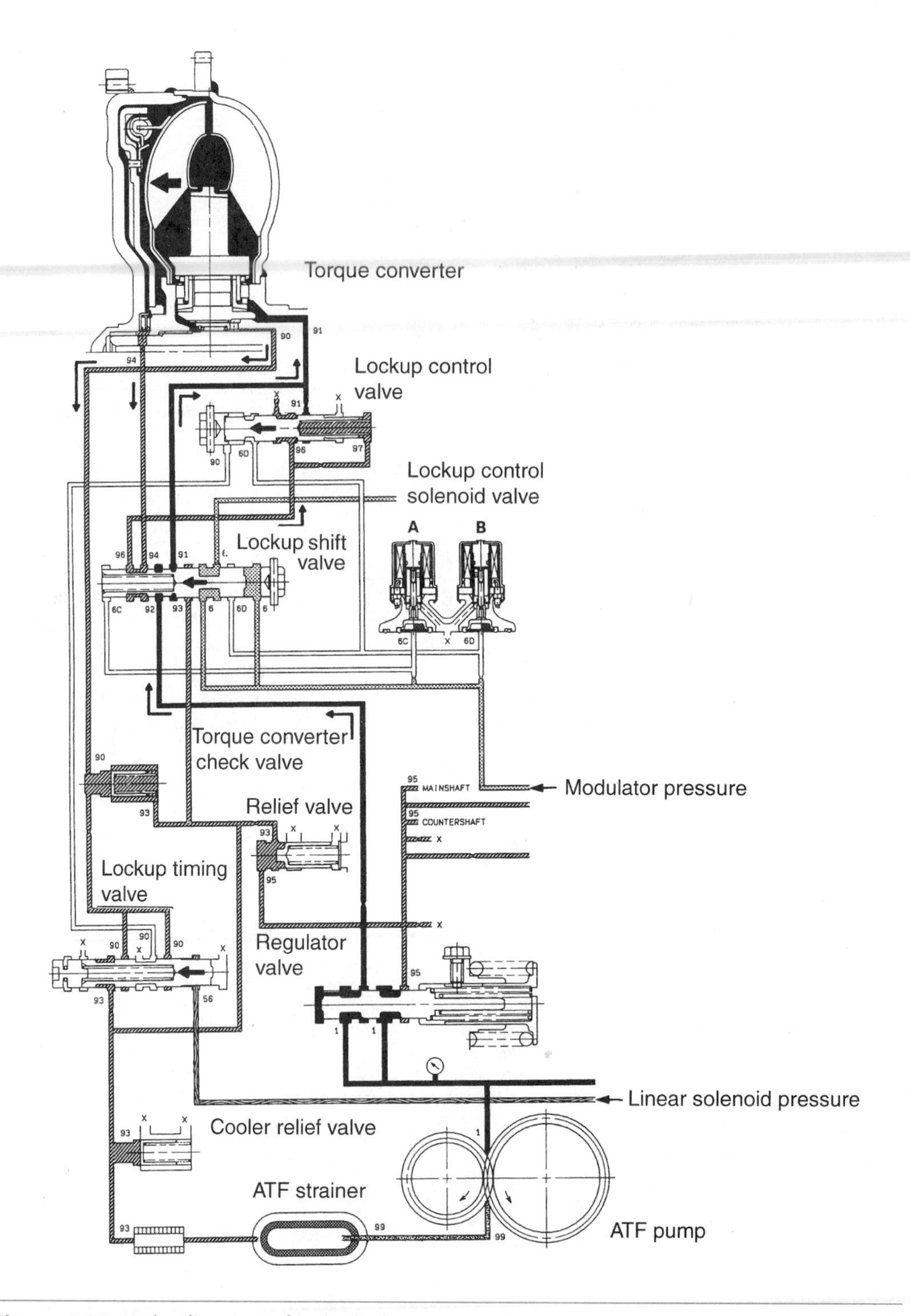

Figure 4-37 Hydraulic system for the lockup torque converter during full lockup. (Courtesy of the Honda Motor Co., Ltd.)

valve overcomes the tension of its spring and moves to one side, closing the fluid port leading to the converter's check valve. The lockup control valve moves to the other side because the throttle pressure is greater than modulator pressure. The movement of the control valve exhausts all of the converter's back pressure and allows the clutch to engage fully.

During deceleration, one solenoid remains on and the other is operating on a duty cycle that causes it to be turned rapidly on and off (Figure 4-38). This condition causes partial and half-clutch lockup to occur in a sequence. This sequence allows for engine temperature control and engine braking during deceleration.

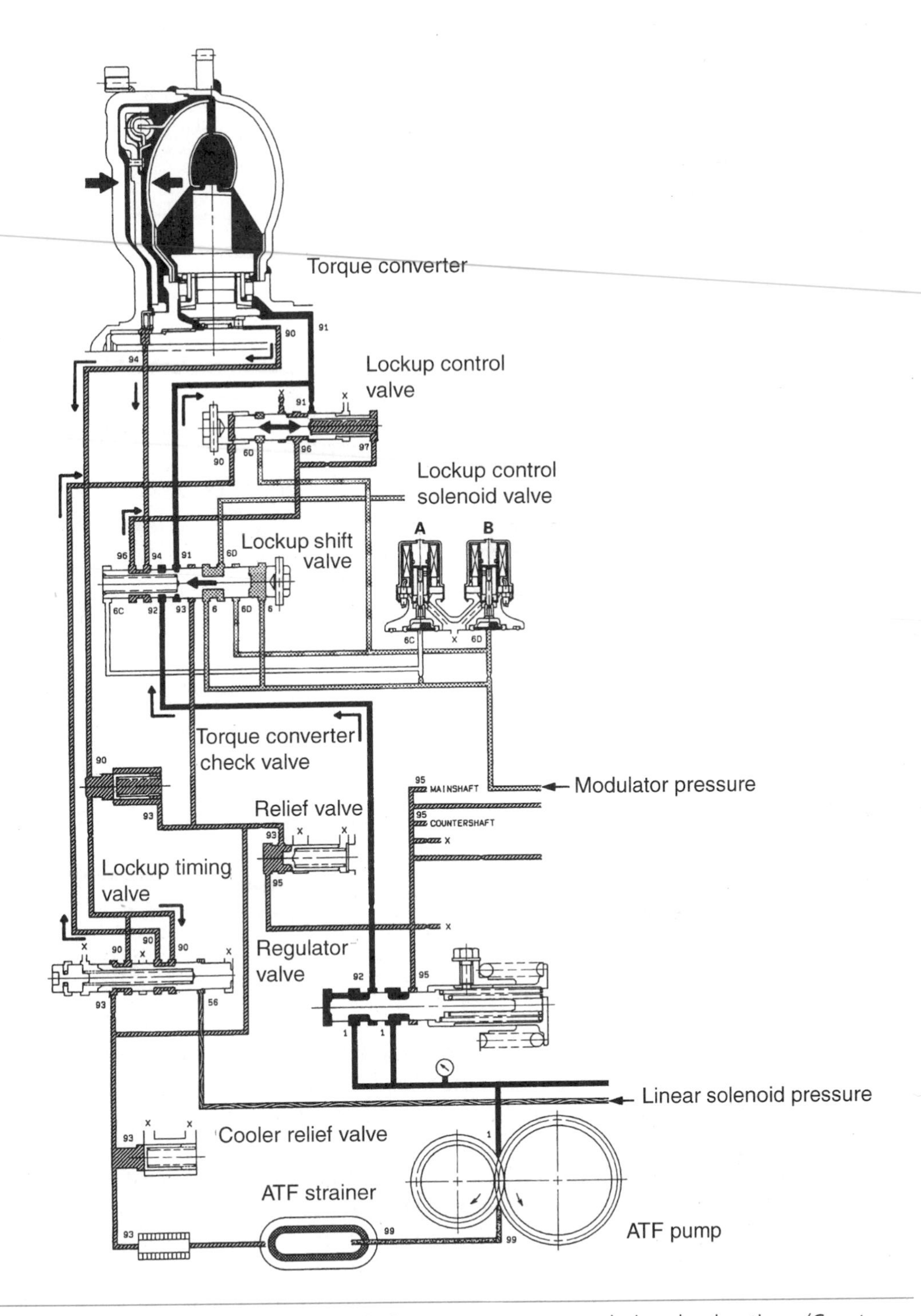

Figure 4-38 Hydraulic system for the lockup torque converter during deceleration. (Courtesy of the Honda Motor Co., Ltd.)

Toyota Transaxles

The PCM in Toyota's TCC system has a program in its memory for converter lockup in each of its driving modes (normal, economy, and power). Based on these programs, the PCM turns the lockup solenoid on or off according to operating conditions (Figure 4-39). The solenoid controls the position of the lockup control valve, which stops or allows fluid pressure within the torque converter to force the piston plate and damper assembly against the cover creating a mechanical link between the engine's crankshaft and the input shaft of the transaxle.

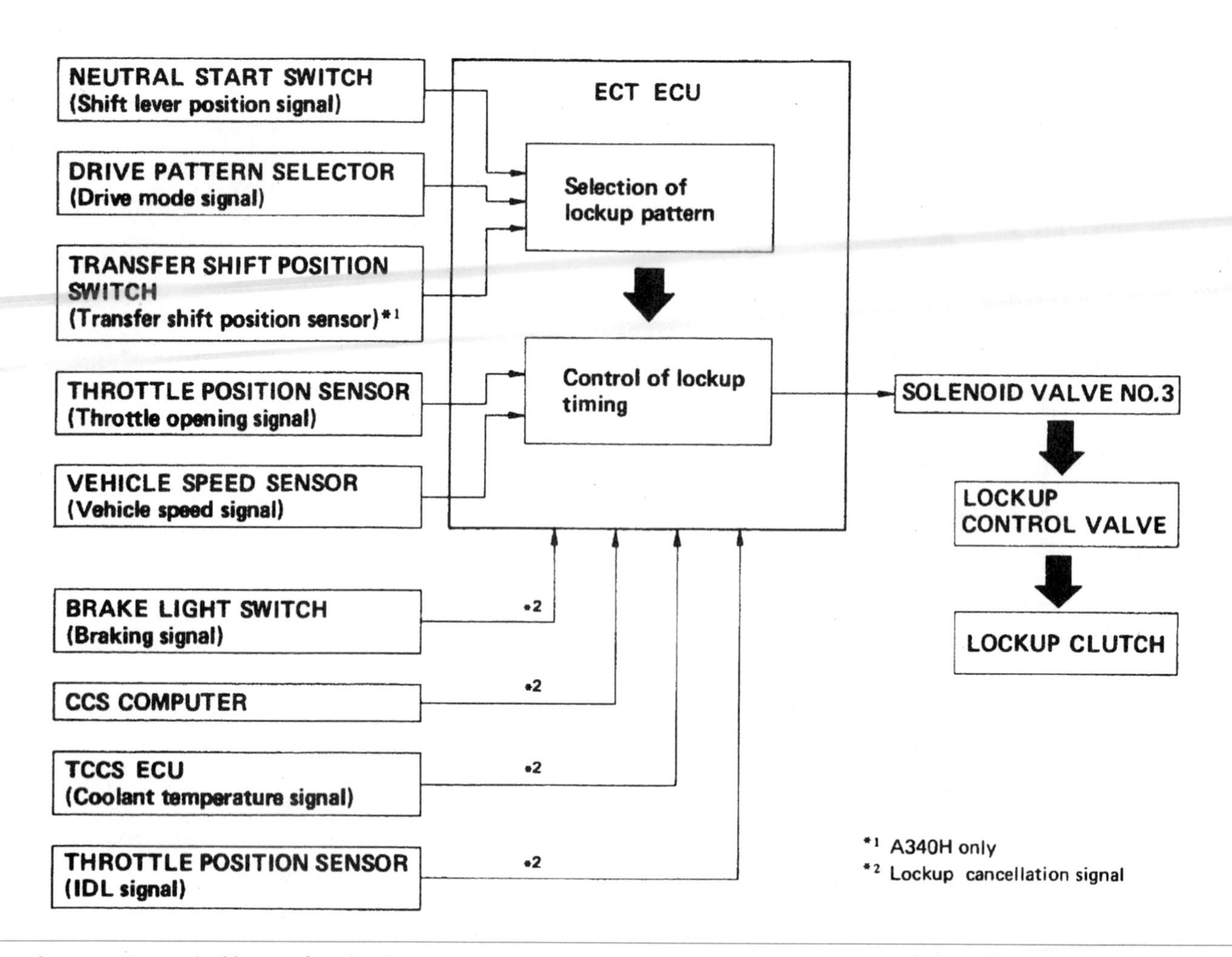

Figure 4-39 Typical layout for clutch control on a late-model Toyota transmission. (Reprinted with permission of Toyota Motor Co.)

The PCM will energize the TCC solenoid when three conditions are present. The vehicle must be traveling in second or third gear of the D range or it must be in Overdrive. The speed of the vehicle must be at or above the specified speed and the throttle opening at or above the specified value. Finally, the PCM must not have received a mandatory lockup cancellation signal.

Mandatory cancellation signals will cause the solenoid to be deenergized when certain conditions exist. Lockup will be canceled if the brake pedal is depressed and the brake light is turned on or the throttle position sensor indicates that the throttle is closed. By preventing lockup during these times, engine stalling is prevented. Lockup will also be canceled when the vehicle's speed drops about 10 mph below the set speed while the cruise control system is operating. Doing this allows the torque converter to serve as a torque multiplier, which aids in acceleration or overcoming heavy loads. If the engine's coolant temperature falls below a specified temperature, lockup will be canceled and prevented. This allows for improved driveability, reduced emissions, and quicker transmission warming. The PCM will also temporarily turn the TCC off during upshifting or downshifting to decrease the harshness of the shifts.

Special Applications

Manufacturers have used different designs to accomplish basically the same result—more efficient torque converter operation. There are currently two common deviations from the normal torque converter and from a normal lockup-type converter. Both of these were introduced by Ford Motor Company. One of these designs involves the use of a planetary gear set in the torque converter. This design allows for the mechanical and hydraulic transferring, sometimes simultaneously, of power from the engine to the transmission. The other design has the same result, but it does it without a torque converter planetary gear set.

The torque converter used in Ford's AOD transmission is not a true lockup converter. Rather, it uses a split torque path to get the same results. The transmission and torque converter are connected by three separate shafts. The input shaft is hollow and transmits hydraulic input torque through the transmission's one-way clutch or direct clutch. Inside the input shaft is an inner or intermediate shaft that transmits a combination of hydraulic and mechanical input to the transmission's intermediate clutch. The third shaft is the pump shaft. This shaft is driven by the converter cover and mechanically drives the oil pump. This setup allows power to enter the transmission along two separate paths, one mechanical and one hydraulic.

Shop Manual
Chapter 4, page 87

In reverse, first, and second gears, only hydraulic drive through the turbine transmits engine torque to the transmission. In third gear, the torque is split between the turbine, about 40 percent, and the intermediate shaft, about 60 percent. In fourth gear, all engine torque is transmitted mechanically. A damper assembly is built into the torque converter and splined to the intermediate shaft to absorb the torsional vibrations of the engine.

Splitter Gear Converters

A BIT OF HISTORY

The Dual-Path transmission used in Buick Specials in the early 1960s also used a planetary converter with an internal multiple-disc clutch assembly.

The torque converters in some Ford ATX transaxles are also a split-torque design. However, it is constructed with an internal planetary gear set (Figure 4-40). The purpose of the planetary gear set is not to multiply torque; rather, it splits the input torque between mechanical and fluid transmittal to reduce converter slip.

Two hollow shafts from the transmission are splined to separate gear set members in the converter. In the center of the converter's gear set is the sun gear, which is splined to the turbine shaft. Because the turbine is driven hydraulically, the turbine and sun gear provide the hydraulic input to the transaxle. The other shaft, the intermediate shaft, is splined to the planetary carrier of the gear set. The carrier and intermediate shaft provide for the mechanical input into the transaxle. The planetary pinion gears are meshed with the internal teeth of the ring gear. The ring gear is splined to the converter's cover and a damper assembly; therefore, the ring gear always rotates at engine speed.

Ford refers to the ATX converter as a splitter torque converter because there are two ways that torque can pass through it.

The **splitter gear converter** provides for staged increases of the mechanical input to the transaxle. In first and reverse gears, the converter operates in a normal manner and all engine torque is transmitted hydraulically to the transaxle (Figure 4-41). When the transaxle is placed into DRIVE and as the transaxle upshifts from first to second (Figure 4-42) and from second to third (Figure 4-43), the planetary gear set within the converter will split engine torque. Part of the torque is transmitted hydraulically through the action of the turbine, while the remainder is transmitted mechanically through the planetary gear set.

The engine's torque is split in second and third gears because the transaxle's intermediate clutch locks the converter's intermediate shaft to the transaxle's gear set. In second gear, the turbine

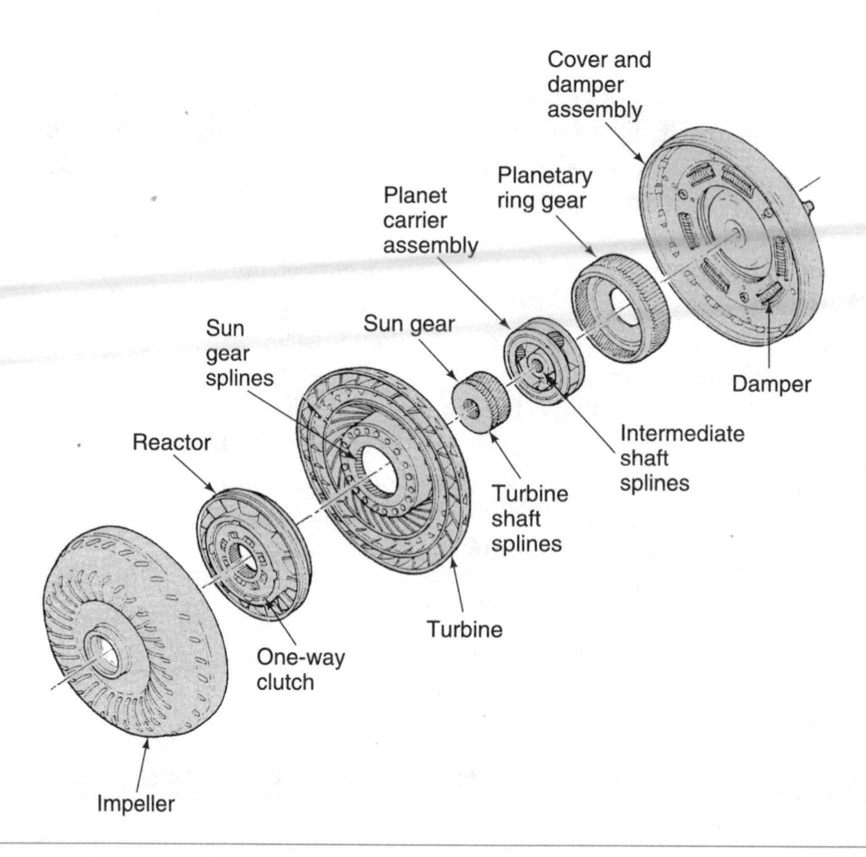

Figure 4-40 Split-type torque converter assembly used in a Ford ATX transaxle. (Reprinted with the permission of Ford Motor Co.)

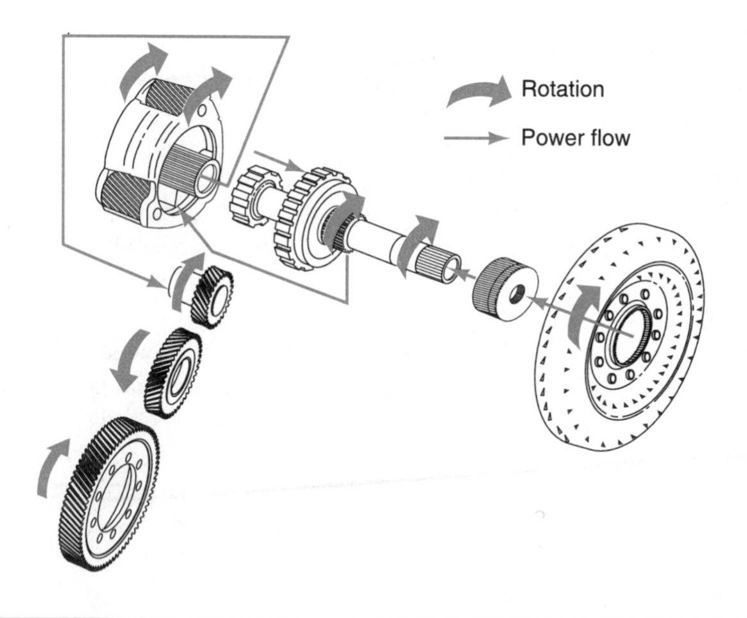

Figure 4-41 Power flow from turbine to transmission when the ATX is in first gear. (Reprinted with the permission of Ford Motor Co.)

shaft hydraulically supplies 38 percent of the torque to the transaxle and 62 percent of the torque is transmitted mechanically by the intermediate shaft.

In third gear, both the intermediate clutch and the direct clutch of the transaxle are applied. The interaction of the converter's gear set with the transmission's causes a 93 percent mechanical input and a 7 percent hydraulic input through the turbine shaft.

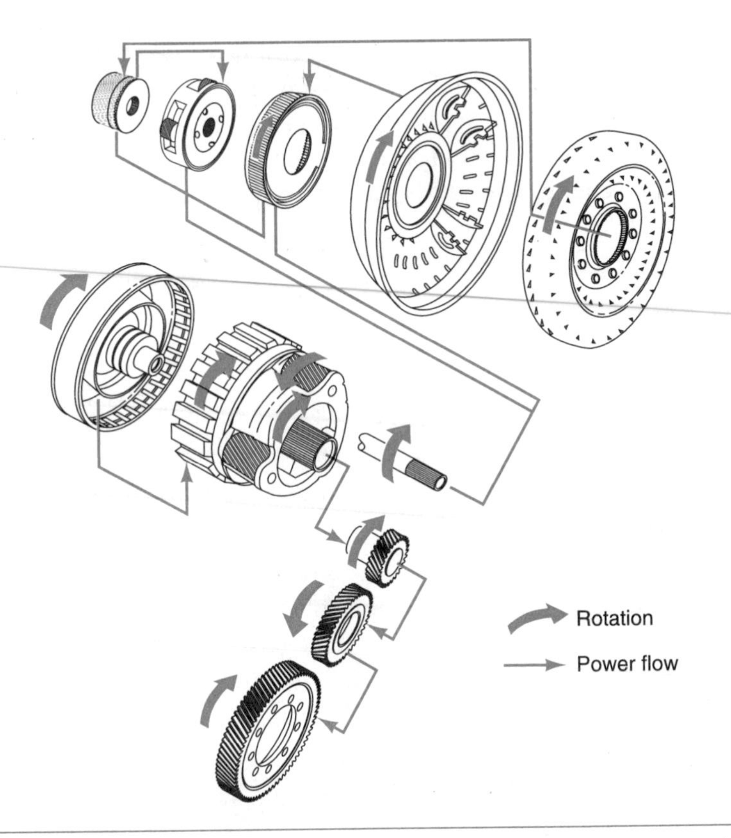

Figure 4-42 Input power flow in an ATX when it is operating in second gear. (Reprinted with the permission of Ford Motor Co.)

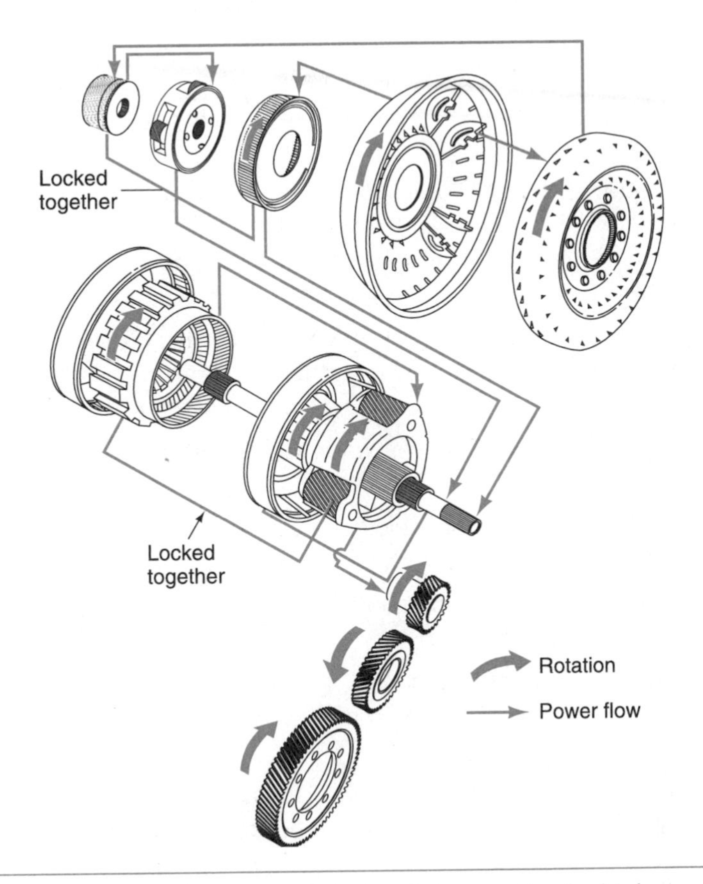

Figure 4-43 Input power flow in an ATX when it is operating in third gear. (Reprinted with the permission of Ford Motor Co.)

The converter cover is fitted with a damper assembly to absorb engine torsional vibrations. Because second and third gear operation includes a percentage of direct mechanical input from the engine through the splitter gear set, engine vibrations must be isolated from the rest of the drive line. A damper assembly with springs and frictional material is welded or riveted to the converter's cover and all mechanical torque is transmitted through the damper before it drives the intermediate shaft.

Summary

- ❑ A torque converter uses fluid to smoothly transfer engine torque to the transmission.
- ❑ All automatic transmissions, except some designs of the constant variable transmission, use a torque converter to transfer power from the engine to the transmission.
- ❑ A simple fluid coupling is composed of three basic members: a housing, an impeller, and a turbine.
- ❑ The impeller acts like a pump and moves the fluid in the direction of its rotation. The moving fluid hits against the turbine vanes. This causes the turbine to rotate, which brings an input into the transmission.
- ❑ The fluid coupling allows the vehicle to stop while in gear without stalling the engine, because at low engine speeds, the impeller doesn't rotate fast enough to drive the turbine.
- ❑ Two types of fluid flow take place inside the fluid coupling: rotary flow and vortex flow.
- ❑ Rotary flow is in the direction of impeller rotation and results from the paddle action of the impeller vanes against the fluid.
- ❑ Vortex oil flow is the fluid flow circulating between the impeller and turbine as the fluid moves from the impeller to the turbine and back to the impeller.
- ❑ On all RWD and most FWD vehicles, the torque converter is mounted in line with the transmission's input shaft. Some transaxles use a drive chain to connect the converter's output shaft with the transmission's input shaft.
- ❑ The torque converter fits between the transmission and the engine and is supported by the transmission's oil pump housing and the engine's crankshaft. The rear hub of the converter's cover fits into the transmission's pump. The hub drives the oil pump whenever the engine is running.
- ❑ The front of a torque converter is mounted to a flexplate. The flexplate is designed to be flexible enough to allow the front of the converter to move forward or backward if it expands or contracts because of heat or pressure.
- ❑ The combined weight of the fluid, torque converter, and flexplate serves as the flywheel for the engine.
- ❑ A typical torque converter consists of three elements sealed in a single housing: the impeller, the turbine, and the stator.
- ❑ The impeller is the drive member of the unit and its fins are attached directly to the converter cover.
- ❑ The turbine is the converter's output member and it is coupled to the transmission's input shaft. The turbine is driven by the fluid flow from the impeller and always turns at its own speed.
- ❑ The stator is the reaction member of the converter. This assembly is about one-half the diameter of the impeller or turbine and is positioned between the impeller and turbine.

Terms to Know

Centrifugal lockup
Converter capacity
Coupling phase
CVT
Damper
EMCC
Fail-safe valve
Fluid coupling
Frequency
Guide rings
Impeller
Isolator springs
Lockup
MCCC
Pulse width
Reactor
Rotary flow
Sheaves
Silicone fluid
Splitter gear converter
Stall speed
Stator
TCC
TCM
Vacuum transducer
VCC
Viscous clutch
Vortex flow

- ❑ All of the fluid returning from the turbine to the impeller must pass through the stator. The stator redirects the fluid leaving the turbine back to the impeller. By redirecting the fluid so that it is flowing in the same direction as engine rotation, it drives the impeller to rotate more efficiently.
- ❑ A stator is supported by a one-way clutch that is splined to the stator support shaft and allows the stator to rotate only in the same direction as the impeller.
- ❑ The outer edge of the stator fins normally forms the inner edge of a three-piece fluid guide ring. This split guide ring directs fluid into a smooth, turbulence-free flow.
- ❑ The fluid in the torque converter is supplied by the transmission's oil pump and enters through the converter's hub, then flows into the passages between the vanes. As the impeller rotates, the fluid is moved outward and upward through the vanes by centrifugal force because of the curved shape of the impeller.
- ❑ The higher the engine speed, the faster the impeller turns, and the more force is transferred from the impeller to the turbine by the fluid.
- ❑ As engine speed increases, the fluid is thrown at the turbine with a greater force, causing the turbine to rotate.
- ❑ As the vortex flow continues, the fluid leaving the turbine to return to the impeller is moving in the opposite direction as crankshaft rotation. If the fluid were allowed to continue in this direction, it would enter the impeller as an opposing force and some of the engine's power would be used to redirect the flow of fluid. To prevent this loss of power, torque converters are fitted with a stator.
- ❑ The redirection of the fluid by the stator not only prevents a torque loss, but also provides for a multiplication of torque.
- ❑ In passing through the stator, the direction of fluid flow is reversed by the curvature of the stator blade. The fluid now moves in a direction that aids in the rotation of the impeller. The impeller accelerates the movement of the fluid and it now leaves with nearly twice the energy and exerts a greater force on the turbine. This action results in a torque multiplication.
- ❑ It is vortex flow that allows for torque multiplication. Torque multiplication occurs when there is high impeller speed and low turbine speed.
- ❑ When the speed of the turbine nearly equals the speed of the impeller, fluid flows against the stator vanes in the same direction as the fluid from the impeller.
- ❑ When there is little vortex flow and the engine's torque is carried through the converter by the rotary flow of the fluid, the coupling phase of the torque converter occurs and no torque multiplication takes place.
- ❑ The condition called stall occurs when the turbine is held stationary and the impeller is spinning. Stall speed is the fastest speed an engine can reach while the turbine is at stall.
- ❑ A low-capacity converter has a high stall speed but allows for a relatively large amount of slippage during the coupling phase.
- ❑ A high-capacity converter allows for less slippage but has a low stall speed.
- ❑ The stall speed of a torque converter changes with a change in the diameter of the converter and the angles of the impeller, turbine, and stator vanes.
- ❑ Torque converter capacity can be changed by changing the angle of the stator vanes. As the angle of the vanes increases, more fluid will return to the impeller. Therefore, the stall speed will increase.
- ❑ The transmission's oil pump continuously delivers pressurized ATF into the torque converter through a hollow shaft in the center of the torque converter assembly.

- ❑ The fluid is circulated through the converter and exits past the turbine through the turbine shaft, which is located within the hollow fluid feed shaft. From there, the fluid is directed to an external oil cooler and back to the transmission's oil circuit.
- ❑ Manufacturers have adapted automatic transmissions to provide a direct mechanical drive for input through the use of centrifugally applied clutches, hydraulically operated clutches, and planetary gear sets in the torque converter.
- ❑ When applied, the lockup clutch locks the turbine to the cover of the converter.
- ❑ Most lockup converters consist of the three basic elements: impeller, turbine, and stator, plus a piston and clutch plate assembly, special thrust washers, and roller bearings.
- ❑ The piston and clutch plate assembly has frictional material on the outer portion of the plate with a spring-cushioned damper assembly in the center. The clutch plate is splined to the turbine shaft and when the piston is applied, the plate locks the turbine to the converter. The thrust washers and roller bearings control the movements of and provide bearing surfaces for the components of the converter.
- ❑ Placed around the outside edge of a centrifugal clutch assembly are several clutch shoes. As turbine speed increases, centrifugal force moves the shoes outward into contact with the converter cover, locking the turbine and the cover together. The amount of locking depends on the amount of force pushing the shoes onto the cover.
- ❑ The clutch shoes are designed to slip when the vehicle demands extra torque from the engine. This slippage allows for some torque multiplication.
- ❑ The converter clutch is applied when the fluid flow through the torque converter is reversed by a valve. When the torque converter clutch control valve moves, the fluid begins to flow in a reversed direction. This forces the clutch disc or pressure plate against the front of the torque converter's cover.
- ❑ To engage the clutch, mainline pressure is directed between the plate and the turbine. This forces the plate into contact with the front inner surface of the cover and locks the turbine to the impeller. With the engagement of the clutch, the fluid in front of the clutch is squeezed out before the clutch is totally engaged. The presence of this fluid softens the engagement of the clutch and acts much like an accumulator.
- ❑ A viscous converter clutch assembly consists of a rotor, body, clutch cover, and silicone fluid.
- ❑ Engine power is transmitted from the pressure plate through the fluid coupling to the transaxle's input shaft. The clutch's fluid coupling uses the viscous properties of thick silicone fluid between the closely spaced pressure plate and cover plate to transmit the power.
- ❑ A typical hydraulic converter clutch has a plate that acts like a clutch piston and is splined to the front of the turbine.
- ❑ Frictional material is bonded to either the forward face of the clutch plate or the inner front surface of the converter cover.
- ❑ A converter clutch disc is splined to the turbine so that it can drive the turbine when the frictional material is forced against the torque converter's cover.
- ❑ All converter clutch assemblies are fitted with a damper assembly that directs the power flow through a group of coil springs. The damper springs are either located at the center of the disc or they are grouped around the outer edge of the disc. These springs are designed to absorb the normal torsional vibrations of an engine, which vary with changes in speed and load.
- ❑ Early Chrysler electronic controlled lockup converter systems use an electrical solenoid mounted on the valve body to engage and disengage the converter clutch. The operation of this solenoid is controlled by the engine control computer through a relay. The computer determines when to lock or unlock the torque converter based on information from the coolant temperature sensor, vacuum transducer, vehicle speed sensor, and carburetor or throttle ground switch.

- The computer energizes the clutch relay, which grounds the lockup solenoid. The activation of the solenoid moves the solenoid check ball against its seat. This prevents the solenoid valve from exhausting line pressure. Line pressure increases and forces the switch valve to move against coil spring tension. Pressure is then directed to the impeller drive hub and stator support to fill the torque converter with fluid. Line pressure flows from the impeller and turbine to fill the space behind the torque converter clutch piston. This high pressure forces the engagement of the clutch.
- Late-model Chryslers with the A-604/41TE transaxle have a torque converter that can lock up in second, third, and overdrive gear. The electronic modulated converter clutch (EMCC) system allows the converter clutch to partially engage between 23 and 47 mph. This feature limits the amount of converter slippage but serves to cushion the engine from the drive line.
- Electronically controlled hydraulic converter clutch circuits in Ford Motor Company products are controlled by the power train control assembly.
- The Ford 4EAT functions with only a lockup control valve and a lockup solenoid.
- The converter clutch of the AXOD and the E4OD is controlled by a converter clutch bypass solenoid, which vents line pressure to the oil pan when the converter clutch is released. The solenoid also redirects line pressure into an apply passage when the solenoid is energized to apply the clutch.
- An AXOD-E/AX4S uses a pulse width modulated solenoid to allow the power train control module (PCM) to vary clutch operation from full release to controlled slip to full apply.
- The converter control system in an AOD-E/4R70W also uses a pulse width modulated solenoid called the modulated converter clutch control (MCCC) solenoid. It operates in the same way as the AXOD-E/AX4S.
- The converter clutch in an A4LD transmission is applied and released hydraulically, but it can be overridden electronically. The A4LD override solenoid is normally energized to prevent fluid flow from unlocking the converter clutch.
- The torque converter clutch assembly's computer controls the application of the clutch by providing a ground circuit for the torque converter clutch (TCC) solenoid circuit.
- When the torque converter clutch solenoid's ground circuit is completed by the PCM, fluid pressure is exhausted from between the converter's pressure plate and the converter cover, and converter clutch apply pressure from the converter clutch regulator valve pushes the converter pressure plate against the converter cover to apply the converter clutch.
- One of two types of vehicle speed sensors is used to monitor vehicle speed for the system. A light emitting diode (LED) type is used in the instrument cluster on some models. Other models use a PM generator mounted in the transmission housing. Vehicle speed must be greater than a certain value before the TCC can be applied.
- A Honda torque converter can operate in various degrees of lockup: zero, partial, half, full, and cycling. Lockup is controlled by the transmission's hydraulic system and the transmission control module.
- The torque converter used in Ford's AOD transmission is not a true lockup converter. Rather, it uses a split-torque path to get the same results. The transmission and torque converter are connected by three separate shafts. The input shaft is hollow and transmits hydraulic input torque through the transmission's one-way clutch or direct clutch. Inside the input shaft is an inner or intermediate shaft that transmits a combination of hydraulic and mechanical input to the transmission's intermediate clutch. The third shaft is the pump shaft. This shaft is driven by the converter cover and mechanically drives the oil pump. This setup allows power to enter the transmission along two separate paths, one mechanical and one hydraulic.

- ❑ The torque converter in some Ford ATX transaxles are a split-torque design with an internal planetary gear set.
- ❑ Two hollow shafts from the transmission are splined to separate gear set members in the converter. In the center of the converter's gear set is the sun gear, which is splined to the turbine shaft. Because the turbine is driven hydraulically, the turbine and sun gear provide the hydraulic input to the transaxle. The other shaft, the intermediate shaft, is splined to the planetary carrier of the gear set. The carrier and intermediate shaft provide for the mechanical input into the transaxle. The planetary pinion gears are meshed with the internal teeth of the ring gear. The ring gear is splined to the converter's cover and a damper assembly; therefore, the ring gear always rotates at engine speed.
- ❑ In second gear, the turbine shaft hydraulically supplies 38 percent of the torque to the transaxle and 62 percent of the torque is transmitted mechanically by the intermediate shaft.
- ❑ In third gear, both the intermediate clutch and the direct clutch of the transaxle are applied. The interaction of the converter's gear set with the transmission's causes a 93 percent mechanical input and a 7 percent hydraulic input through the turbine shaft.

Review Questions

Short Answer Essays

1. Describe the operation of a fluid coupling.
2. What is the difference between rotary and vortex oil flow?
3. Name the three major components of a torque converter and describe the purpose of each.
4. What allows for torque multiplication in a torque converter?
5. What is meant by stall speed?
6. What can change the capacity of a torque converter?
7. Describe how a centrifugal torque converter clutch works.
8. What is the purpose of the damper in a lockup clutch?
9 How does the computer control the converter clutch in most Chrysler transmissions?
10. What is a split-path torque converter? Give an example.

Fill in the Blanks

1. The combined weight of the ____________________ , ____________________ ____________________ , and ____________________ serve as the flywheel for the engine on cars equipped with an automatic transmission.
2. The ____________________ the engine speed, the ____________________ the impeller turns, and ____________________ force is transferred from the impeller to the turbine by the fluid.
3. A simple fluid coupling is composed of three basic members: a ____________________ , an ____________________ , and a ____________________ .

4. A Honda torque converter can operate in various degrees of lockup: ____________________, ____________________, ____________________, ____________________, and ____________________.

5. One of two types of vehicle speed sensors is used to monitor vehicle speed for the system. A ____________________ ____________________ ____________________ type is used in the instrument cluster on some models. Other models use a ____________________ ____________________ mounted in the transmission housing.

6. The torque converter in some of Ford's ATX transaxles is a ____________________-____________________ design with an internal ____________________ ____________________.

7. Most lockup converters consist of the three basic elements: ____________________, ____________________, and ____________________, plus a ____________________ and ____________________ ____________________ assembly, special ____________________ ____________________, and ____________________ ____________________.

8. Two types of fluid flow take place inside the fluid coupling: ____________________ and ____________________ flow.

9. A viscous converter clutch assembly consists of a ____________________, ____________________, ____________________ ____________________, and ____________________ ____________________.

10. It is ____________________ flow that allows for torque multiplication. Torque multiplication occurs when there is ____________________ impeller speed and ____________________ turbine speed.

ASE Style Review Questions

1. *Technician A* says all RWD and most FWD vehicles have the torque converter mounted in line with the transmission's input shaft.
   Technician B says some transaxles use a drive chain to connect the converter's output shaft with the transmission's input shaft.
   Who is correct?
   A. A only
   B. B only
   C. Both A and B
   D. Neither A nor B

2. While discussing typical electronic controlled converter clutch systems:
   Technician A says most use an electrical solenoid mounted on the valve body to engage and disengage the converter clutch.
   Technician B says the operation of the clutch is controlled by a relay, which is activated by the coolant temperature sensor.
   Who is correct?
   A. A only
   B. B only
   C. Both A and B
   D. Neither A nor B

3. *Technician A* says a typical torque converter clutch assembly's computer controls the application of the clutch by providing a ground circuit for the clutch solenoid circuit.
Technician B says the clutch solenoid simply redirects fluid flow to activate the clutch.
Who is correct?
A. A only **C.** Both A and B
B. B only **D.** Neither A nor B

4. *Technician A* says two-element fluid couplings are no longer used because their engagement was sudden and jerky.
Technician B says two-element fluid couplings are no longer used because they had high impeller speeds, which caused engines to stall while in gear and at idle.
Who is correct?
A. A only **C.** Both A and B
B. B only **D.** Neither A nor B

5. While discussing clutch lockup solenoids:
Technician A says some systems use a PM generator-type solenoid, which allows for full-time, but controlled, slip.
Technician B says some converter control systems use a pulse width modulated solenoid.
Who is correct?
A. A only **C.** Both A and B
B. B only **D.** Neither A nor B

6. *Technician A* says an ATX transaxle has two hollow shafts from the transmission splined to separate gear set members in the converter.
Technician B says an ATX uses a solenoid to force converter clutch lockup.
Who is correct?
A. A only **C.** Both A and B
B. B only **D.** Neither A nor B

7. *Technician A* says when the speed of the turbine nearly equals the speed of the impeller, fluid flows against the stator vanes in the same direction as the fluid from the impeller.
Technician B says when there is little vortex flow and the engine's torque is carried through the converter by the rotary flow of the fluid, the coupling phase of the torque converter occurs and no torque multiplication takes place.
Who is correct?
A. A only **C.** Both A and B
B. B only **D.** Neither A nor B

8. While discussing the construction of typical converter clutch assemblies:
Technician A says the clutch plate is splined to the turbine shaft, and when the piston is applied, the plate locks the turbine to the converter.
Technician B says the piston and clutch plate assembly has frictional material on the outside of the plate with a spring-cushioned damper assembly in the center.
Who is correct?
A. A only **C.** Both A and B
B. B only **D.** Neither A nor B

9. *Technician A* says the rear hub of the torque converter is bolted to the flexplate.
Technician B says the flexplate is designed to be flexible enough to allow the front of the converter to move forward or backward if it expands or contracts because of heat or pressure.
Who is correct?
A. A only **C.** Both A and B
B. B only **D.** Neither A nor B

10. While discussing the operation of the stator:
Technician A says the stator redirects the fluid leaving the turbine back to the impeller.
Technician B says only a portion of the fluid returning from the turbine to the impeller passes through the stator.
Who is correct?
A. A only **C.** Both A and B
B. B only **D.** Neither A nor B

Planetary Gears and Shafts

Upon completion and review of this chapter, you should be able to:

- ❑ Describe how different gear ratios are obtained from a single planetary gear set.
- ❑ Identify the components in a basic planetary gear set and describe their operation.
- ❑ Describe the different designs of compound planetary gear sets.
- ❑ Describe the construction and operation of typical Simpson gear train-based transmissions.
- ❑ Describe the construction and operation of Ravigneaux gear train-based transmissions.
- ❑ Describe the construction and operation of Ravigneaux gear train-based transmissions with overdrive.
- ❑ Describe the construction and operation of transmissions that use planetary gear sets in tandem.
- ❑ Describe the purpose of a differential.
- ❑ Identify the major components of a differential and explain their purpose.
- ❑ Describe the various gears in a differential assembly and state their purpose.
- ❑ Explain the operation of a four-wheel-drive differential and its drive axles.
- ❑ Describe the different designs of four-wheel-drive systems and their applications.
- ❑ Compare and contrast the components of part- and full-time four-wheel-drive systems.
- ❑ Discuss the purpose, operation, and application of a viscous coupling in four-wheel-drive systems.

Introduction

An automatic transmission is based on the principle of levers, automatically allowing the engine to move heavy loads with little effort. As the heavy load decreases or the vehicle begins to move, less leverage or lower gear ratios are required to keep the vehicle moving. By providing different gear ratios, a transmission provides for performance and economy over the entire driving range. Most automatic transmissions use planetary gear sets to provide for the different gear ratios. The gear ratios are selected manually by the driver or automatically by the hydraulic control system, which engages and disengages the clutches and bands used to shift gears. A single planetary gear set consists of a sun gear, a **planet carrier** with two or more **planet pinion gears,** and a ring gear. Planet gears are always in constant mesh; therefore, they allow quick, smooth, and precise gear changes without the worry of clashing or partial engagement.

Torque is multiplied and speed is decreased according to the gear ratio.

The planet carrier is the bracket in a planetary gear set on which the planet pinion gears are mounted on pins and are free to rotate.

Simple Planetary Gear Sets

Planet gears provide for the different gear ratios needed to move a vehicle in the desired direction at the correct speed. A simple planetary gear set consists of a sun gear, planet gears, and a ring gear. These gears are typically helically cut gears, which offer quiet operation.

At the center of the planetary gear set is the sun gear. The placement of this gear in the gear set is the reason for its name; the earth's sun is at the center of our solar system. Planet gears surround the sun gear, just like the earth and other planets surround the sun in our solar system. These gears are mounted and supported by the planet carrier and each gear turns on its own separate shaft. The planet gears are in constant mesh with the sun and ring gears. The ring gear is the outer gear of the gear set. It has internal teeth and surrounds the rest of the gear set. Its gear teeth are in constant mesh with the planet gears. The number of planet gears used in a planetary gear set varies according to the loads the transmission is designed to face. For heavy loads, the number of planet gears is increased to spread the work load over more gear teeth.

The planet gears can be referred to as the **pinion gears.**

The ring gear is also known as the **annulus gear** or **internal gear.**

The planetary gear set can provide a gear reduction or overdrive, direct drive or reverse, or a neutral position. Because the gears are in constant mesh, gear changes are made without engaging or disengaging gears, as is required in a manual transmission. Rather, clutches and bands are used to either hold or release different members of the gear set to get the proper direction of rotation or gear ratio (Figure 5-1).

A basic understanding of gears is essential to an understanding of planetary gear sets. The basic principles of gear action that you need to remember are:

1. When a small gear drives a larger gear, output torque is increased and output speed is decreased.
2. When a large gear drives a smaller gear, output torque is decreased and output speed is increased.

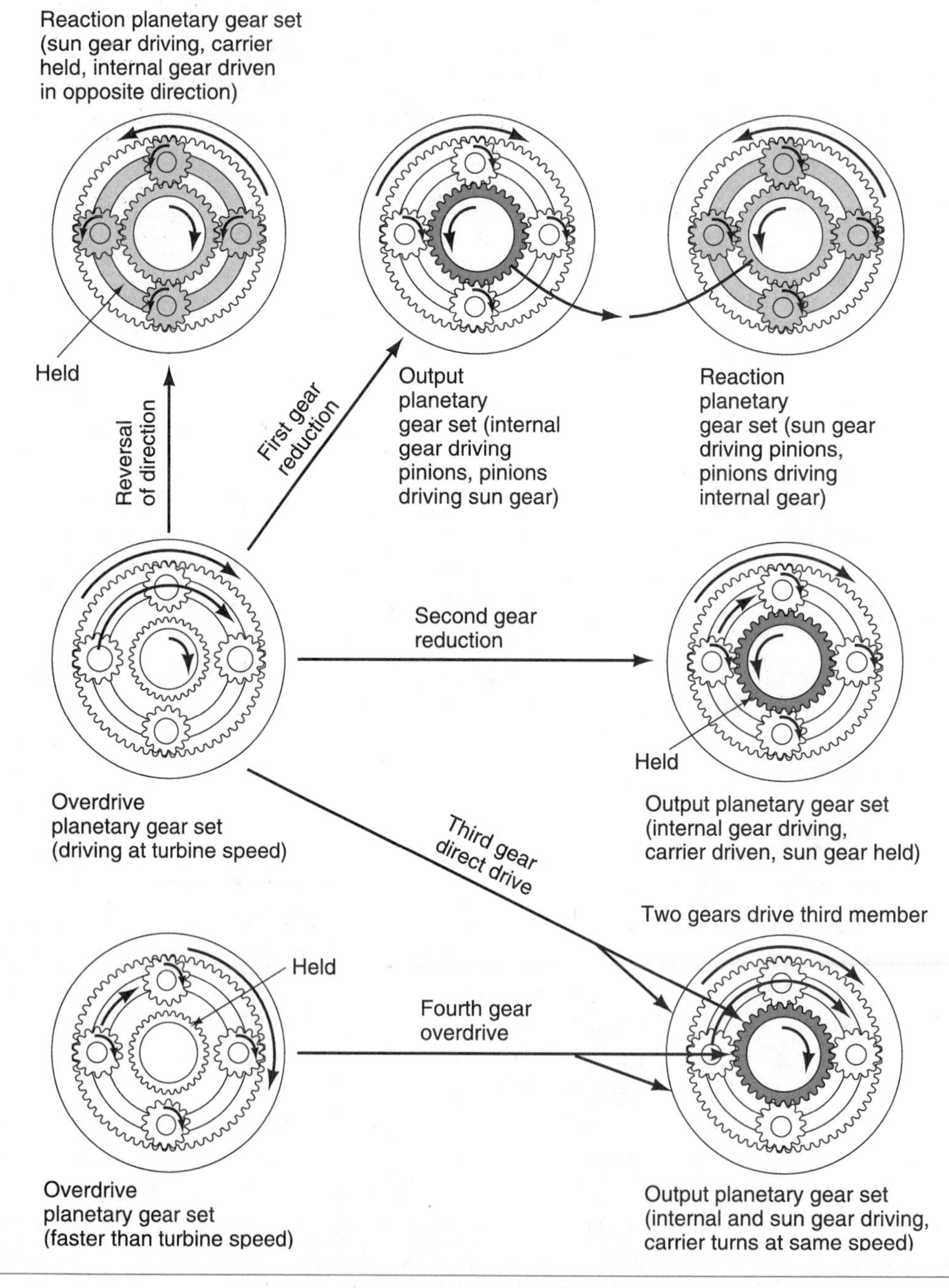

Figure 5-1 Power flow through a simple planetary gear set. (Courtesy of the Hydra-Matic Division of General Motors Corp.)

3. When an external gear drives another external gear, the output gear will rotate in the opposite direction to the input gear.
4. When an external gear is in mesh with an internal gear, the two gears will rotate in the same direction.

Gear Reduction

When the output gear turns at a slower speed than the input gear, gear reduction has occurred and this results in increased torque on the output gear. This is accomplished in a planetary gear set anytime the carrier is the output member. If power is applied to the ring gear, the planet gears will rotate on their shafts in the planet carrier (Figure 5-2). Because the sun gear is being held, the rotating planet gears will walk around the sun gear and carry the planet carrier with them. This causes a gear reduction because one complete revolution of the ring gear will not cause one complete revolution of the planet carrier. The planet carrier is turning at a slower speed and therefore is increasing the torque output. By holding the sun gear, the planet carrier moves in the same direction as the ring gear.

Gear reduction can also occur if the ring gear is held and the sun gear is the input gear. The planet carrier will rotate around the sun gear, but will turn slower than it did when the ring gear drove the planet carrier. This results in greater gear reduction and therefore greater torque. Whenever the planet carrier becomes the output member of the gear set, the planetary gear assembly works in reduction and output torque increases.

Increasing the engine's torque is generally known as operating in reduction because there is always a decrease in the speed of the output member that is proportional to the increase in the output torque.

Holding a member of the gear set prevents it from turning.

Gear reduction is a condition that exists in a gear set when the input driving gear turns faster than the output driven gear.

Overdrive

When the planet carrier is the driving member of the gear set, overdrive occurs (Figure 5-3). As the planet carrier rotates, the pinion gears are forced to walk around the held sun gear, which drives the ring gear faster. One complete rotation of the planet carrier causes the ring gear to rotate more than one complete revolution in the same direction. This provides more output speed but less torque or overdrive.

When the ring gear and carrier pinions are free to rotate at the same time, the pinions will always follow the same direction as the ring gear.

Overdrive is a condition in which the output gear of a member of a gear set rotates at a greater speed than the input gear. The output speed is increased, but the torque is decreased.

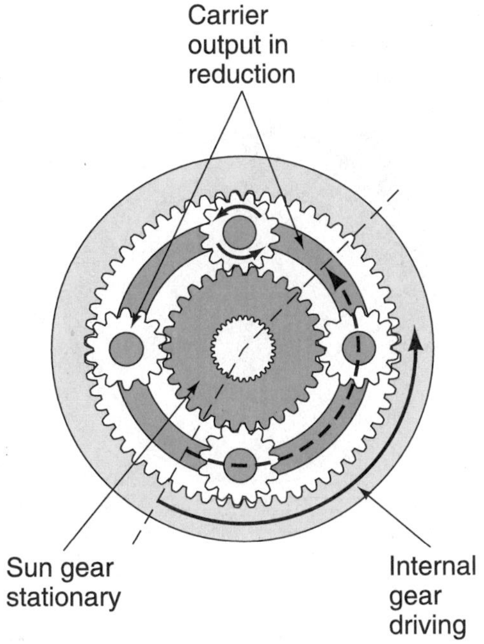

Carrier output in reduction
Sun gear stationary
Internal gear driving

Figure 5-2 Planetary gear set providing for gear reduction. (Courtesy of the Hydra-Matic Division of General Motors Corp.)

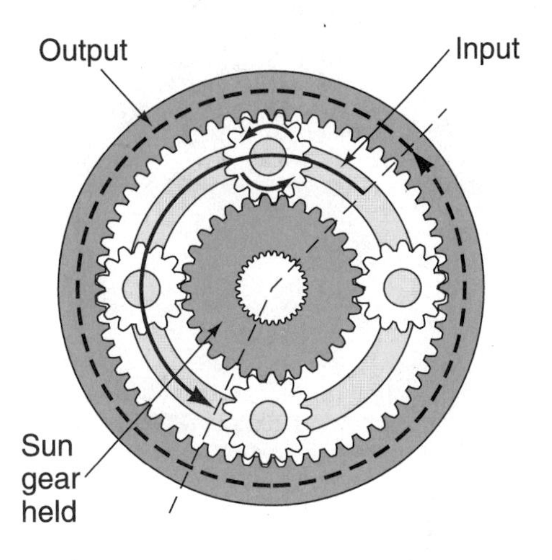

Figure 5-3 Planetary gear set providing for overdrive. (Courtesy of the Hydra-Matic Division of General Motors Corp.)

A higher speed overdrive is possible by holding the ring gear stationary. With input on the planet carrier, the pinion gears are forced to walk around the inside of the ring gear, driving the sun gear clockwise. The planetary carrier rotates much less than one turn to rotate the sun gear one complete revolution. The result is a great reduction in torque output and a maximum increase in output speed.

Direct Drive

Direct drive is a condition in which the speed and torque of the output are the same as those of the input.

If any two of the planetary gear set members receive power in the same direction at the same speed, the third member is forced to move with the other two (Figure 5-4). If the ring gear and the sun gear are the input members, the internal teeth of the ring gear will try to rotate the planetary pinions in one direction while the external teeth of the sun gear will try to drive them in the opposite direction. This action locks the planetary pinions between the other members and the entire planetary gear set rotates as a single unit. The input is now locked to the output, which results in direct drive. One input revolution equals one output revolution.

Reverse

If the planet carrier is held and the sun gear is rotated in a clockwise rotation (Figure 5-5), the ring gear will rotate in a counterclockwise direction. The planet gears cannot travel around the teeth of an enmeshed gear. Rather, the carrier is held in place and the planet gears rotate on their shafts. The sun gear spins the planet gears, which drive the ring gear in the opposite direction but at a slower speed. Therefore, the gear set is providing reverse gear with reduction.

An idler gear is a gear that does not contribute to the input or output torque. It merely transmits torque from one gear to another, in the opposite direction.

Reverse with overdrive is possible by putting the input on the ring gear and using the sun gear as the output member. A reversal of direction is obtained whenever the planet carrier is stopped from turning and power is applied to either the sun gear or the ring gear. This causes the pinion gears to act as idler gears, thus driving the output member in the opposite direction to the input.

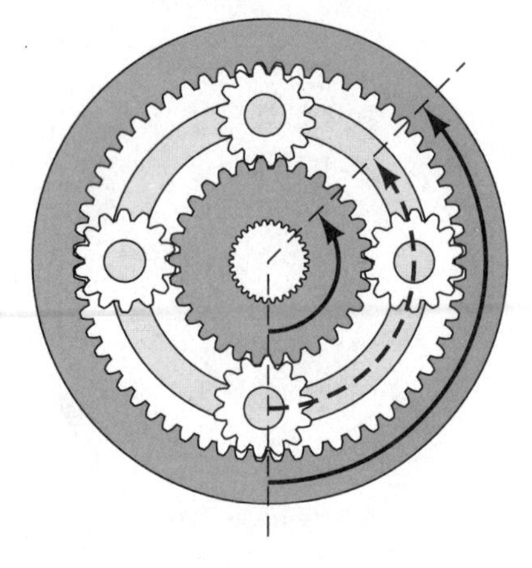

Figure 5-4 Planetary gear set providing for direct drive. (Courtesy of the Hydra-Matic Division of General Motors Corp.)

Carrier restrained from turning

Counter-clockwise input on sun gear

Clockwise output on internal gear

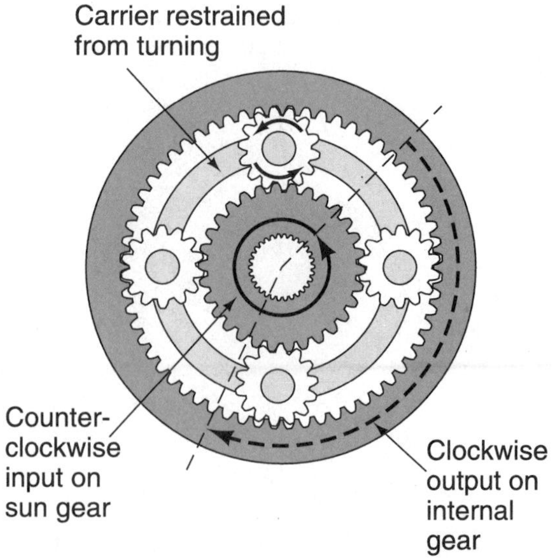

Figure 5-5 Planetary gear set providing for reverse gear rotation. (Courtesy of the Hydra-Matic Division of General Motors Corp.)

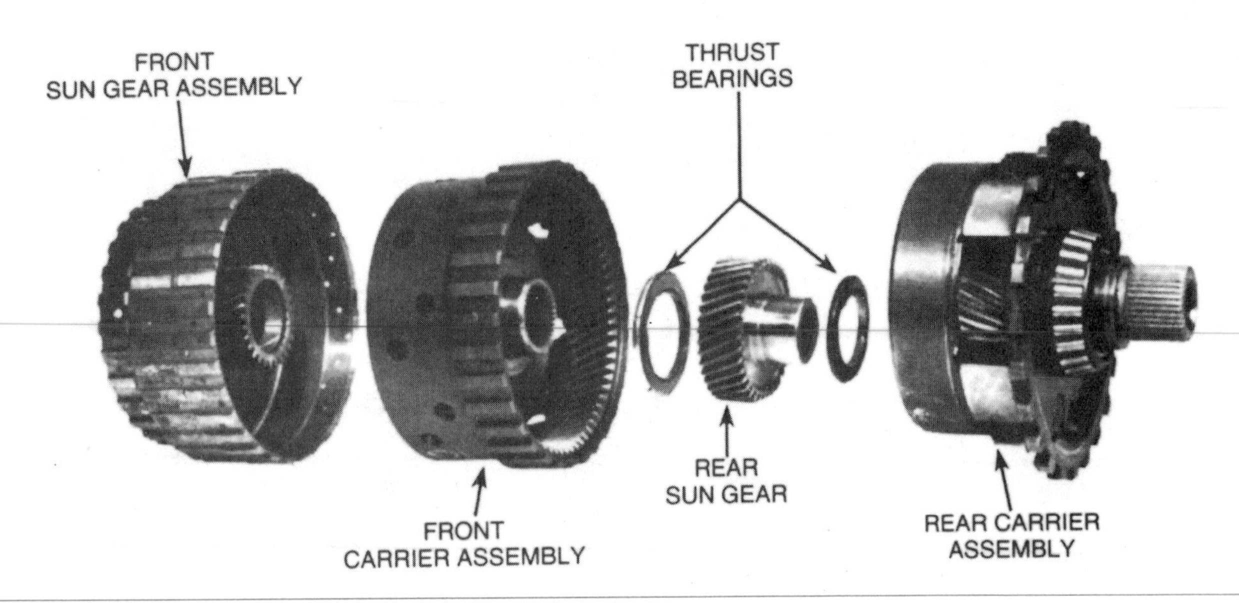

Figure 5-6 A compound planetary gear set. (Courtesy of Chrysler Corp.)

Neutral

When no member of the planetary gear set is held stationary or locked to another, there will be no output regardless of which member receives the input. This is the normal NEUTRAL position of a planetary gear set.

When there is input into a planetary gear set, but a member is not held, NEUTRAL results.

Compound Planetary Gear Sets

A limited number of gear ratios are available from a simple or single planetary gear set. To increase the number of available gear ratios, gear sets can be added. The typical automatic transmission with three or four forward speeds has at least two planetary gear sets.

In automatic transmissions, two or more planetary gear sets (Figure 5-6) are connected together to provide the various gear ratios needed to efficiently move a vehicle. There are two common designs of compound gear sets, the Simpson gear set in which two planetary gear sets share a common sun gear and the Ravigneaux gear set, which has two sun gears, two sets of planet gears, and a common ring gear. Some transmissions are fitted with an additional single planetary gear set, which is used to provide an "add-on" overdrive gear.

Simpson Gear Train

The Simpson gear train is an arrangement of two separate planetary gear sets with a common sun gear, two ring gears, and two planet pinion carriers (Figure 5-7). A Simpson gear train is the most commonly used compound planetary gear set. It is used to provide three forward gears. Half of the compound set or one planetary unit is referred to as the front planetary and the other planetary unit is the rear planetary (Figure 5-8). The two planetary units do not need to be the same size or have the same number of teeth on their gears. The size and number of gear teeth determine the actual gear ratios obtained by the compound planetary gear assembly.

Shop Manual
Chapter 5, page 145

Gear ratios and direction of rotation are the result of applying torque to one member of either planetary units, holding at least one member of the gear set, and using another member as the output. For the most part, each automobile manufacturer uses different parts of the planetary assemblies as input, output, and reaction members. This also varies with the different transmission models from the same manufacturer. There are also many different apply devices used in the various transmission designs.

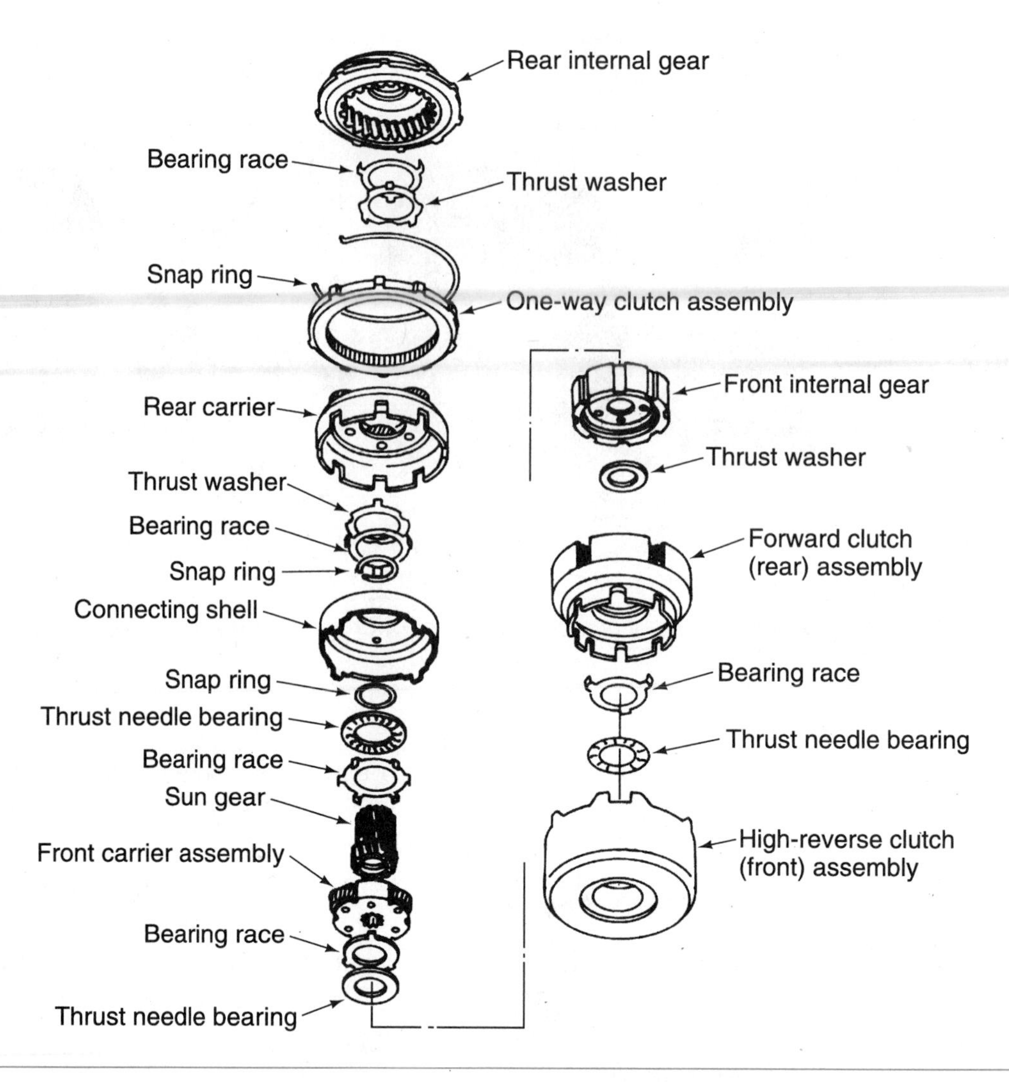

Figure 5-7 A Simpson planetary gear set. (Courtesy of Nissan Motor Co., Ltd.)

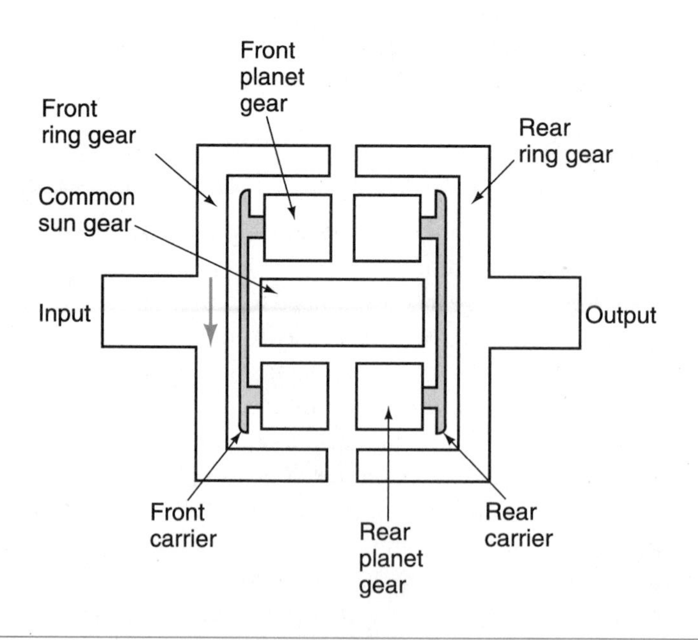

Figure 5-8 Components of a Simpson gear set.

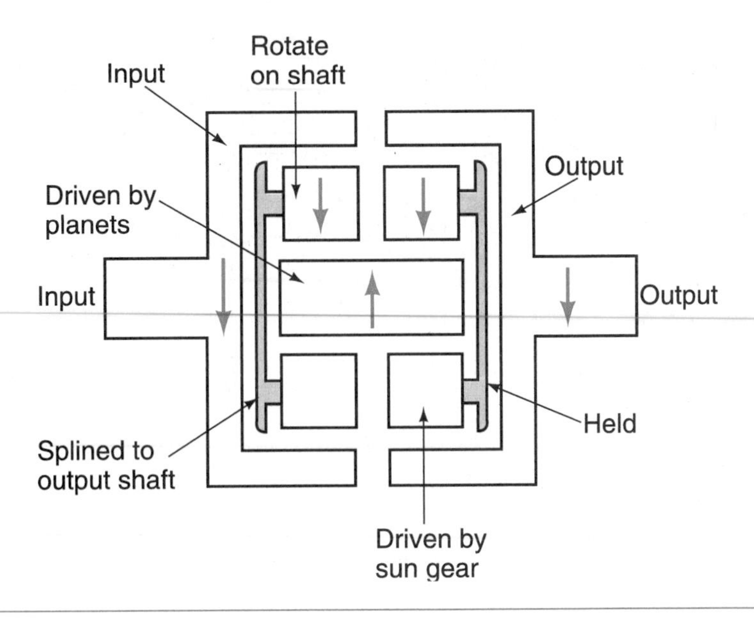

Figure 5-9 Power flow in first gear.

A Simpson gear set can provide the following gear ranges: neutral, first reduction gear, second reduction gear, direct drive, and reverse. The typical power flow through a Simpson gear train when it is in neutral has engine torque being delivered to the transmission's input shaft by the torque converter's turbine. No planetary gear set member is locked to the shaft; therefore, engine torque enters the transmission but goes nowhere else.

When the transmission is shifted into first gear (Figure 5-9), engine torque is again delivered into the transmission by the input shaft. The input shaft is now locked to the front planetary ring gear, which turns clockwise with the shaft. The front ring gear drives the front planet gears, also in a clockwise direction. The front planet gears drive the sun gear in a counterclockwise direction. The rear planet carrier is locked; therefore, the sun gear spins the rear planet gears in a clockwise direction. These planet gears drive the rear ring gear, which is locked to the output shaft, in a clockwise direction. The result of this power flow is a forward gear reduction, typically with a ratio of 2.5:1 to 3:1.

When the transmission is operating in second gear (Figure 5-10), engine torque is again delivered into the transmission by the input shaft. The input shaft is locked to the front planetary ring

The sun gear always rotates opposite the rotation of the pinion gears.

When the planet carrier is the output, it always follows the direction of the input member.

Forward gear reduction always occurs when the planet carrier is the output member of the gear set.

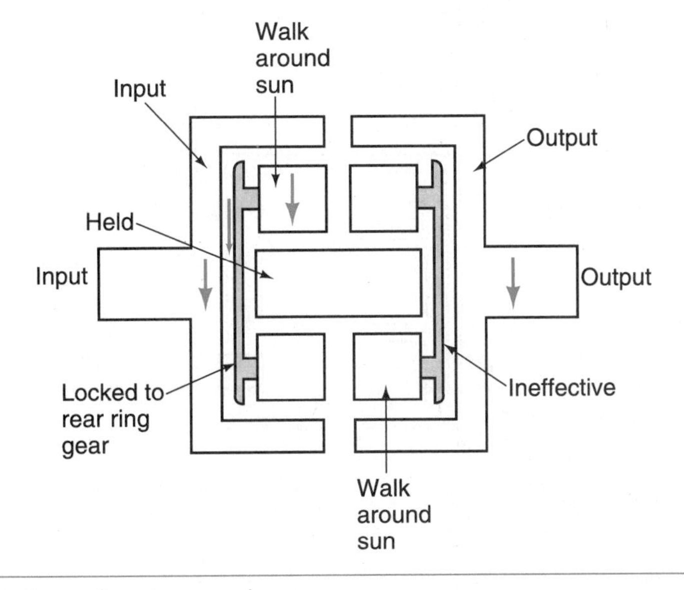

Figure 5-10 Power flow in second gear.

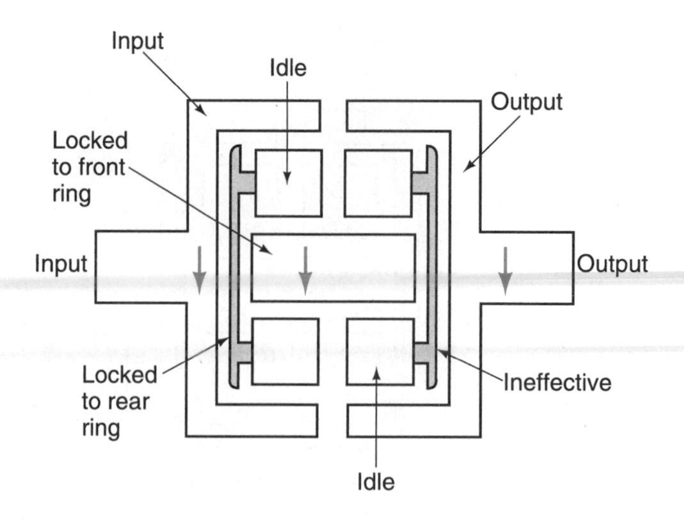

Figure 5-11 Power flow in third gear or direct drive.

gear, which turns clockwise with the shaft. The front ring gear drives the front planet gears, also in a clockwise direction. The front planet gears walk around the sun gear because it is held. The walking of the planets forces the planet carrier to turn clockwise. Since the carrier is locked to the output shaft, it causes the shaft to rotate in a forward direction with some gear reduction. A typical second gear ratio is 1.5:1.

When the planet carrier is held, the output of the gear set will be in the opposite direction to the input and reverse gear will be produced.

When operating in third gear (Figure 5-11), the input is received by the front ring gear as in the other forward positions. However, the input is also received by the sun gear. Since the sun and ring gears are rotating at the same speed and in the same direction, the front planet carrier is locked between the two and is forced to move with them. Since the front carrier is locked to the output shaft, direct drive results.

To obtain a suitable reverse gear in a Simpson gear set, there must be a gear reduction but in the opposite direction to the input torque (Figure 5-12). The input is received by the sun gear, as in the third gear position, and rotates in a clockwise direction. The sun gear then drives the rear planet gears in a counterclockwise direction. The rear planet carrier is held; therefore, the planet gears drive the rear ring gear in a counterclockwise direction. The ring gear is locked to the output

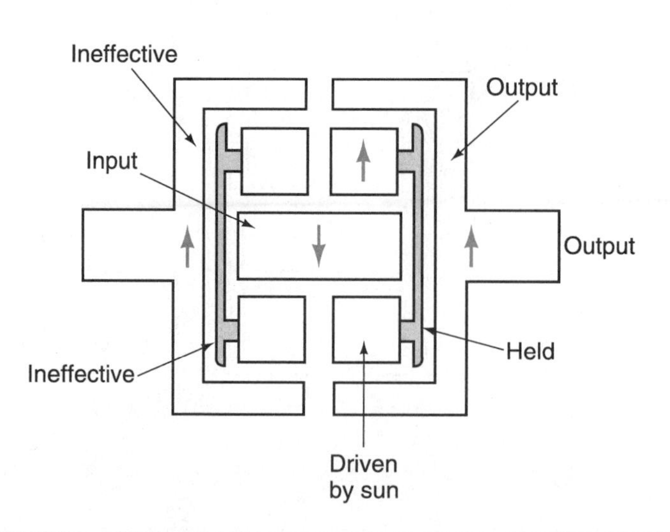

Figure 5-12 Power flow in reverse gear.

shaft, which turns at the same speed and direction as the rear ring gear. The result is a reverse gear with a ratio of 2.5:1 to 2:1.

Typically, when the transmission is in neutral or park, no apply devices are engaged, allowing only the input shaft and the transmission's oil pump to turn with the engine. In park, a pawl is mechanically engaged to a parking gear that is splined to the transmission's output shaft, locking the drive wheels to the transmission's case.

Ravigneaux Gear Train

Shop Manual
Chapter 5, page 145

The Ravigneaux gear train, like the Simpson gear train, provides forward gears with a reduction, direct drive, overdrive, and a reverse operating range. The Ravigneaux gear train offers some advantages over a Simpson gear train: it is very compact, it can carry large amounts of torque because of the great amount of tooth contact, and it can have three different output members. However, it has a disadvantage to students and technicians—it is more complex and therefore its actions are more difficult to understand.

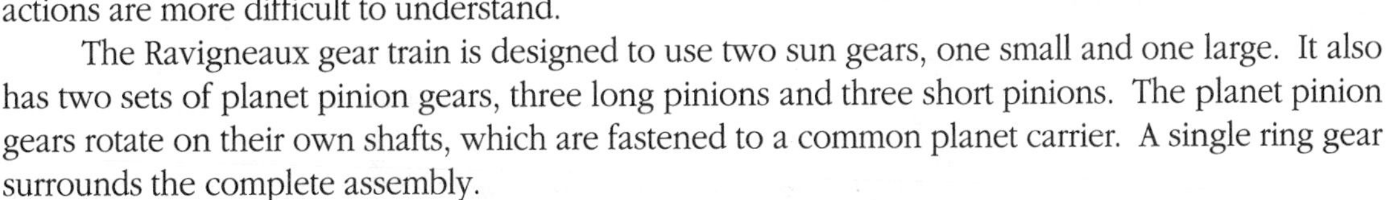

The Ravigneaux gear train is designed to use two sun gears, one small and one large. It also has two sets of planet pinion gears, three long pinions and three short pinions. The planet pinion gears rotate on their own shafts, which are fastened to a common planet carrier. A single ring gear surrounds the complete assembly.

The small sun gear is meshed with the short planet pinion gears. These short pinions act as idler gears to drive the long planet pinion gears. The long planet pinion gears mesh with the large sun gear and the ring gear.

Typically, when the gear selector is in neutral position, engine torque, through the converter turbine shaft, drives the forward clutch drum. Since the forward clutch is not applied, the power is not transmitted through the gear train and there is no power output.

When the transmission is operating in first gear (Figure 5-13), engine torque drives the forward clutch drum. The forward clutch is applied and drives the forward sun gear clockwise. The planet carrier is prevented from rotating counterclockwise by the one-way clutch; therefore, the for-

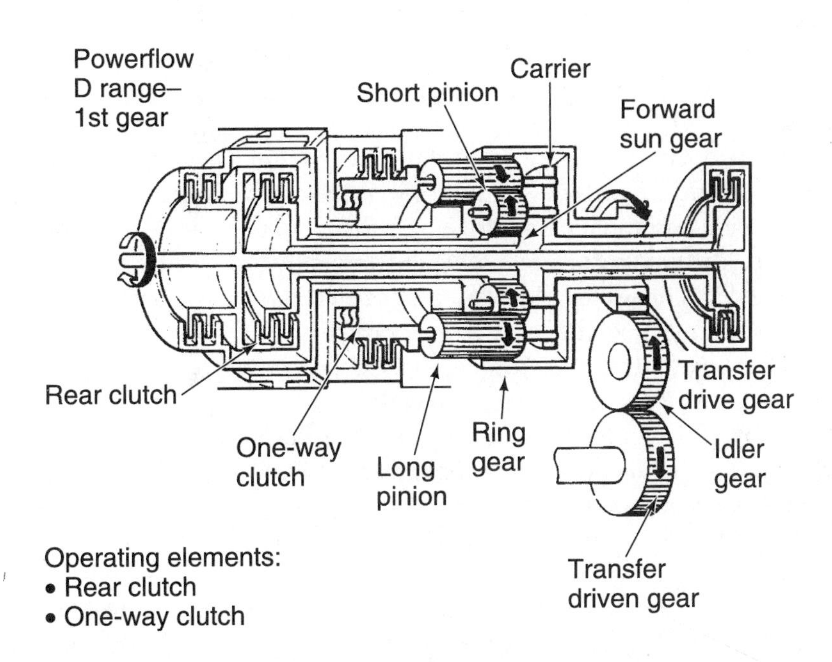

Figure 5-13 Power flow through a Ravigneaux gear train while operating in first gear. (Courtesy of Chrysler Corp.)

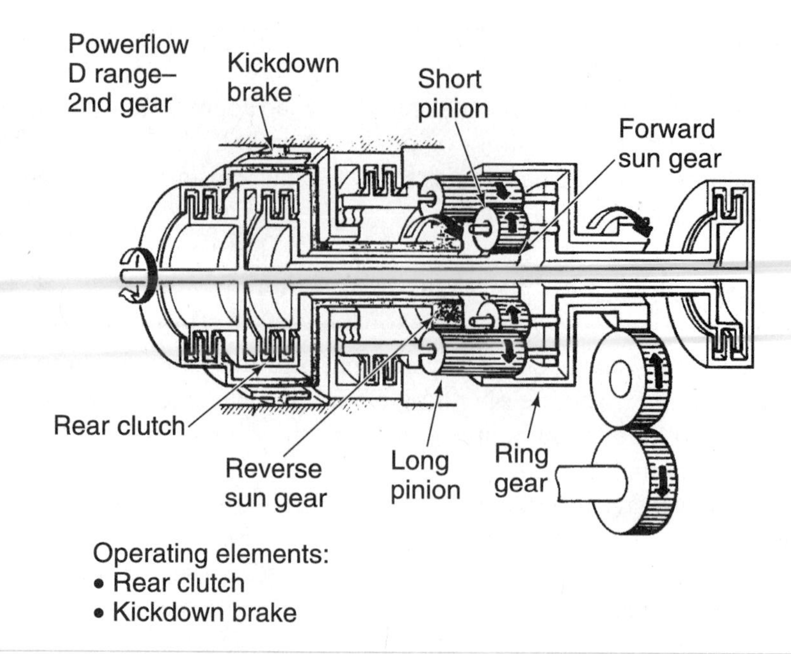

Figure 5-14 Power flow through a Ravigneaux gear train while operating in second gear. (Courtesy of Chrysler Corp.)

ward sun gear drives the short planet gears counterclockwise. The direction of rotation is reversed as the short planet gears drive the long planet gears, which drive the ring gear and output shaft in a clockwise direction but at a lower speed than the input. This results in a gear reduction and a ratio of approximately 2.5:1.

In second gear operation (Figure 5-14), the intermediate clutch is applied and locks the outer **race** of the one-way clutch. This prevents the reverse clutch drum, shell, and reverse sun gear from turning counterclockwise. The forward clutch is also applied and locks the input to the forward sun gear and it rotates in a clockwise direction. The forward sun gear drives the short planet gears counterclockwise. The direction of rotation is reversed as the short planet gears drive the long planet gears, which walk around the stationary reverse sun gear. This walking drives the ring gear and output shaft in a clockwise direction and at a reduction with a gear ratio of approximately 1.5:1.

During third gear operation (Figure 5-15), there are two inputs into the planetary gear train. As in other forward gears, the turbine shaft of the torque converter drives the forward clutch drum. The forward clutch is applied and drives the forward sun gear in a clockwise direction. Input is also received by the direct clutch, which is driven by the torque converter cover. The direct clutch is applied and drives the planetary gear carrier. Since two members of the gear train are being driven at the same time, the planetary gear carrier and the forward sun gear rotate as a unit. The long planet gears transfer the torque, in a clockwise direction, through the gear set to the ring gear and output shaft. This results in direct drive.

To operate in overdrive or fourth gear, input is received only by the direct clutch from the torque converter cover. The direct clutch is applied and drives the planet carrier in a clockwise direction. The long planet gears walk around the stationary reverse sun gear in a clockwise direction and drive the ring gear and output shaft. This results in an overdrive condition with an approximate ratio of 0.75:1.

During reverse gear operation, input is received through the torque converter's turbine shaft to the reverse clutch. The reverse clutch, when applied, connects the turbine shaft to the reverse sun gear. The low/reverse band is applied and holds the planetary gear carrier. The clockwise rotation of the reverse sun gear drives the long planet gears in a counterclockwise direction. The

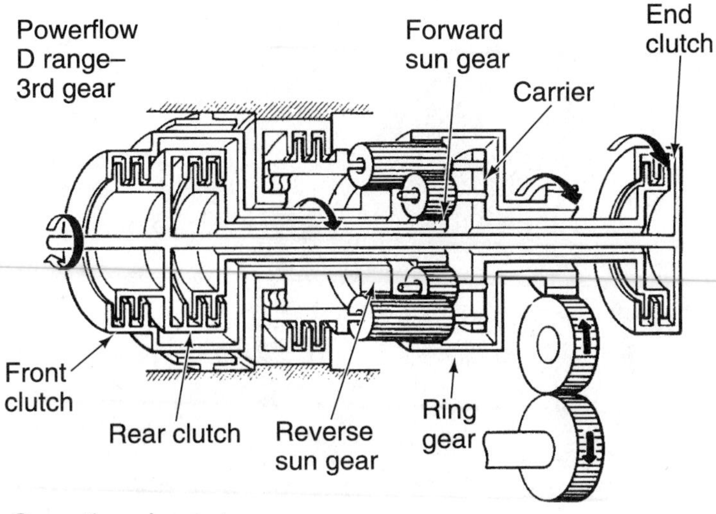

Figure 5-15 Power flow through a Ravigneaux gear train while operating in third gear. (Courtesy of Chrysler Corp.)

long planets then drive the ring gear and output shaft in a counterclockwise direction with a speed reduction and a gear ratio of approximately 2:1.

Planetary Gear Sets in Tandem

Rather than relying on the use of a compound gear set, some automatic transmissions use two simple planetary units in series. In this type of arrangement, gear set members are not shared. The holding devices are used instead to lock different members of the planetary units together.

Two common transaxles use this planetary gear arrangement, the THM 440-T4/4T60 and the Ford AXOD family. Both of these will be covered in detail, although they are quite similar in design and power flow.

Although the gear train is based on two simple planetary gear sets operating in tandem, the combination of the two planetary units does function much like a compound unit. The two tandem units do not share a common member, rather certain members are locked together or are integral with each other. The front planet carrier is locked to the rear ring gear and the front ring gear is locked to the rear planet carrier. The transaxle houses a third planetary unit, which is used only as the final drive unit and not for overdrive.

Honda's Nonplanetary-Based Transmission

The Honda CA, F4, and G4 transaxles are used in many Honda and Acura cars. These transmissions are unique in that they do not use a planetary gear set to provide for the different gear ranges. Constant-mesh helical and square-cut gears (Figure 5-16) are used in a manner similar to that of a manual transmission.

These transaxles have a mainshaft and countershaft on which the gears ride. To provide the four forward and one reverse gears, different pairs of gears are locked to the shafts by hydraulically controlled clutches (Figure 5-17). Reverse gear is obtained through the use of a shift fork, which slides the reverse gear into position (Figure 5-18). The power flow through these transaxles is also similar to that of a manual transaxle.

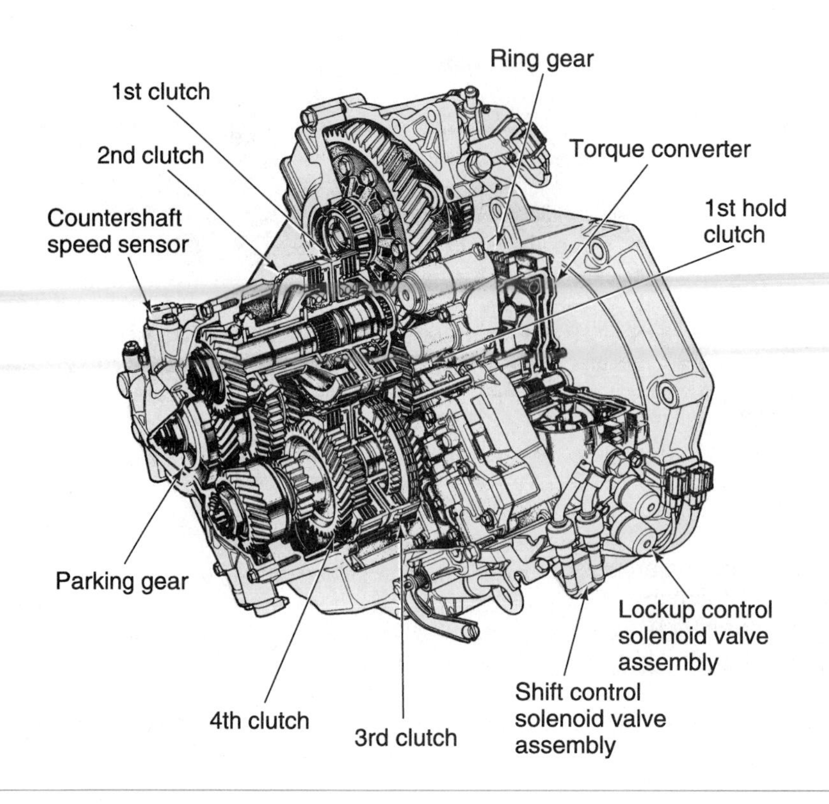

Figure 5-16 Honda automatic transmissions use constant-mesh helical gears instead of planetary gear sets. (Courtesy of the Automatic Transmissions Rebuilders Association—ATRA)

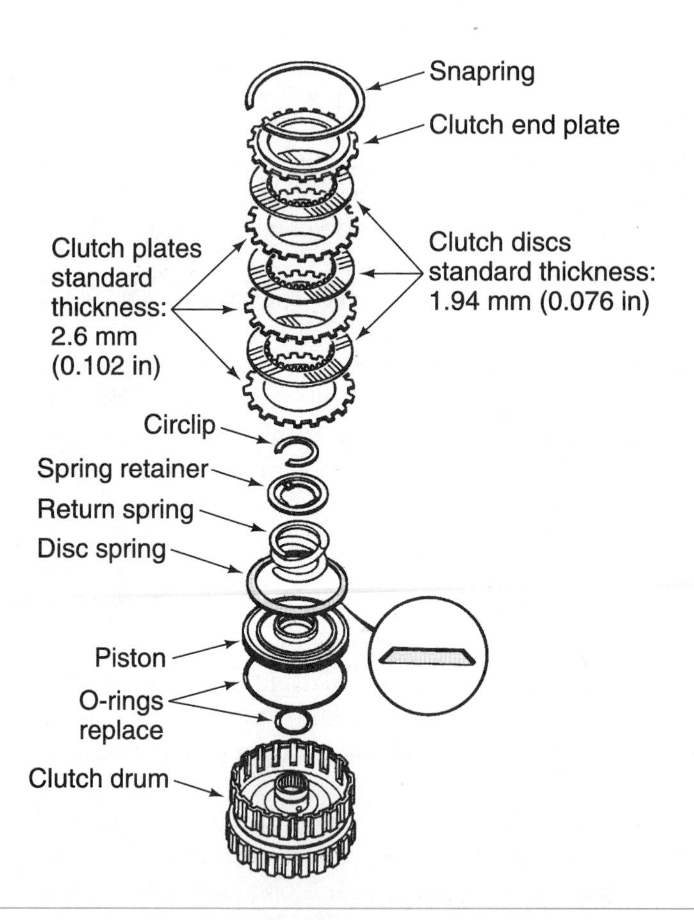

Figure 5-17 Different pairs of gears are locked to the shafts by hydraulically controlled clutches. (Courtesy of the Automatic Transmissions Rebuilders Association—ATRA)

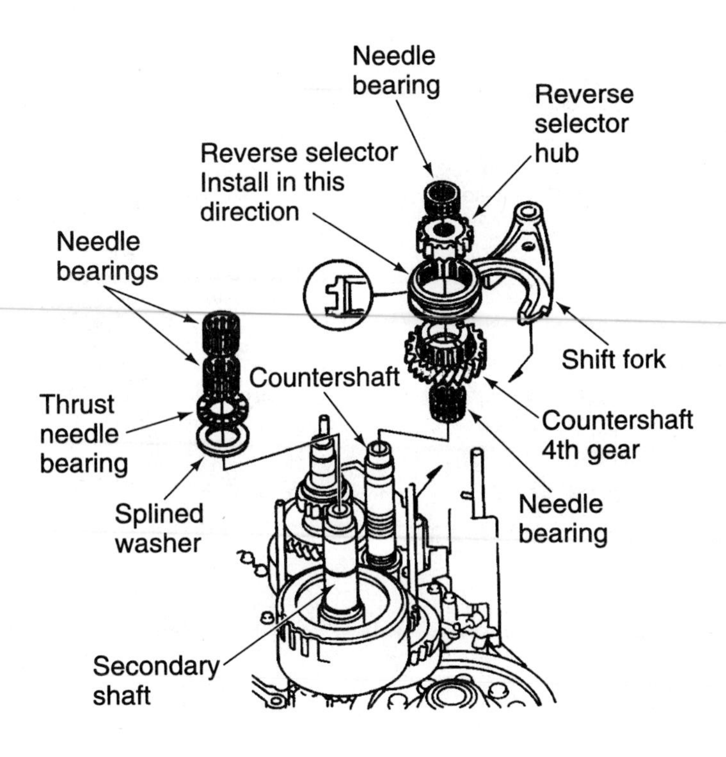

Figure 5-18 Reverse gear is obtained by sliding a gear into position. (Courtesy of the Automatic Transmissions Rebuilders Association—ATRA)

Bearings, Bushings, and Thrust Washers

Shop Manual
Chapter 5, page 140

Rolling bearings are typically either ball or roller-type bearings.

Torrington bearings are a commonly used type of thrust washer. The sides of these thrust washers are fitted with rollers to reduce friction and wear.

When a component slides over or rotates around another part, the surfaces that contact each other are called bearing surfaces. A gear rotating on a fixed shaft can have more than one bearing surface. It is supported and held in place by the shaft in a **radial** direction. Also the gear tends to move along the shaft in an **axial** direction as it rotates and is therefore held in place by some other components. The surfaces between the sides of the gear and the other parts are bearing surfaces.

A **bearing** is a device placed between two bearing surfaces to reduce friction and wear. Most bearings have surfaces that either slide or roll against each other. In automatic transmissions, sliding bearings are used where one or more of the following conditions prevail: low rotating speeds, very large bearing surfaces compared to the surfaces present, and low use. Rolling bearings are used in the following applications: high speed, high load with relatively small bearing surfaces, and high use.

Transmissions use sliding bearings that are composed of a relatively soft bronze alloy. Many are made from steel with the bearing surface bonded or fused to the steel. Those that take radial loads are called **bushings** and those that take axial loads are called **thrust washers** (Figure 5-19). The bearing's surface usually runs against a harder surface such as steel to produce minimum friction and heat-wear characteristics.

Bushings are cylindrically shaped and usually held in place by press-fit. Since bushings are made of a soft metal, they act like a bearing and support many of the transmission's rotating parts (Figure 5-20). They are also used to precisely guide the movement of various valves in the transmission's valve body (Figure 5-21). Bushings can also be used to control fluid flow; some restrict the flow from one part to another whereas others are made to direct fluid flow to a particular point or part in the transmission.

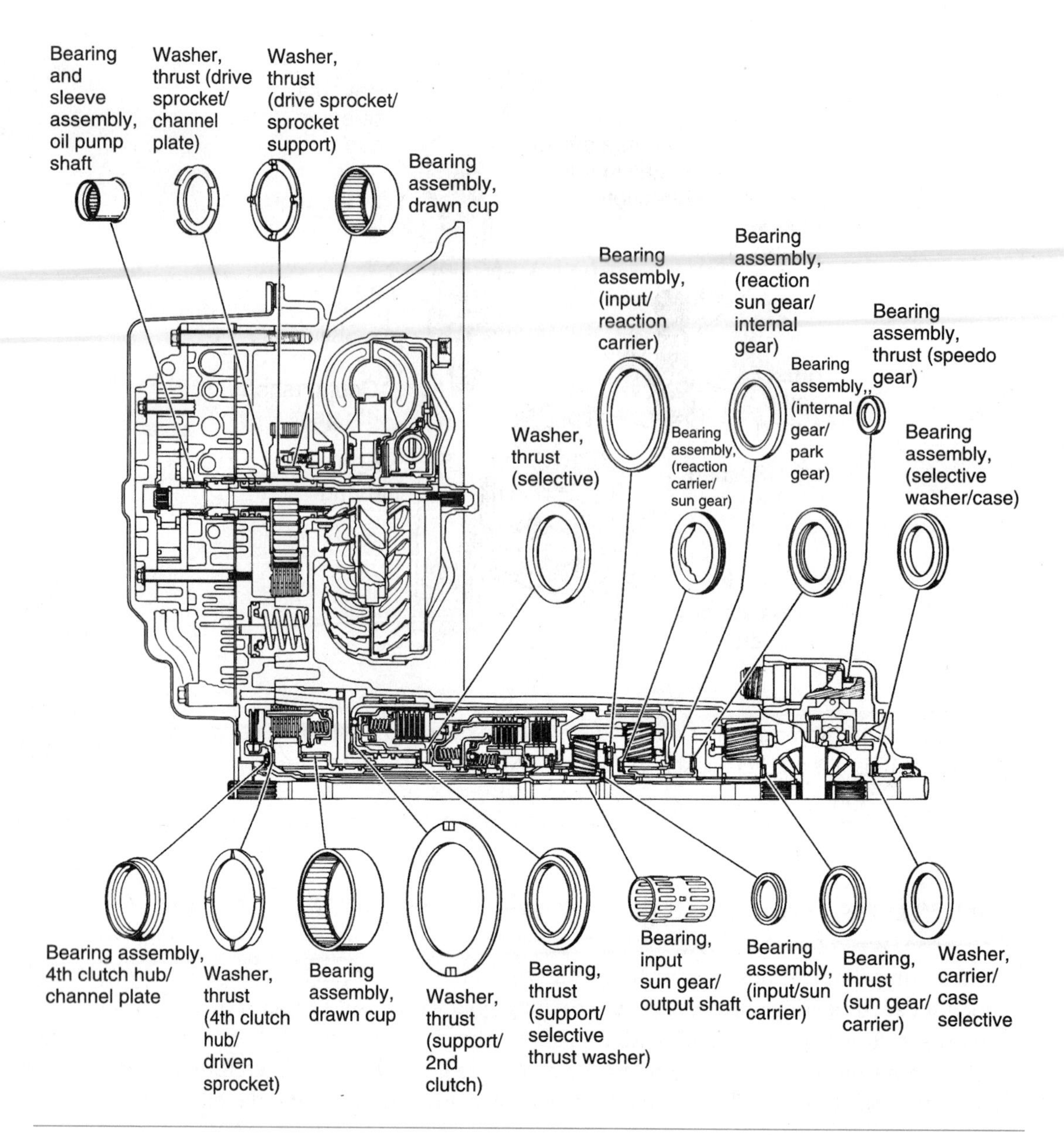

Figure 5-19 Location of various bearing and thrust washers in a typical transmission. (Courtesy of the Hydra-Matic Division of General Motors Corp.)

Items held in place by press-fit have their outside diameter a few thousandths of an inch larger than the inside diameter of the hole into which they will be inserted. Press-fit bushings must be pressed into the bore.

Shop Manual
Chapter 5, page 131

Often serving both as a bearing and a spacer, thrust washers are made in various thicknesses. They may have one or more tangs or slots on the inside or outside circumference that mate with the shaft bore to keep them from turning. Some thrust washers are made of nylon or Teflon, which are used when the load is low. Others are fitted with rollers to reduce friction and wear (Figure 5-22).

Free axial movement or endplay is normally controlled by thrust washers (Figure 5-23). Since some endplay is necessary in all transmissions because of heat expansion, proper endplay is often accomplished through selective thrust washers. These thrust washers are inserted between various parts of the transmission. Whenever endplay is set, it must be set to the manufacturer's specifications. Thrust washers work by filling the gap between two components and become the primary wear item because they are made of softer materials than the parts they protect. Normally, thrust washers are made of copper-faced soft steel, bronze, nylon, or plastic.

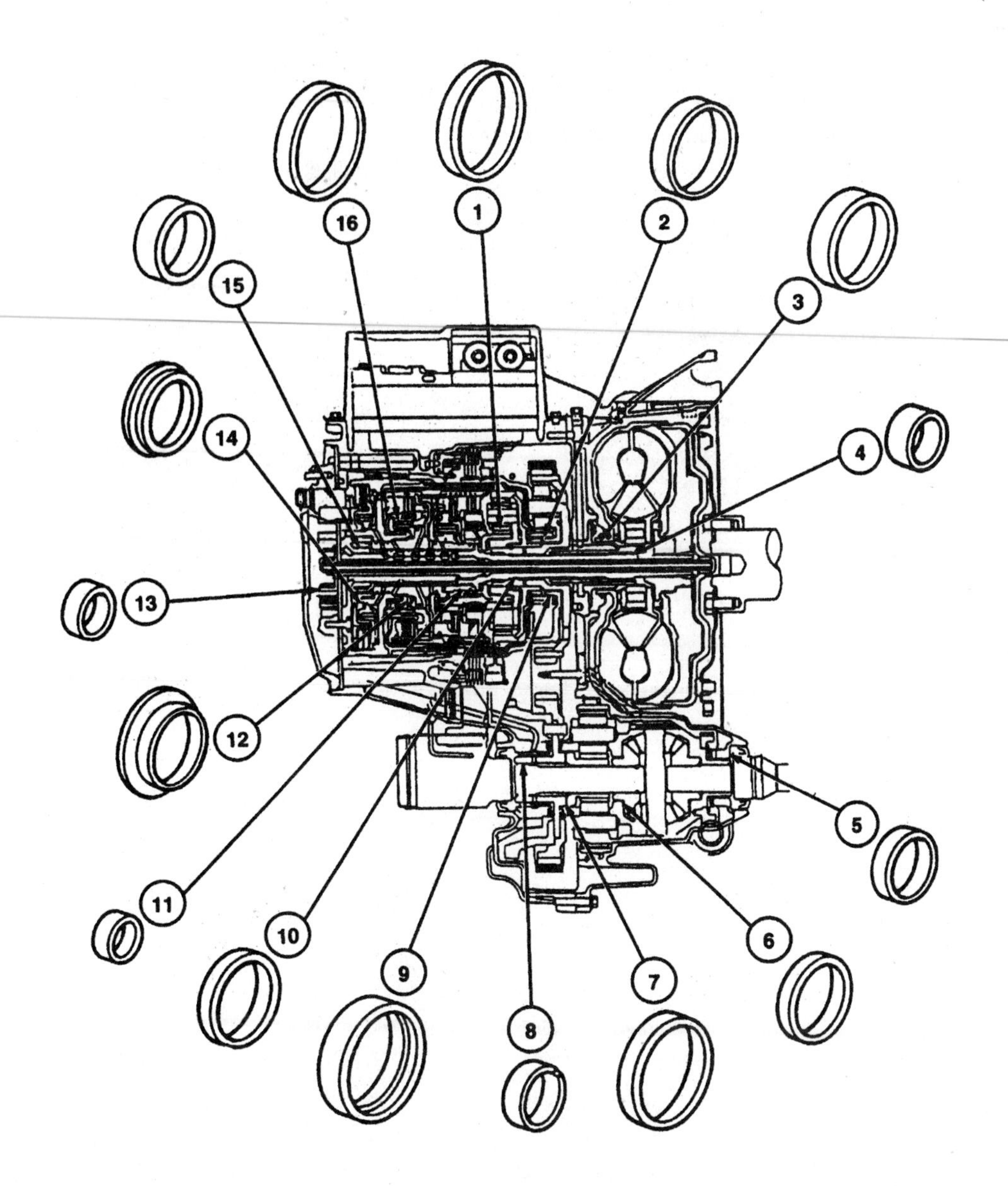

Item	Part Number	Description
1	7F366	Low-Intermediate Sun Gear Bushings
2	7H043	Reverse / Overdrive Hub Plate Bushing
3	7N805	Converter Impeller Hub Bushing
4	7N817	Stator Support Bushing
5	7E110	Case Bushing
6	7H041	Final Drive Differential Rear Bushing
7	7H040	Final Drive Differential Front Bushing
8	7G344	Final Drive Sun Gear Bushing
9	7H033	Reverse / Overdrive Drum Gear Bushing
10	7H034	Low-Intermediate Gear Hub Bushing
11	7H025	Turbine Shaft Bushing
12	7H240	Direct Clutch Bushing
13	7B259	Pump Assembly Drive Bushing
14	7F218	Reverse Clutch Drum Support Bushing

(Continued)

Figure 5-20 Bushings are used throughout a transmission. (Reprinted with the permission of Ford Motor Co.)

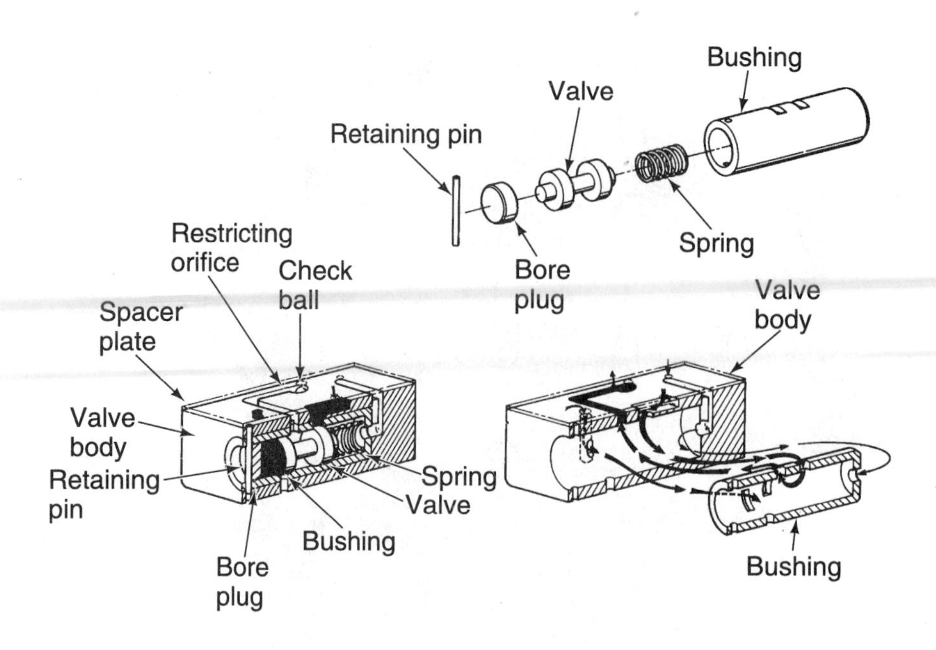

Figure 5-21 Typical bushing and valve assembly. (Courtesy of the Hydra-Matic Division of General Motors Corp.)

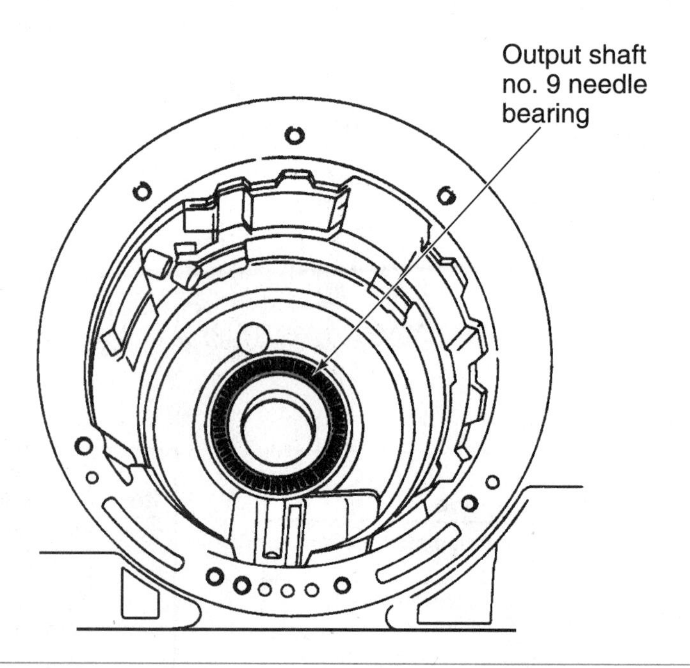

Figure 5-22 Typical application of a Torrington bearing. (Reprinted with the permission of Ford Motor Co.)

Torrington bearings are thrust washers fitted with roller bearings. These thrust bearings are primarily used to limit endplay but also to reduce the friction between two rotating parts. Most often Torrington bearings are used in combination with flat thrust washers to control endplay of a shaft or the gap between a gear and its drum.

Some sliding bearings have holes drilled in them to match oil passages so they can be lubricated under pressure. Others simply are lubricated by splashing. The bearing surface is often grooved to promote improved lubrication.

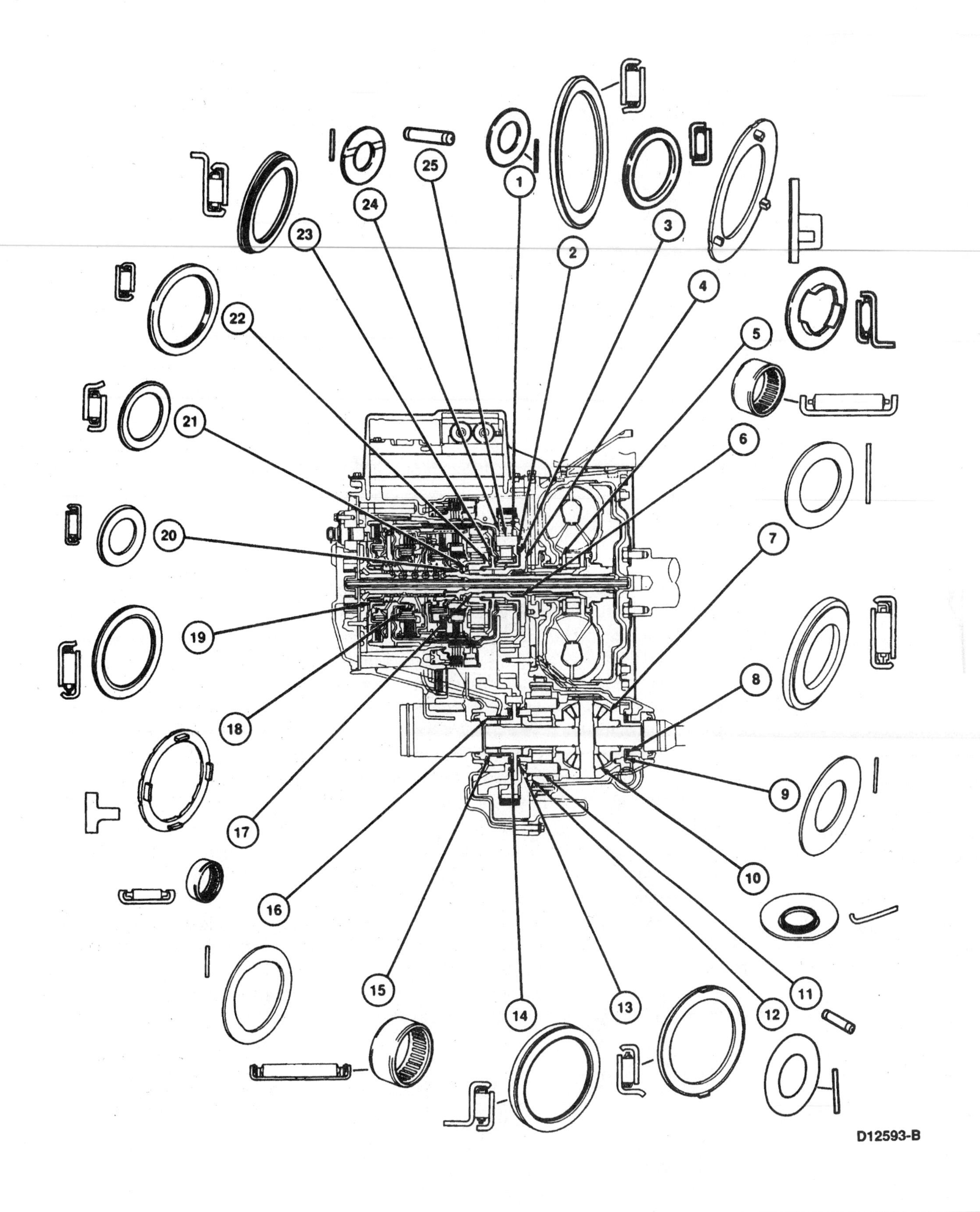

Figure 5-23 Location of the various thrust washers and bearings in a transaxle. (Reprinted with the permission of Ford Motor Co.)

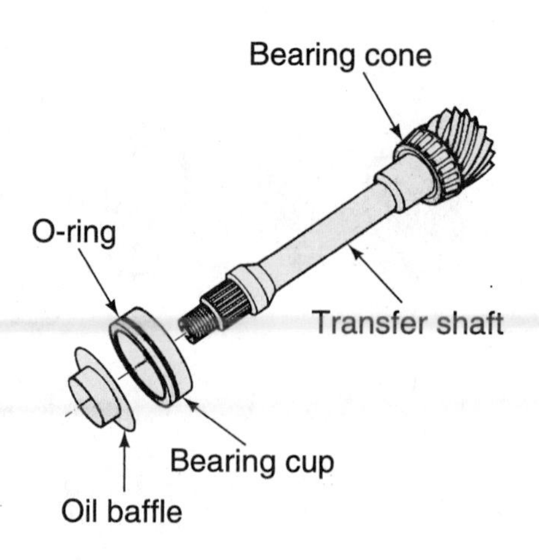

Figure 5-24 Use of a tapered roller bearing in a transaxle. (Courtesy of Chrysler Corp.)

A race is a ring-shaped track that controls the travel of the rollers.

A cage is a device used to hold the rollers in place and to keep them separated from each other.

The outer race of a tapered bearing is generally referred to as a **cup.** The inner race is the **cone.**

Preload is simply a tightening process accomplished by using shims or by tightening a shaft nut to a specified torque that limits endplay.

The bearing surface is greatly reduced through the use of roller bearings. The simplest roller bearing design leaves enough clearance between the bearing surfaces of two sliding or rotating parts to accept some rollers. Each roller's two points of contact between the bearing surfaces are so small that friction is greatly reduced. The bearing surface is more like a line than an area.

If the ratio of roller length to diameter is about 5:1 or more, the roller is called a needle and such a bearing is called a needle bearing. The needles can be loose or they can be held in place by a steel cylinder or by rings at each end. The latter are often drilled to accept pins at the ends of each needle that act as an axle. These small assemblies help save the agony of losing one or more loose needles and the delay caused by searching for them.

Many other roller bearings are designed as assemblies. The assemblies consist of an inner and outer race, the rollers, and a **cage.** There are roller bearings designed for radial loads and others designed for axial loads.

A tapered roller bearing is designed to accept both radial and axial loads. Its rollers turn on an angle to the bearing assembly's axis rather than parallel to it. The rollers are also slightly tapered to fit the angle of the inner and outer races. The bearing assembly consists of an inner race, the rollers, and the cage, and the outer race (Figure 5-24). Because the outer race is separate, bearing clearance or **preload** can be adjusted during assembly to meet both axial and radial specifications. Tapered roller bearings are used in pairs and are rarely used in automatic transmissions. They are commonly used in final drive units.

The heaviest radial loads in automatic transmissions are carried by either roller or ball bearings. Ball bearings are constructed similarly to roller bearings, except that the races are grooved to accept the balls. The groove radius is slightly larger than the ball radius, which reduces the bearing surface area more than the roller bearing does. A ball bearing can also withstand light axial loads. **Lip seals** are sometimes built into ball bearings to retain lubricants.

Snaprings

Many different sizes and types of snaprings (Figure 5-25) are used in today's transmissions. External and internal snaprings are used as retaining devices throughout the transmission. Internal snaprings are used to hold servo assemblies and clutch assemblies together. In fact, snaprings are also available in several thicknesses (Figure 5-26) and may be used to adjust the clearances in multiple-disc clutch packs. Some snaprings for clutch packs are waved to smooth clutch application. External snaprings are used to hold gear and clutch assemblies to their shafts.

Figure 5-25 Various sizes and types of snaprings used in a single Saturn transaxle. (Courtesy of Saturn Corp.)

A gasket is typically used to seal the space between two parts that have irregular surfaces.

A hard gasket is one that will compress less than 20% when it is tightened in place.

A soft gasket is one that will compress more than 20% when it is tightened in place.

Shop Manual
Chapter 5, page 151

A composition gasket is one that is made from two or more different materials.

The oil pan gasket is most often called the *pan gasket.*

RTV stands for room temperature vulcanizing, which means this sealant will begin to solidify at room temperature and form a seal prior to being heated by the operation of the transmission.

RTV is also known as an aerobic; it requires air and humidity to cure and form a seal (anaerobic is the opposite; it cures in the absence of air).

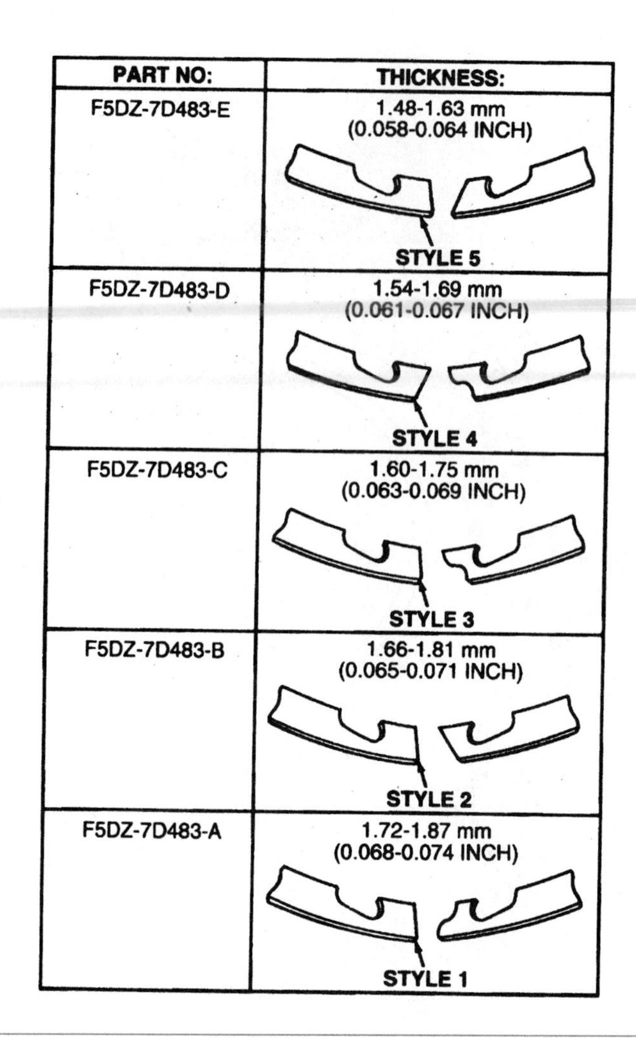

PART NO:	THICKNESS:
F5DZ-7D483-E	1.48-1.63 mm (0.058-0.064 INCH) STYLE 5
F5DZ-7D483-D	1.54-1.69 mm (0.061-0.067 INCH) STYLE 4
F5DZ-7D483-C	1.60-1.75 mm (0.063-0.069 INCH) STYLE 3
F5DZ-7D483-B	1.66-1.81 mm (0.065-0.071 INCH) STYLE 2
F5DZ-7D483-A	1.72-1.87 mm (0.068-0.074 INCH) STYLE 1

Figure 5-26 Various selectively sized snaprings. (Reprinted with the permission of Ford Motor Co.)

Gaskets and Seals

The gaskets and seals of an automatic transmission help contain the fluid within the transmission and prevent the fluid from leaking out of the various hydraulic circuits. Various types of seals are used in automatic transmissions. They can be made of rubber, metal, or Teflon (Figure 5-27). Transmission gaskets are made of rubber, cork, paper, synthetic materials, or plastic.

Gaskets

Gaskets are used to seal two parts together or to provide a passage for fluid flow from one part of the transmission to another (Figure 5-28). Gaskets are easily divided into two separate groups, hard and soft, depending on their application. **Hard gaskets** are used whenever the surfaces to be sealed are smooth. This type of gasket is usually made of paper. A common application of a hard gasket is the gasket used to seal the valve body and oil pump against the transmission case. Hard gaskets are also often used to direct fluid flow or seal off some passages between the valve body and the separator plate.

Gaskets that are used when the sealing surfaces are irregular or in places where the surface may distort when the component is tightened into place are called **soft gaskets.** A typical location of a soft gasket is the oil pan gasket, which seals the oil pan to the transmission case. Oil pan gaskets are typically a **composition gasket** made with rubber and cork. However, some late-model transmissions use an **RTV** sealant (Figure 5-29) instead of a gasket to seal the oil pan.

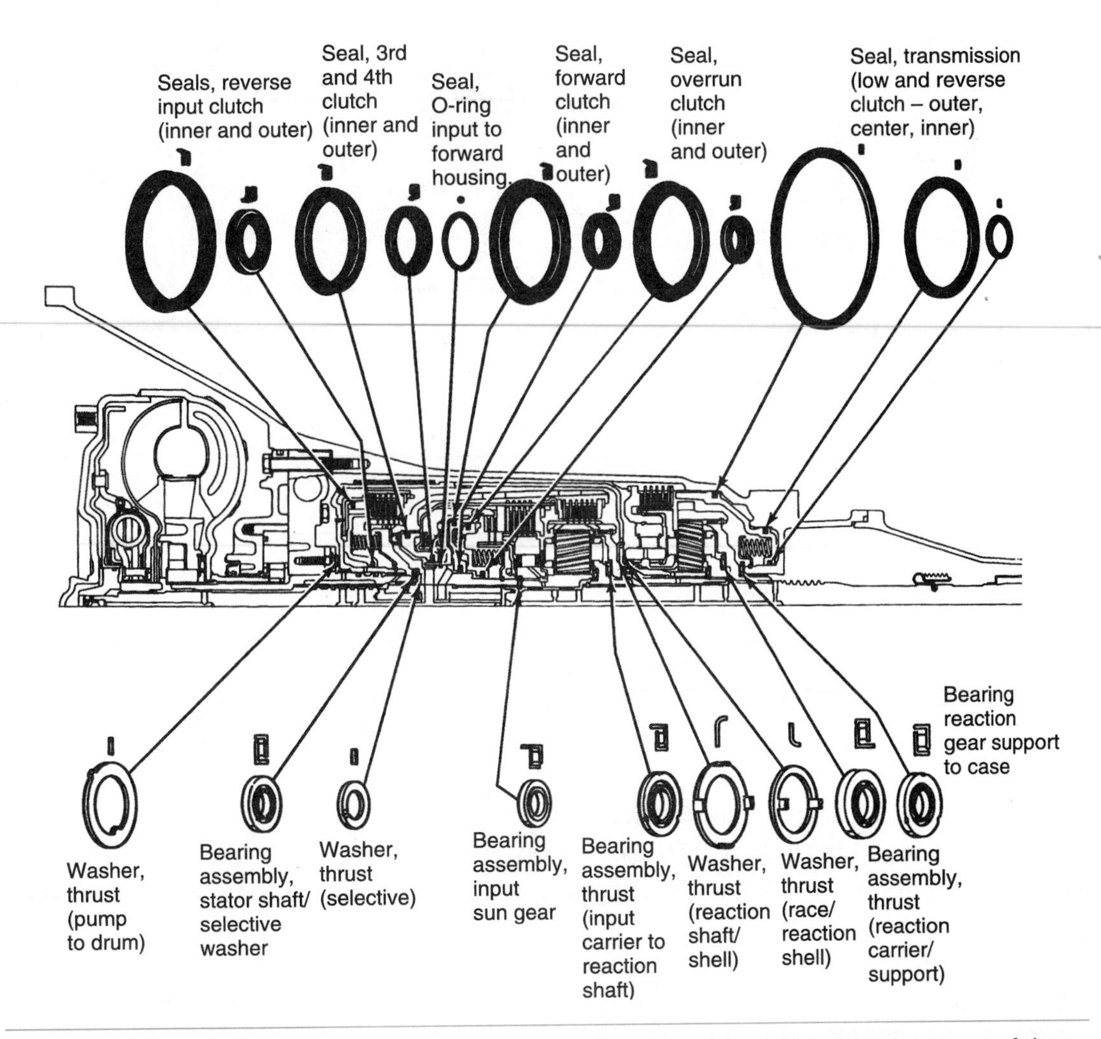

Figure 5-27 Location of various seals and gaskets in a typical transmission. (Courtesy of the Hydra-Matic Division of General Motors Corp.)

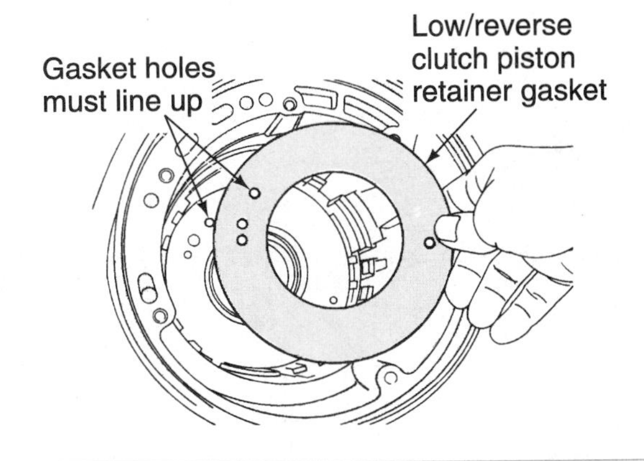

Figure 5-28 Typical application of a gasket in an automatic transmission. (Courtesy of Chrysler Corp.)

Shop Manual
Chapter 5, page 130

A static seal prevents fluid from passing between two or more parts that are always in the same relationship with each other.

Seals

As valves and transmission shafts move within the transmission, it is essential that the fluid and pressure be contained within its bore. Any leakage would decrease the pressure and result in poor transmission operation. Seals are used to prevent leakage around valves, shafts, and other moving parts. Rubber, metal, or Teflon materials are used throughout a transmission to provide for **static**

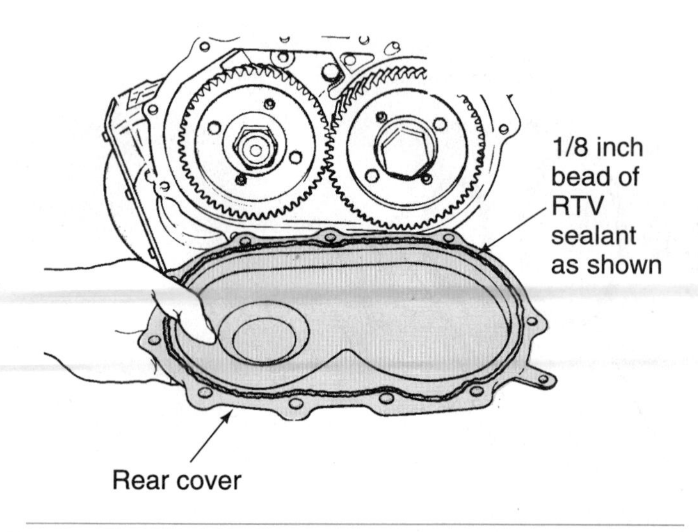

Figure 5-29 Use of RTV to seal a sheet metal cover to the transmission housing. (Courtesy of Chrysler Corp.)

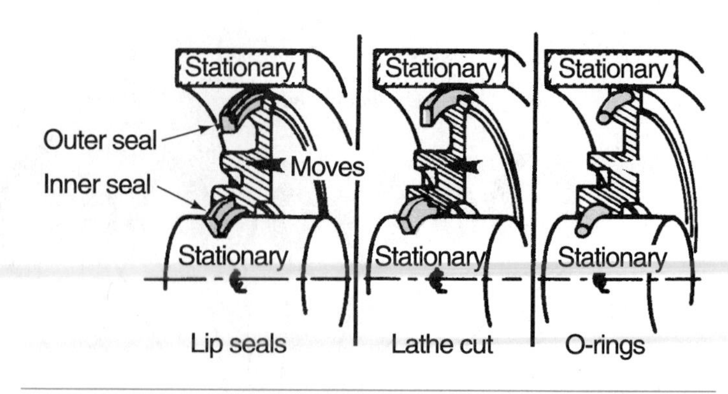

Figure 5-30 Three types of seals and mountings shown in their typical position. (Reprinted with permission ©1962 Society of Automotive Engineers, Inc., from *Design Practices: Passenger Car Automatic Transmissions.*)

A dynamic seal is one that prevents leaks between two or more parts that do not have a fixed position and move in relation to each other.

Synthetic rubber is made from neoprene, nitrile, silicone, fluoroelastomers, and polyacrylics.

Shop Manual
Chapter 5, page 152

and **dynamic sealing.** Both static and dynamic seals can provide for positive and nonpositive sealing. A definition of each of the basic classifications of seals follows:

Static—A seal used between two parts that do not move in relation to each other.

Dynamic—A seal used between two parts that move in relation to each other. This movement is either a rotating or reciprocating (up and down) motion.

Positive—A seal that prevents all fluid leakage between two parts.

Nonpositive—A seal that allows a controlled amount of fluid leakage. This leakage is typically used to lubricate a moving part.

Three major types of rubber seals are used in automatic transmissions: the O-ring, the lip seal, and the **lathe-cut seal** or square-cut seal (Figure 5-30). Rubber seals are made from synthetic rubber rather than natural rubber.

O-rings are round seals with a circular cross section. An O-ring is normally installed in a groove cut into the inside diameter of one of the parts to be sealed. When the other part is inserted into the bore and through the O-ring, the O-ring is compressed between the inner part

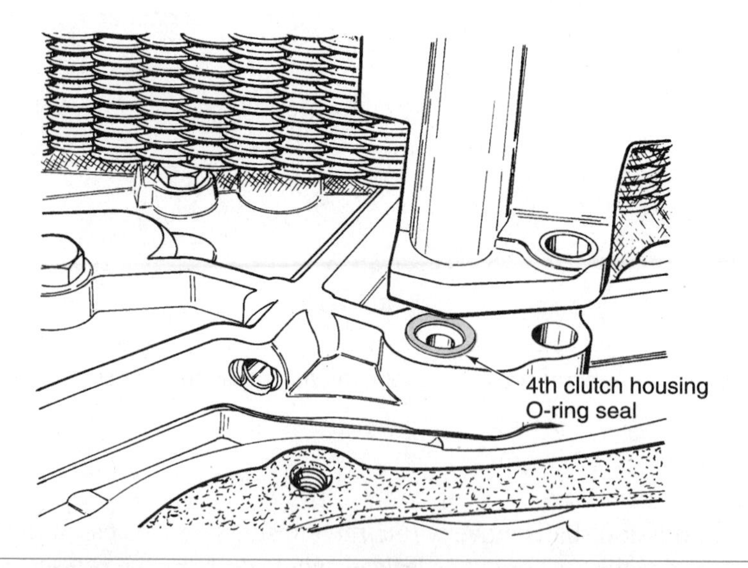

Figure 5-31 Typical application of an O-ring seal. (Courtesy of the Hydra-Matic Division of General Motors Corp.)

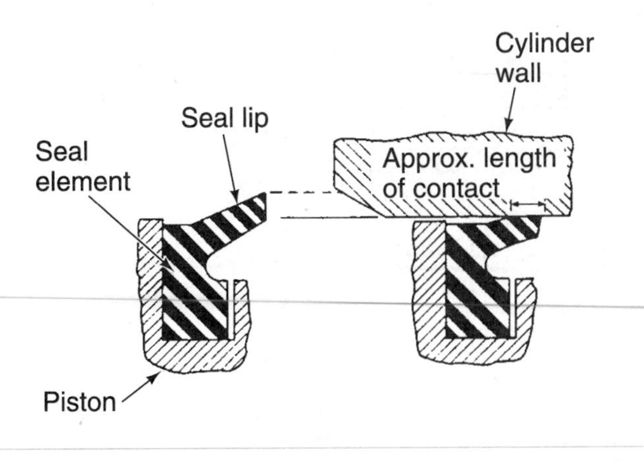

Figure 5-32 Sealing action of a lip seal. (Reprinted with permission ©1962 Society of Automotive Engineers, Inc., from *Design Practices: Passenger Car Automatic Transmissions.*)

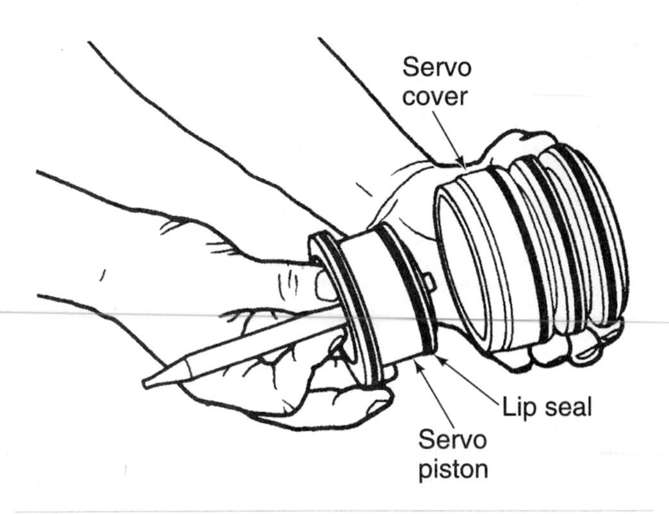

Figure 5-33 Typical application of a lip seal. (Reprinted with the permission of Ford Motor Co.)

and the groove. This pressure distorts the O-ring and forms a tight seal between the two parts (Figure 5-31).

O-rings can be used as dynamic seals, but they are most commonly used as static seals. An O-ring can be used as a dynamic seal when the parts have relatively low amounts of axial movement. If there is a considerable amount of axial movement, the O-ring will quickly be damaged as it rolls within its groove. O-rings are never used to seal a shaft or part that has rotational movement.

Axial movement is movement along, or parallel to, the centerline of a shaft.

Lip seals are used to seal parts that have axial or rotational movement. They are round to fit around a shaft, but the entire seal does not serve as a seal, rather the sealing part is a flexible lip (Figure 5-32). The flexible lip is normally made of synthetic rubber and shaped so that it is flexed when it is installed to apply pressure at the sharp edge of the lip. Lip seals are used around input and output shafts to keep fluid in the housing and dirt out. Some seals are double-lipped.

Shop Manual
Chapter 5, page 153

When the lip is around the outside diameter of the seal, it is used as a piston seal (Figure 5-33). Piston seals are designed to seal against high pressures and the seal is positioned so that the lip faces the source of the pressurized fluid. The lip is pressed firmly against the cylinder wall as the fluid pushes against the lip. This forms a tight seal. The lip then relaxes its seal when the pressure on it is reduced or exhausted.

Lip seals are also commonly used as shaft seals. When used to seal a rotating shaft, the lip of the seal is around the inside diameter of the seal and the outer diameter is bonded to the inside of a metal housing. The outer metal housing is pressed into a bore. To help maintain good sealing pressure on the rotating shaft, a garter spring is fitted behind the lip. This **toroidal** spring pushes on the lip to provide for uniform contact on the shaft. Shaft seals are not designed to contain pressurized fluid, rather they are designed to prevent fluid from leaking over the shaft and out of the housing. The tension of the spring and lip is designed to allow an oil film of about 0.0001 inch. This oil film serves as a lubricant for the lip. If the tolerances increase, fluid will be able to leak past the shaft, and if the tolerances are too small, excessive shaft and seal wear will result.

Shaft lip seals are often called metal clad seals.

The word *toroidal* infers that the spring is doughnut-shaped.

A **square-cut seal** is similar to an O-ring; however, a square-cut seal can withstand more axial movement than an O-ring can. Square-cut seals are also round seals, but they have a rectangular or square cross section. They are designed this way to prevent the seal from rolling in its groove when there are large amounts of axial movement. Added sealing comes from the distortion of the seal during axial movement. As the shaft inside the seal moves, the outer edge of the seal moves more than the inner edge, causing the diameter of the sealing edge to increase, which creates a tighter seal (Figure 5-34).

A square-cut seal is often called a lathe-cut seal.

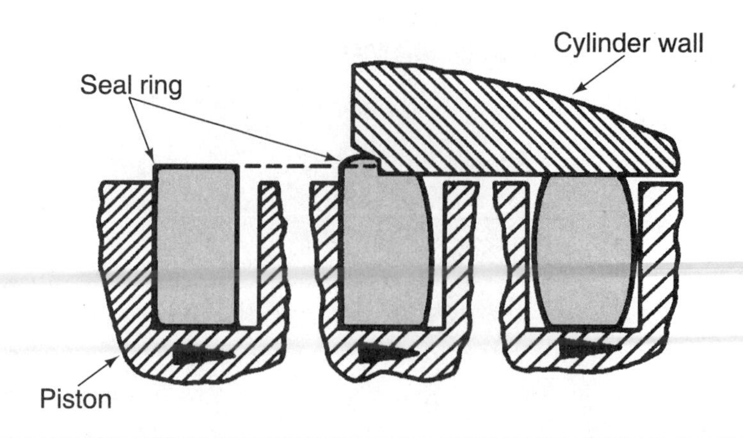

Figure 5-34 Sealing action of a rubber seal as a piston moves in its bore. (Reprinted with permission ©1962 Society of Automotive Engineers, Inc., from *Design Practices: Passenger Car Automatic Transmissions.*)

Shop Manual
Chapter 5, page 152

Metal seals are often called **steel rings**, although they are made of cast iron.

Hook-end seals are also referred to as locking-end seals.

Teflon is a soft, durable plastic material that has high heat resistance.

Teflon locking-end seals are normally called **scarf-cut rings.**

Metal Sealing Rings

There are some parts of the transmission that do not require a positive seal and where some leakage is acceptable. These components are sealed with ring seals that fit into a groove on a shaft (Figure 5-35). The outside diameter of the ring seals slide against the walls of the bore into which the shaft is inserted. Most ring seals in a transmission are placed near pressurized fluid outlets on rotating shafts to help retain pressure. Ring seals are made of cast iron, nylon, or Teflon.

Three types of metal seals are used in automatic transmissions: butt-end seals, open-end seals, and hook-end seals. In appearance, butt-end and open-end seals are much the same; however, when an **open-end seal** is installed, there is a gap between the ends of the seal. When a **butt-end seal** is installed, the square-cut ends of the seal touch, or butt, against each other. **Hook-end seals** (Figure 5-36) have small hooks at their ends, which are locked together during installation to provide better sealing than the open-end or butt-end seals.

Teflon Seals or Rings

Some transmissions use Teflon seals instead of metal seals. Teflon provides for a softer sealing surface, which results in less wear on the surface that it rides on and therefore a longer-lasting seal. Teflon seals are similar in appearance to metal seals except for the hook-end type. The ends of

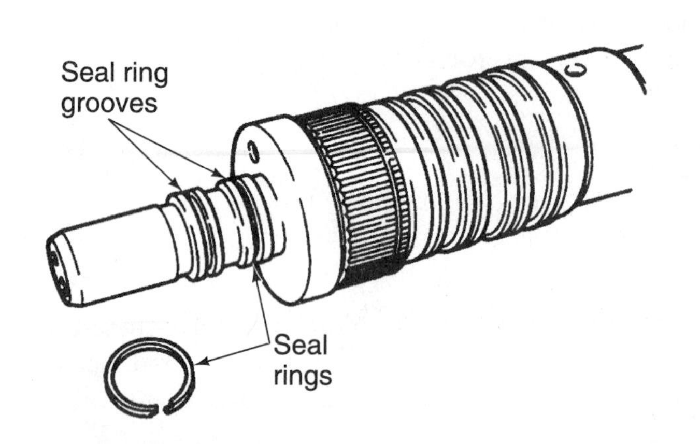

Figure 5-35 Metal sealing rings are fit into grooves on a shaft. (Reprinted with the permission of Ford Motor Co.)

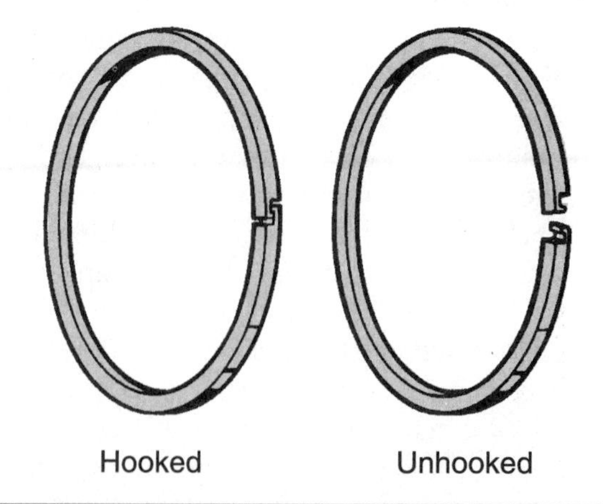

Figure 5-36 Hook-end sealing rings. (Courtesy of the Oldsmobile Division of General Motors Corp.)

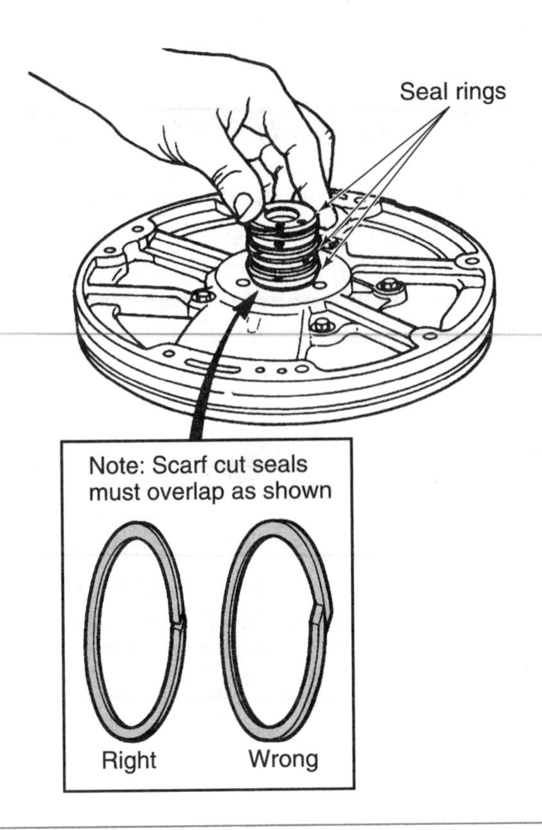

Figure 5-37 Scarf-cut seals. Notice that the ends of the seal are cut at opposing angles. (Reprinted with the permission of Ford Motor Co.)

locking-end Teflon seals are cut at an angle (Figure 5-37), and the locking hooks are somewhat staggered.

Many late-model transmissions are equipped with solid one-piece Teflon seals. Although one-piece seals require some special tools for installation, they provide for a near positive seal. These Teflon rings seal much better than other metal rings.

Shop Manual
Chapter 5, page 152

GM uses a different type of synthetic seal on some late-model transmissions. The material used in these seals is Vespel, which is a flexible but highly durable plastic-like material. Vespel seals are found on 4T60-E and 4T80-E transaxles.

Final Drives and Differentials

The last set of gears in the drive train is the final drive. In most RWD cars, the final drive is located in the rear axle housing. On most FWD cars, the final drive is located within the transaxle. Some FWD cars with longitudinally mounted engines have the differential and final drive in a separate case that bolts to the transmission.

Final drive is the final set of reduction gears the engine's power passes through on its way to the drive wheels.

In RWD systems, the housing is in three parts. The center part houses the final drive and differential gears. The outer parts support the axles by providing attachments for the axle bearings. These parts also serve as suspension components and attachment points for the steering gear and/or brakes. In FWD applications, the differential and final drive are either enclosed in the same housing as the transmission or in a separate housing bolted directly to the transmission housing.

Shop Manual
Chapter 5, page 135

RWD final drives normally use a hypoid final drive gear set that turns the power flow 90 degrees from the drive shaft to the drive axles. On FWD cars with transversely mounted engines, the power flow axis is naturally parallel to that of the drive axles; therefore, the power doesn't need to turn. Simple gear connections can be made to connect the output of the transmission to the final drive.

Differentials

A BIT OF HISTORY

Early automobiles were driven by means of belts and ropes around pulleys mounted on the driving wheels and engine shaft or transmission shaft. Since there was always some slippage of the belts, one wheel could rotate faster than the other when turning a corner. When belts proved unsatisfactory, automobile builders borrowed an idea from the bicycle and applied sprockets and chains. Since this was a positive driving arrangement, it was necessary to provide differential gearing to permit one wheel to turn faster than the other.

The term *differential* means relating to or exhibiting a difference or differences.

The differential (Figure 5-38) is a geared mechanism located between two driving axles. It rotates the driving axles at different speeds when the vehicle is turning, and at the same speed when the vehicle is traveling in a straight line. The drive axle assembly directs drive-line torque to its driving wheels. The gear ratio between the pinion and ring gear is used to increase torque and improve driveability. The gears within the differential create a state of balance between the forces or torques between the drive wheels, and allow the drive wheels to turn at different speeds as the vehicle turns a corner.

When a car turns a corner, the outside wheels must travel farther and faster than the inside wheels. If compensation is not made for this difference in speed and travel, the wheels could skid and slide, causing poor handling and excessive tire wear. Compensation for the variations in wheel speeds is made by the differential assembly.

Power is delivered to the differential assembly through the pinion gear (Figure 5-39). The pinion teeth engage the ring gear, which is mounted upright at a 90-degree angle to the pinion. Therefore, as the drive shaft turns, so do the pinion and ring gears.

The ring gear in a transaxle is sometimes referred to as the differential drive gear.

The ring gear is fastened to the differential case with several hardened bolts or rivets. Holes machined through the center of the differential housing support the differential pinion shaft. The pinion shaft is retained in the housing case by clips or a specially designed bolt. Two beveled

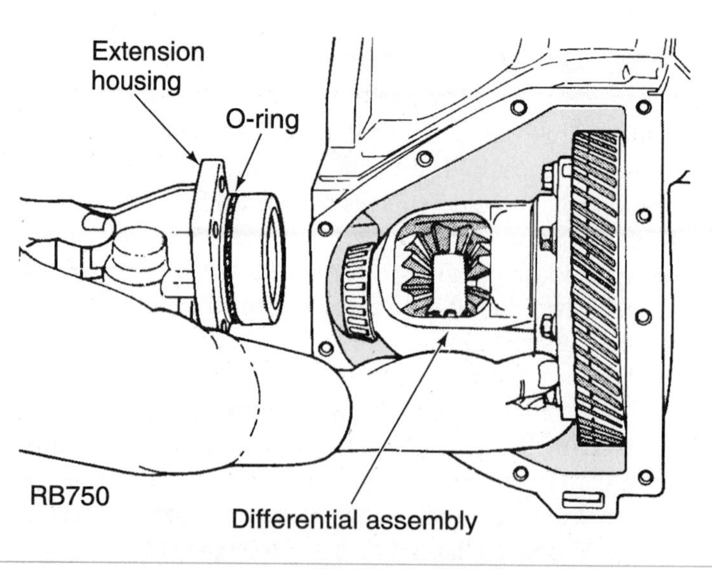

Figure 5-38 A differential assembly in a transaxle housing. (Courtesy of Chrysler Corp.)

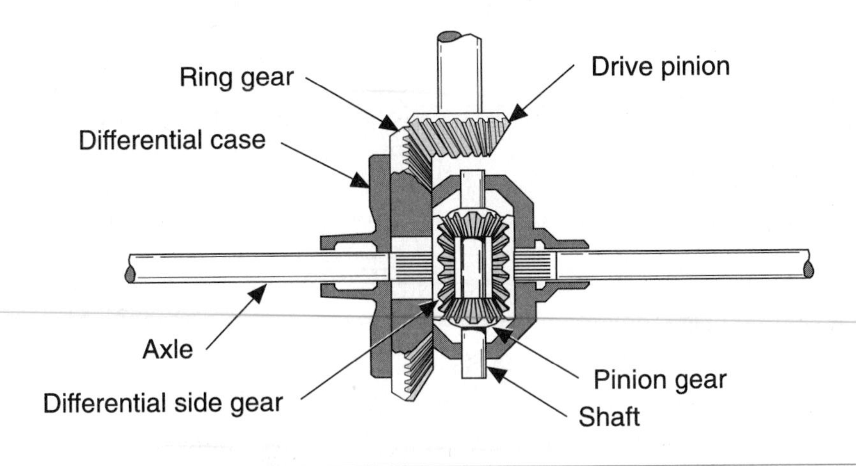

Figure 5-39 Major components of a basic differential.

differential pinion gears and thrust washers are mounted on the differential pinion shaft. In mesh with the differential pinion gears are two axle side gears splined internally to mesh with the external splines on the left and right axle shafts. Thrust washers are placed between the differential pinions, axle side gears, and differential case to prevent wear on the inner surfaces of the differential case.

The differential case is located between the side gears and is mounted on bearings so that it is able to rotate independently of the drive axles. Pinion shafts, with small pinion gears, are fitted inside the differential case. The pinion gears mesh with the side gears.

The differential of a RWD vehicle is normally housed with the drive axles in a large casting called the rear axle assembly. Power from the engine enters into the center of the rear axle assembly and is transmitted to the drive axles. The drive axles are supported by bearings and are attached to the wheels of the car. The power entering the rear axle assembly has its direction changed by the drive axle gears.

Engine torque is delivered to the drive pinion gear, which is in mesh with the ring gear and causes it to turn. Power flows from the pinion gear to the ring gear. The ring gear drives the differential case, which is bolted to the ring gear. The differential case extends from the side of the ring gear and normally houses the pinion gears and the side gears. The side gears are mounted so they can slip over splines on the ends of the axle shafts.

There is a gear reduction between the drive pinion gear and the ring gear, causing the ring gear to turn about one-third to one-fourth the speed of the drive pinion. The pinion gears are located between and meshed with the side gears, thereby forming a square of gears inside the differential case. Differentials have two or four pinion gears that are in mesh with the side gears. The differential pinion gears are free to rotate on their own centers and can travel in a circle as the differential case and pinion shaft rotate. The side gears are meshed with the pinion gears and are also free to rotate on their own centers. However, since the side gears are mounted at the centerline of the differential case, they do not travel in a circle with the differential case as do the pinion gears.

The small pinion gears are mounted on a pinion shaft that passes through the gears and the case (Figure 5-40). The pinion gears are in mesh with the axle side gears, which are splined to the axle shafts. In operation, the rotating differential case causes the pinion shaft and pinion gears to rotate end over end with the case. Since the pinion gears are in mesh with the side gears, the side gears and axle shafts are also forced to rotate.

When a car is moving straight ahead, both drive wheels are able to rotate at the same speed. Engine power comes in on the pinion gear and rotates the ring gear. The differential case is rotated with the ring gear. The pinion shaft and pinion gears are carried around by the ring gear and all of the gears rotate as a single unit. Each side gear rotates at the same speed and in the same plane as

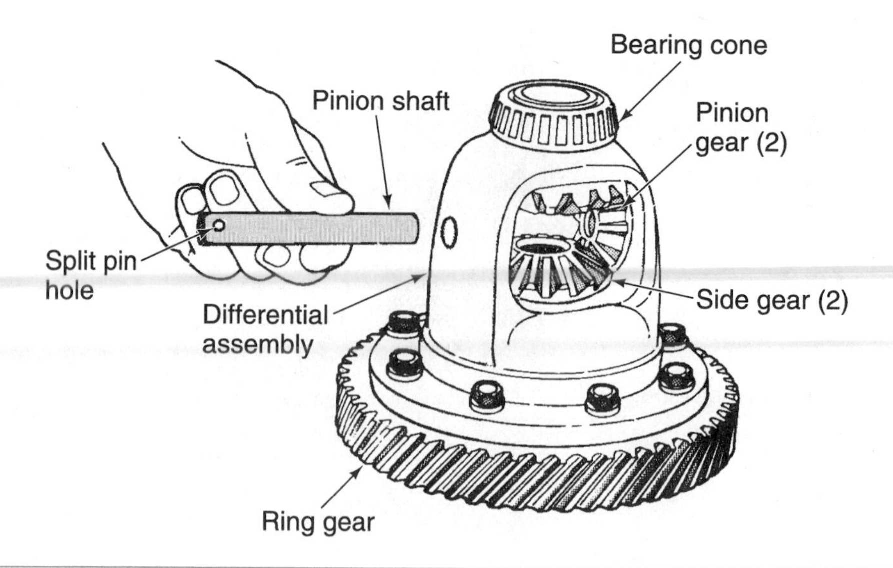

Figure 5-40 Small pinion gears are held in the differential case by pinion shafts. (Courtesy of Chrysler Corp.)

does the case and they transfer their motion to the axles. Each axle rotates at the same speed and the vehicle moves straight.

As the vehicle goes around a corner, the inside wheel travels a shorter distance than the outside wheel. The inside wheel must therefore rotate slower than the outside wheel. In this situation, the differential pinion gears will walk forward on the slower turning or inside side gear. As the pinion gears walk around the slower side gear, they drive the other side gear at a greater speed. An equal percentage of speed is removed from one axle and given to the other; however, an equal amount of torque is applied to each wheel.

Only the outside wheel is free to rotate when a car is making a very sharp turn; therefore, only one side gear rotates freely. Since one side gear is nearly stationary, the pinion gears now turn on their own centers as they walk around the stationary side gear. As they walk around that side gear, they drive the other side gear at twice their own speed. The moving wheel is now turning at nearly twice the speed of the differential case, but the torque applied to it is about half of the torque applied to the differential case.

When one of the driving wheels has little or no traction, the torque required to turn the wheel without traction is very low. The wheel with good traction is, in effect, holding the axle gear on that side stationary. This causes the pinions to walk around the stationary side gear and drive the other wheel at twice the normal speed but without any vehicle movement. With one wheel stationary, the other wheel turns at twice the speed shown on the speedometer.

Final Drive Assemblies

A transaxle's final drive gears provide a way to transmit the transmission's output to the differential section of the transaxle. The pinion and ring gears and the differential assembly are normally located within the transaxle housing. There are four common configurations used as the final drives on FWD vehicles: helical gear, planetary gear, hypoid gear, and chain drive. The helical, planetary, and chain final drive arrangements are found with transversely mounted engines. Hypoid final drive gear assemblies are normally found in vehicles with a longitudinally placed engine. The hypoid assembly is basically the same unit as would be used on RWD vehicles and it is mounted directly to the transmission.

Common transaxles that use a planetary final drive are the THM 125/125C and 440-T4, and the Ford AXOD.

The differential section of the transaxle has the same components as the differential gears in a RWD axle and basically operates in the same way. The drive pinion gear is connected to the

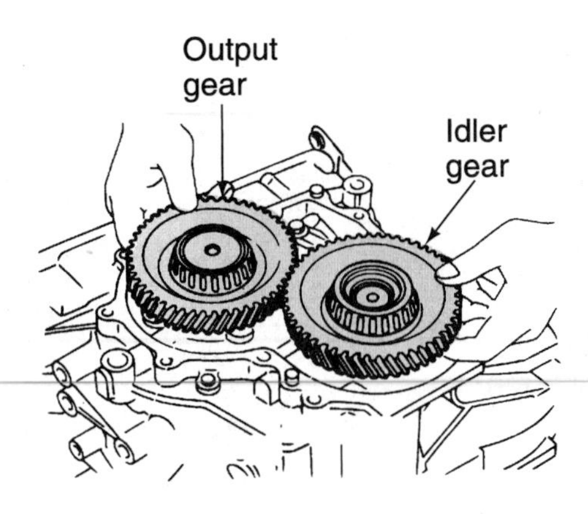

Figure 5-41 Some transaxles use an input gear and an idler gear to transfer torque from the transmission to the output shaft and final drive. (Courtesy of Nissan Motor Co., Ltd.)

transmission's output shaft and the ring gear is attached to the differential case. The pinion and ring gear set provides for a multiplication of torque.

Some transaxles route power from the transmission through two helical-cut gears to a transfer shaft. A helical-cut pinion gear attached to the opposite end of the transfer shaft drives the differential ring gear and carrier. The differential assembly then drives the axles and wheels.

The final drive assembly in a Ford ATX consists of an input gear, an idler gear, and an output gear. The use of an idler gear allows the input gear and the output gear to rotate in the same direction (Figure 5-41). The input gear is driven by the output of the transmission and drives the idler gear, which rotates on a shaft. The idler gear then drives the output gear, which is riveted to the differential case.

The differential case rotates with the final drive output gear and is supported by tapered roller bearings on each side. The differential case contains four bevel gears: two pinion gears and two side gears. The pinion gears are installed on a pinion shaft that is retained by a pin in the differential case. The pinion gears transmit power from the case to the side gears. Each side gear is splined to one of the halfshafts that transmits power to the front wheels.

Rather than use helical-cut or spur gears in the final drive assembly, some transaxles use a simple planetary gear set for their final drive (Figure 5-42). The sun gear of this planetary unit is driven by the final drive sun gear shaft, which is splined to the front carrier and rear ring gear of the transmission's gear set. The final drive sun gear meshes with the final drive planet pinion gears, which rotate on their shafts in the planet carrier. The pinion gears mesh with the ring gear, which is splined to the transaxle case. The planet carrier is part of the differential case, which contains typical differential gearing, two pinion gears and two side gears.

The ring gear of a planetary final drive assembly has lugs around its outside diameter that fit into grooves machined inside the transaxle housing. These lugs and grooves hold the ring gear stationary. The transmission's output is connected to the planetary gear set's sun gear. In operation, the transmission's output drives the sun gear, which, in turn, drives the planet pinion gears. The pinion gears walk around the inside of the stationary ring gear. The rotating planet pinion gears drive the planet carrier and differential case. This combination provides maximum torque multiplication from a simple planetary gear set.

Chain-drive final drive assemblies use a multiple-link chain to connect a drive sprocket, which is connected to the transmission's output shaft, to a driven sprocket, which is connected to the differential case. This design allows for remote positioning of the differential within the transaxle housing. Final drive gear ratios are determined by the size of the driven sprocket compared to the drive sprocket.

Shop Manual
Chapter 5, page 138

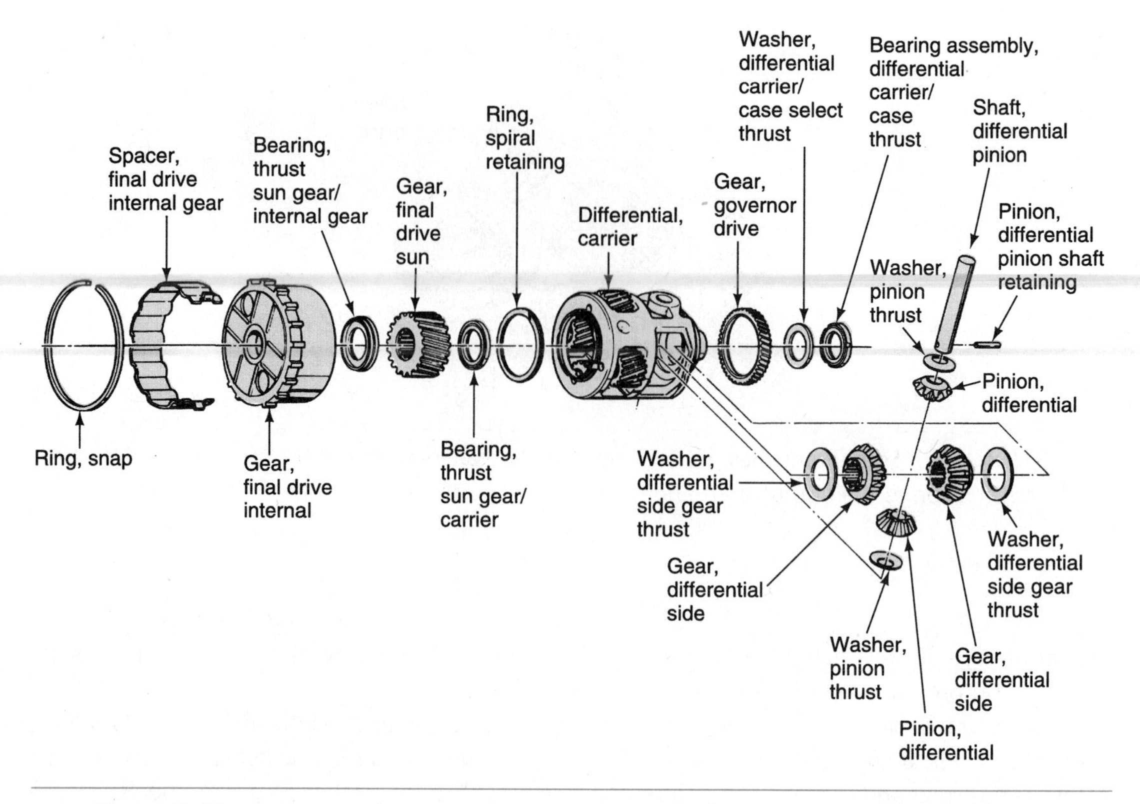

Figure 5-42 Some transaxles use a planetary gear set as the final drive and differential unit. (Courtesy of the Hydra-Matic Division of General Motors Corp.)

Four-Wheel-Drive Design Variations

The common acronym for four-wheel drive is 4WD.

Four-wheel-drive (4WD) vehicles normally use some sort of **transfer case** to distribute the transmission's output to two- or four-drive wheels. Some transfer cases are mounted directly onto the transmission housing, others are separate and mounted to the frame between the transmission and the axles.

A BIT OF HISTORY

The first known gasoline-powered four-wheel-drive automobile was the Spyker, which was built in the Netherlands in 1902.

With four-wheel drive, engine power can flow to all four wheels. This action can greatly increase a vehicle's traction when traveling in adverse conditions and can also improve handling since side forces generated by the turning of a vehicle or by wind gusts will have less of an effect on a vehicle that has power applied to the road on four wheels.

Transfer Cases

Four-wheel drive is most useful when a vehicle is traveling off the road or in deep mud or snow. However, some high-performance cars are equipped with four-wheel drive to improve the handling

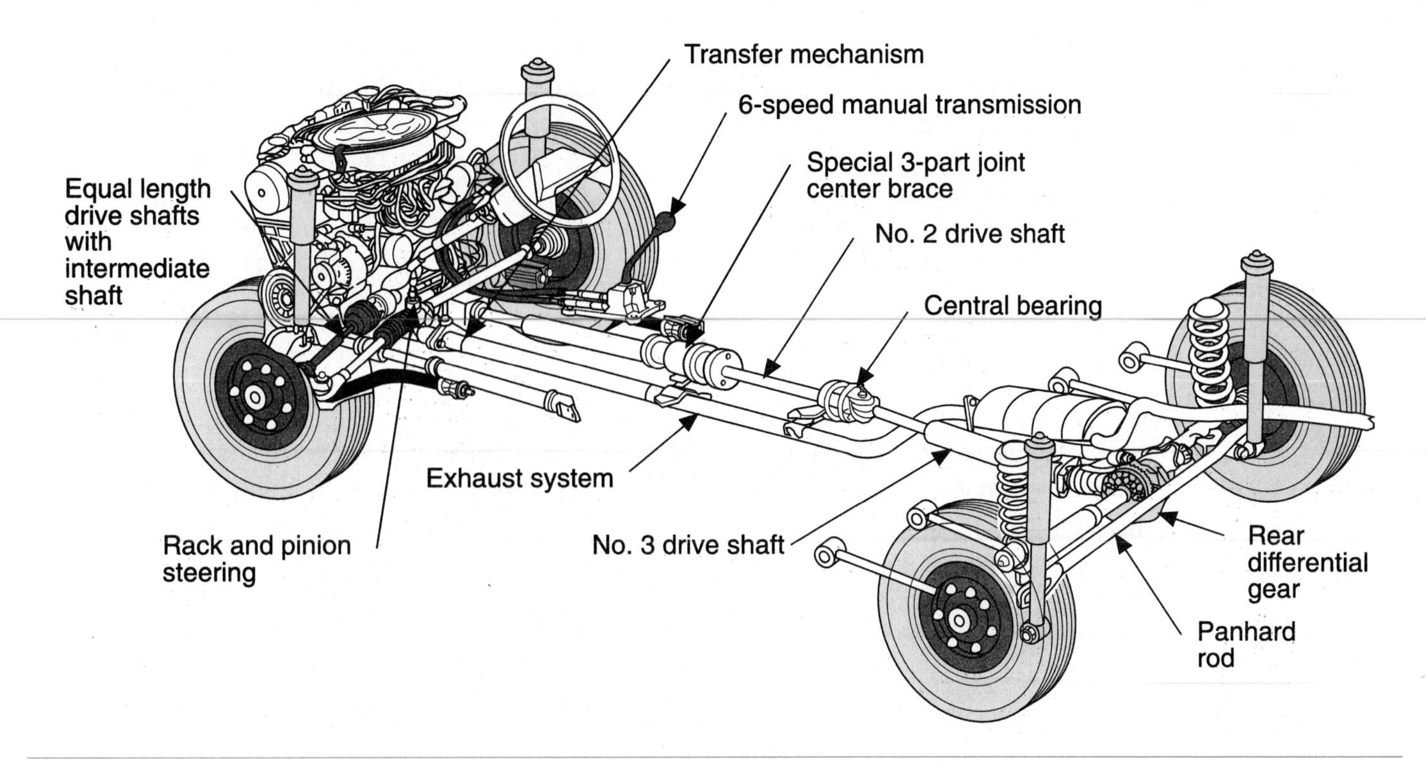

Figure 5-43 A front-wheel-drive car modified to an all-wheel-drive car. (Courtesy of Honda Motor Company, Ltd.)

characteristics of the car. Nearly all of these cars are front-wheel-drive models converted to four-wheel drive. Normally, FWD cars are modified by adding a transfer case, a rear drive shaft, and a rear axle with a differential (Figure 5-43). Although this is the typical modification, some cars are equipped with a center differential in place of the transfer case. This differential unit allows the rear and front wheels to turn at different speeds.

The typical 4WD system consists of a front-mounted, longitudinally positioned engine, either an automatic or manual transmission, front and rear drive shafts, front and rear drive axle assemblies, and a transfer case.

The transfer case is usually mounted to the side or rear of the transmission. When a drive shaft is not used to connect the transmission to the transfer case, a chain or gear drive within the transfer case receives the transmission's output and transfers it to the drive shafts leading to the front and rear drive axles.

The drive shafts from the transfer case shafts connect to differentials at the front and rear drive axles. As on 2WD vehicles, these differentials are used to compensate for road and operating conditions by altering the speed of the wheels connected to the axles.

An electric switch or shift lever, located in the passenger compartment, controls the transfer case so that power is directed to the axles selected by the driver. Power can typically be directed to all four wheels, two wheels, or none of the wheels. On many vehicles, the driver can also select a low-speed range for extra torque while traveling in very adverse conditions.

While most 4WD trucks and utility vehicles are design variations of basic RWD vehicles, most passenger cars equipped with 4WD are based on FWD designs. These modified FWD systems consist of a transaxle and differential to drive the front wheels, plus some type of mechanism for connecting the transaxle to a rear driveline. In many cases, this mechanism is a simple clutch or differential. Some vehicles are fitted with a compact transfer case bolted to the front-drive transaxle. A drive shaft assembly carries the power to the rear differential. The driver can switch from 2WD to 4WD by pressing a dashboard switch. This switch activates a solenoid vacuum valve, which applies vacuum to a diaphragm unit in the transfer case. The diaphragm unit linkage locks the output of the transaxle to the input shaft of the transfer case.

The center differential is commonly referred to as the **interaxle differential.**

Full-time 4WD systems use a center differential that accommodates speed differences between the two axles, which is necessary for highway operation.

Integrated full-time 4WD systems use computer controls to enhance full-time operation, adjusting the torque split depending on which wheels have traction.

On-demand 4WD systems power a second axle only after the first begins to slip.

Part-time 4WD systems can be shifted in and out of 4WD.

The driver cannot select between 2WD or 4WD in an all-wheel-drive system. These systems always drive four wheels. AWD vehicles are not designed for off-road operation; rather, they are designed to increase vehicle performance in poor traction situations, such as icy or snowy roads. AWD allows for maximum control by transferring a large portion of the engine's power to the axle with the most traction. Most AWD designs use a center differential to split the power between the front and rear axles. On some designs, the center differential locks automatically or the driver can manually lock it with a switch. AWD systems may also use a viscous coupling to allow variations in axle speeds.

Viscous Clutch

The handling characteristics of four-wheel-drive vehicles are excellent in all driving conditions except turning on a dry pavement. While turning a corner, the front wheels cannot rotate at the same speed as the rear wheels. This causes one set of wheels to scuff along the pavement. To overcome this tendency, many full-time 4WD and AWD vehicles use a viscous coupling between the front and rear axles (Figure 5-44).

A viscous coupling is basically a drum filled with a thick fluid that houses several closely fitted, thin steel discs. One set of the discs is connected to the front wheels and the other to the rear.

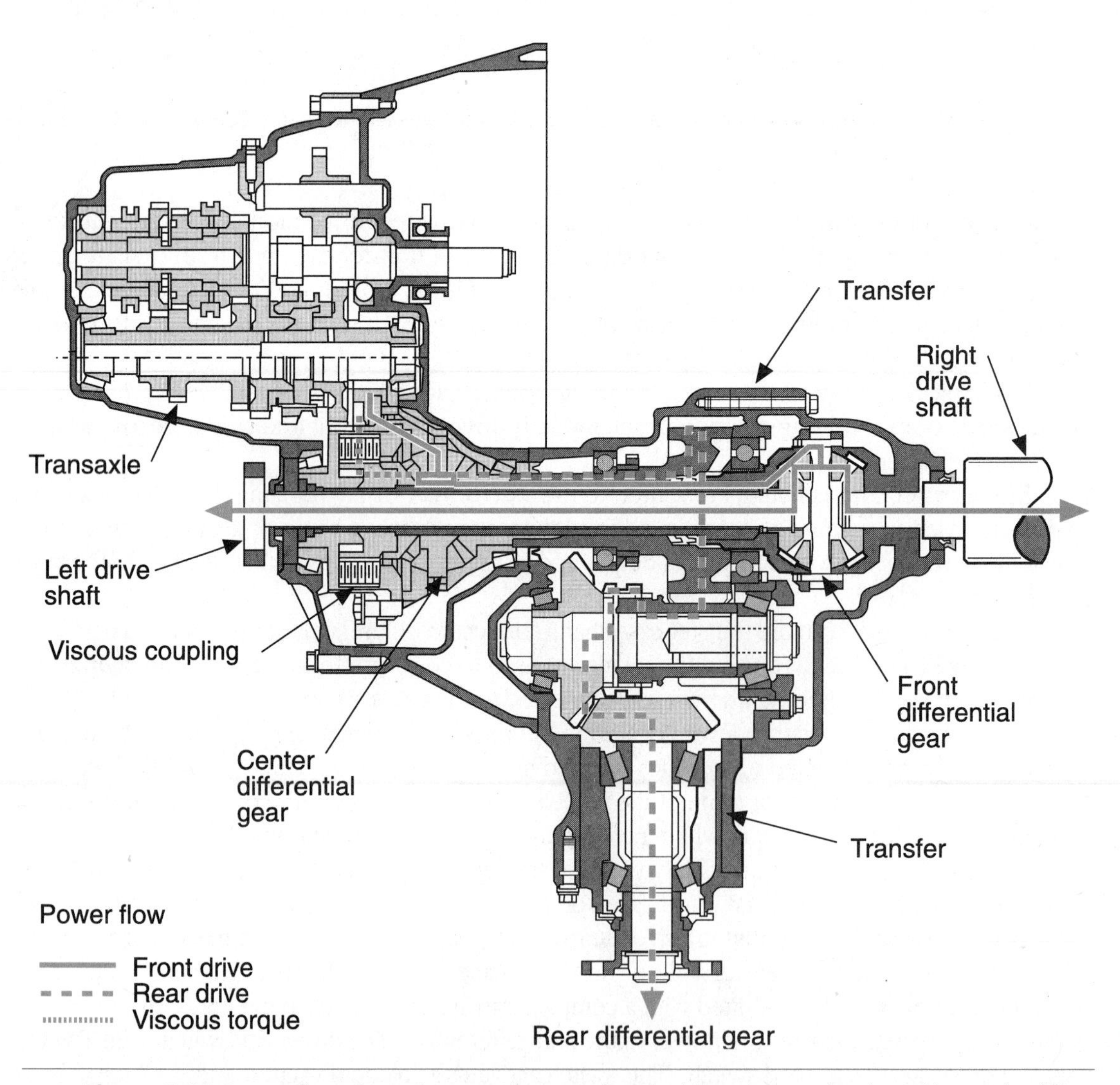

Figure 5-44 Power flow through a transfer case fitted with a viscous clutch and interaxle differential. (Courtesy of Nissan Motor Co., Ltd.)

As the demand for more torque to one axle is evident, the fluid heats up and immediately changes viscosity. This change in viscosity reacts on the discs, and torque is split according to the actual needs of the axles.

Viscous couplings may also be used in the front or rear axle differentials. They provide a holding force between two shafts as they rotate under force. Like a limited slip differential, a viscous coupling transfers torque to the drive wheel with more traction when the other wheel has less traction. A **viscous clutch** often takes the place of the interaxle differential. Viscous clutches operate automatically as soon as it becomes necessary to improve wheel traction.

High-performance AWD vehicles use a viscous coupling in the center and rear differentials to improve cornering and maneuvering at higher speeds. The combination of a center differential viscous coupling with open front and rear differentials improves the vehicle's distribution of braking forces and is compatible with antilock braking systems.

Summary

- ❑ The sun gear in a planetary gear set is the central gear around which other gears revolve.
- ❑ The planet gears are in constant mesh with the sun and ring gears.
- ❑ The planet carrier is the bracket in a planetary gear set on which the planet pinion gears are mounted on pins and are free to rotate.
- ❑ The ring gear is the outer gear of the gear set. It has internal teeth, which are in constant mesh with the planet gears.
- ❑ Gear reduction is accomplished by holding the sun gear and applying power to the ring gear.
- ❑ Gear reduction can also occur if the ring gear is held and the sun gear is the input gear.
- ❑ When the planet carrier is the driving member of the gear set and the sun gear is held, overdrive occurs.
- ❑ If any two of the planetary gear set members receive power in the same direction at the same speed, the third member is forced to move with the other two and direct drive results.
- ❑ When two or more planetary gear sets are used in a transmission, they are collectively called a compound planetary gear system.
- ❑ When a large gear drives a smaller gear, output torque is decreased and output speed is increased.
- ❑ When an external gear drives another external gear, the output gear will rotate in the opposite direction to the input gear.
- ❑ When an external gear is in mesh with an internal gear, the two gears will rotate in the same direction.
- ❑ When there is input into a planetary gear set but a member is not held, NEUTRAL results.
- ❑ When the ring gear and carrier pinions are free to rotate at the same time, the pinions will always follow the same direction as the ring gear.
- ❑ The sun gear always rotates opposite of the rotation of the pinion gears.
- ❑ When the planet carrier is the output, it always follows the direction of the input member and results in a gear reduction.
- ❑ When there is a reaction member and the planet carrier is the input, overdrive results.
- ❑ Gear reduction is accomplished in a planetary gear set by holding the sun gear. If power is applied to the ring gear, the planet gears will rotate on their shafts in the planet carrier. Because the sun gear is being held, the rotating planet gears will walk around the sun gear and carry the planet carrier with them. This causes a gear reduction.

Terms to Know

Annulus gear
Axial
Bearing
Bushings
Butt-end seals
Cage
Composition gasket
Cone
Cup
Dynamic seal
Gasket
Hard gasket
Hook-end seals
Interaxle differential
Internal gear
Lathe-cut seal
Lip seals
Open-end seals
O-rings
Pinion gear
Planet carrier
Planet pinion gears
Preload
Race
Radial
RTV
Scarf-cut rings
Soft gasket

Terms to Know (continued)

Square-cut seal

Static seal

Steel rings

Thrust washer

Toroidal

Torrington bearing

Transfer case

Viscous clutch

- ❑ Gear reduction can occur if the ring gear is held and the sun gear is the input gear. The planet carrier rotates around the sun gear, but more slowly than it did when the ring gear drove the planet carrier.
- ❑ When the planetary carrier is the input member, the gear set produces an overdrive condition.
- ❑ When the planetary carrier is held, the output of the gear set is in the opposite direction to the input and reverse gear is produced.
- ❑ There are two common designs of compound gear sets, the Simpson gear set, in which two planetary gear sets share a common sun gear, and the Ravigneaux gear set, which has two sun gears, two sets of planet gears, and a common ring gear.
- ❑ A Simpson gear train is the most commonly used compound planetary gear set and is used to provide three forward gears.
- ❑ The Honda CA, F4, and G4 transaxles do not use a planetary gear set. They use constant-mesh helical and square-cut gears in a manner similar to that of a manual transmission.
- ❑ As the vehicle goes around a corner, the inside wheel travels a shorter distance than the outside wheel. The inside wheel must therefore rotate slower than the outside wheel. An equal percentage of speed is removed from one axle and given to the other; however, an equal amount of torque is applied to each wheel.
- ❑ There are four common configurations used as the final drives on FWD vehicles: helical gear, planetary gear, hypoid gear, and chain drive.
- ❑ The transfer case is usually mounted to the side or rear of the transmission.
- ❑ A viscous coupling is basically a drum filled with a thick fluid that houses several closely fitted, thin steel discs. One set of the discs is connected to the front wheels and the other to the rear.
- ❑ Bearings that take radial loads are called bushings and those that take axial loads are called thrust washers.
- ❑ The gaskets and seals of an automatic transmission help contain the fluid within the transmission and prevent the fluid from leaking out of the various hydraulic circuits. Different types of seals are used in automatic transmissions. They can be made of rubber, metal, or Teflon.
- ❑ Transmission gaskets are made of rubber, cork, paper, synthetic materials, or plastic.
- ❑ Three major types of rubber seals are used in automatic transmissions: the O-ring, the lip seal, and the square-cut seal.
- ❑ Three types of metal seals are used in automatic transmissions: butt-end seals, open-end seals, and hook-end seals.
- ❑ A bearing is a device placed between two bearing surfaces to reduce friction and wear.
- ❑ Bearings that take radial loads are called bushings and those that take axial loads are called thrust washers.

Review Questions

Short Answer Essays

1. How does direct drive result in a planetary gear set?
2. What is expressed by a gear ratio?
3. What are the two common designs of compound planetary gear sets and how do they differ?
4. How is reverse gear accomplished in a planetary gear set?

5. How can a gear reduction be obtained from a planetary gear set?
6. How does a differential help a vehicle go around a turn?
7. What is the purpose of a transfer case?
8. Explain how Honda's helical-gear, constant-mesh automatic transmission works.
9. What happens in a planetary gear set when no member is held? Explain your answer.
10. When a transmission is described as having two planetary gear sets in tandem, what does this mean?

Fill in the Blanks

1. The four common configurations used as the final drives on FWD vehicles are the ________ gear, ________ gear, ________ gear, and ________ ________.
2. Gear reduction can occur if the ________ gear is held and the ________ gear is the input gear.
3. If any two of the planetary gear set members receive power in the same ________ at the same ________, the third member is forced to move with the other two and ________ drive results.
4. The major components of a planetary gear set are the: ________ ________, ________ ________, and ________ ________.
5. Constant-mesh, helical-gear-based transmissions are also referred to as ________ ________ transmissions.
6. When a small gear drives a larger gear, torque is ________ while speed is ________.
7. Three major types of rubber seals are used in automatic transmissions: the ________, the ________ ________, and the ________ seal.
8. The three types of metal seals used in automatic transmissions are the ________, ________, and ________ seals.
9. Transmission gaskets are made of ________, ________, ________, ________ ________, or ________.
10. A Ravigneaux gear set has two ________ gears, two sets of ________ gears, and a common ________ gear.

ASE Style Review Questions

1. *Technician A* says when the planet carrier is the input member, the gear set produces a gear reduction.
   Technician B says when the planet carrier is held, the output of the gear set will be in the opposite direction to the input and reverse gear will be produced.
   Who is correct?
   A. A only **C.** Both A and B
   B. B only **D.** Neither A nor B

2. *Technician A* says bearings that take radial loads are called Torrington washers.
   Technician B says bearings that take axial loads are called thrust washers.
   Who is correct?
   A. A only **C.** Both A and B
   B. B only **D.** Neither A nor B

3. While discussing reverse gear in a planetary gear set:
   Technician A says that when the ring gear is held, the output will rotate in the opposite direction to the input.
   Technician B says that when the sun gear is held, the output will rotate in the opposite direction to the input.
   Who is correct?
   A. A only **C.** Both A and B
   B. B only **D.** Neither A nor B

4. While discussing gear ratios available from a planetary gear set:
   Technician A says a single planetary gear set can provide only one reverse gear.
   Technician B says a single planetary gear set can provide only one gear reduction forward gear.
   Who is correct?
   A. A only **C.** Both A and B
   B. B only **D.** Neither A nor B

5. *Technician A* says that when a small gear drives a larger gear, output torque is increased and output speed is decreased.
   Technician B says that when a large gear drives a smaller gear, output torque is decreased and output speed is increased.
   Who is correct?
   A. A only **C.** Both A and B
   B. B only **D.** Neither A nor B

6. *Technician A* says when the planet carrier is the output, it always follows the direction of the input member.
   Technician B says when the planet carrier is the input, the output gear member always moves in the opposite direction to the carrier.
   Who is correct?
   A. A only **C.** Both A and B
   B. B only **D.** Neither A nor B

7. *Technician A* says a viscous coupling for 4WD uses a gear set similar to a differential to provide different amounts of torque to the drive axles.
   Technician B says a viscous coupling is basically a drum filled with a thick fluid that houses several closely fitted, thin steel discs that are connected to the front and rear axles.
   Who is correct?
   A. A only **C.** Both A and B
   B. B only **D.** Neither A nor B

8. *Technician A* says when an external gear drives another external gear, the output gear will rotate in the opposite direction to the input gear.
   Technician B says when an external gear is in mesh with an internal gear, the two gears will rotate in the same direction.
   Who is correct?
   A. A only **C.** Both A and B
   B. B only **D.** Neither A nor B

9. *Technician A* says when the ring gear and carrier pinions are free to rotate at the same time, the pinions will always follow the same direction as the ring gear.
   Technician B says the sun gear always rotates opposite of the rotation of the pinion gears.
   Who is correct?
   A. A only **C.** Both A and B
   B. B only **D.** Neither A nor B

10. *Technician A* says a Simpson gear set is two planetary gear sets that share a common sun gear.
    Technician B says a Ravigneaux gear set has two sun gears, two sets of planet gears, and a common ring gear.
    Who is correct?
    A. A only **C.** Both A and B
    B. B only **D.** Neither A nor B

Hydraulic Systems and Apply Devices

Upon completion and review of this chapter, you should be able to:

- ❑ Explain the basic operation of hydraulic machines.
- ❑ State Pascal's Law and explain how it applies to the operation of an automatic transmission.
- ❑ Identify the major components in a transmission's hydraulic circuit and describe how they provide fluid flow and pressurization.
- ❑ Describe the various configurations of a spool valve and explain how it can be used to open and close various hydraulic circuits.
- ❑ Describe the design and operation of the hydraulic controls and valves used in modern transmissions and transaxles.
- ❑ Explain the role and operation of the following components of the transmission control system: pressure regulation valve, throttle valve, governor assembly, manual valve, shift valves, and kickdown assembly.
- ❑ Describe the operation of the main pressure regulator valve.
- ❑ Describe the different designs of a governor and explain the operation of each type.
- ❑ List and describe the various load-sensing devices used in automatic transmissions.
- ❑ Identify the various pressures in the transmission and explain their purposes in the total functioning of the transmission.
- ❑ Describe the purpose of a transmission's valve body.
- ❑ Identify the components in a hydraulic multiple-disc clutch and describe their functions.
- ❑ Identify the basic components in a hydraulic servo and describe their functions.
- ❑ Describe the purpose and operation of the common reaction members.
- ❑ Explain the purpose and operation of an accumulator and modulator valve.
- ❑ Describe the basic construction of a hydraulic servo and explain its basic operation.

Introduction

An automatic transmission receives engine power through a torque converter, which is indirectly attached to the engine's crankshaft. Hydraulic pressure in the converter allows power to flow from the engine to the transmission's input shaft. The input shaft drives a planetary gear set, which provides the different forward gears, a neutral position, and one reverse gear. Power flow through the gears is controlled by multiple-disc clutches, one-way clutches, or friction bands. These hold or drive a member of the gear set when hydraulic pressure (Figure 6-1) is applied to them. Hydraulic pressure is routed to the correct apply device by the transmission's valve body, which controls the pressure and direction of the hydraulic fluid.

An understanding of the transmission's hydraulic control circuit is essential for proper diagnostics.

The various gear ratios are selected by the transmission according to engine and vehicle speeds, engine load, and other operating conditions. Both upshifts and downshifts occur automatically. The transmission can also be manually shifted into a lower forward gear, REVERSE, NEUTRAL, or PARK.

To better understand how an automatic transmission works, a good understanding of how basic hydraulic circuits work is needed.

The basic operation of an automatic transmission involves the controlled use of hydraulic pressures. The hydraulic fluid used, ATF, is a special oil designed for proper transmission operation. The transmission's oil pump is the source of all pressures within the hydraulic system. It also provides a constant supply of oil under pressure to operate, lubricate, and cool the transmission. **Pressure-regulating valves** change the pressure of the oil to control the shift points of the transmission. Flow-directing valves direct the pressurized oil to the appropriate reaction members, which cause a change in gear ratios.

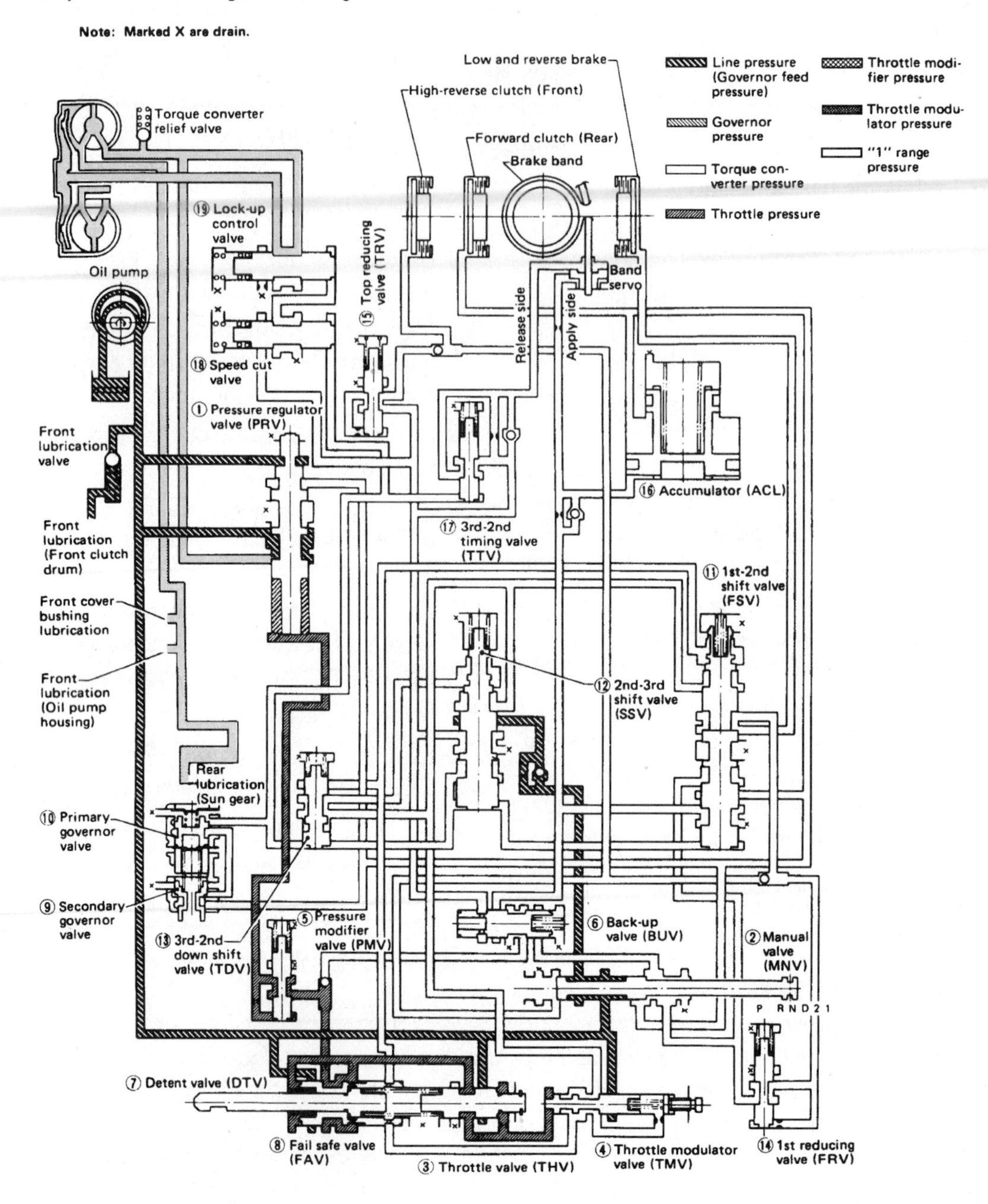

Figure 6-1 Typical hydraulic system for a late-model transmission. (Courtesy of Nissan Motor Co., Ltd.)

The operation of an automatic transmission is based on fluid power, an important aspect of hydraulics that represents a liquid's ability to transmit energy.

The valve body assembly is machined from aluminum or iron and has many precisely machined bores and fluid passages fitted with many different valves. These valves are used to start, stop, or control the flow of fluid through the transmission.

Laws of Hydraulics

An automatic transmission has complex hydraulic circuits. All hydraulic systems have a liquid, a pump, lines, control valves, and an output device. The liquid must be continuously available from an oil pan or sump. An oil pump moves and pressurizes the liquid through the system. The lines carry and direct the liquid. The lines may be pipes, hoses, or a network of internal bores or passages in a housing. Control valves are used to regulate hydraulic pressure and direct and control the flow of the liquid. The output device is the unit that uses the pressurized liquid to do work.

Hydraulics is the study of liquids in motion. Everything in our universe exists as solid, liquid, or gas. Liquids and gases are considered fluids because they do not have a definite shape. All fluids conform to the shape of their container. A major difference between a gas and a liquid is that a gas will always fill a sealed container whereas a liquid may not. A gas will readily expand or compress according to the pressure exerted on it. Liquids will respond predictably to pressures exerted on them. This fact allows hydraulics to do work.

Pascal's Law

Over 300 years ago, French scientist Blaise Pascal determined that if you had a liquid-filled container with only one opening and applied force to the liquid through that opening, the force would be evenly distributed throughout the liquid. This explains how pressurized liquid is used to operate and control an automatic transmission. An oil pump pressurizes the transmission fluid and the fluid is delivered to the various apply devices. When the pressure increases enough to activate the apply device, it holds a gear set member until the pressure decreases. The valve body is responsible for the distribution of the pressurized fluid to the appropriate reaction member according to the operating conditions of the vehicle.

Pascal constructed the first known hydraulic device, which consisted of two sealed containers connected by a tube. The pistons inside the cylinders seal against the walls of each cylinder and prevent the liquid from leaking out of the cylinder and air from entering the cylinder. When the piston in the first cylinder has a force applied to it, the pressure moves everywhere within the system. The force is transmitted through the connecting tube to the second cylinder. The pressurized fluid in the second cylinder exerts force on the bottom of the second piston, moving it upward and lifting the load on the top of it. By using this device, Pascal found he could increase the force available to do work (Figure 6-2), just as could be done with levers or gears.

Pascal determined that force applied to liquid creates pressure (Figure 6-3), or the transmission of force through the liquid. These experiments revealed two important aspects of a liquid when it is confined and put under pressure. The pressure applied to it is transmitted equally in all directions and this pressure acts with equal force at every point in the container.

Air or gas of any kind in a hydraulic system renders it ineffective because the air can be compressed.

If we put a 1000-pound weight on top of a 10-square inch table, there are 100 pounds of pressure per square inch, abbreviated as 100 psi.

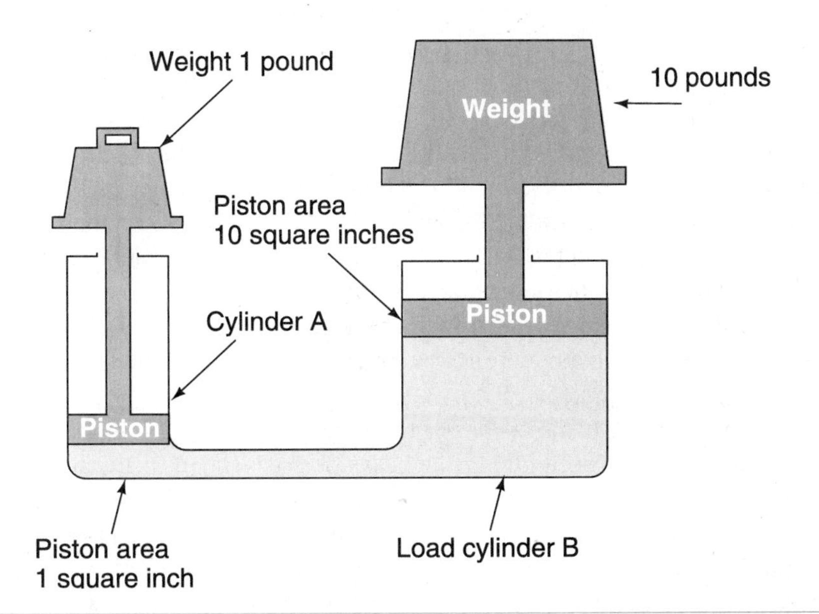

Figure 6-2 Hydraulics can be used to multiply the amount of force available to do work. (Courtesy of Chrysler Corp.)

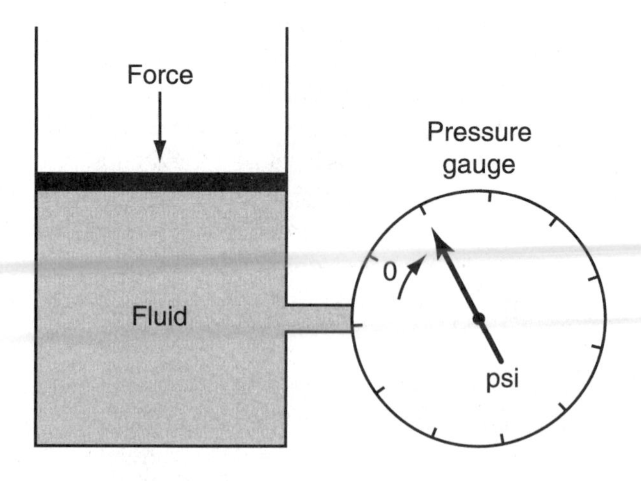

Figure 6-3 A force on a contained fluid causes it to pressurize.

Fluid Characteristics

If a liquid is confined and a force applied, pressure is produced. In order to pressurize a liquid, the liquid must be in a sealed container. Any leak in the container will decrease the pressure.

Fluids are not totally incompressible. If 32 tons of force are applied to one cubic inch of water, the water will be compressed by 10%.

The basic principles of hydraulics are based on certain characteristics of liquids. Liquids have no shape of their own, they acquire the shape of the container they are put in. They also always seek a common level. Therefore, oil in a hydraulic system will flow in any direction and through a passage, regardless of size or shape. Liquids are basically incompressible, which gives them the ability to transmit force. The pressure applied to a liquid in a sealed container is transmitted equally in all directions and to all areas of the system and acts with equal force on all areas. As a result, liquids can provide great increases in the force available to do work. A liquid under pressure may also change from a liquid to a gas in response to temperature changes.

Mechanical Advantage With Hydraulics

Work is equal to force multiplied by distance.

In the metric measuring system, amounts of force are given in units called newtons (N). A unit of work is expressed as a newton-meter or joule.

Hydraulics are used to do work in the same way a lever or gear does. All of these systems transmit energy. Since energy cannot be created or destroyed, these systems only redirect energy to perform work and do not create more energy. **Work** is actually the amount of force applied and the distance over which it is applied.

When a pressure is applied to a confined liquid, the pressure of the liquid is the same everywhere within the hydraulic system. If the hydraulic pump provides 100 psi, there will be 100 pounds of pressure on every square inch of the system (Figure 6-4). If the system includes a piston with an area of 30 square inches, each square inch receives 100 pounds of pressure. This means there will be 3,000 pounds of force applied to that piston (Figure 6-5). The use of the larger piston gives the system a **mechanical advantage** or increase in the force available to do work. The multiplication of force through a hydraulic system is directly proportional to the difference in the piston sizes throughout the system.

By changing the size of the pistons in a hydraulic system, force is multiplied, and as a result, low amounts of input force are needed to move heavy objects. The mechanical advantage of a hydraulic system can be further increased by the use of levers to increase the force applied to a piston.

Although the force available to do work is increased by using a larger piston in one cylinder, the total movement of the larger piston is less than that of the smaller one. A hydraulic system with two cylinders, one with a one-square-inch piston and the other with a two-square-inch piston, will double the force at the second piston. However, the total movement of the larger piston will be half the distance of the smaller one (Figure 6-6).

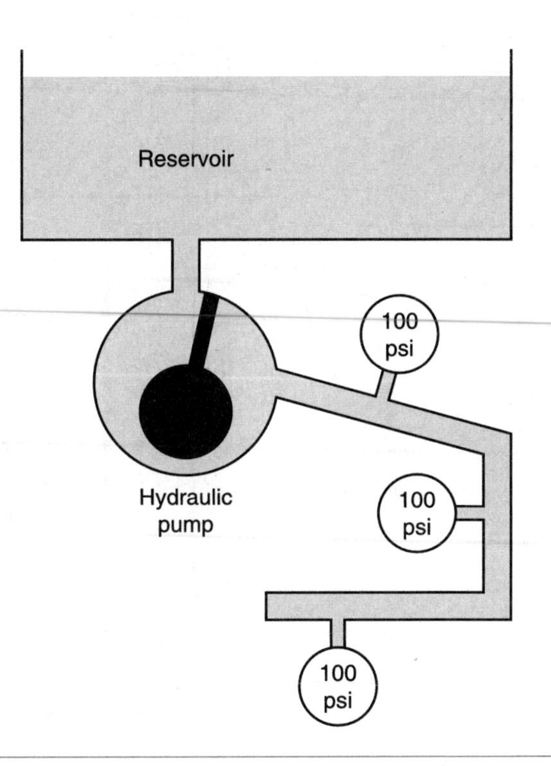

Figure 6-4 While contained, the pressure of a fluid is the same throughout the container.

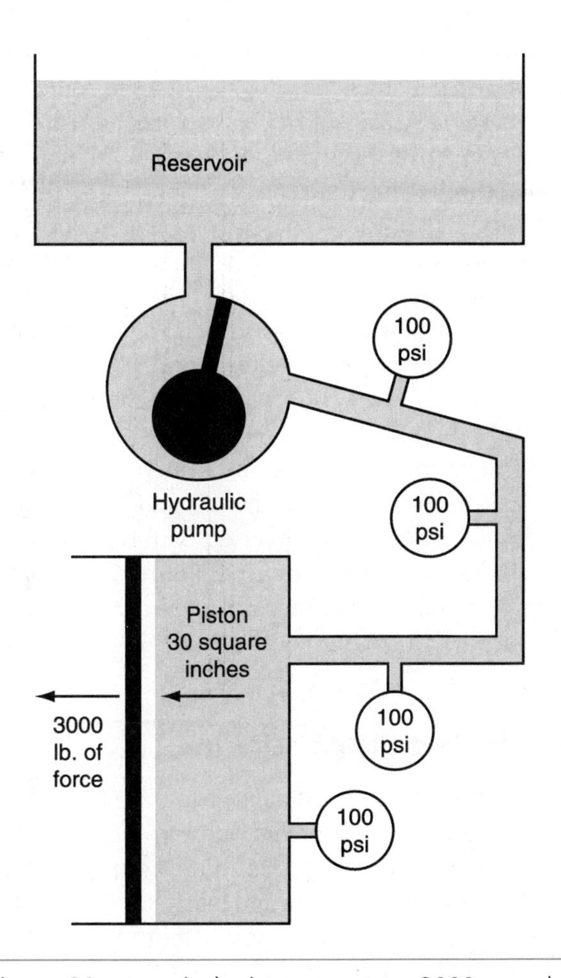

Figure 6-5 100 psi on a 30-square-inch piston generates 3000 pounds of force.

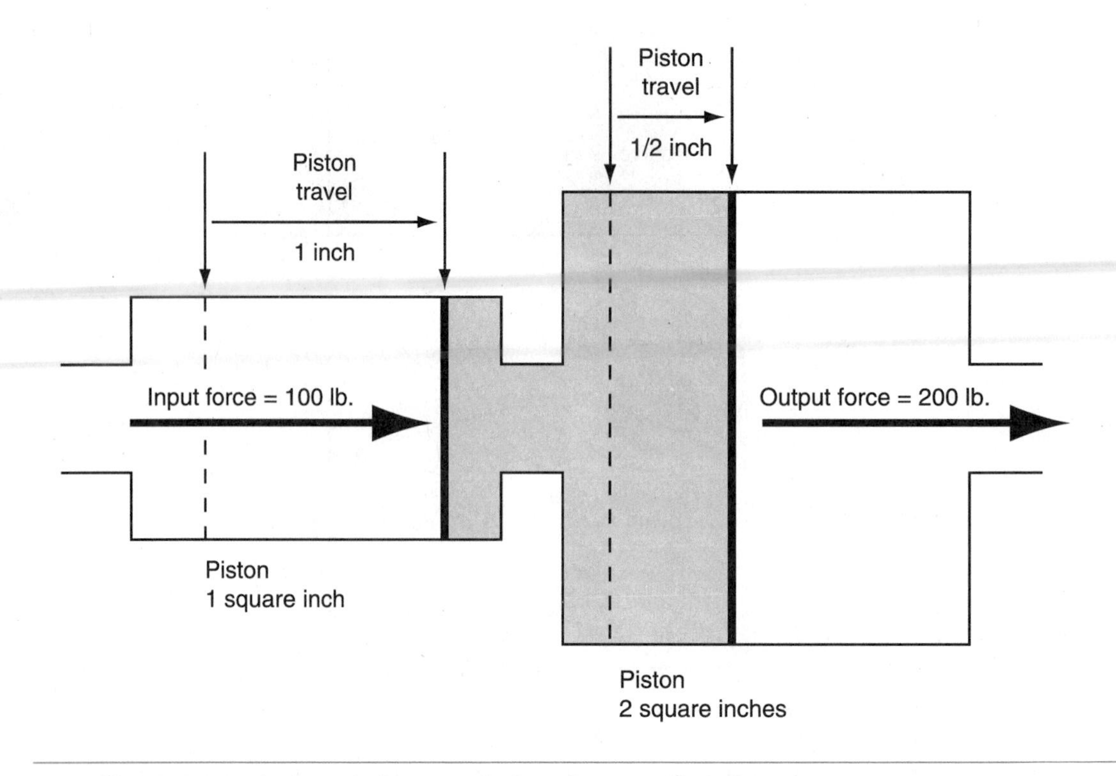

Figure 6-6 An increase in force results in a decrease of total movement.

The use of hydraulics to gain a mechanical advantage is similar to the use of levers or gears. All these systems can increase force, but with an increase of force comes a decrease in the distance moved. Although hydraulic systems, gears, and levers can accomplish the same results, hydraulics are preferred when size and shape of the system is of concern. In hydraulics, the force applied to one piston will be transmitted through the fluid causing the opposite piston to have the same force on it. The distance between the two pistons in a hydraulic system does not affect the operation of the system. Therefore, the force applied to one piston can be transmitted without change to another piston located somewhere else.

A hydraulic system responds to the pressure or force applied to it. The mere presence of different-sized pistons does not always result in fluid power. Either the pressure applied to the pistons or the size of the pistons must be different in order to cause fluid power. If an equal amount of pressure is exerted onto both pistons in a system and both pistons are the same size, neither piston will move. The system is balanced or is at equilibrium. The pressure inside the hydraulic system is called static pressure because there is no fluid motion.

When an unequal amount of pressure is exerted on the pistons, the piston with the least amount of pressure on it will move in response to the difference between the two pressures. Likewise, if the size of the two pistons is different and an equal amount of pressure is exerted on the pistons, the fluid will move. The pressure of the fluid while it is in motion is called dynamic pressure.

Automatic Transmission Applications

The fluid used in an automatic transmission's hydraulic system is ATF. ATF is a hydraulic oil designed specifically for automatic transmissions. Its primary purpose is to transmit pressure to activate the transmission's bands and clutches. It also serves as a fluid coupling between the engine and the transmission, removes heat from the transmission, and lubricates the transmission's moving parts.

A common hydraulic system within most automatic transmissions is the servo assembly. The servo assembly is used to control the application of a band. The band must tightly hold the drum it surrounds when it is applied. The holding capacity of the band is determined mainly by the construction of the band and the pressure applied to it. This pressure or holding force is the result of the action of a servo. The servo multiplies the force through hydraulic action.

If a servo has an area of 10 square inches and a pressure of 70 psi applied to it, the apply force of the servo is 700 pounds. The force exerted by the servo is increased further by its lever-type linkage and the self-energizing action of the band. The total force applied by the band is sufficient to stop and hold the rotating drum connected to a planetary gear set member.

A multiple-disc clutch assembly is also used to stop and hold gear set members. This assembly also uses hydraulics to increase its holding force. If the fluid pressure applied to the clutch assembly is 70 psi and the diameter of the clutch piston is 5 inches, the force applying the clutch pack is 1374 pounds. If the clutch assembly uses a **belleville spring,** which adds a mechanical advantage of 1.25, the total force available to engage the clutch will be 1374 pounds multiplied by 1.25, or 1717 pounds.

The outward force from a servo can be calculated by multiplying the force (oil pressure) applied to the piston by its area (force × area).

The area of a clutch assembly with a 5-inch diameter is 19.6 square inches (3.14 × 2.5 × 2.5).

Fluid

All hydraulic systems use a liquid to perform work. ATF is the liquid used in automatic transmissions. ATF can be a petroleum-based, partially synthetic, or totally synthetic oil. All domestic automobile manufacturers require a petroleum-based fluid in their automatic transmissions; however, some imported vehicles require the use of a partially synthetic fluid.

Shop Manual
Chapter 6, page 165

ATF is a compound liquid that contains special additives that allow the lubricant to better meet the flow and friction requirements of an automatic transmission. ATF is normally dyed red, primarily to help distinguish it from engine oil when determining the source of fluid leaks.

Petroleum-based ATF typically has a clear red color that will darken when it is burned or will become milky when contaminated by water. Synthetic ATF is normally a darker red than petroleum-based fluid. Synthetic fluids tend to look and smell burned after normal use; therefore, the appearance and smell of these fluids is not a good indicator of the fluid's condition.

The various chemicals added to ATF ensure the durability and overall performance of the fluid. Zinc, phosphorous, and sulfur are commonly added to reduce friction. Detergent additives are added to ATF to help keep the transmission parts clean. Also added are dispersants, which keep contaminants suspended in the fluid so they can be trapped by the filter.

All certified types and brands of ATF have been tested to ensure they meet the criteria set by the manufacturers. Some of these test standards apply to all types of ATF, such as oxidation resistance, corrosion and rust inhibition, flash and flame points, and resistance to foaming. Other standards are specific for a particular fluid rating or type.

The following is a summary of the common additives blended into the various types of ATF.

- *Antifoam agents:* These minimize foaming caused by the movement of the planet gears and the movement inside the torque converter. These movements tend to cause fluid foaming.
- *Antiwear agents:* Zinc is blended into the fluid to minimize gear, bushing, and thrust washer wear.
- *Corrosion inhibitors:* These are added to prevent corrosion of the transmission's bushings and thrust washers, and to prevent fluid cooler line corrosion.
- *Dispersants:* These keep dirt suspended in the fluid, which helps prevent the buildup of sludge inside the transmission.
- *Friction modifiers:* Additives are blended into the base fluid to provide a certain amount of clutch and band slippage. This improves or softens the feel of the shift.
- *Oxidation stabilizers:* To control oxidation of the fluid, additives are used to allow the ATF to absorb and dissipate heat. If the fluid is not designed to handle the high heat normally present in a transmission, the fluid will burn or oxidize. Oxidized fluid will cause severely damaged frictional materials, clogged fluid filters, and sticky valves.

- *Seal swell controllers:* These additives control the swelling and hardening of the transmission's seals, while maintaining their normal pliability and tensile strength.
- *Viscosity index improvers:* These are blended into the fluid as an attempt to maintain the viscosity of fluid regardless of its temperature.

Because some chemicals in the transmission fluid may adversely react with the fibers or synthetic materials in the seals of the transmission, the compatibility of the fluids with specific transmissions is also tested. Incompatibility can result in external and internal transmission fluid leaks due to deterioration, swelling, or shrinking of the seals.

Miscibility is the property of a fluid that allows it to mix and blend with other similar fluids.

All brands of ATF are tested for their miscibility or compatibility with other brands of ATF. Although the different brands of transmission fluids must meet the same set of standards, they may differ in their actual chemical composition and be incompatible with one another. There should be no fluid separation, color change, or chemical breakdown when two different brands of ATF are mixed together. This level of compatibility is important to the service life of a transmission as it allows for the maintenance of fluid levels without the worry about switching to or mixing different brands of fluid.

Recommended Applications

There are several ratings or types of ATF available. Each type is designed for a specific application. The different classifications of transmission fluid have resulted from the inclusion of new or different additives that enhance the operation of the different transmission designs. Each automobile manufacturer specifies the proper type of ATF that should be used in its transmissions. Both the design of the transmission and the desired shift characteristics are considered when a specific ATF is chosen.

Viscosity index improvers are additives that allow an oil to maintain its viscosity over a wide range of temperatures.

To reduce wear and friction inside a transmission, most commonly used transmission fluids are mixed with friction modifiers. Fluid types A, CJ, H, DEXRON, and MERCON have friction modifiers added to the ATF. Transmission fluids with these additives allow for the use of lower clutch and band application pressures, which in turn provide for a very smooth feeling shift. Transmission fluids without a friction modifier, such as Types F and G, tend to have a firmer shift because higher clutch and band application pressures are required to avoid excessive slippage during gear changes.

If an ATF without friction modifiers is used in a transmission designed for friction-modified fluid, the service life of the transmission is not normally affected. However, firmer shifting will result and the driver may not welcome this change in shifting quality. Transmission durability is affected by using friction-modified fluid in a transmission designed for nonmodified fluids. This incorrect use of fluid will cause slippage, primarily when the vehicle is working under a load. Any amount of slippage can cause the clutches and bands to wear prematurely. Also, because of the high heat generated by the slippage, the fluid may overheat and lose some of its lubrication and cooling qualities, which could cause the entire transmission to fail.

Viscosity is the rating given to the degree that a fluid resists flow.

The formulation of an ATF must also be concerned with the viscosity of the fluid. Although the fluids are not selected according to viscosity numbers, proper flow characteristics of the fluid are important in the operation of a transmission. If the viscosity is too low, the chances of internal and external leaks increase, parts can prematurely wear due to a lack of adequate lubrication, system pressure will be reduced, and overall control of the hydraulics will be less effective. If the viscosity is too high, internal friction will increase, resulting in an increase in the chance of building up sludge, hydraulic operation will be sluggish, and the transmission will require more engine power for operation.

The viscosity of a fluid is directly affected by temperature. Viscosity increases at low temperatures and decreases with higher temperatures. A transmission operates at many different temperatures. Since the fluid is used for lubricating and for shifting it must be able to flow well at any temperature. ATF has a low viscosity but is viscous enough to prevent deterioration at higher temperatures. High temperature performance is improved by many additives, such as friction modifiers. Low temperature fluid flow of ATF is enhanced by mixing pour point depressants into the base fluid. These additives are normally referred to as viscosity index improvers.

The use of correct ATF is critical to the operation and durability of automatic transmissions. We have already discussed the differences between friction-modified and nonmodified fluids. Certainly this is one aspect of ATF that must be considered when putting fluid into a transmission. There are other considerations. Each type of ATF is specifically blended for a particular application. Each has a unique mixture of additives that makes it suitable for certain types of transmissions. Always fill a transmission with the fluid recommended by the transmission's manufacturer. Through the years, automatic transmissions have changed and so have their fluid requirements. In many cases, the development of new fluids has allowed automobile manufactures to improve their transmission designs. In other cases, changes in the transmission has mandated the development of new fluid types.

Each manufacturer recommends the use of a particular type of ATF. Each of these is specially blended for its transmissions. Some manufacturers, such as Volkswagen, have their own brand of fluid that should be used in their transmissions. To give a feel for the history of the development of automatic transmissions and ATF, the following discussion is arranged by the year a particular type of fluid was introduced.

- Prior to 1949: ATF was basically engine oil dyed red. Some manufacturers added chemicals to improve the quality of the fluid, but no particular blend type existed during th0se years.
- 1949: General Motors introduced Type A fluid. This was a friction-modified fluid with additives that made it compatible with the frictional materials used in clutches and bands. It also contained additives that resisted oxidation and fluid aeration. This type of fluid was recommended by all American manufacturers until 1957. Type A fluid is no longer available.
- 1957: Type A fluid was modified to further resist oxidation. The new fluid, Type A, Suffix A, was the recommended fluid for GM and other transmissions until 1967. This blend of fluid is no longer recommended for any automatic transmission.
- 1959: Ford Motor Company introduced a fluid for its transmissions, Type F. This fluid was a nonmodified ATF. Compared to the Type A it replaced, Type F offered improved viscosity stabilization and oxidation resistance.
- 1967: An improved version of Type F was introduced, Type F-2P. This blend was designed to replace the previous versions of Type F and had improved frictional characteristics, as well as more resistance to oxidation.
- 1967: DEXRON fluid was introduced by General Motors. This blend was designed to better withstand the high temperatures of the current transmissions. It was a friction-modified fluid designed to replace Type A fluid.
- 1973: General Motors released a new blend of DEXRON, called DEXRON-II. This fluid had a lower coefficient of friction, improved oxidation resistance, and improved low temperature performance. But the original blend of DEXRON-II was found to cause corrosion in transmission cooler lines and was soon replaced by DEXRON-IID.
- 1976: DEXRON-IID was introduced to replace DEXRON-IIC and became the recommended fluid for GM transmissions. Other manufacturers did not recommend this change and stayed with DEXRON-B as their choice of fluids.
- 1977: Ford introduced Type CJ fluid for use in their new C-6 transmission. Type CJ fluid is a friction-modified fluid and was recommended for all transmissions requiring a friction-modified fluid. In 1980, Ford dropped the Type CJ fluid in favor of DEXRON-II for those same transmissions.
- 1982: Type H fluid was released by Ford. This was a friction-modified fluid designed for use in the C-5 transmission with a lockup torque converter. Type H was discontinued in 1988 when it was replaced by MERCON fluid.
- 1987: Prior to this time, Chrysler had the same fluid recommendations as GM. With the introduction of torque converter clutches, a new fluid needed to be developed. That fluid, MOPAR ATF-Plus, was designed to prevent shudder when the torque converter clutch was partially locked. This is the recommended fluid for all late-model Chrysler transmissions.

DEXRON fluid came out after Type A fluid and is sometimes referred to as DEXRON-B.

The first blend of DEXRON-II came out after DEXRON-B and was often called DEXON-IIC.

- 1988: Ford introduced MERCON fluid. The primary purpose of this fluid was to replace all other fluids recommended for Ford transmissions that required a friction-modified fluid. This fluid remains the recommended fluid for Ford transmissions. Although the fluid was modified in 1992, the name of the blend did not change.
- 1990: GM introduced DEXRON-IIE to replace all previous friction-modified fluids for its transmissions. This new fluid was designed for electronically controlled transmissions. These transmissions required a lower viscosity than previous designs. This blend of fluid has a very low viscosity and contains additives to stabilize the fluid at high temperatures.
- 1995: DEXRON-III was introduced. This new blend is compatible with GM transmissions produced since 1949. It was designed to meet the ever-changing needs of electronic transmissions and offers better oxidation resistance, compatibility, and flow than the blend it replaced.
- The Future: There are many new blends of fluids scheduled to be released in the near future. All of these are designed to improve low temperature viscosity, high temperature oxidation resistance, and durability, and to reduce friction. Always follow the recommendations of the manufacturer for the fluid to be used in a specific transmission.

Filtering

Shop Manual
Chapter 6, page 166

To trap dirt and metal particles from the circulating ATF, automatic transmissions have a fluid filter (Figure 6-7), which is normally located inside the transmission case between the pump pickup and the bottom of the pan. If dirt, metal, and frictional materials are allowed to circulate, they can cause valves to stick and can cause premature transmission wear.

WARNING: Some automatic transmissions are equipped with an extra deep sump, which allows for improved cooling and increased capacity. If the transmission is equipped with a deep pan, a special filter that will reach into the bottom of the pan must be used.

Screen filters sometimes are made of polyester or nylon.

Current automatic transmissions are fitted with one of three types of filters: a screen filter, paper filter, or felt filter. Screen filters use a fine wire mesh to trap the contaminants in the ATF. This type of filter is considered a surface filter because it traps the contaminants on its surface. As a screen filter traps dirt, metal, and other materials, fluid flow through the filter is reduced. The openings in the screen are relatively large so that only larger particles are trapped and small particles remain in the fluid. Although this does not remove all of the contaminants from the fluid, it does prevent quick clogging of the screen and helps maintain normal fluid flow.

Cellulose is a resin-based substance that serves as the basis for producing paper.

Paper filters are more efficient than the screen type because they can trap smaller particles. Paper filters are typically made from a cellulose or Dacron fabric. Although this type of filter is quite efficient, it can quickly clog and cause a reduction in fluid flow through the transmission. Therefore, some older transmissions equipped with a paper filter have a bypass circuit that allows contaminated fluid to circulate through the transmission if the filter becomes clogged and greatly restricts fluid flow.

Felt filters are the most commonly used filters in current model transmissions. These are not surface filters; rather they are considered depth filters because they trap contaminants within the filter and not just on its surface. Normally made from randomly spaced polyester materials, felt filters are able to trap both large and small particles and are less susceptible to clogging.

To protect vital transmission circuits and components, most transmissions are equipped with a secondary filter located in a hydraulic passage that helps keep dirt out of the pump, valves, and solenoids. Secondary filters are simply small screens fitted into a passage or bore.

CUSTOMER CARE: Customers should be made aware that transmission filters should be replaced at the mileage intervals recommended by the manufacturer. Failure to do this will result in a loss of fluid flow and pressure, which can cause premature transmission failure.

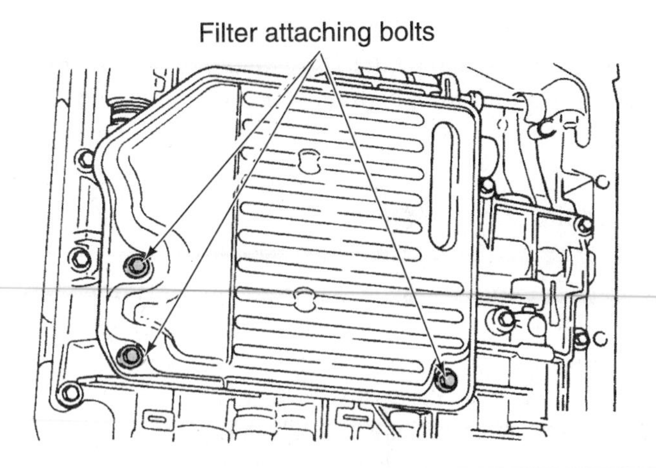

Figure 6-7 Typical oil filter mounting to valve body. (Reprinted with the permission of Ford Motor Co.)

Reservoir

All hydraulic systems require a reservoir to store fluid and to provide a constant source of fluid for the system. In an automatic transmission, the reservoir is the pan, typically located at the bottom of the transmission case (Figure 6-8). Transmission fluid is forced out of the pan by atmospheric pressure into the pump and returned to it after it has circulated through the selected circuits. The level of ATF in a transmission must be maintained and a transmission dipstick and filler tube are used to check the level of the fluid and to add ATF to the transmission. The tube and dipstick are normally located at the front of the transaxle housing and the right front of a transmission housing.

The pan is often called the oil sump.

Venting

In order to allow the fluid to be pumped through the transmission by the oil pump, all reservoirs must have an air vent that allows atmospheric pressure to force the fluid into the pump when the oil pump creates a low pressure at its inlet port. The pans of many automatic transmissions vent

When a fluid becomes contaminated with trapped air, it is aerated.

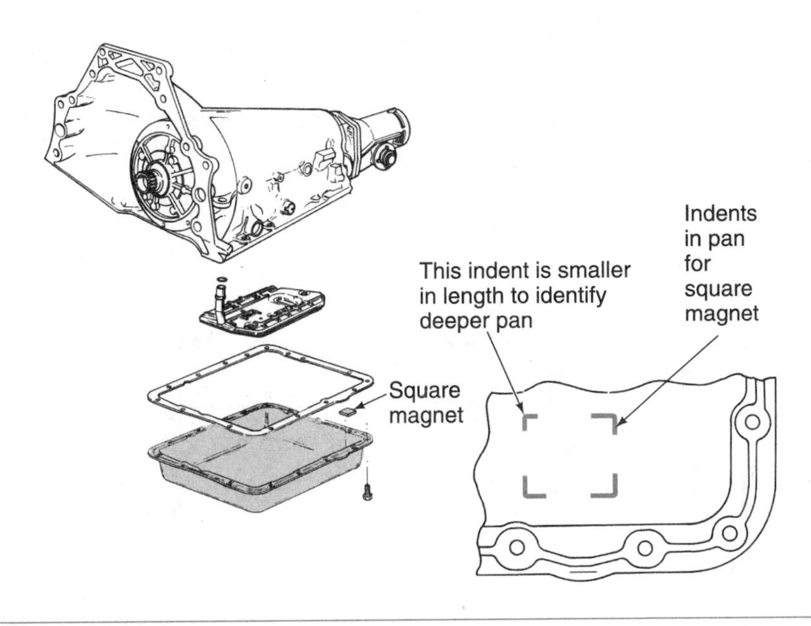

Figure 6-8 Typical pan location. (Courtesy of the Hydra-Matic Division of General Motors Corp.)

through the handle of the dipstick. Transmissions must also be vented to allow for the exhaust of built up air pressure that results from the moving components inside the transmission. The movement of these parts can force air into the ATF, which would not allow it to cool or lubricate the transmission properly.

Transmission Coolers

Shop Manual
Chapter 6, page 211

Keeping the ATF cool ensures that it will be able to adequately lubricate and cool the transmission. As ATF circulates through the transmission, it absorbs some of the heat from the transmission's gears, clutches, and bands and the torque converter. To remove the heat from the fluid, the ATF is directed through an oil cooler where it dissipates heat.

The removal of heat from the ATF is extremely important to the durability of the transmission. Excessive heat causes the ATF to break down. Once broken down, ATF no longer lubricates well and has poor resistance to oxidation. When a transmission is operated for some time with overheated ATF, varnish is formed inside the transmission. Varnish buildup on valves can cause them to stick or move slowly. The result is poor shifting and glazed or burned friction surfaces. Continued operation can lead to the need for a complete rebuilding of the transmission.

It is important to note that the ATF is designed to operate at 17 ° F (80° C). At this temperature, the fluid should remain effective for 100,000 miles. However, when the operating temperature increases, the useful life of the fluid quickly decreases. A 20-degree increase in operating temperature will decrease the life of ATF by half!

A transmission's oil pump is commonly called the front pump because of its location.

Transmission housings are fitted with ATF cooler lines (Figure 6-9). These lines direct the hot fluid from the torque converter to the transmission cooler, located in the vehicle's radiator. The heat of the fluid is reduced by the cooler and the cool ATF returns to the transmission. In some transmissions, the cooled fluid flows directly to the transmission's bushings, bearings, and gears. Then, the fluid is circulated through the rest of the transmission. The cooled fluid in other transmissions is returned to the pan where atmospheric pressure forces it into the oil pump and is circulated throughout the transmission.

The flow of pressurized fluid from the oil pump to the rest of the transmission is typically referred to as the supply fluid.

Some vehicles are equipped with an auxiliary fluid cooler in addition to the one in the radiator. This cooler serves to remove additional amounts of heat from the fluid before it is sent back to the transmission. Auxiliary coolers are most often found on heavy duty and performance vehicles.

Maintenance

The level and condition of ATF should be checked on a regular basis. If the fluid is low, more fluid should be added to bring the level up to normal. If the fluid is contaminated or burned, both the fluid and the filter should be replaced. Whenever fluid is added to a transmission, it is important that only the recommended fluid be used.

Figure 6-9 Typical oil cooler lines. (Courtesy of the Buick Motor Division of General Motors Corp.)

Transmission fluid also needs to be changed periodically. The recommended interval for changing the fluid or filter varies with the transmission design. It also varies with the operating conditions of the vehicle.

Heat is generated by the operation of a transmission and the torque converter. Heat causes ATF to break down. The control of heat is not a problem if the vehicle is always operated on a flat highway in cool temperatures and with no load. This is not where and how most vehicles are driven. City traffic, heavy loads, hard acceleration, and summer heat all contribute to increases in the operational temperature of ATF.

We have already seen that ATF is designed to operate at 175 degrees F (80 degrees C). At this temperature, the fluid should remain effective for 100,000 miles. As a result, most manufacturers recommend fluid and filter changes at 100,000 mileswhen the vehicle is operating under normal conditions. Some manufacturers do not have a recommended ATF change interval and claim the fluid is good for the service life of the vehicle. However, all manufacturers recommend more frequent changes when the vehicle undergoes worse than normal or severe conditions. Some recommend that the fluid be changed once a year if the transmission has seen severe duty. Therefore it is important to consider the operating conditions of the vehicle when determining the correct interval for fluid and filter change. Of course, if the fluid appears contaminated or has an abnormal smell, it should be changed immediately.

Oil Pumps

The transmission relies on the oil pump to provide the circulation of the ATF and to provide the pressure needed to activate the apply devices. Although the oil pump is the source of all fluid flow through the transmission, the valve body regulates and directs the fluid flow to provide for the gear changes. The oil pump is driven by the hub of the torque converter; therefore, it operates whenever the engine is running. Three types of oil pumps are commonly used: the gear-type, rotor-type, and vane-type. The gear-type is the most common.

Gear-type oil pumps and **rotor-type oil pumps** are commonly used in automatic transmissions. They are considered **positive displacement pumps** because they create and maintain the flow of fluid. As the fluid flows out of the pump, it is prevented from backing up into the inlet of the pump. Oil pumps are considered positive displacement because they move oil in at one location and out at another. The pump has a continuous delivery of fluid and the rate of delivery is in direct proportion to the speed at which the pump is driven. The pump's output sometimes exceeds the hydraulic needs of the transmission and the excessive fluid is bled off by the pressure regulator. If this excessive pressure is not exhausted, the transmission could become damaged. Because positive displacement pumps move fluid when it is not needed, these pumps waste energy and horsepower. Although they are the most common type of pump, they are not the preferred type of pump.

Positive displacement pumps are sometimes called positive delivery pumps.

Displacement is the volume of fluid moved or displaced during each cycle of an oil pump.

An oil pump changes the rotating force of a motor or engine into hydraulic energy.

Pump Drives

Automatic transmission oil pumps are driven by the engine's crankshaft through the torque converter housing (Figure 6-10). The oil pumps in all rear-wheel-drive and some front-wheel-drive transmissions are driven externally by the torque converter's drive hub. The hub has two slots or flats machined in it that fit into the oil pump drive. As the converter rotates, the pump is driven directly by the hub. Many front-wheel-drive transaxles use a hex-shaped or splined shaft fitted in the center of the converter cover to drive the oil pump (Figure 6-11). This system is called an internal drive. Many older and a few late-model transmissions use a second pump mounted at the rear of the transmission case and driven by the output shaft. This second pump allows for pump operation whenever the output shaft is rotating, such as when the vehicle is being towed or push started.

Shop Manual
Chapter 6, page 182

During pump operation, ATF moves from an area of high pressure to an area of low pressure.

Since the oil pump operates whenever the engine is running, its output is dependent on engine speed. Therefore, the output of the pump is considered a variable output. At some speeds,

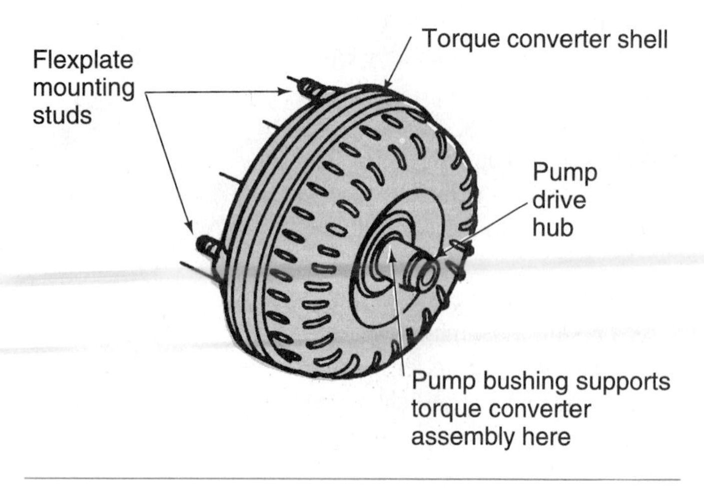

Figure 6-10 Oil pump drive on a typical torque converter. (Reprinted with the permission of Ford Motor Co.)

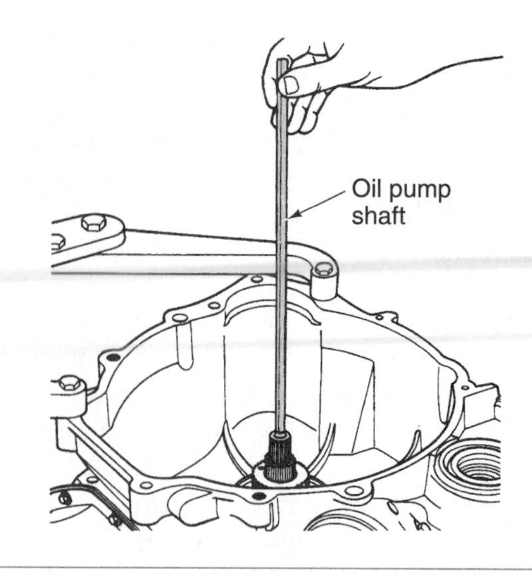

Figure 6-11 Oil pump shaft used on some transmissions and transaxles. (Reprinted with the permission of Ford Motor Co.)

Atmospheric pressure is approximately 14.7 psi at sea level, but registers 0 psi on a pressure gauge.

A vacuum is best defined as any pressure lower than atmospheric pressure.

the pressure from the oil pump can exceed the pressure requirements of the transmission. A pressure regulator valve is used to limit the pressure generated by the oil pump.

Operation

While the pump is operating, the pump creates a low-pressure area at its inlet port. The oil pan is vented to the atmosphere and atmospheric pressure forces the fluid into the inlet port of the pump. The fluid is now moved by the pump and forced out at a higher pressure through the outlet port.

ATF leaves the oil pan and enters the oil pump because atmospheric pressure is pressing down on the fluid in the pan. Because the pump creates a low pressure, the pressure of the atmosphere, which is greater than the low pressure, pushes the fluid into the pump inlet. Before ATF will flow into the pump, the pump must create a vacuum at the inlet port. Then atmospheric pressure can push the fluid into the pump to fill it.

Shop Manual
Chapter 6, page 182

Gear-Type Pumps

Gear-type pumps consist of two gears in mesh to pressurize the fluid. The gears rotate on their own shafts and in opposite directions. The gears are assembled in a housing that surrounds and totally encloses the gears. The shafts may be sealed in the housing by bushings and the housing is normally sealed with a cover. In order for the pump to create low and high pressures, the pump housing must be sealed.

Some gear-type pumps use a wear plate instead of a pump cover to seal the gears in the housing.

One pump gear is driven by the torque converter and drives the other gear. As the gears rotate, the gear teeth move in and out of mesh. As the teeth move out of mesh, inlet fluid is drawn into the gear teeth and the walls of the housing. The trapped fluid is carried around with the teeth until the gear teeth again mesh. At this time, the meshing of the teeth forces the fluid out. The continuous release of trapped fluid provides for a buildup of oil pressure as the fluid is pushed out of the pump's outlet port. The meshing of the gears also forms a seal that stops the fluid from moving out the inlet port. The initial meshing of the gear teeth creates the high pressure and the lack of fluid in the teeth after the gears mesh causes the low pressure.

This type of pump is made with close tolerances. Excessive wear or play between the teeth of the gears or between the gears and the housing or pump cover will reduce the output and efficiency of the pump.

Another common type of gear pump is the **crescent** pump, which also uses two gears. One gear of this pump has internal teeth and the other has external teeth. The smaller gear with external teeth is in mesh with one part of the larger gear. In the gap where the teeth are not meshed is a crescent-shaped separator (Figure 6-12). The small gear is driven by the torque converter and it drives the larger gear. The gears' teeth mesh tightly together. As the gears rotate, the teeth mesh and then separate. As they separate, a low pressure is created between the gear teeth. This causes a suction on the inlet side of the crescent. The fluid is pushed by atmospheric pressure into this void until it is full.

A crescent is a half-moon-shaped part that isolates the inlet side of the pump from the outlet side of the pump.

As the gears rotate, fluid is trapped between the teeth and the crescent. The fluid is then carried around the pump housing toward the outlet port. As the gear teeth get close to the outlet and the gear teeth begin to mesh, the gap between the gear teeth begins to narrow. This narrowing of the gap continues until the gear teeth fully mesh. The narrowing of the gap squeezes the fluid and forces it through the outlet port as the gears move into mesh. A continuous flow of fluid toward the outlet port pushes the fluid out into the transmission's hydraulic circuit. The crescent blocks the pressurized fluid and prevents it from leaking back toward the pump's inlet port.

Gear-type pumps are sometimes referred to as **IX pumps.** IX is an abbreviation that refers to the internal/external design of the gears used in this style pump.

Crescent pumps are made with close tolerances and can deliver the same amount of fluid each time the gear makes one complete revolution. The rate of fluid delivery does change with changes in engine speed.

Shop Manual
Chapter 6, page 182

Rotor-Type Pump

A rotor-type pump is a variation of the gear-type pump. However, instead of gears, an inner and outer rotor turns inside the housing (Figure 6-13). A rotor utilizes rounded lobes instead of teeth. The lobes of one rotor mesh with the recess area between the lobes of the other rotor. The torque converter drives the inner rotor, which rotates inside the outer rotor or rotor ring. The inner rotor has one less lobe than the ring, so that only one lobe is engaged with the outer ring at any one time.

Rotor-type pumps are also sometimes referred to as **IX pumps** because of the internal/external design of the lobes.

Fluid is carried in the recess between the lobes and squeezed out toward the outlet port as the lobe moves into the recess of the outer ring (Figure 6-14). A constant supply of fluid being forced out of the outlet port supplies pressurized fluid for the operation of the transmission. Fluid is prevented from backing up into the inlet port by the action of the lobes sliding over the lobes of the outer ring. The lobe-to-lobe contact causes them to seal against each other and creates small

The rotor-type pump is sometimes referred to as the gerotor-type pump.

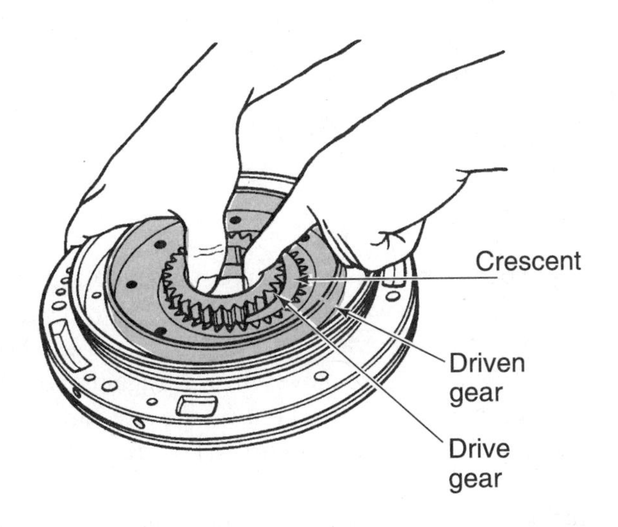

Figure 6-12 Typical gear-type pump. (Reprinted with the permission of Ford Motor Co.)

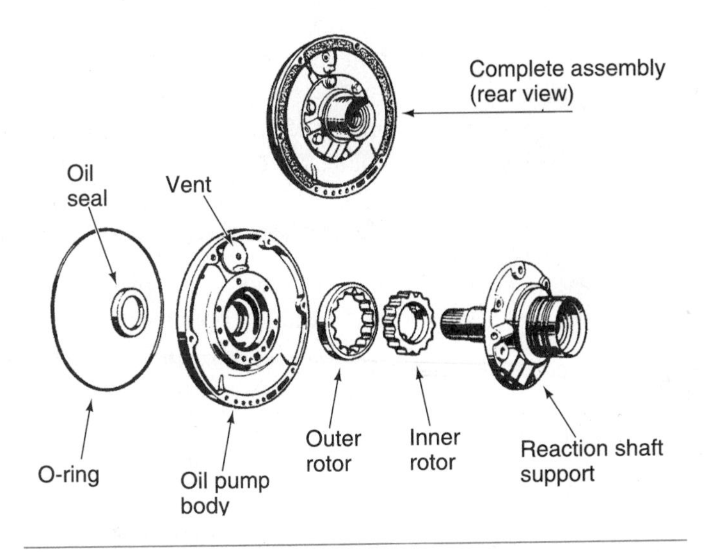

Figure 6-13 Typical rotor-type pump. (Courtesy of Chrysler Corp.)

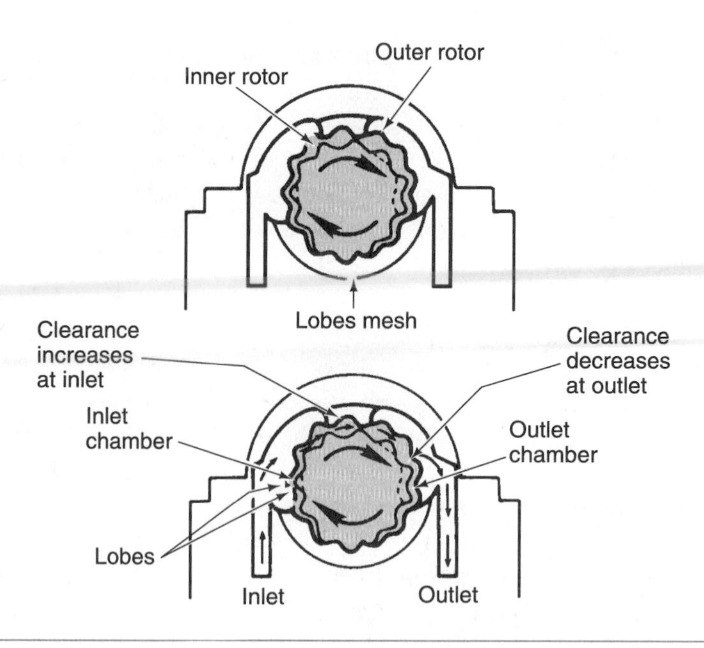

Figure 6-14 Action of rotors that move to pressurize the fluid. (Courtesy of Chrysler Corp.)

fluid chambers whose volumes increase as the lobes separate on the inlet side. This creates the low pressure that allows fluid to be pushed into the inlet of the pump. As the lobes move into the recesses of the outer ring, fluid is squeezed out. The fluid must move or it is forced to the outlet side of the pump where it is pressurized.

Shop Manual
Chapter 6, page 186

Vane-Type Pump

In a **vane-type pump,** a vane ring is driven by the torque converter. The vane ring contains several sliding vanes that seal against a slide mounted in the pump housing (Figure 6-15). As in a rotor-type pump, the vanes create small fluid chambers whose size increases on the inlet side and decreases on the outlet side. The size increase on the inlet side creates a low pressure that becomes filled with fluid, and the size decrease supplies pressurized fluid to the hydraulic system of the transmission.

As the vane ring rotates, the sliding vanes are thrown out against the inside surface of the housing by centrifugal force. As the vanes follow the contour of the housing, crescent-shaped areas between the vanes and the housing are formed. These chambers are constantly being increased and decreased in size. As the size of the fluid chamber begins to decrease, the trapped fluid is forced out at the outlet port.

Variable Capacity Pumps

Many late-model automatic transmissions are equipped with vane-type pumps. This type of pump is a **variable capacity pump** or variable displacement pump whose output can be reduced when high fluid volumes are not necessary. To monitor and control output pressure, a sample of the output is applied to the backside of the pump's slide. The pressure moves the slide against spring pressure and changes its position in relation to the rotor. This action controls the size variations between the chamber at the inlet and the outlet ports, which in turn control the output volume of the pump.

The rotor and vanes of the pump are contained within the bore of a movable slide that is able to pivot on a pin. The position of the slide determines the output of the pump (Figure 6-16). When the slide is in the fully extended position, the slide and rotor vanes are at maximum output.

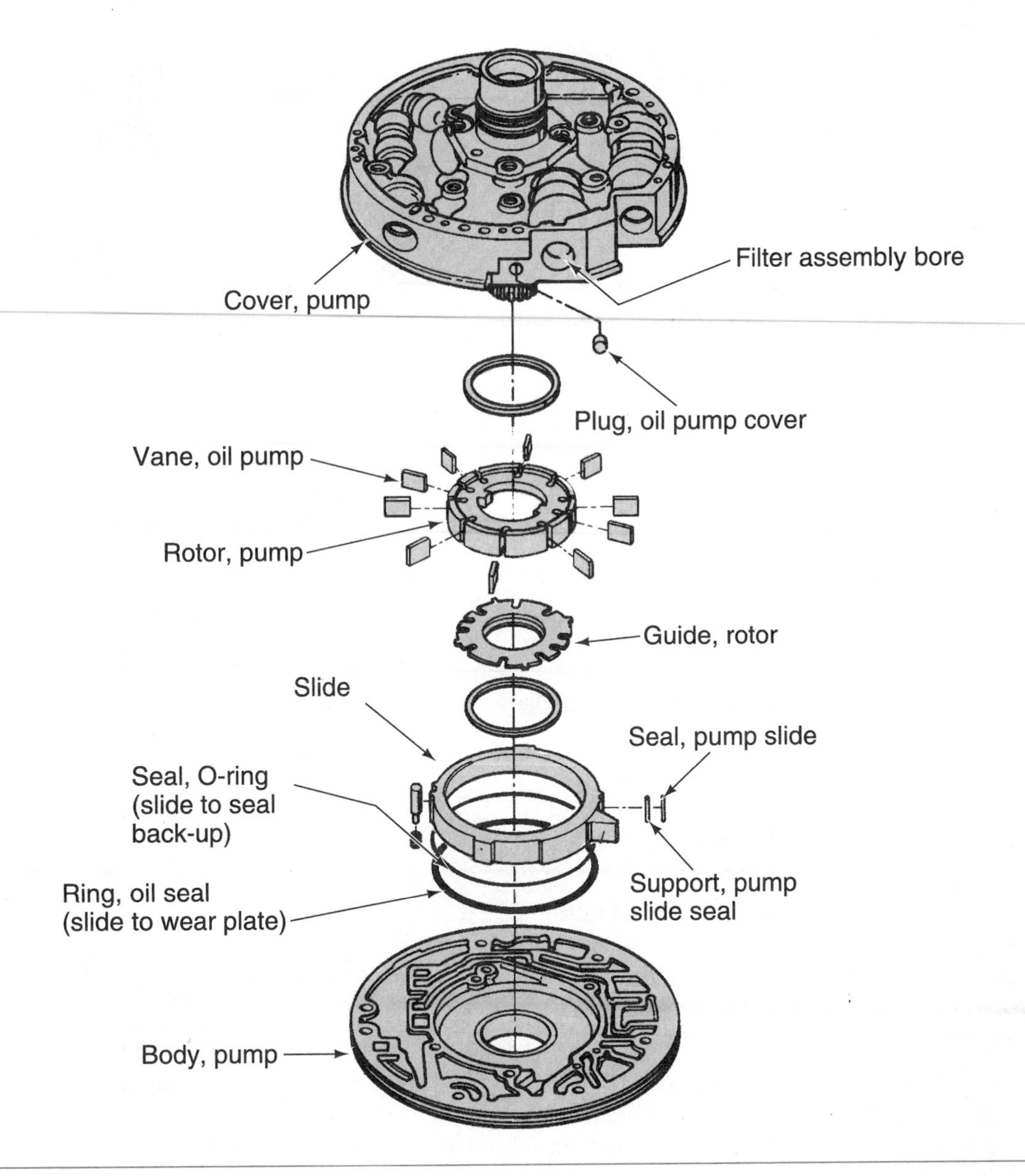

Figure 6-15 Typical vane-type pump. (Courtesy of the Hydra-Matic Division of General Motors Corp.)

As the rotor and vanes rotate within the bore of the slide, the changes in the chamber size form low and high pressure areas. Fluid trapped between the vanes at the inlet port is moved to the outlet side. As the slide pivots toward the center or away from the fully extended position, a greater amount of fluid is allowed to move from the outlet side back to the inlet side. When the slide is centered, there is no output from the pump. A neutral or no-output condition exists. Since the slide moves in response to the sample of pressure sent to it, the slide can be at an infinite number of positions.

The output of a variable-capacity pump will vary according to the needs of the transmission, not according to the engine's speed. Therefore, it does not waste energy as a positive displacement pump does. A variable capacity pump can deliver a large volume of fluid when the need is great, especially at low pump speeds. At high pump speeds, the volume requirements are normally low and the variable capacity pump can reduce its output accordingly. Once the needs of the transmission are met, the pump delivers only the amount of fluid that is needed to maintain the regulated pressure.

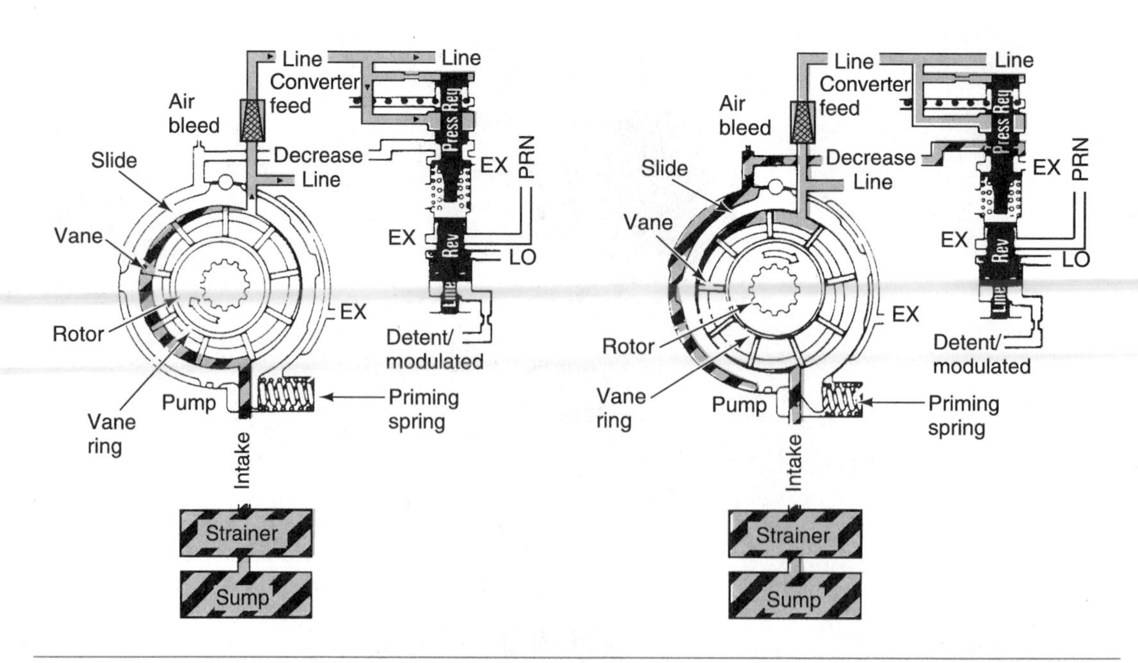

Figure 6-16 Action of vane-type pump: maximum output on the left and minimum output on the right. (Courtesy of the Hydra-Matic Division of General Motors Corp.)

Control Devices

Although the pump is the source of all fluid flow through the transmission, the valve body regulates and directs the fluid flow to provide for the gear changes. Many different valves and hydraulic passages are used to regulate, direct, and control the pressure and movement of the fluid after it leaves the pump (Figure 6-17). The basic types of valves used in automatic transmissions are ball, poppet, or needle check valves, and relief valves, orifices, and spool valves. A simple check valve is used to block a hydraulic passageway. It normally stops fluid flow in one direction, while allowing it in the opposite direction. A **relief valve** prevents or allows fluid flow until a particular pressure is reached, then it either opens or closes the passageway. Relief valves are used to control maximum pressures in a hydraulic circuit. **Orifices** are used to regulate and control fluid pressures. **Spool valves** are normally used as flow-directing valves. They are the most commonly used type of valve in an automatic transmission.

Check Valves

Shop Manual
Chapter 6, page 173

Check valves are used to hold fluid in cylinders and to prevent fluid from returning to the reservoir. A check valve opens when fluid is flowing and closes when the flow stops. The valve also closes when fluid pressure is applied to the outlet side of the valve. Check valves can also serve as a one-way valve. The direction of the fluid flow controls and operates the check valve (Figure 6-18). **Ball-type** check **valves** are commonly used to redirect fluid flow (Figure 6-19) or to stop fluid from backflowing to the reservoir. The seats for the ball-type check valves are often holes cut into the valve body's **separator plate** (Figure 6-20).

Needle-type and poppet-type check valves are also used to prevent backflow. **Needle check valves** are not used very frequently in automatic transmissions, but they operate in the same way as ball-type valves.

A return spring is used with a **poppet valve** and allows the valve to totally stop backflow. The valve is forced closed by the spring as soon as spring pressure is greater than the pressure of the fluid. Ball-type check valves may also be fitted with a return spring to accomplish the same

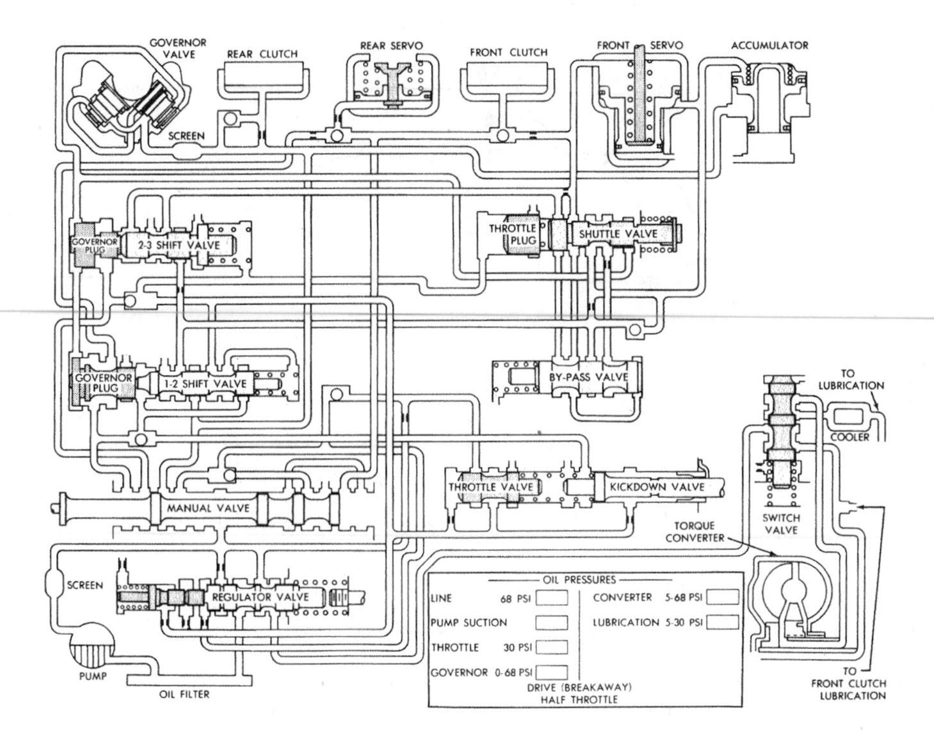

Figure 6-17 Typical fluid passages fed by the output of a pump. (Courtesy of Chrysler Corp.)

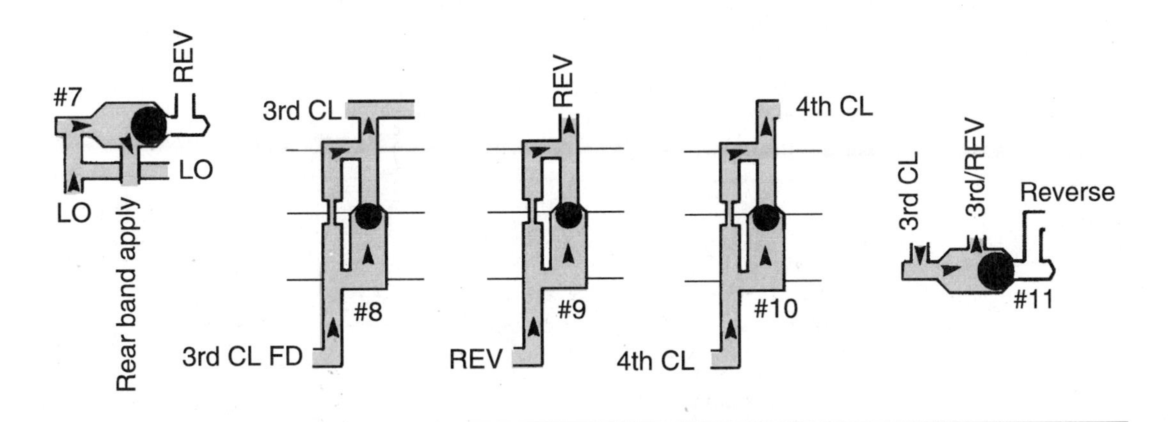

Figure 6-18 Action of a check ball valve. (Courtesy of the Hydra-Matic Division of General Motors Corp.)

task as a poppet valve. A spring-loaded check valve will open whenever fluid pressure on the inlet side of the valve exceeds the tension of the spring. The valve will remain open until the fluid's pressure drops below the spring's tension. Check valves without a return spring will not seat until there is a small amount of backflow.

A ball-type check valve without a return spring can be used as a **two-way check valve** (see Figure 6-19). This type valve is used where hydraulic pressure from two different sources is sent to the same outlet port. When hydraulic pressure on one side of the valve is stronger than the pressure on the other side of the valve, the ball moves to the weaker side and closes that port. The ball will toggle between the ports in response to differing pressures. Both ports are open at the same time unless the pressures on both sides of the ball are equal. Normally, if the pressures were equal, the ball would be centered and would block off the outlet port.

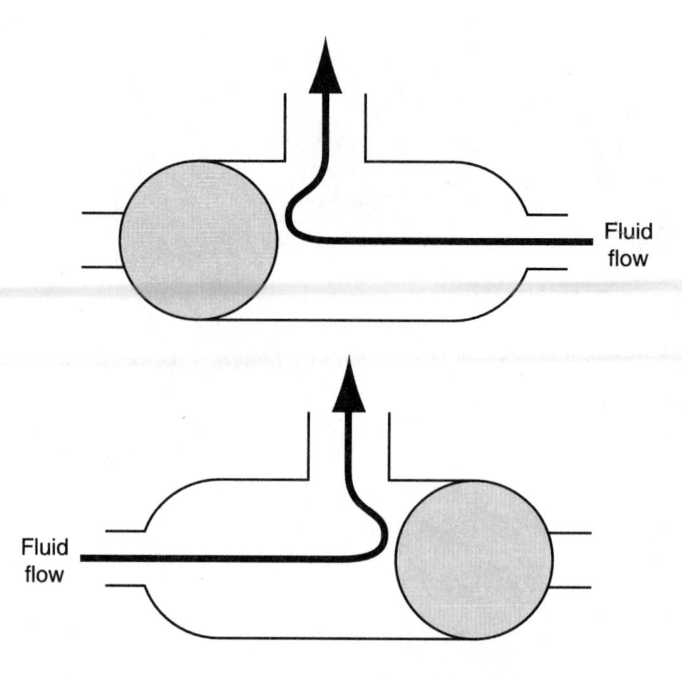

Figure 6-19 Fluid is directed by the movement of the check ball.

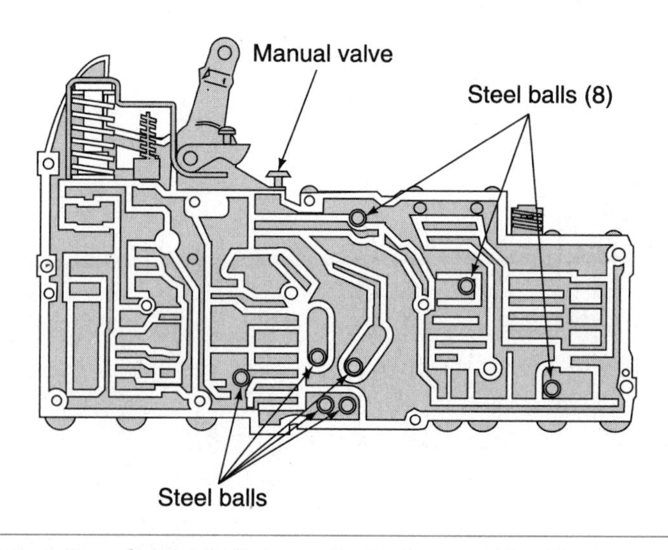

Figure 6-20 Location of check balls in a valve body assembly. (Courtesy of Chrysler Corp.)

Relief Valves

A check valve fitted with a spring designed to prevent the valve from opening until a specific pressure is reached is called a pressure relief valve. A pressure relief valve is used to protect the system from damage due to excessive pressure. When the pressure builds beyond the rating of the spring, the valve will open and reduce the pressure by exhausting some fluid back to the reservoir. After the pressure decreases, the valve again closes, allowing normal fluid flow.

Orifices

Shop Manual
Chapter 6, page 175

The dynamic pressure of a fluid is affected by the fluid's movement from one point to another through a restriction. The hydraulic system in a transmission has many restrictions, such as connecting lines, small bores, and orifices. These restrictions cause the pressure of the fluid to increase

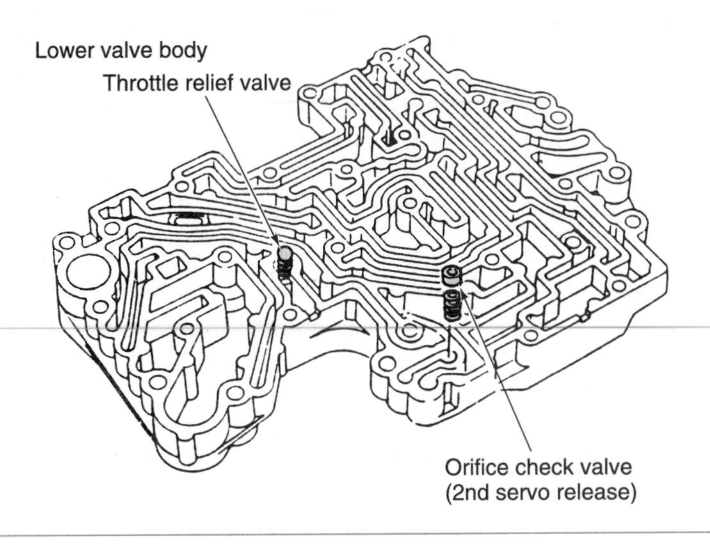

Figure 6-21 Location of a fixed orifice to restrict fluid flow. (Courtesy of Nissan Motor Co., Ltd.)

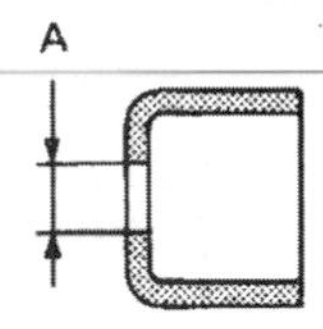

Figure 6-22 The size of the opening in an orifice determines its flow rate. (Courtesy of Nissan Motor Co., Ltd.)

at the inlet side of the restriction and decrease at the outlet side. Orifices are used in transmissions to control dynamic pressures. An orifice is a passage in a hydraulic system or item that has been placed in a hydraulic circuit that restricts and slows down fluid flow (Figure 6-21).

The flow is slowed down, which results in higher pressure on the inlet side of the restriction than on the outlet side of the restriction. The opening or size of the orifice will determine the amount of pressure decrease. Pressure drops only when fluid flows through the orifice. When the flow stops, the pressure becomes equalized. Orifice sizes are specifically selected to meet the needs of the transmission (Figure 6-22). Orifices can be cast as part of the hydraulic passages in the transmission's case or valve body, or they can be line plugs with precisely drilled holes.

A number of orifices are often placed in series to provide a cushioning effect on the hydraulic system. Any number of restrictions may be present in a line and these allow for a gradual activation of an apply device, which improves shift quality.

In the separator plate of a valve body, the smaller holes are orifices. The larger rectangular holes in the plate do not act as orifices.

Spool Valves

The most commonly used valve in an automatic transmission is the spool valve (Figure 6-23), which is so named because it looks similar to a sewing thread spool. The large circular portions of the

Shop Manual
Chapter 6, page 175

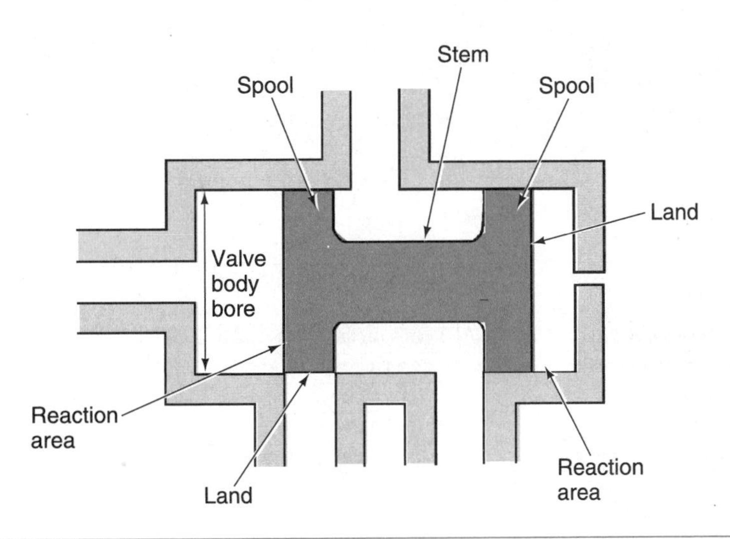

Figure 6-23 Typical spool valve. (Courtesy of Chrysler Corp.)

The spools or sealing areas of the valve are frequently referred to as the lands of the valve.

Some spool valves are fitted inside a sleeve or bushing that fits into the valve body.

Spool valves that rely on spring tension and hydraulic force for movement are sometimes called **hybrid valves.**

valve are called spools. These spools are precisely machined to fit into a bore and are connected together by the valve's stem. The stem has a smaller diameter than the spools but is not a precisely machined part of the valve. The area or space between the spools is called the **valley.** When the valve is installed in its bore, the valleys form a fluid pressure chamber between the spools and the bore.

The spools ride on a very thin film of fluid in the valve's bore. As the spool valve moves in its bore, the spools cover and uncover ports in the valve's bore. This opens and closes hydraulic passages. The valley of the valve forms a **reaction area.** This area is made smaller to allow fluid to pass between the **lands** and through the valve. The movement of the valve allows fluid to flow from various inlets to various outlets. Fluid passages in the bore are closed and opened as the lands of the spool cover and uncover the passage openings.

By the positioning of the spools in the valve's bore, the valve is able to direct fluid flow. The movement of the valve is controlled by either mechanical or hydraulic forces. The forces may be a result of mechanical linkages or hydraulic pressures applied to the reaction area of the valve.

Some hydraulically operated spool valves have different-sized spools and are fitted with a spring (Figure 6-24). When hydraulic pressure on the valve causes it to move toward the larger spool, spring tension is effectively increased. Hydraulic pressure on the other side of the valve must become greater before it can overcome the tension of the spring and move the spool valve.

Hydraulic pressure can move the valves against spring pressure or move the valve back and forth in its bore. The latter is the case for a regulating spool valve. If the pressure of the fluid increases to levels that exceed the tension of the spring, the valve will move to open an exhaust port, which allows some of the fluid to return to the reservoir. Once the pressure has been reduced

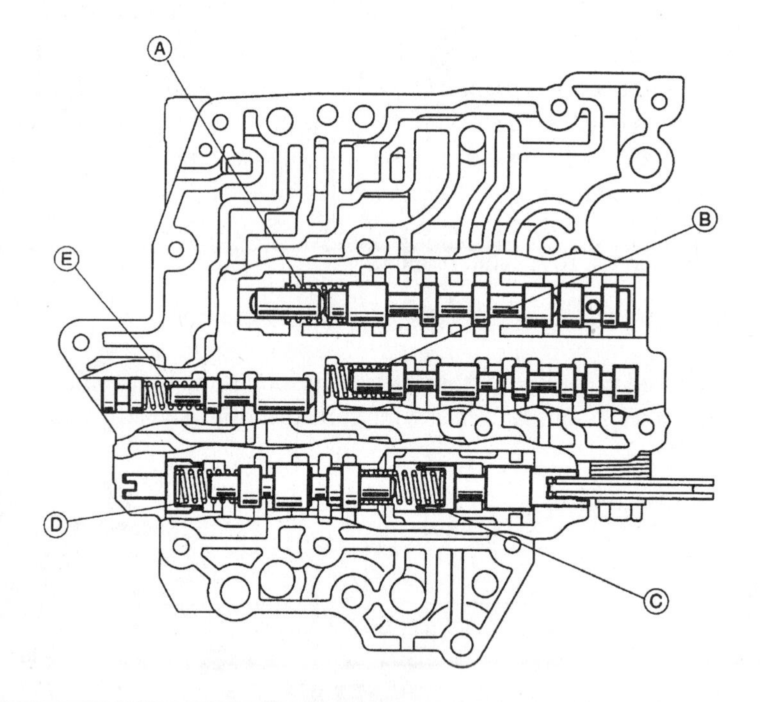

Mark	Name (Color)	Free Length / Outer Diameter mm (in.)	Total No. of Coils
Ⓐ	Lock–Up Relay Valve (Yellow)	26.8 (1.055) / 10.2 (0.402)	10.8
Ⓑ	B_1 Orifice Control Valve (White)	24.8 (0.976) / 6.4 (0.252)	12.0
Ⓒ	Throttle Valve (Green)	31.5 (1.240) / 7.0 (0.276)	11.4
Ⓓ	Down–Shift Plug (None)	15.0 (0.591) / 11.0 (0.433)	7.0
Ⓔ	Low Coast Modulator Valve (Purple)	20.2 (0.795) / 7.9 (0.311)	11.9

Figure 6-24 Some spool valves are set in place by springs that must be overcome by hydraulic pressure before the spool valve will move. (Reprinted with permission.)

to a pressure below the tension of the spring, the valve will move back to its original position. This toggling back and forth controls the fluid pressure at the valve.

Control Valves

A **control valve** is used as a flow-directing valve. A constant supply of fluid is fed to the valve and the valve's position directs the fluid to the appropriate hydraulic passageway. If a control valve is fitted in a bore with one inlet and three separate outlets, the position of the valve will determine where the fluid will be directed. When the valve is in its neutral position, fluid flows directly through the valve to a passageway that sends it back into the reservoir. When the valve is moved down in its bore, fluid is directed to the bottom outlet of the bore, where it may activate a piston. Any fluid present at the top of the valve flows through the center outlet, which directs that fluid back to the reservoir. When the valve moves up, fluid is directed to the top outlet. Any fluid present at the bottom of the valve will flow back to the reservoir.

Shop Manual
Chapter 6, page 175

Relay Valves

A **relay valve** is a spool valve with several spools, lands, and reaction areas. It is used to control the direction of fluid flow; a relay valve does not control pressure. A relay valve is held in position in a bore of the valve body by spring tension, auxiliary fluid pressure, or a mechanical linkage. Forces from hydraulic pressure or mechanical linkages oppose spring tension and move the relay valve to a different position in its bore. In each position, the valve's valley aligns with outlet ports. Fluid pressure flows from an inlet port across the valve's valley to an outlet port. A relay valve may block fluid flow in one position. When moved to the other positions, fluid is directed through outlet ports.

Regulator Valves

Regulator valves are valves that change the pressure of the fluid as it passes through the valve. The pressure change occurs as the position of the valve allows some of the fluid to exhaust. These valves are widely used in transmissions to control line pressure as well as the pressures used to control or signal shift points.

Regulator valves have a spring that holds the valve in a position until the fluid pressure is great enough to compress the spring. At that time, the valve moves and allows some fluid to exhaust. Once the fluid begins to exhaust, the pressure on the valve decreases. The valve remains in the exhausted position until the fluid's pressure is not great enough to work against the tension of the spring.

Valve Bodies

For efficient operation of the transmission, the bands and clutches must be released and applied at the proper time. It is the responsibility of the valve body to control the hydraulic pressure being sent to the different reaction members in response to engine and vehicle load, as well as to meet the needs and desires of the driver.

Shop Manual
Chapter 6, page 172

The fluid passages inside the valve body are often called worm tracks.

The valve body is machined from aluminum or iron castings and has many precisely machined bores and fluid passages (Figure 6-25). Fluid flow to and from the valve body is routed through passages and bores. Although many of these are located in the transmission case and pump housing, most will be found in the valve body. Most of the transmission's valves are also housed in the valve body as well as many different check balls.

Valve bodies are normally fitted with three different types of valves: poppet valves, check ball valves, and spool valves (Figure 6-26). The purpose of these valves is to start, stop, or regulate and direct the flow of fluid throughout the transmission. Fluid flow through the passages and bores is controlled by either a single valve or a series of valves. Most of the control valves operate automatically to direct the fluid as needed to perform a certain function.

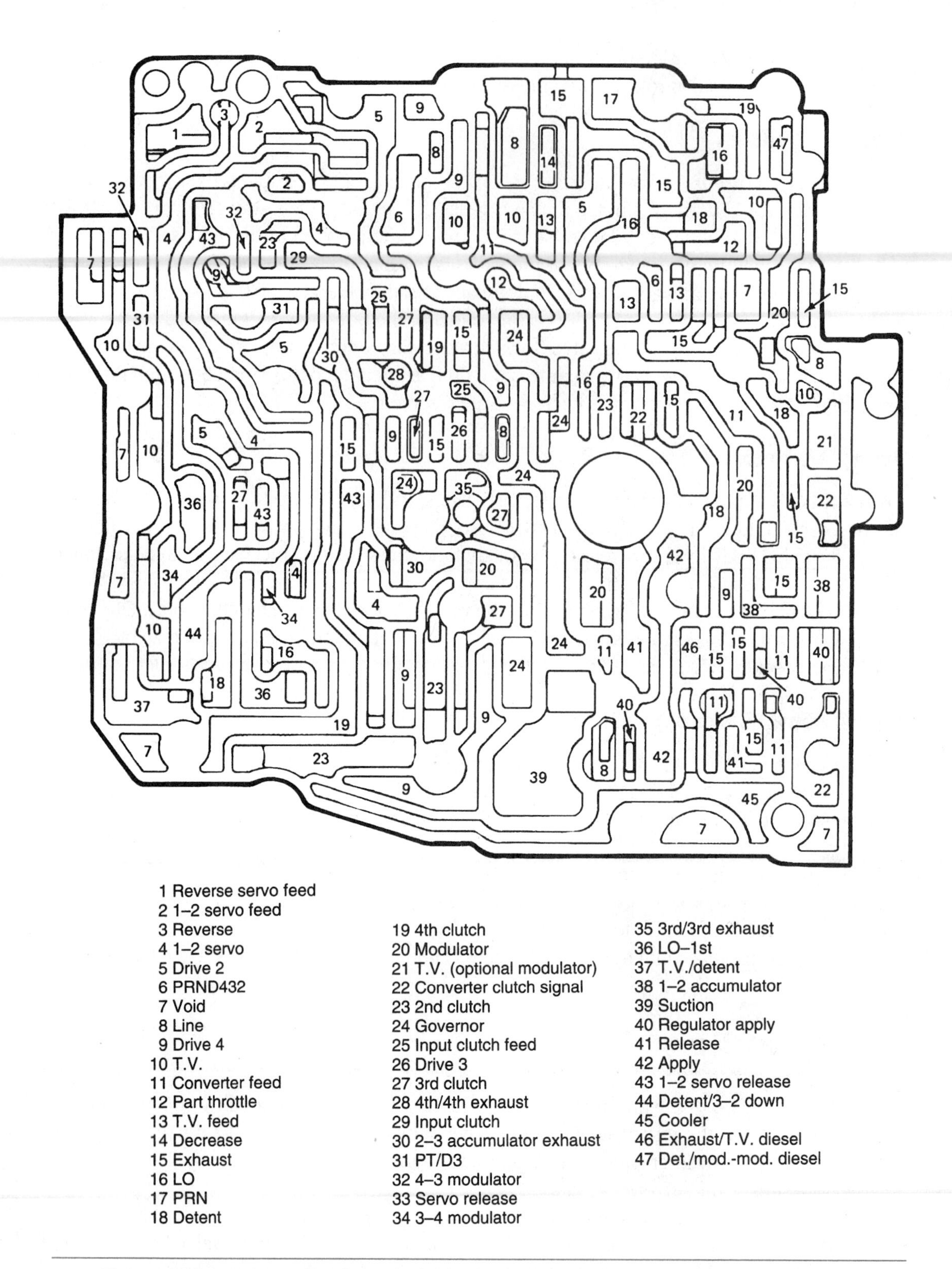

Figure 6-25 Worm tracks of a typical valve body. (Courtesy of the Hydra-Matic Division of General Motors Corp.)

The valve body is composed of two or three main parts: a valve body, separator plate, and transfer plate (Figure 6-27). The **separator plate** and **transfer plate** are designed to seal off some of the passages and they contain some openings that help control and direct fluid flow through specific passages. The three parts are typically bolted together and mounted as a single unit to the transmission housing.

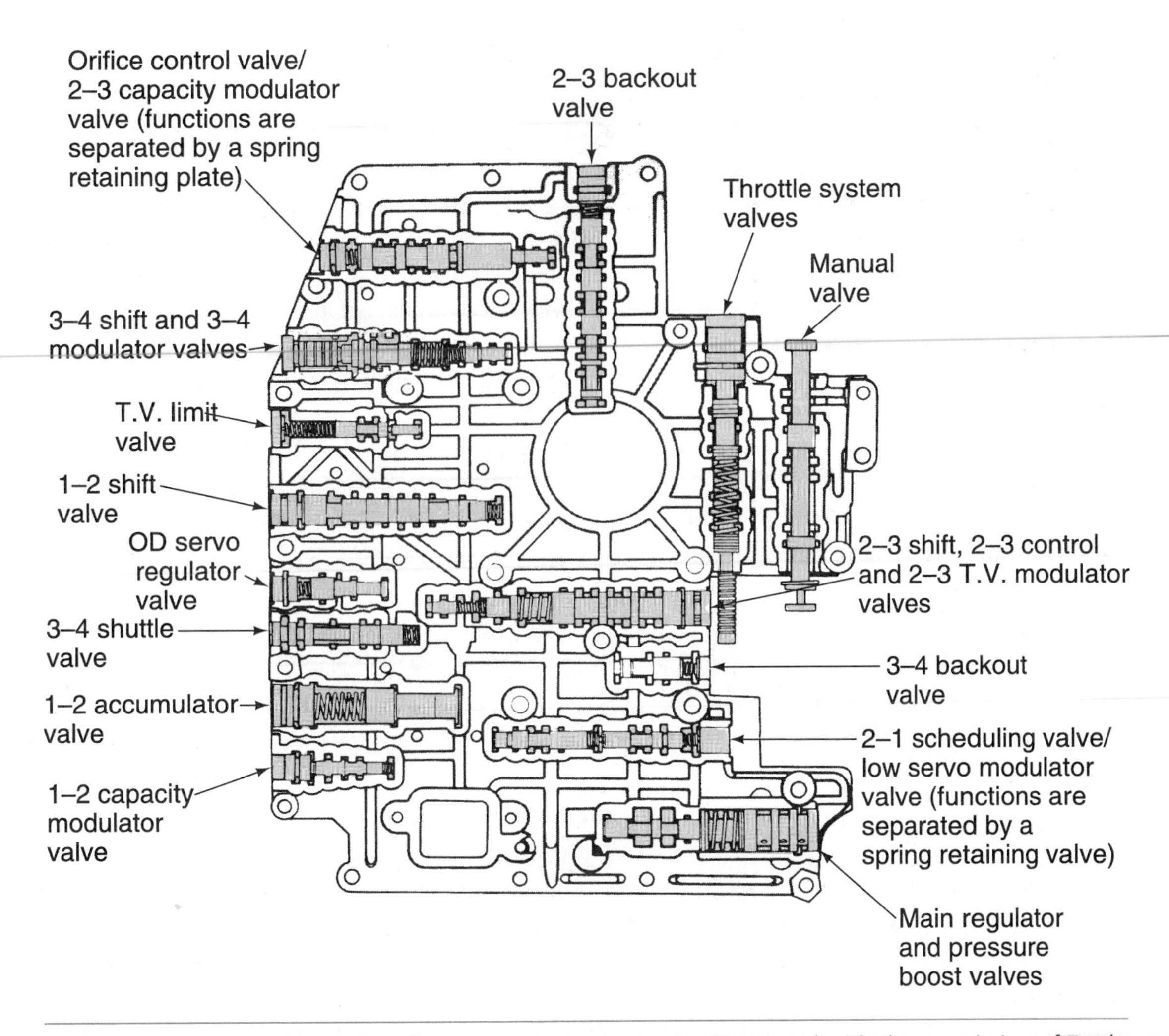

Figure 6-26 Various valves fit into a typical valve body. (Reprinted with the permission of Ford Motor Co.)

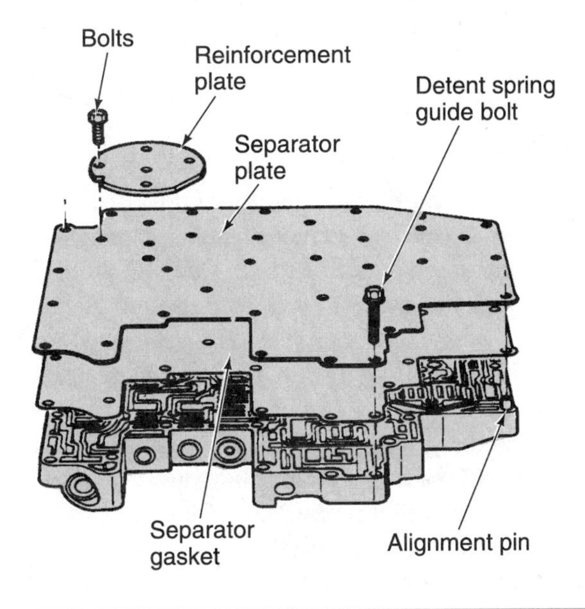

Figure 6-27 Separator plate mounting to a valve body. (Reprinted with the permission of Ford Motor Co.)

Shop Manual
Chapter 6, page 166

Mainline pressure is sometimes referred to as baseline, line pressure, or working pressure.

Pressures

All automatic transmissions use three basic pressures to control their operation: mainline pressure, throttle pressure, and governor pressure. The valves that regulate these pressures are the pressure regulator valve, **throttle valve,** and governor valve. Mainline pressure is a regulated pump pressure that is the source of all other pressures in the transmission. Mainline pressure typically fills the torque converter, lubricates the transmission, supplies fluid to the valve body, and is used to apply the bands and clutches. Governor pressure is a regulated hydraulic pressure that varies with vehicle speed. Throttle pressure is a regulated pressure that varies with engine load or throttle position. Throttle pressure interacts with governor pressure to control shifting.

Although most discussions of an automatic transmission focus on the fluid flows and pressures used to provide for gear selection, it is important to remember that ATF is also used to lubricate the transmission. Part of the pump's output is always directed to lubricate the moving parts of the transmission. ATF flows through various circuits and passages in the case, valve body, shafts, and gear sets to reach these moving parts. Normally, the fluid flowing from the oil cooler is used as the lubricator.

Mainline Pressure

The oil pump provides mainline pressure. This pressure is limited by engine speed and the pressure regulator valve. Mainline pressure is the source of all other pressures used by the transmission. It is the pressure used to engage or apply the clutches and bands within the transmission.

The pressure regulator valve controls line pressure to meet the needs of the transmission regardless of operating speeds. As engine speed increases, the pump works faster and delivers higher pressures than when the engine is running at low speeds. The pressure regulator keeps line pressure from building to a point where it could damage the transmission. Line pressure also increases as the pump fills all of the circuits in the hydraulic system. Again, the pressure regulator prevents excessive pressure buildup by exhausting some fluid when the pressure reaches a predetermined limit. Pump output pressure is line pressure until it reaches that point.

Some of the oil pump's output is routed to the torque converter. Although line pressure is used to direct the ATF into the torque converter, the converter is not pressurized. In fact, a converter works best when it is filled with very low pressure ATF. A special valve is used in the torque converter circuit to limit the amount of pressure in the converter.

Throttle Pressure

Throttle pressure is a signal pressure. The amount of throttle pressure depends directly on the engine's load. Engine load is sensed by mechanical linkages relaying the amount of throttle opening, the effect of engine vacuum on a modulator, a combination of these two, or various sensors tied to a PCM. Many late-model transmissions use a pulse-modulated solenoid, controlled by the computer, to control throttle pressure in response to inputs to the computer. Throttle pressure is a boosted line pressure. The amount that throttle pressure is greater than line pressure represents the amount of load on the engine.

Throttle pressure opposes governor pressure to control shift timing. Basically, governor pressure is supplied to one side of a spool-type shift valve. Throttle pressure is supplied to the other side of the valve. When governor pressure, which builds with vehicle speed, is greater than throttle pressure, the **shift valve** moves and the transmission shifts into the next gear. Under heavy load conditions, throttle pressure opposes the movement of the shift valve. This delays upshifting and allows the transmission to stay in the lower gear for a longer period of time. Shifting will not occur until governor pressure is greater than throttle pressure.

Governor Pressure

Governor pressure increases with an increase of vehicle speed. When governor pressure is higher than throttle pressure, an upshift occurs. Governor pressure is a boosted line pressure. Early transmission designs used a mechanical governor made of springs, weights, and a spool valve to create this signal. Late-model transmissions use a computer-controlled solenoid to control this pressure.

A transmission's change of gears is actually caused by the movement of a shift valve. The shift valve simply controls the direction of line pressure. When the shift valve moves, line pressure is directed to the appropriate apply device. Since throttle and governor pressures only control the movement of the shift valve, they are signal pressures. Signal pressures control the direction of the line pressure.

Boost Pressures

During certain operating conditions, fluid pressure must be increased above its mainline pressure. Increased pressure is needed to hold bands and clutches more tightly. Increasing the pressure above normal mainline pressures allows the vehicle to overcome heavy loads, such as pulling a trailer or climbing a steep hill.

When the vehicle is placed under heavy load, throttle pressure is applied to a boost valve at the pressure regulator. This pressure acting on the boost valve assists the pressure regulator valve's spring in pushing the regulator valve up against the line pressure. Line pressure is able to continue to increase until the pressure on the regulator valve overcomes the spring in the pressure regulator and throttle pressure. The pressure regulator valve now opens its exhaust port with a boosted line pressure, which is used to hold the reaction members tightly to resist slippage. Some transmissions are equipped with two boost valves. The second boost valve is used for improved shifting into reverse gear and for greater holding power when operating in reverse.

When one pressure is designed to work against another, we are use the balanced valve principle, which is common in pump pressure regulator valves, governor valves, and throttle valves.

The THM 4L60 and THM 200-4R transmissions are examples of current transmissions that use two boost valves.

Pressure Regulator Valves

Automatic transmissions use a positive displacement oil pump. As the pump delivers fluid to the transmission, the pressure increases as the engine's speed increases. Enough pressure can be generated to stall the oil pump. To prevent this stalling, a pressure regulator valve is normally located in the valve body (Figure 6-28). The pressure regulator valve serves as a relief valve in the pump output circuit to develop, regulate, and maintain mainline pressure.

Shop Manual
Chapter 6, page 169

Pressure-regulating valves use the principles of both a pressure relief valve and a spool valve. As a relief valve, they respond and move to pressures that are either greater or less than the tension of the valve's spring. As a spool valve, pressure regulator valves toggle back and forth in their bore to open and close an exhaust port. Some pressure regulator valves incorporate a boost valve, which increases the regulated pressure.

The fluid pressure from the regulator is sometimes referred to as the main control pressure.

The pressure regulator valve has three primary purposes and modes of operation: filling the torque converter, exhausting fluid pressure, and establishing a balanced condition (Figure 6-29). When the pump begins to turn and send fluid out through the transmission, there is little resistance to fluid flow in the system; therefore, pressure does not build up. During this time the springs hold the pressure regulator valve in the closed position and fluid flows throughout the transmission. As the pressure increases beyond a predetermined level, normally about 60 psi, the regulator valve is moved against the tension of the springs. This movement opens an additional outlet port, which sends fluid out to the torque converter. The opening of this port also decreases the pressure in the system.

The pressure regulator valve may block off fluid flow to the converter if there is a large decrease in pressure.

Fluid flows into the converter circuit and the converter becomes pressurized. The torque converter pressure regulator valve has several spools and lands, so when different oil pressures are applied, converter pressure can vary in proportion to torque requirements. Converter pressure is

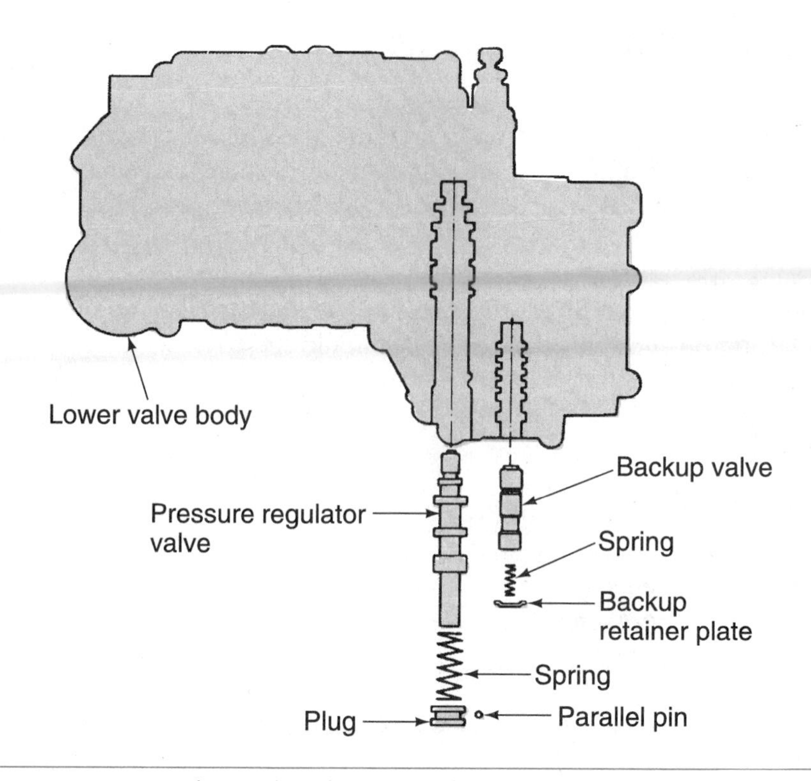

Figure 6-28 Pressure regulator valve. (Courtesy of Nissan Motor Co., Ltd.)

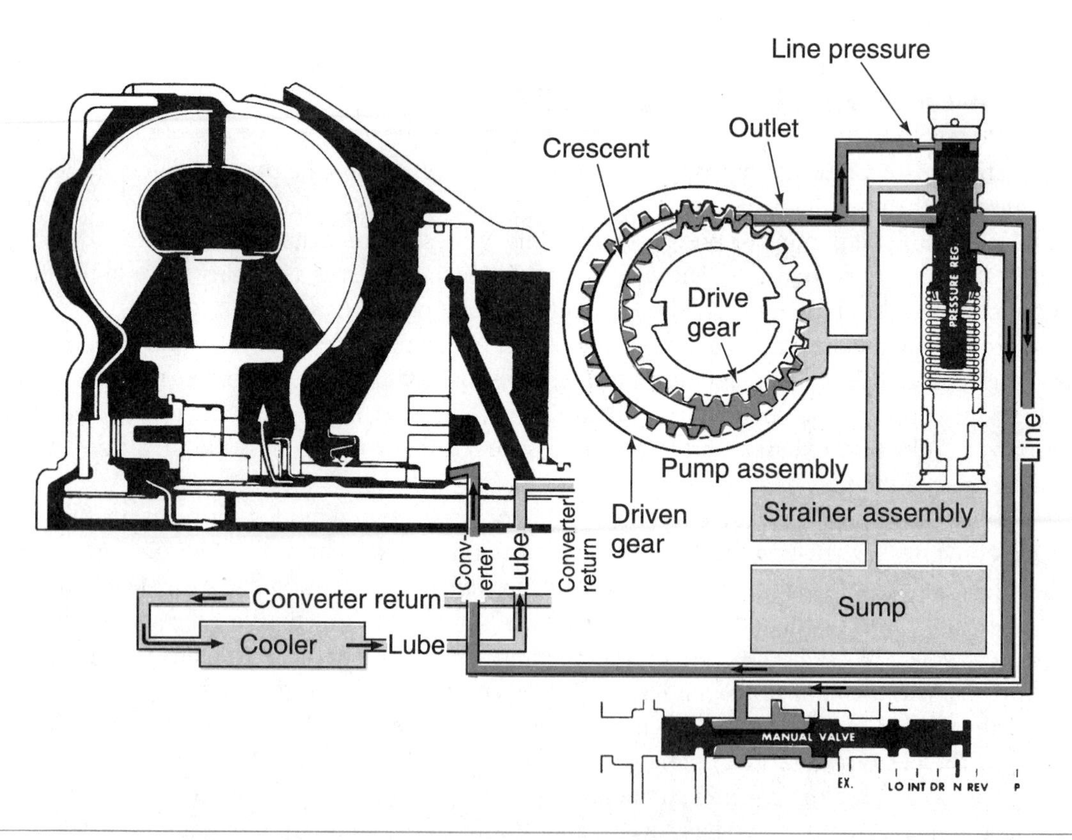

Figure 6-29 Oil pump output is regulated by the pressure regulator valve. (Courtesy of the Hydra-Matic Division of General Motors Corp.)

used to transmit torque and keep ATF circulating in and out of the torque converter. This action reduces the possibility of forming air bubbles in the fluid and helps keep the fluid cool. Converter pressures typically do not vary when the converter is in its lockup mode of operation.

Once the converter is pressurized, the pressures of the fluid can again begin to increase. Therefore, an additional exhaust port is needed to regulate the fluid flow from the pump. The high pressure from the converter moves the pressure regulator valve and uncovers the exhaust port. All fluid flow not needed for transmission operation is exhausted back into the fluid reservoir.

With fluid pressure acting at one end of the valve and spring tension at the opposite end, the pressure regulator valve is in a balanced position. As soon as the fluid pressure is less than the tenssion of the spring, the valve moves and allows pressure to build up again. This process is ongoing while the transmission is operating. The constant toggling of the pressure regulating valve maintains a constant pressure in the system. The valve's spring controls the mainline pressure. If the pressure begins to decrease, the spring moves the valve and prevent some of the fluid from flowing to the reservoir, thus maintaining the desired mainline pressure. If the pressure increases, the valve again moves and reopens the exhaust outlet to the reservoir.

Governors

A governor is a control device driven by the transmission's output shaft. It senses road speed and sends a fluid pressure signal to the valve body to either upshift or downshift. Basically, the governor valve controls transmission shift points and quality based on vehicle speed. A governor converts mainline pressure into a variable governor pressure according to the speed of the vehicle. Governor pressure increases with increasing vehicle speeds and this pressure is directed to the shift valve.

Shop Manual
Chapter 6, page 179

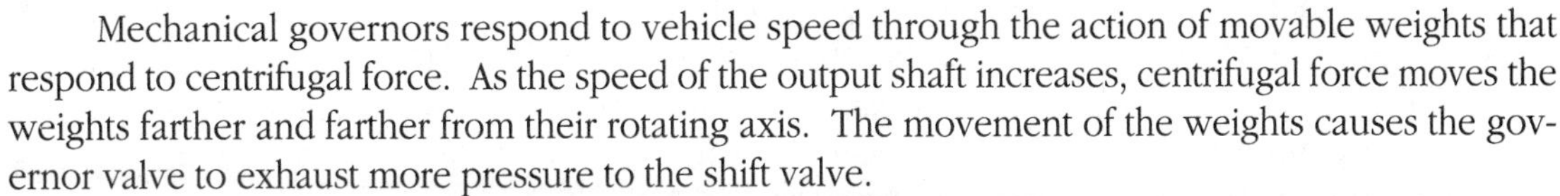

Mechanical governors respond to vehicle speed through the action of movable weights that respond to centrifugal force. As the speed of the output shaft increases, centrifugal force moves the weights farther and farther from their rotating axis. The movement of the weights causes the governor valve to exhaust more pressure to the shift valve.

As the pressure increases, the spring tension on the shift valve is eventually overcome and the valve moves to the upshifted position. In this position, the inlet and outlet ports of the valve bore are in the same valley of the shift valve. This results in an upshift as fluid flows out of the outlet port and through the connecting worm tracks to engage the apply device for the next higher gear. A decrease in output shaft speed will result in a decrease in pressure and cause a downshift.

Many different designs of governors have been used in automatic transmissions. All governors, regardless of design, regulate a variable pressure in relation to road speed. The governors in automatic transmissions either rotate with or are turned by the transmission's output shaft. Governors are either mounted in the transmission case and driven off the output shaft by a worm gear or mounted directly on the output shaft. On FWD vehicles, the governor is driven by a gear on the final drive unit.

The governor assembly normally consists of a small separate valve body with fluid passages for line pressure, governor pressure, and an exhaust port to send fluid back to the reservoir. The governor's valve body has one or two valves controlled by the weights and spring tension. When the vehicle is stopped, fluid to the governor circuit is blocked. As the vehicle begins to move, the governor rotates and centrifugal force will begin to move the valves and weights. This allows a regulated mainline pressure to enter the governor circuit, which sends fluid to the shift valves.

There are three common designs of governors: shaft-mounted, gear-driven with check balls, and gear-driven with a spool valve (Figure 6-30). All these designs have primary and secondary weights, springs, and valves connected to the output shaft by a valve shaft. The heavier primary weights move out at low speeds and the lighter secondary weights move out at higher speeds.

The valve shaft is referred to as a governor pin by some manufacturers.

The shaft-mounted governor has a spring-loaded spool valve attached to the output shaft of the transmission. The governor's weights are mounted so that the primary weight moves against spring tension while the secondary weight moves against mainline pressure. The combined force

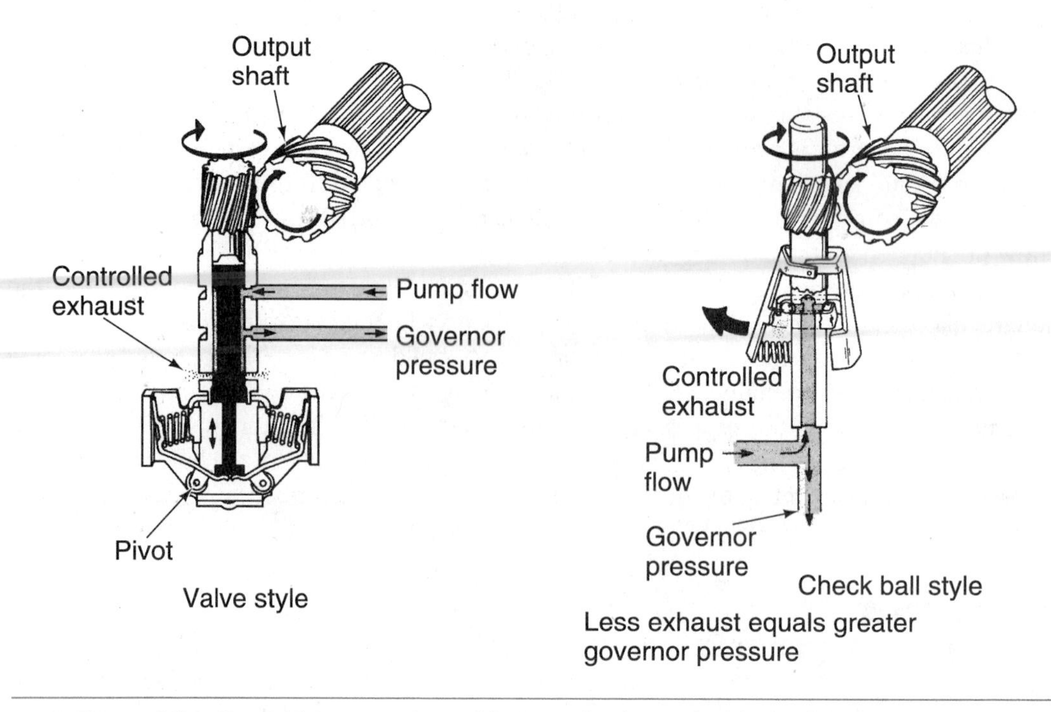

Figure 6-30 Gear-driven governors with a spool valve and with check balls. (Courtesy of the Hydra-Matic Division of General Motors Corp.)

of the weights and springs on one side of the output shaft is much greater than the force of the valve on the other side. Therefore, as the output shaft rotates faster, the valve is pulled inward by the valve shaft. As speed increases, fluid is exhausted out through a port that feeds the shift valves.

As the output shaft's speed increases, the governor gradually opens the circuit and increases governor pressure until the governor pressure equals mainline pressure. At this point, fluid is exhausted and the pressure decreases. After the pressure is exhausted, the valve once again is pulled inward by the rotating weights, allowing mainline pressure into the governor passage, and the cycle starts again.

It is not unusual for a governor to still be regulating at 60 or 70 mph. Some may regulate at even higher speeds depending on the weights and the spring tension calibration. When the vehicle is stopped, the governor spring forces the valve closed, which blocks mainline pressure from the governor circuit. If the pressure of the governor is supplied by a rear oil pump, it will be mainline pressure only after the vehicle is in motion.

Shop Manual
Chapter 6, page 179

Gear-driven governors with check balls mount to the transmission at a right angle to the output shaft. The centrifugal weights act directly on the check balls that open or close two fluid passages. The check balls are forced open by mainline pressure to exhaust fluid to the valve body. When the vehicle is stopped, the check balls unseat and mainline pressure fed to the governor is allowed to bleed off. As vehicle speed increases, the weights apply force to seat the check balls and restrict the flow of escaping fluid. This increases governor pressure, up to the point where no fluid escapes and governor pressure equals mainline pressure. At this point, governor pressure increases and exerts force on the shift valves.

The gear-driven governor with a spool valve is driven by, and mounted at a right angle to, the transmission output shaft (Figure 6-31). Hydraulically, this governor operates much like the shaft-mounted governor; however, its mechanical operation is different. The weights, which move by centrifugal force, indirectly control the spool valve through levers attached to the weights. The weights also serve as levers. They act against a spool valve located in the governor shaft, which is located in the valve body. As vehicle speed increases, centrifugal force moves the weights out, causing the spool valve to move farther into the valve body. This closes the exhaust port and increases governor pressure.

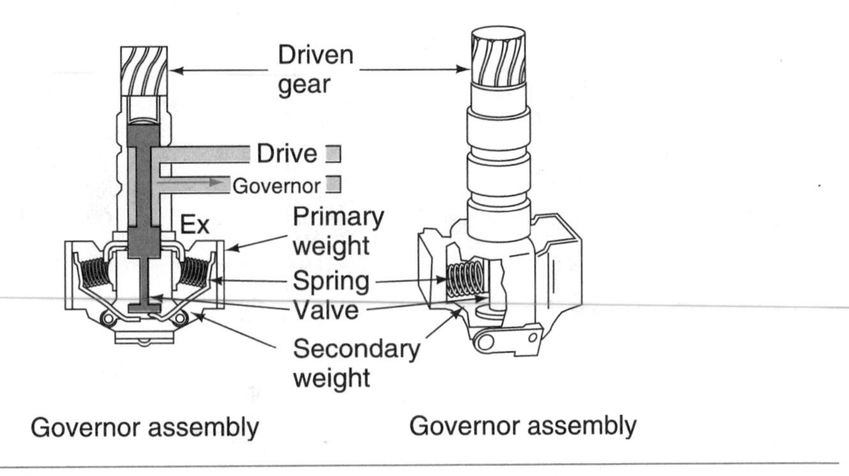

Figure 6-31 Components of a gear-driven governor. (Courtesy of the Hydra-Matic Division of General Motors Corp.)

Vehicle Speed Sensors

Electronically controlled transmissions do not rely on hydraulic signals from a governor to determine when to shift, nor are they fitted with a speedometer cable. Instead, they use permanent magnetic (PM) generators to sense vehicle speeds.

Shop Manual
Chapter 6, page 179

PM generators are often referred to as pulse generators.

A PM generator is an electronic device that utilizes a magnetic pickup sensor located on the transmission/transaxle housing and a trigger wheel mounted on the output shaft. The trigger wheel has a number of projections evenly spaced around it. As the wheel rotates, the projections move by the magnetic pickup. The passing of the projections produces a voltage in a coil inside the pickup assembly. This voltage is actually ac voltage, but it is changed to a digital dc voltage by a converter before it is sent to the transmission control computer. The frequency of the pulsations and the amount of voltage generated by the pickup assembly are translated into vehicle speed by the computer.

Load Sensors

Domestic and import transmissions typically have similar components in function, construction, and operation.

Shift timing and quality should vary with engine load, as well as with vehicle speed. Throttle pressure increases the fluid pressure applied to the apply devices of the planetary units to hold them tightly to reduce the chance of slipping while the vehicle is operating under heavy load. Throttle pressure interacts with mainline pressure to coordinate shift points and shift quality with vehicle and engine load. On some transmissions, a vacuum modulator is used to control throttle pressure. Transmissions that are not equipped with a vacuum modulator use a throttle cable or electronic devices to sense engine load and change fluid pressures. It is not safe to assume that no electronic transmissions are equipped with a vacuum modulator; some are. For example, GM's THM 4T60-E uses a vacuum modulator, in addition to electronic inputs and shift solenoids, to control shift timing and quality.

The terms modulator and throttle are used to describe the same function and pressure in a transmission. Although a vacuum modulator is very different from a throttle pressure sensor, they both generate a hydraulic engine torque signal based on engine load. As a result of this signal, the transmission has many different possible shift points. Each of those shift points is dictated by engine speed and load. This feature allows for a delay in shifting when load demands are high.

Vacuum Modulators

Shop Manual
Chapter 6, page 166

A vacuum modulator measures engine vacuum to sense the load placed on the engine and the drive train. A vacuum modulator is a load-sensing device that increases or decreases fluid pressure in response to a vacuum signal, which varies with throttle opening and vehicle load (Figure 6-32). Engine

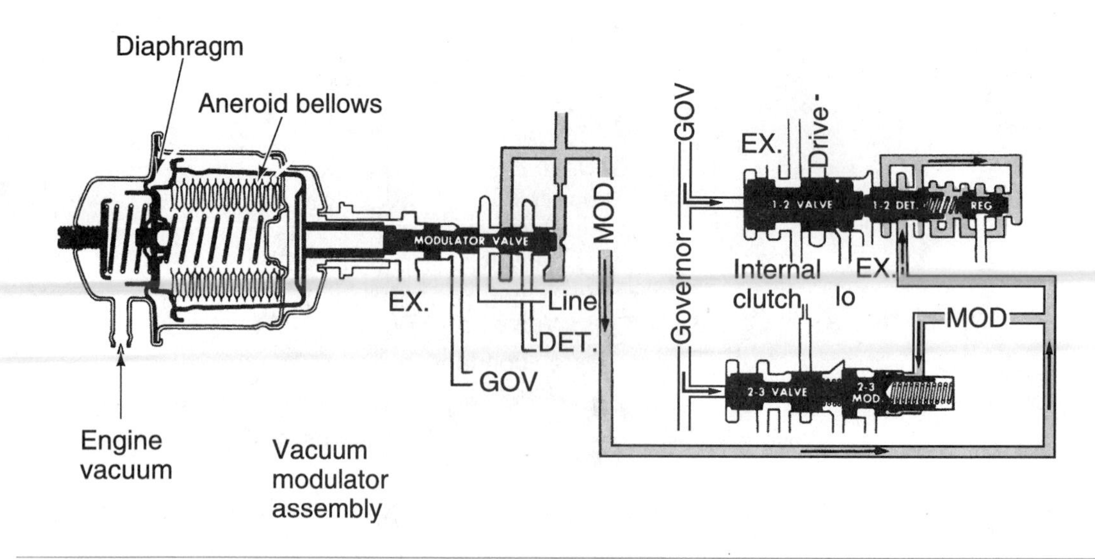

Figure 6-32 Action of a vacuum modulator and the modulated pressure it regulates. (Courtesy of the Hydra-Matic Division of General Motors Corp.)

The vacuum modulator is also called the vacuum diaphragm, vacuum control diaphragm, or vacuum capsule.

vacuum is low when there is a heavy load and high when the load is light. The vacuum modulator allows for an increase in pressure when vacuum is low and a decrease when vacuum is high. When high vacuum is present at the modulator, the pressure regulator valve works normally and maintains normal pressure. However, when the vacuum is low, the modulator allows fluid flow to enter onto the side of the spool valve in the pressure regulator, which allows for an increase in pressure.

The vacuum modulator is normally a small round metal canister that is threaded or push-fit into the rear of the transmission's housing. A vacuum line connects the vacuum modulator to the intake manifold of the engine. This supplies vacuum to operate the modulator.

The vacuum modulator is basically a canister divided into two chambers by a diaphragm. The diaphragm forms a seal between the two chambers. The chamber closest to the transmission housing is open to atmospheric pressure and the other chamber is closed to atmospheric pressure. A pushrod is connected to the open side of the diaphragm. This rod connects the diaphragm to the **modulator valve.** The closed chamber of the vacuum modulator contains a coil spring positioned between the diaphragm and the end of the canister. Engine vacuum is sensed on the spring side of the diaphragm.

The throttle valve is often referred to as the modulator valve.

The throttle valve, or modulator valve, directs fluid flow to other control devices at low vacuum and withholds it at high vacuum. When engine load is low and the throttle is closed, high manifold vacuum retracts the diaphragm and pushrod against spring tension to reduce throttle pressure. As engine load increases and the throttle is opened wider, manifold vacuum drops and the **diaphragm spring** forces the pushrod into firmer contact with the throttle valve to increase throttle pressure.

The movement of the diaphragm can be best explained by a law of nature: a high pressure always moves toward a lower pressure. The diaphragm will move in response to pressure differences on either side of it.

A vacuum diaphragm responds to differences in atmospheric pressure on one side and vacuum on the other side. Normal vacuum modulators are less effective at higher altitudes because atmospheric pressure and engine vacuum are less than they are at sea level. Some vacuum modulators use an altitude-compensating vacuum valve with a spring-type bellows added to the atmospheric side of the valve. When the bellows units are made, a vacuum is drawn in the bellows, which causes atmospheric pressure to compress the bellows. The bellows unit is then sealed to hold the vacuum. As altitude increases and atmospheric pressure decreases, the bellows unit expands because of the decrease in pressure differential. This allows the bellows unit to exert pressure against the atmospheric side of the diaphragm to compensate for the lower atmospheric pressure at higher altitudes.

Some vacuum modulators use two diaphragms; one diaphragm is exposed to intake vacuum, while the other receives a ported vacuum signal. Ported vacuum increases with throttle plate opening; therefore, ported vacuum is available to assist the lower engine vacuum during times of heavy load.

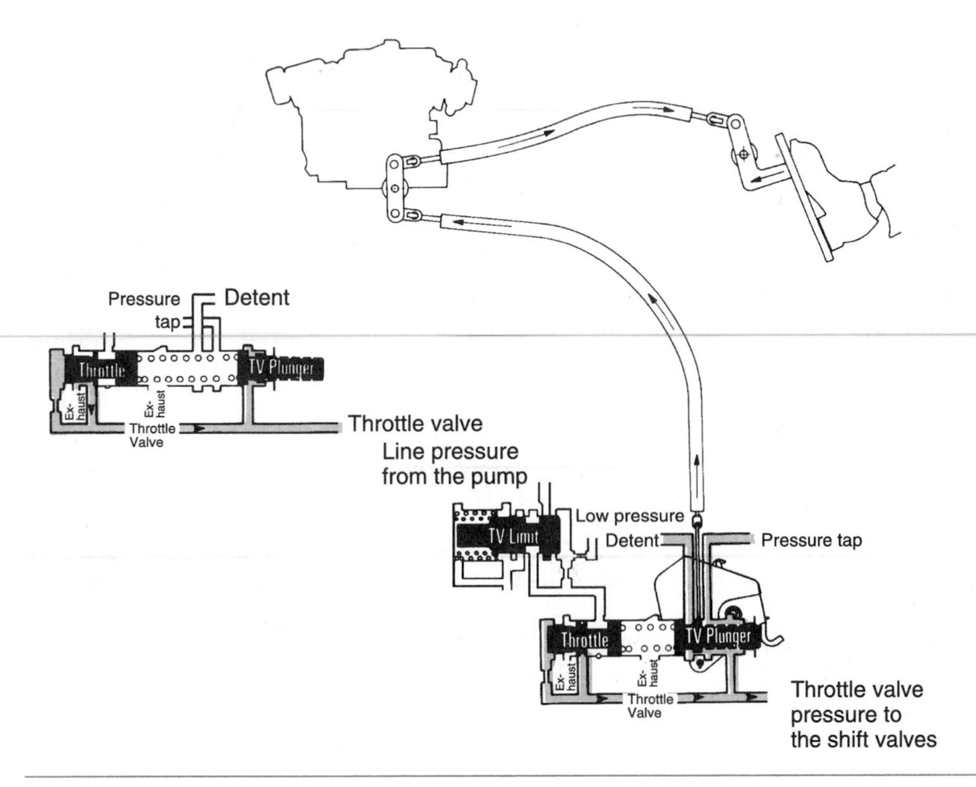

Figure 6-33 Typical cable-type throttle pressure control system. (Courtesy of the Hydra-Matic Division of General Motors Corp.)

Throttle Linkages

Shop Manual
Chapter 6, page 166

Some transmissions use mechanical linkages to control throttle valve pressure and increase line pressure. Throttle pedal movement is used to control the operation of the throttle valve. Therefore, the throttle valve converts mainline pressure into a variable throttle pressure based on the position of the throttle plate (Figure 6-33). The linkage moves the throttle valve against spring pressure—the more the throttle is opened, the higher the throttle pressure—until throttle pressure equals mainline pressure. The throttle cable is connected between the fuel injection throttle body or the throttle linkage at the carburetor and a lever located at the transmission housing (Figure 6-34). The throttle lever moves with the movement of the throttle pedal.

Some automatic transmissions use a rod-type linkage assembly, instead of a cable, to move the throttle lever.

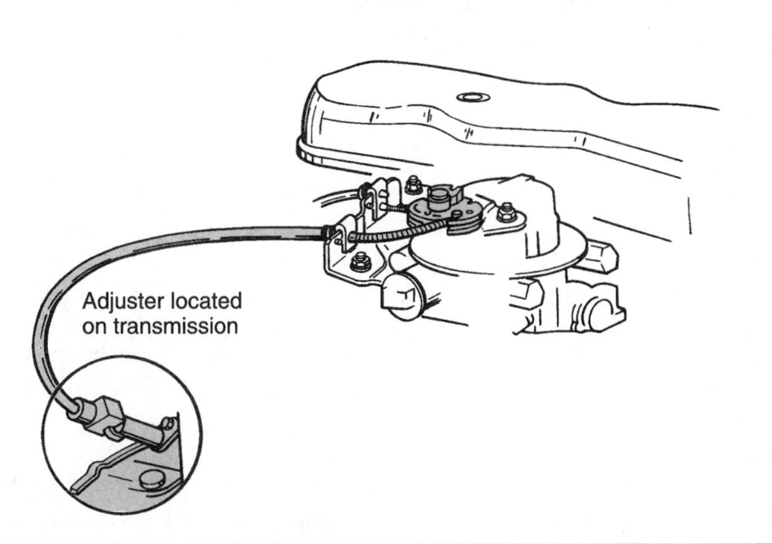

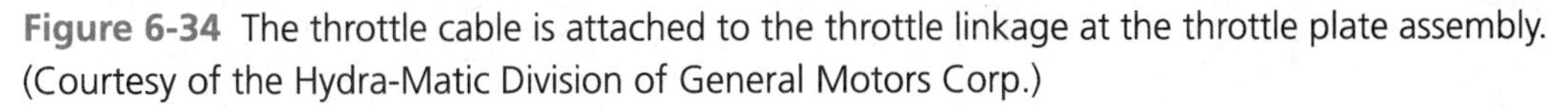

Figure 6-34 The throttle cable is attached to the throttle linkage at the throttle plate assembly. (Courtesy of the Hydra-Matic Division of General Motors Corp.)

When the throttle pedal is depressed, the throttle valve opens to produce throttle pressure, which is directed to the pressure regulator valve. This pressure helps the pressure regulator valve spring hold the pressure regulator valve in position to close the exhaust port, which causes the pressure to increase. This increased pressure is used to hold the apply devices and causes a delay in the upshift. As the throttle pedal is released, throttle pressure is decreased, as is mainline pressure. This decrease in pressure allows for a downshift when vehicle speed reaches a particular point.

If the throttle pedal is depressed quickly, there is a sudden increase in throttle pressure and mainline pressure. Throttle pressure becomes higher than governor pressure. The high throttle pressure and coil spring tension force the shift valve to move to the downshifted position against governor pressure. The transmission automatically downshifts to the next lower gear. The position of the shift valve blocks fluid pressure at its inlet port, which prevents an upshift during this time. Often, the hydraulic circuit is fitted with a kickdown valve, which applies additional pressure to the shift valve and provides for a quick and positive downshift.

A BIT OF HISTORY

Oldsmobile introduced the Hydramatic in 1940. This was the first fully automatic transmission in a mass-produced U.S. car. Two planetary gear sets were used and controlled by opposing throttle and governor pressures to provide four forward speeds. Cadillac offered the Hydramatic in 1941 and Pontiac offered it in 1948.

MAP Sensor

Engine load can be monitored electronically through the use of various electronic sensors that send information to an electronic control unit, which in turn controls the pressure at the valve body. The most commonly used sensor is the **manifold absolute pressure (MAP) sensor.** The MAP sensor senses air pressure or vacuum in the intake manifold. The control unit uses this information as an indication of engine load. A pressure-sensitive ceramic or silicon element and electronic circuit in the sensor generate a voltage signal that changes in direct proportion to pressure. A MAP sensor measures manifold air pressure against a precalibrated absolute pressure; therefore, the readings from these sensors are not adversely affected by changes in operating altitudes or barometric pressures.

Shift Timing

Shift timing is determined by throttle position and governor pressure acting on opposite ends of the shift valve. When a vehicle is accelerating from a stop, throttle pressure is high and governor pressure is low. As vehicle speed increases, the throttle pressure decreases and the governor pressure increases. When governor pressure overcomes throttle pressure and the spring tension at the shift valve, the shift valve moves to direct pressure to the appropriate apply device and the transmission upshifts.

Shop Manual
Chapter 6, page 165

Shift Valves

Shift valves control the upshifting and downshifting of the transmission by controlling the flow of pressurized fluid to the apply devices that engage the different gears. Transmissions have several shift valves that provide for control of all possible gear changes. Most shift valves are spool valves operated by two different hydraulic pressure sources that oppose each other. As one pressure gains strength over the other, the valve moves in the direction of the lower pressure.

For example, when the governor sends a signal that would normally force an upshift, throttle pressure may delay the shift, which allows for continued operation in a lower gear when the vehicle still needs the gear reduction. While operating under a load, governor pressure must overcome the throttle pressure plus the spring tension on the shift valve before it can force an upshift.

Manual Valve

The **manual valve** is a spool valve operated manually by the gear selector linkage. When the gear selector is placed in a forward or reverse gear, the manual valve directs mainline pressure to the circuits that apply the reaction members involved in that operating range. The gear selector linkage positions the manual valve in the valve body. If the driver selects the drive range, the manual valve is moved into a position that allows fluid pressure to flow across the manual valve valley through its outlet port to activate the forward circuit. If the gear selector is moved to the reverse position, the manual valve is moved to open the reverse inlet and outlet ports to activate the reverse circuit.

Kickdown is a forced downshift accomplished by increasing throttle pressure using a kickdown valve.

Kickdown Valve

A valve body may be fitted with a kickdown circuit to provide a downshift when additional torque is needed. When the throttle is opened quickly, throttle pressure increases rapidly and is directed to the kickdown valve. This moves the kickdown valve to open a port, which allows mainline pressure to flow against the shift valve. The spring tension on the shift valve, kickdown pressure, and throttle pressure push on the end of the shift valve causing it to move to the downshift position, forcing a quick downshift.

The automatic downshifting of the transmission is often referred to as hitting passing gear.

A transmission can also automatically downshift when operating under a load, such as climbing a hill. During this time, throttle pressure exceeds governor pressure and forces a downshift.

Some transmissions use a solenoid kickdown valve operated electrically by a switch on the linkage of the throttle pedal. This switch senses when the throttle plate is wide open and allows the kickdown valve to increase throttle pressure, which forces an immediate downshift.

Electronic Shift-Timing Controls

Many current transmissions use electronics to control shift timing and quality. Shifting electronically allows for shifting at better times than does shifting by hydraulics. The use of electronics also reduces the occurrence of many common transmission problems, such as dirty, seized, or worn springs, check balls, valves, and orifices.

Shop Manual
Chapter 3, page 166

Shifting is based on inputs to a control computer from various sensors, such as engine temperature, engine speed, engine load, vehicle speed, throttle position, and gear selector position. The computer may be the vehicle's engine control computer or one designated just for transmission control. The computer compares the information from the sensors against the shifting instructions programmed into it. The computer then controls the appropriate solenoid valves to provide for optimal shift timing.

Solenoid-Operated Valves

Most solenoid-operated valves are ball-type valves that open and close a hydraulic passage. They are designed to block fluid flow when voltage is applied to the solenoid and to allow fluid flow when voltage is not applied. Most transmissions have two solenoid valves that control shifting through all forward gears. By controlling which solenoid is energized, the computer controls the shift timing. Many transmissions energize one solenoid for first gear, the other solenoid for third gear, both solenoids for second gear, and neither solenoid for fourth gear.

Oil Circuits

Shop Manual
Chapter 6, page 165

An oil circuit is actually a schematic that shows the fluid path and valves necessary to perform a specific function.

Boosted throttle compensates for lower rate of vacuum change at a throttle opening of more than 50%.

The valve body is the master flow control for an automatic transmission. It contains the worm tracks and numerous bores fitted with many spool-type and other type valves. Each passage, bore, and valve forms a specific oil circuit with a specific function. These oil circuits can be traced and followed through the use of flow diagrams. These diagrams can also be used to see clearly how each valve functions to allow the transmission to operate in any particular gear range.

By tracing through an oil circuit, the fluid flow through passages, bores, and valves to the components that make the actual shift can be seen. An automatic transmission has a circuit for each of its functions.

To fully understand how a particular transmission works, locate a flow diagram for that transmission and trace the oil circuit for each gear. Automatic transmission hydraulic systems vary with each design and model, as well as model year. However, by carefully analyzing the oil circuits given in Figure 6-35, you will not only gain a better understanding of the operation of an automatic transmission, but also be able to understand the oil circuits of other transmissions.

Following are some guidelines that will make tracing through an oil circuit's flow diagram easier:

1. Trace through one circuit at a time. Flow diagrams are usually available for each of the possible gear ranges plus the following circuits: supply, main control pressure, converter and cooler, governor, throttle valve, boosted throttle, modulated throttle, **accumulator,** and converter clutch.
2. Begin at the oil pump.
3. Determine the type and amount of pressure the manual shift valve circuit and the modulator valve circuit are receiving.
4. Trace the main path of the circuit first, then trace the effect of the alternate circuits.
5. Since not all diagrams show the actual position of the valves, follow the flow based on your knowledge.
6. Flow diagrams do not show the exact location of the components so don't be misled by them. They are simply a summary of the flow.
7. Trace through the fluid passages as they are drawn.
8. Pay particular attention to the direction of the fluid flow and the placement of the orifices and check valves.

Using these guidelines and the oil circuits shown in Figure 6-35, take a simple look at the shifting of a GM THM 4T60-E transmission. Because this is an electronically controlled four-speed automatic transmission with lockup torque converter, it is a good example using oil circuits to understand transmission operation. Use a pencil or pen to trace the oil circuit diagrams and to follow along with the discussion. The operational overview will be presented according to the various gear selector positions available.

PARK and NEUTRAL. With the engine running, fluid pushed from the transmission's oil pan, through the filter, and into the oil pump. The oil pump supplies the line pressure according to the calibration of the pressure regulator. The manual valve, which is controlled by the gear selector, directs line pressure into the PRN passage. LIne pressure is not sent elsewhere in the transmission and the transmission is effectively disconnected from the engine.

REVERSE. The gear selector moves the manual valve, which allows line pressure to move from the PRN passage into the reverse fluid passage. Reverse fluid seats the #5 check ball and directs fluid to the reverse servo boost valve and through the reverse servo feed orifice. After passing through the orifice, the fluid enters the reverse servo passage and forces the servo's piston to move against the servo's spring pressure. The movement of the piston applies the reverse band. During hard acceleration, while in reverse, the reverse boost valve can allow the reverse fluid to bypass the reverse feed orifice and enter directly into the reverse servo passage. This allows for a quick fill of the servo passage and a quick apply of the reverse band. Both of these actions prevent band slippage during abusive shifts from PARK or NEUTRAL into REVERSE.

OVERDRIVE. The manual valve is moved by the gear selector. Its position allows line pressure to fill the D-4 passage. D-4 pressure seats the #6 check ball and sends the fluid through the forward servo feed orifice into the servo apply passage. Servo apply fluid applies the forward band. The transmission is now operating in *first gear.* D-4 fluid also feeds the servo apply passage through the thermo element. D-4 pressure is also directed, by the manual valve, to the 1–2 shift valve, 2–3 shift valve, 1–2 accumulator valve, 2–3 accumulator valve, and 3–4 accumulator valve.

When the PCM receives the correct information from its sensors, it will shut down the "A" shift solenoid. This allows the line pressure at the 1–2 shift valve to exhaust through the solenoid. The spring pressure on the 1–2 shift valve causes the valve to move once the line pressure is exhausted. The movement of the valve allows the fluid to enter the 2nd fluid passage. The 2nd fluid feeds to the #2 check ball and seats it, forcing the fluid through a feed orifice before applying the 2nd clutch. At the same time, 2nd clutch fluid is fed to the 1–2 accumulator piston and moves the piston up against spring pressure and 1–2 accumulator fluid pressure, cushioning 2nd clutch apply. *Second gear* is engaged. Notice that the 1–2 accumulator fluid is then forced through the 1–2 accumulator passage and exhausts at the 1–2 secondary accumulator valve.

When ideal operating conditions are met, the PCM turns off the "B" shift solenoid allowing the fluid at the solenoid to exhaust. This allows spring force to move the 4–3 manual downshift valve. Also, line pressure at the end of the 2–3 shift valve moves the 3–2 manual downshift valve against spring pressure. Line pressure at the 3–2 downshift valve also enters the "B" shift solenoid passage and helps prevent the 1–2 shift valve from moving. When the 2–3 shift valve moves, D-4 fluid enters the 3rd fluid passage. The 3rd fluid passage feeds fluid to the converter clutch solenoid (where it exhausts through the solenoid), the 1–2 shift valve, and against the #9 check ball to seat it. This last action forces fluid through an orifice and into the 3rd clutch passage. The 3rd clutch passage directs fluid to seat the #4 check ball, sending the fluid into the 3rd clutch/lo–1st passage to apply the 3rd clutch. The 3rd clutch fluid is also sent to the 2–3 accumulator piston and forces the piston against the spring and 2–3 accumulator fluid to cushion the 3rd clutch apply. *Third gear* is now engaged. The 2–3 accumulator fluid is then forced to exhaust at the 2–3 accumulator valve. In third gear, the input clutch is released to allow the input clutch apply fluid to exhaust through the 3–4 shift valve into the input clutch feed passage. At the 2–3 shift valve, the exhausting input clutch apply fluid is directed into the D-3 passage and out through the manual valve.

The torque converter clutch can be applied in third gear. This will happen only when the PCM receives the appropriate input signals. At this time, the TCC solenoid is turned on and is no longer exhausting the TCC signal fluid. This moves the converter clutch valve against line pressure and the strength of the valve's spring. With the valve in this position, release fluid from the converter clutch is allowed to exhaust; converter feed pressure passes through the clutch valve, enters the TCC accumulator passage, and is sent to the TCC accumulator piston and the converter clutch regulator valve; and regulated apply fluid enters the apply passage. The apply fluid seats the #1 check ball against the release fluid and applies the converter clutch. Apply fluid is also at the torque converter blow-off valve to exhaust excess fluid pressure. Converter feed fluid at the converter clutch valve is allowed to feed into the cooler passage where it passes through the cooler check valve and cooler and into the transaxle's lubrication system.

To obtain fourth gear, the PCM turns on the "A" shift solenoid. This directs line pressure into the "A" shift solenoid passage to the 3–4 shift valve. "A" shift solenoid pressure shifts the valve against spring pressure allowing 3rd fluid to enter the 4th fluid passage and the 4th clutch fluid passage. The 4th fluid is directed to the 4th clutch discrete pressure switch to close the switch and seat the #10 check ball, which forces the fluid through a feed orifice to move the 3–4 accumulator piston while applying the 4th clutch. *Fourth gear* is now engaged. The 3–4 accumulator fluid on the spring side of the piston is forced through the 3–4 accumulator pin into the 3–4 accumulator passage and exhausts at the 3–4 accumulator valve.

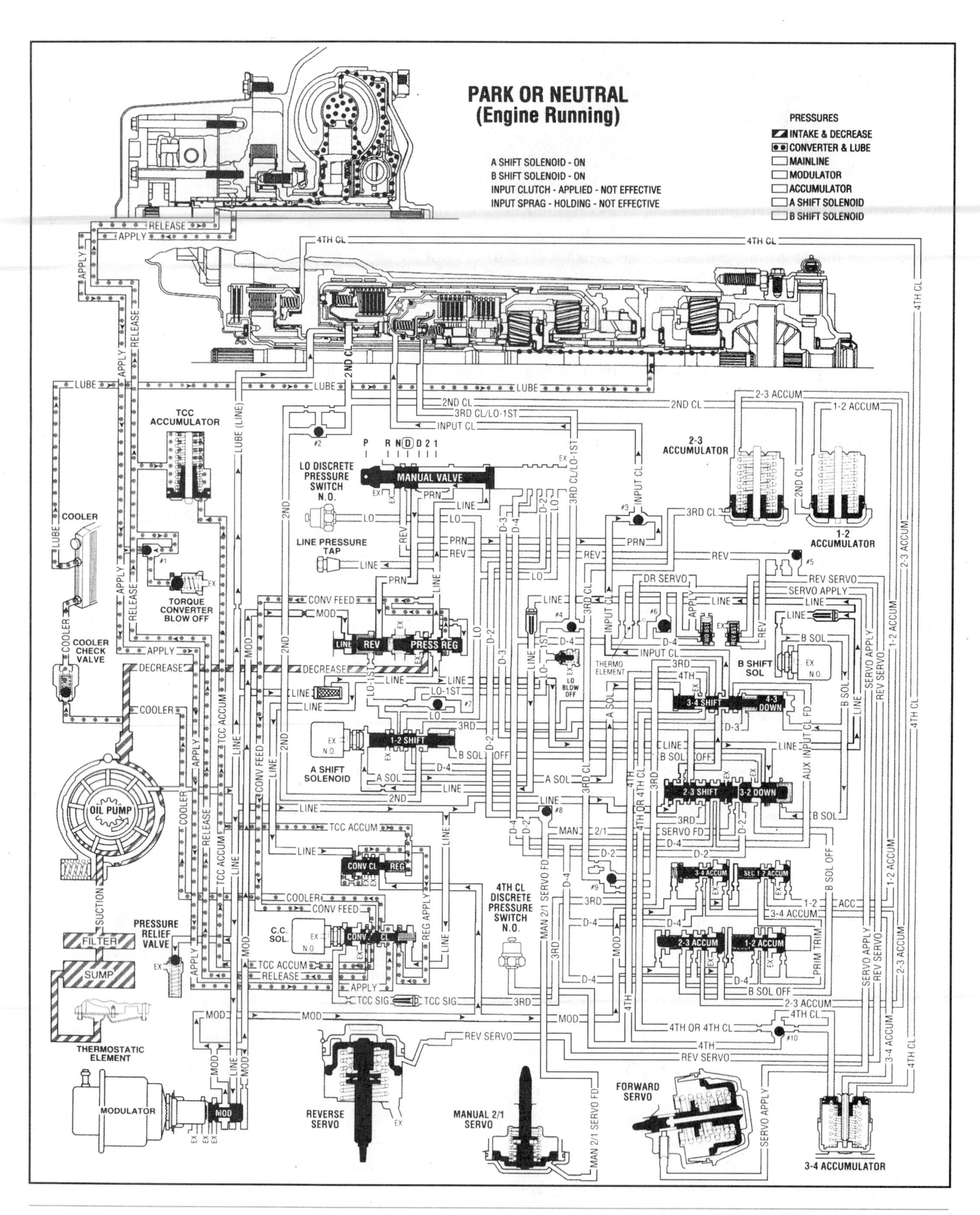

Figure 6-35 Oil circuit for a THM 4T60-E transaxle. (Courtesy of the Chevrolet Motor Division of General Motors Corp.)

PARK OR NEUTRAL
ENGINE RUNNING

"A" SHIFT SOLENOID - ON

"B" SHIFT SOLENOID - ON

INPUT CLUTCH - APPLIED (NOT EFFECTIVE)

INPUT SPRAG - HOLDING (NOT EFFECTIVE)

In Park (P) range with the engine running, fluid is pulled from the sump, through the filter (100) and into the oil pump (200). Pump output is regulated by the pressure regulator valve (313) and a "decrease" passage which creates a calibrated output pressure called "line pressure". In PARK (P) range, line pressure is directed to the following components:

Pressure Regulator Valve (313): Regulates pump output in response to modulator pressure, PRN pressure at the reverse boost valve (310) and spring force from the pressure regulator valve springs (311, 312).

Manual Valve (404): Manually controlled through the gear selector lever, it directs line pressure from the pressure regulator valve (313) into the PRN passage in PARK, REVERSE or NEUTRAL. The manual valve (404) also provides for a hydraulic override of the electronic shift control system in the transaxle for manual third and manual second gear selection.

Modulator Valve (34): Controlled by the vacuum modulator, it regulates modulator pressure inversely to engine vacuum. Modulator pressure is directed to the line boost valve (304), 1-2 accumulator valve (341), secondary 1-2 accumulator valve (347), 2-3 accumulator valve (343) and 3-4 accumulator valve (350).

2-3 Shift Valve (357): In PARK range, line pressure passes through the 2-3 shift valve (357) and feeds the input clutch feed passage. Line pressure then passes through the 3-4 shift valve (362), around the #3 checkball (372, located in the channel plate) to apply the input clutch. Although the input clutch is applied and the input sprag is holding, they are not effective because neither the forward band assembly (688) or the reverse band assembly (615) are applied.

Converter Clutch Valve (335): Line pressure and spring force holds the valve against the converter clutch solenoid (315) allowing converter feed pressure into the release passage and the TCC accumulator passage. Release fluid is then directed to the spring side of the TCC accumulator piston (416) and to the release side of the converter clutch plate. TCC accumulator fluid is directed to the TCC accumulator piston (416) located in the channel plate assembly (400).

Converter Clutch Regulator Valve (330): Biased by modulator and TCC accumulator pressure, it uses line pressure to provide regulated apply fluid to the converter clutch valve (335).

"A" Shift Solenoid (315): Controlled by the Powertrain Control Module (PCM), it is energized (On) in PARK range. Line pressure moves the 1-2 shift valve (318) against spring pressure and directs line fluid into the "A" shift solenoid passage. "A" shift solenoid fluid is then directed to the 3-4 shift valve (362).

"B" Shift Solenoid (315): Controlled by the Powertrain Control Module (PCM), it is energized (On) in PARK range and directs line pressure into the "B" shift solenoid passage. "B" shift solenoid fluid is directed to: the 4-3 manual downshift valve (360) and moves the valve against spring pressure, and also the 3-2 manual downshift valve (356).

Pressure Relief Valve (321): Exhausts line pressure above 1,690-2,480 kPa (245-360 psi).

Line Pressure Tap (38): Allows line pressure to be monitored with a gage.

Figure 6-35 Oil circuit for a THM 4T60-E transaxle. (Courtesy of the Chevrolet Motor Division of General Motors Corp.) *(continued)*

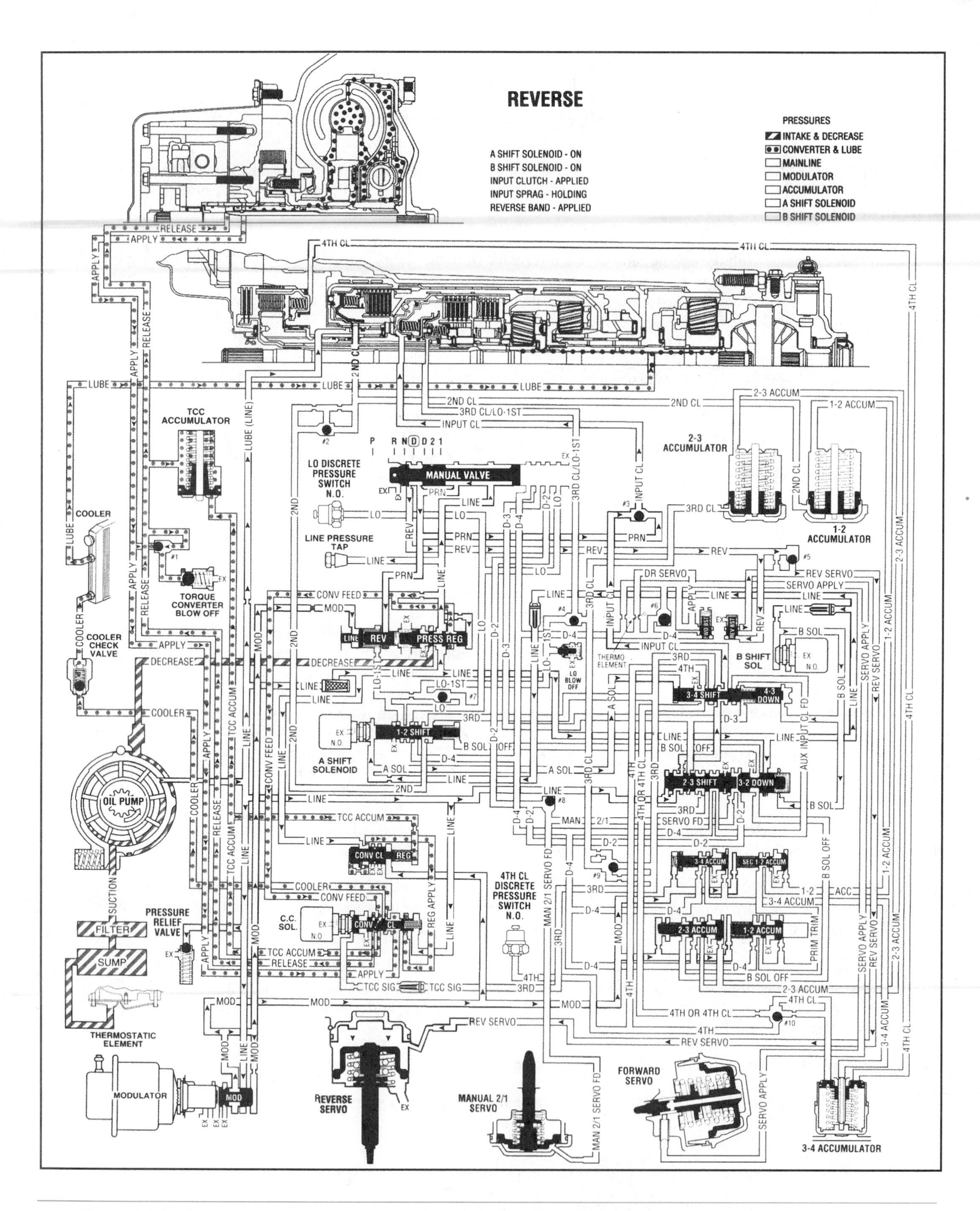

Figure 6-35 Oil circuit for a THM 4T60-E transaxle. (Courtesy of the Chevrolet Motor Division of General Motors Corp.) *(continued)*

REVERSE

"A" SHIFT SOLENOID - ON | INPUT CLUTCH - APPLIED | REVERSE BAND - APPLIED

"B" SHIFTSOLENOID - ON | INPUT SPRAG - HOLDING

When the gear selector lever is moved to REVERSE (R), the manual valve allows line pressure to be fed through the PRN passage into the reverse fluid passage. Reverse fluid seats #5 checkball (372) and directs fluid to the reverse servo boost valve (365) and through the reverse servo feed orifice. Fluid passing through the orifice enters the reverse servo passage and forces the reverse servo piston (44) to overcome spring pressure to apply the reverse band (615). Depending on throttle position, engine manifold vacuum supply to the vacuum modulator can boost to full line pressure, 2100 kPa (305 psi) at 0 kPa (0" psi) engine manifold vacuum.

Manual Valve (404): Is moved manually to the right through the gear selector lever and allows line pressure to enter the reverse fluid passage through the PRN fluid passage.

#5 Checkball (373): Located in the valve body, (300), it blocks the reverse servo feed passage forcing reverse fluid through an orifice in the spacer plate (370) into the reverse servo passage. When the manual valve (404) is moved out of reverse, the checkball unseats allowing reverse servo fluid to exhaust through the ball seat instead of through the orifice.

Reverse Boost Valve (367): Opens under hard acceleration (high line pressure/high throttle position) to allow reverse fluid to by-pass the feed orifice and enter the reverse servo passage. This provides for a quick fill of the servo passage and quick apply of the reverse band to prevent band slippage during abusive shifts from Park or Neutral to Reverse.

Reverse Servo Assembly (39-49): Applies the reverse band (615) in response to reverse servo fluid pressure feeding into the servo cover (40) side of the reverse servo piston (44).

Reverse Band Assembly (615): Wraps around the second clutch housing (617) and holds the input carrier (672), through the reverse reaction drum (669), allowing the vehicle to move in reverse.

Figure 6-35 Oil circuit for a THM 4T60-E transaxle. (Courtesy of the Chevrolet Motor Division of General Motors Corp.) *(continued)*

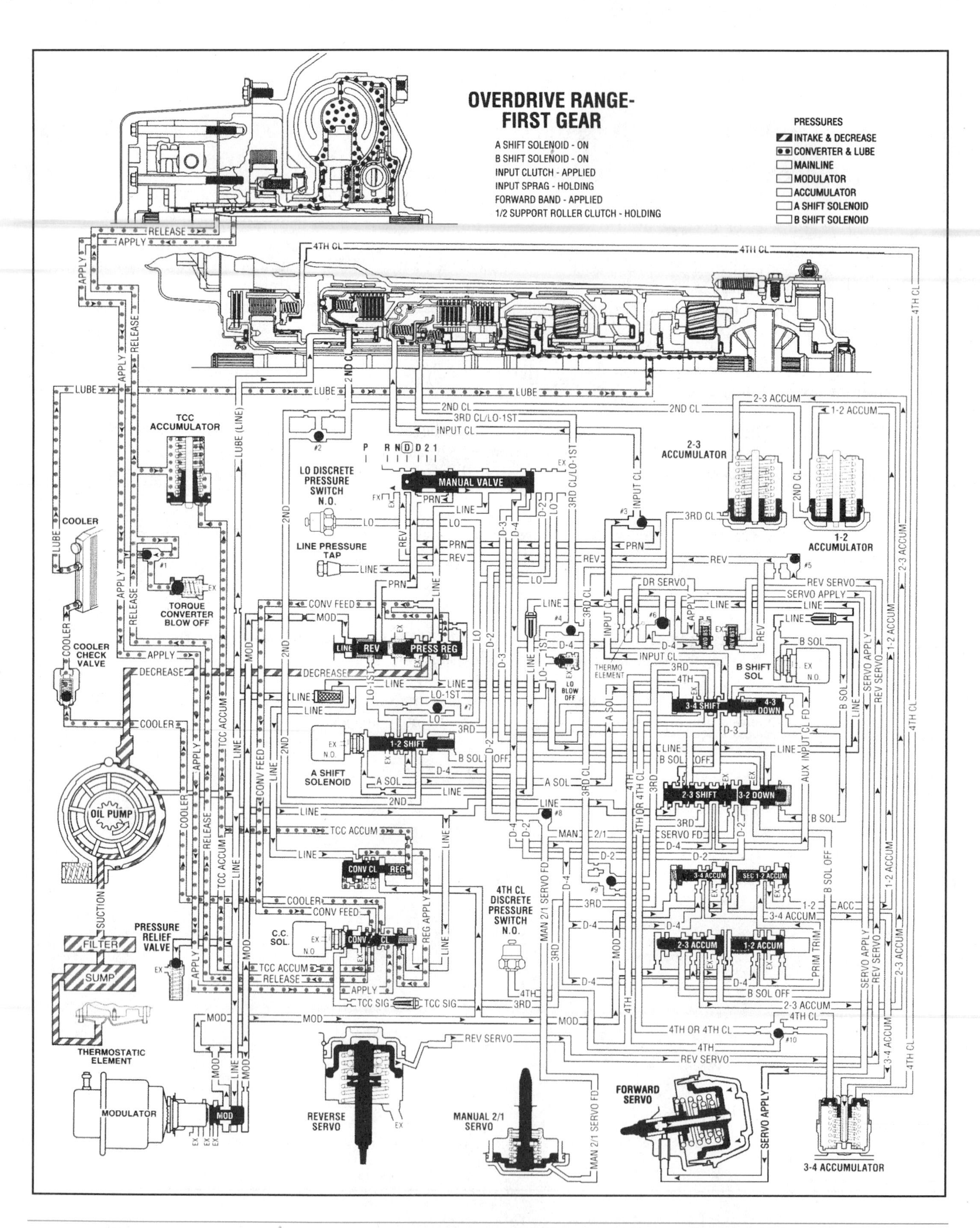

Figure 6-35 Oil circuit for a THM 4T60-E transaxle. (Courtesy of the Chevrolet Motor Division of General Motors Corp.) *(continued)*

OVERDRIVE RANGE - FIRST GEAR

"A" SHIFT SOLENOID - ON | **INPUT CLUTCH - APPLIED** | **FORWARD BAND - APPLIED**

"B" SHIFT SOLENOID - ON | **INPUT SPRAG - HOLDING** | **1/2 SUPPORT ROLLER CLUTCH - HOLDING**

When the gear selector lever is moved to the OVERDRIVE RANGE [D] the manual valve (404) moves and allows line pressure to fill the D-4 passage. D-4 pressure seats the No. 6 checkball (372) and sends fluid through the forward servo feed orifice into the servo apply passage. Servo apply fluid strokes the forward servo assembly (15-22) to apply the forward band (688). D-4 fluid also feeds the servo apply passage through the thermo element located on the spacer plate (370). D-4 fluid from the manual valve directs fluid to the following valves:

- 1-2 accumulator valve (341) into the primary trim passage to the secondary 1-2 accumulator valve (347) and into the 1-2 accumulator passage.
- 2-3 accumulator valve (343) and into the 2-3 accumulator passage.
- 3-4 accumulator valve (350) and into the 3-4 accumulator passage.
- 2-3 shift valve (357) and into the auxiliary input clutch feed passage.
- 1-2 shift valve (318)

Manual Valve (404): Is moved by the gear selector lever and allows line pressure to enter the D-4 passage.

No. 6 Checkball (372): Located in the valve body (300), blocks forward servo apply passage forcing D-4 pressure to the forward servo boost valve (367), the forward servo feed orifice and thermo element on the spacer plate (370). When the manual valve (404) is moved from Drive to Park or Neutral, the checkball unseats to allow for a quick exhaust of servo apply fluid and release of the forward band assembly (688).

Forward Servo Boost Valve (367): Opens under high line pressure (high throttle position) allowing D-4 fluid to by-pass the feed orifice and thermo element to enter the servo apply passage. This provides for a quick fill of the servo passage and quick apply of the forward band assembly (690) to prevent slippage during abusive shifts from Park or Neutral to Drive.

Forward Servo Assembly (15-22): Applies and holds the forward band (688) during all forward gear drive ranges.

Forward Band Assembly (688): Wraps around and holds the 1-2 support outer race (687) during all forward gear drive ranges.

1-2 Accumulator Valve (341): Fed by D-4 pressure, it regulates primary trim fluid pressure to the secondary 1-2 accumulator valve (347) in proportion to changes in modulator pressure.

Secondary 1-2 Accumulator Valve (347): Fed by primary trim fluid pressure, it regulates 1-2 accumulator pressure in proportion to modulator pressure. When primary trim reaches its breakpoint pressure, it is regulated lower by the secondary 1-2 accumulator valve.

2-3 Accumulator Valve (343): Fed by D-4 pressure, it regulates 2-3 accumulator pressure in proportion to modulator and "B" shift solenoid OFF pressure.

3-4 Accumulator Valve (350): Fed by D-4 pressure, it regulates 3-4 accumulator pressure in proportion to modulator pressure.

2-3 Shift Valve (357): Allows D-4 pressure to enter the auxiliary input clutch feed passage.

1-2 Shift Valve (318): Held against spring pressure by "A" shift solenoid pressure, D-4 fluid stops at this valve until a 1-2 shift is made.

Figure 6-35 Oil circuit for a THM 4T60-E transaxle. (Courtesy of the Chevrolet Motor Division of General Motors Corp.) *(continued)*

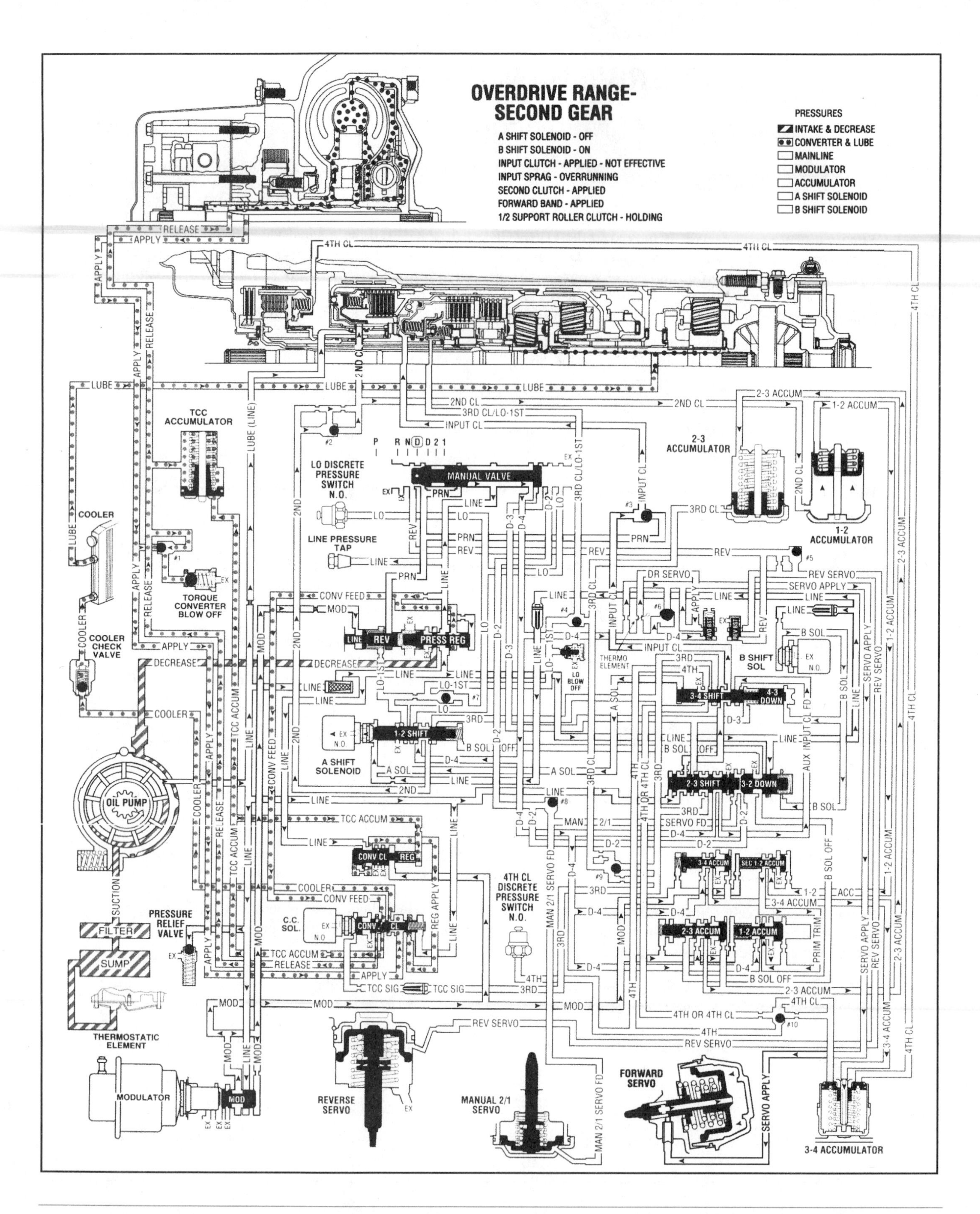

Figure 6-35 Oil circuit for a THM 4T60-E transaxle. (Courtesy of the Chevrolet Motor Division of General Motors Corp.) (*continued*)

OVERDRIVE RANGE - SECOND GEAR

"A" SHIFT SOLENOID - OFF	**INPUT CLUTCH - APPLIED**	**FORWARD BAND - APPLIED**
"B" SHIFT SOLENOID - ON	**INPUT SPRAG - OVERRUNNING**	**1/2 SUPPORT ROLLER CLUTCH - HOLDING**
	SECOND CLUTCH - APPLIED	

To obtain second gear, the PCM receives input signals from the Vehicle Speed Sensor (VS Sensor), Throttle Position Sensor (TP Sensor) and other engine sensors, to determine the precise moment to "de-energize" or "turn off current" to "A" shift solenoid (315). When this occurs, line pressure to the 1-2 shift valve (318) exhausts through the solenoid allowing spring pressure to move the valve. D-4 pressure at the valve during first gear operation can now enter the 2nd fluid passage. 2nd fluid feeds to the No. 2 checkball (372) and seats the checkball forcing the fluid through a feed orifice before applying the 2nd clutch. At the same time, 2nd clutch fluid is fed to the 1-2 accumulator piston (136) and strokes the piston up against spring pressure and 1-2 accumulator fluid pressure, cushioning 2nd clutch apply. 1-2 accumulator fluid is then forced through the 1-2 accumulator passage and exhausts at the 1-2 secondary accumulator valve (347).

"A" Shift Solenoid (315): De-energizes, allowing line pressure to exhaust through the solenoid.

1-2 Shift Valve (318): With "A" shift solenoid off, spring pressure moves the valve allowing D-4 pressure to enter the 2nd fluid passage directing 2nd fluid to the No. 2 checkball.

No. 2 Checkball (372): Located in the channel plate (400), 2nd fluid seats the checkball and is forced through an orifice into the 2nd clutch passage to apply the 2nd clutch.

1-2 Accumulator Piston (136): Cushions the apply of the 2nd clutch using 2nd clutch fluid to force the piston against spring force and 1-2 accumulator fluid.

Secondary 1-2 Accumulator Valve (347): Regulates the exhaust rate of 1-2 accumulator fluid during the apply of the 2nd clutch.

Figure 6-35 Oil circuit for a THM 4T60-E transaxle. (Courtesy of the Chevrolet Motor Division of General Motors Corp.) *(continued)*

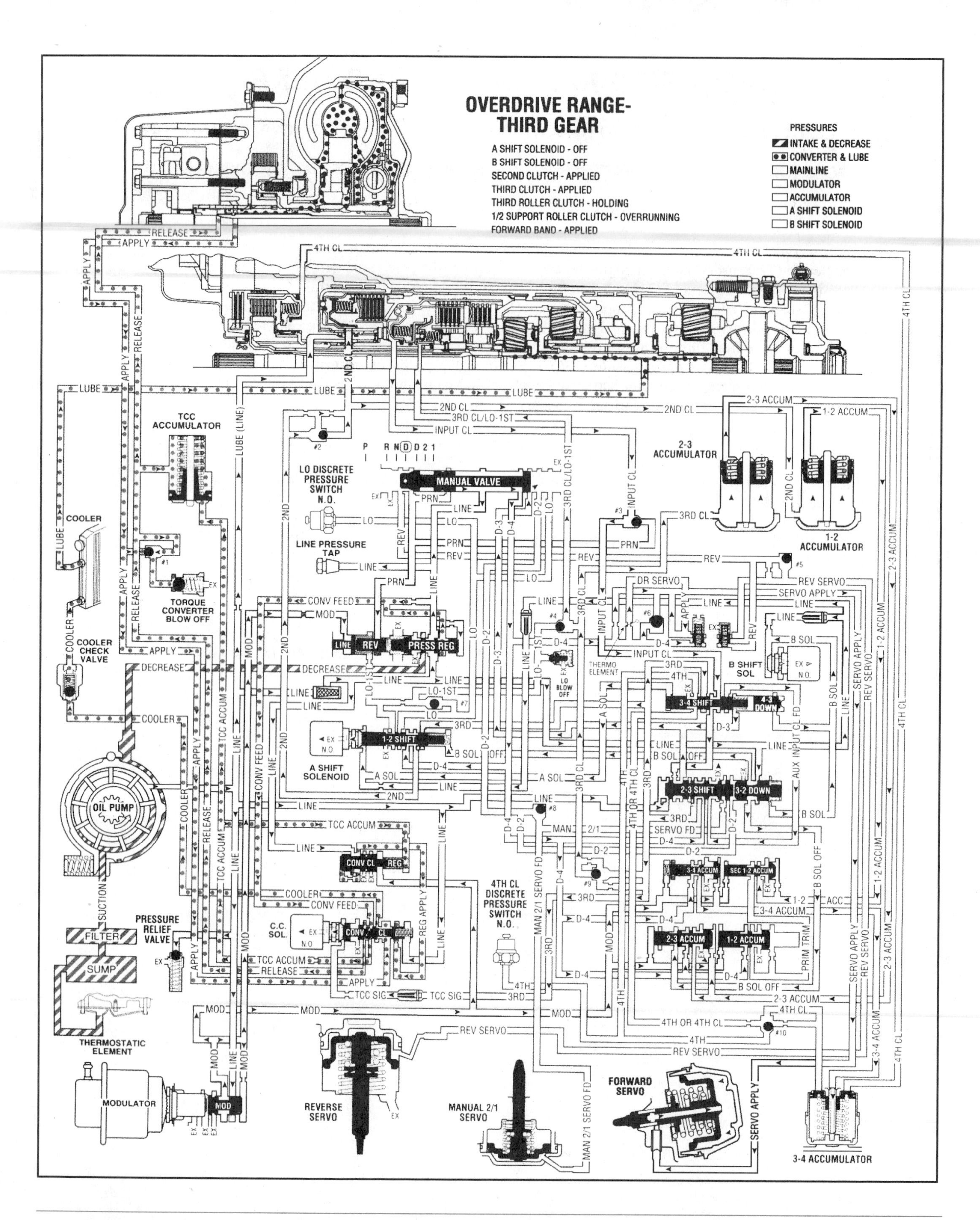

Figure 6-35 Oil circuit for a THM 4T60-E transaxle. (Courtesy of the Chevrolet Motor Division of General Motors Corp.) *(continued)*

OVERDRIVE RANGE - THIRD GEAR

"A" SHIFT SOLENOID - OFF

"B" SHIFT SOLENOID - OFF

SECOND CLUTCH - APPLIED

THIRD CLUTCH - APPLIED

THIRD ROLLER CLUTCH - HOLDING

1/2 SUPPORT ROLLER CLUTCH - OVERRUNNING

FORWARD BAND - APPLIED

To obtain third gear, the PCM receives input signals from the VS Sensor and TP Sensor in order to "de-energize" or "turn off current" supply to "B" shift solenoid (315) allowing "B" shift solenoid fluid to exhaust. When this occurs, spring force moves the 4-3 manual downshift valve (360), and, line pressure at the end of the 2-3 shift valve (357) moves the 3-2 manual downshift valve (356) against spring pressure. Line pressure also at the manual 3-2 downshift valve (356) enters the "B" shift solenoid off passage and assists in preventing the 1-2 shift valve (318) from moving. When the 2-3 shift valve (357) moves, D-4 fluid enters the 3rd fluid passage. The 3rd fluid passage feeds fluid to:

- the converter clutch solenoid (315), where it exhausts through the solenoid
- the 1-2 shift valve (318), and,
- to seat No. 9 checkball (372) forcing fluid through an orifice and into the 3rd clutch passage.

The 3rd clutch passage directs fluid to seat No. 4 checkball (372) sending fluid into the 3rd Clutch/Lo-1st passage to apply the 3rd clutch. 3rd clutch fluid pressure is also sent to the 2-3 accumulator piston (136) and forces the piston against the spring and 2-3 accumulator fluid to cushion the 3rd clutch apply. The 2-3 accumulator fluid is then forced to exhaust at the 2-3 accumulator valve (343). In third gear, the input clutch is released allowing input clutch apply fluid to exhaust through the 3-4 shift valve (362) into the input clutch feed passage. At the 2-3 shift valve (357), exhausting input clutch apply fluid is directed into the D-3 passage and out the manual valve (404).

"B" Shift Solenoid (315): De-energizes, allowing "B" shift solenoid fluid to exhaust through the solenoid. Spring pressure forces the 4-3 manual downshift valve (360) to move while line pressure moves the 2-3 shift valve (357) and 3-2 manual downshift valve (356).

No. 9 Checkball (372): Located in the valve body (300), forces 3rd fluid through a feed orifice and into the 3rd clutch passage.

No. 4 Checkball (372): Located in the channel plate (400), directs 3rd clutch fluid into the 3rd clutch/lo-1st passage to apply the 3rd clutch.

2-3 Shift Valve (357): When shifted, allows D-4 fluid to enter the 3rd passage to: stroke the 2-3 accumulator piston (136); apply the 3rd clutch; and, direct fluid to the converter clutch solenoid (315). The 2-3 shift valve also allows the input clutch apply fluid to exhaust into the D-3 passage and out at the manual valve (404).

2-3 Accumulator Piston (136): Cushions the apply of the 3rd clutch by using 3rd clutch fluid to force the piston against spring force and 2-3 accumulator fluid.

2-3 Accumulator Valve (343): Regulates the exhaust of 2-3 accumulator fluid during 3rd clutch apply.

Figure 6-35 Oil circuit for a THM 4T60-E transaxle. (Courtesy of the Chevrolet Motor Division of General Motors Corp.) *(continued)*

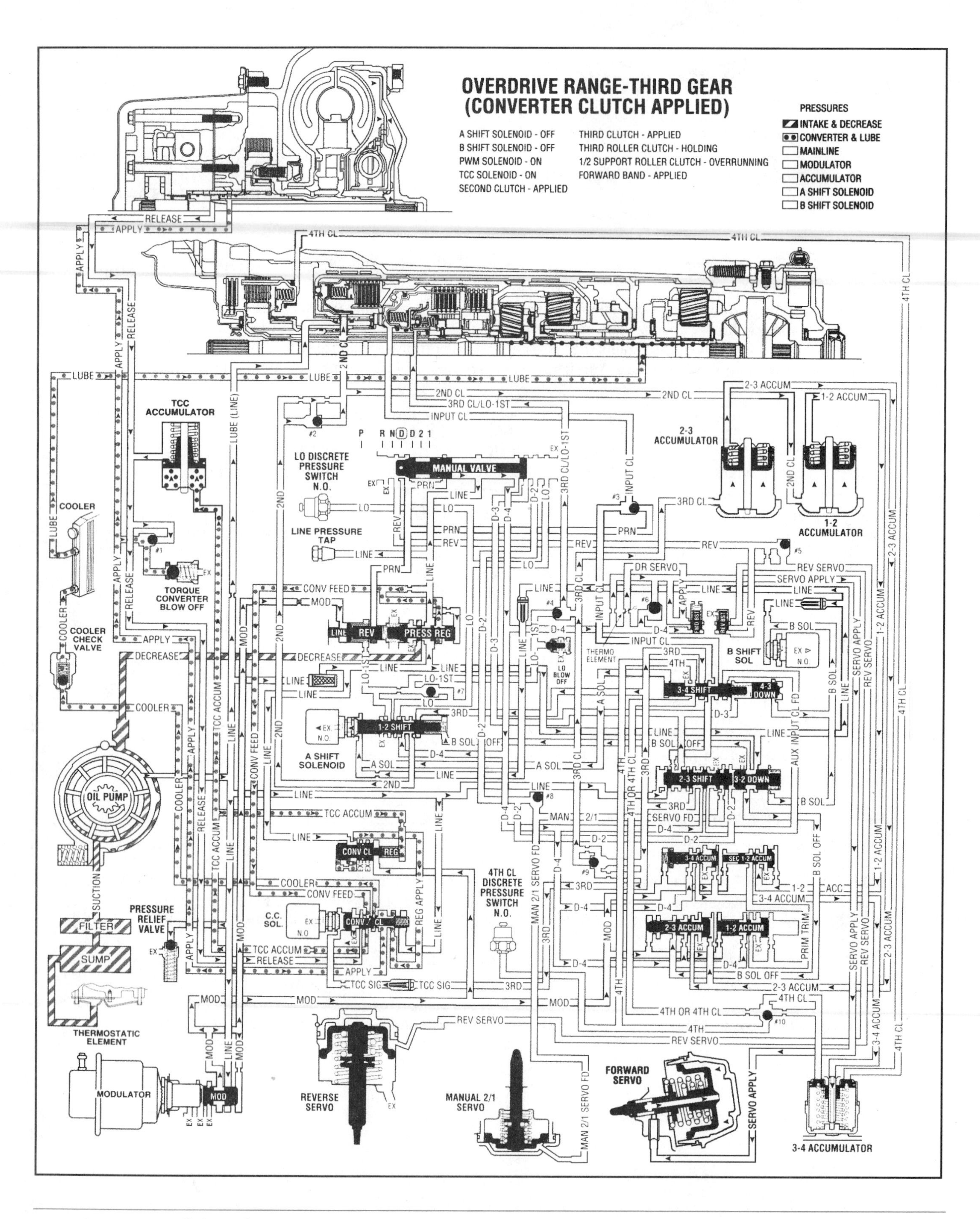

Figure 6-35 Oil circuit for a THM 4T60-E transaxle. (Courtesy of the Chevrolet Motor Division of General Motors Corp.) (*continued*)

OVERDRIVE RANGE - THIRD GEAR (CONVERTER CLUTCH APPLIED)

"A" SHIFT SOLENOID - OFF

"B" SHIFT SOLENOID - OFF

TCC SOLENOID - ON

SECOND CLUTCH - APPLIED

THIRD CLUTCH - APPLIED

THIRD ROLLER CLUTCH - HOLDING

1/2 SUPPORT ROLLER CLUTCH - OVERRUNNING

FORWARD BAND - APPLIED

The torque converter clutch applies during third gear operation after the PCM receives appropriate input signals from the VS Sensor, TP Sensor and engine temperature sensor. When the proper vehicle operating conditions have been met, the "on/off" TCC (315) is "energized". TCC signal fluid at the converter clutch solenoid (315) no longer exhausts and moves the converter clutch valve (335) against line pressure and spring force. With the valve in this position, the following occurs:

- release fluid from the converter clutch plate is allowed to exhaust
- converter feed pressure passes through the valve, enters the TCC accumulator passage and is sent to the TCC accumulator piston (416) and the converter clutch regulator valve (330)
- regulated apply fluid, from the converter clutch regulator valve (330), enters the apply passage

Apply fluid seats the No. 1 checkball (372) against release fluid and applies the converter clutch. Apply fluid is also at the torque converter blow off valve (417-20) to exhaust excess fluid pressure. Converter feed fluid at the converter clutch valve (335) is allowed to feed into the cooler passage where it passes through the cooler check valve (28), through the cooler and into the lube system of the transaxle.

TCC Accumulator Assembly (413-416): Absorbs converter feed fluid through the TCC accumulator passage and uses spring force to provide an increasing TCC accumulator bias pressure for the converter clutch regulator valve (330) during the apply of the torque converter clutch plate.

Converter Clutch Regulator Valve (330): Biased by modulator and TCC accumulator pressure, it uses line pressure to provide regulated apply fluid to the converter clutch valve (335).

Torque Converter Clutch Solenoid (315): When energized, TCC signal fluid shifts the converter clutch valve (335) against line pressure and spring force.

Converter Clutch Valve (335): When shifted, release fluid from the torque converter clutch exhausts. Converter feed fluid enters into the TCC accumulator passage to stroke the TCC accumulator piston (416) and also feeds into the cooler passage. Regulated apply fluid from the converter clutch regulator valve (330) can now enter the apply passage to the torque converter clutch.

No. 1 Checkball (372): Located in the channel plate (400), blocks release fluid while sending apply fluid to the torque converter clutch blow off valve (417-20).

Cooler Check Valve (28): Allows cooler fluid to pass through the cooler and provide lubrication for the transaxle. It also prevents converter drain back when the engine is off.

Figure 6-35 Oil circuit for a THM 4T60-E transaxle. (Courtesy of the Chevrolet Motor Division of General Motors Corp.) *(continued)*

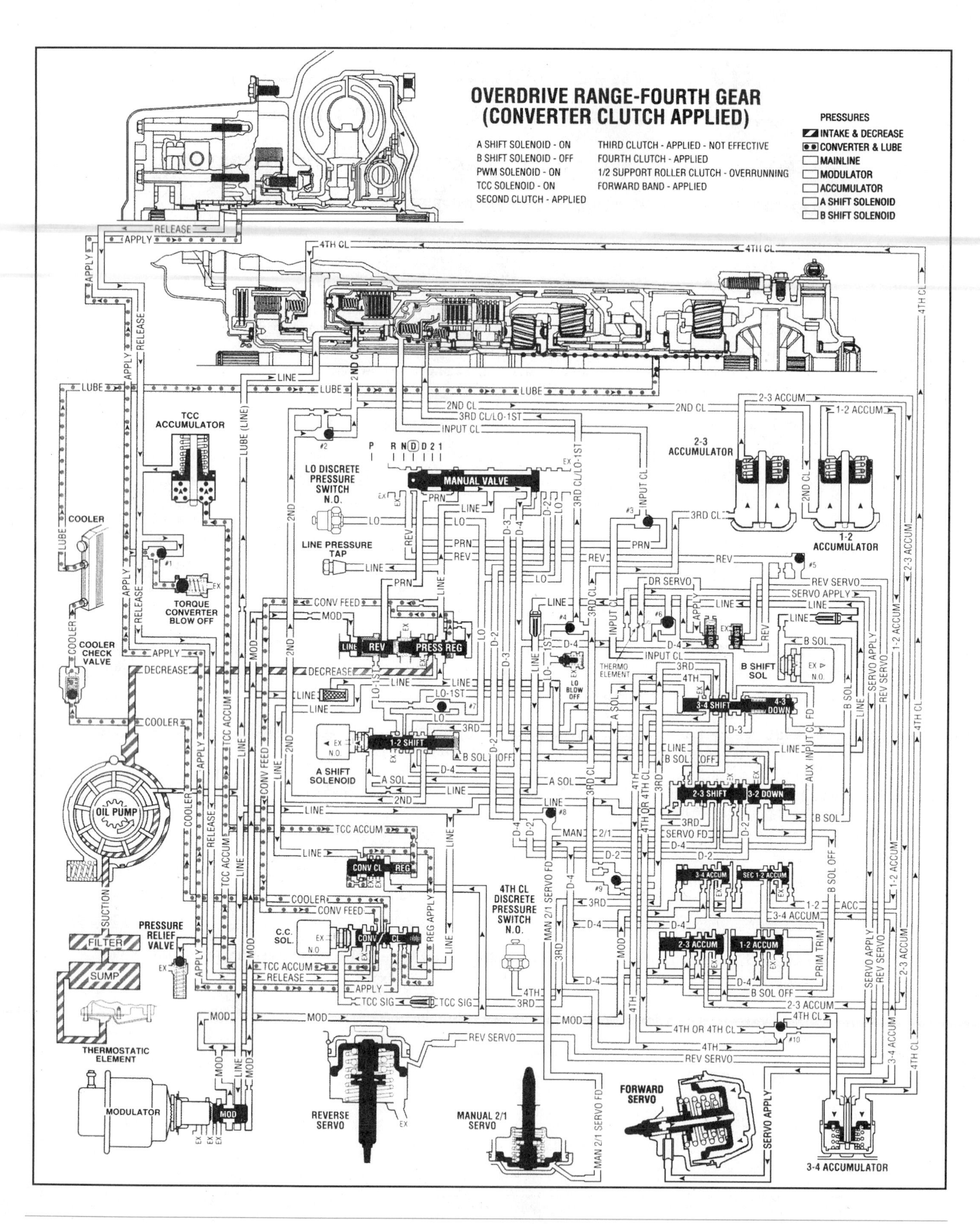

Figure 6-35 Oil circuit for a THM 4T60-E transaxle. (Courtesy of the Chevrolet Motor Division of General Motors Corp.) (*continued*)

OVERDRIVE RANGE - FOURTH GEAR (CONVERTER CLUTCH APPLIED)

"A" SHIFT SOLENOID - ON

"B" SHIFT SOLENOID - OFF

TCC SOLENOID - ON

SECOND CLUTCH - APPLIED

THIRD CLUTCH - APPLIED/ NOT EFFECTIVE

THIRD ROLLER CLUTCH - OVERRUNNING

FOURTH CLUTCH - APPLIED

1/2 SUPPORT ROLLER CLUTCH - OVERRUNNING

FORWARD BAND - APPLIED

To obtain fourth gear, the PCM receives input signals from the VS Sensor and TP Sensor in order to "energize" by providing a ground for "A" shift solenoid (315). When energized, line pressure is directed into the "A" shift solenoid passage to the 3-4 shift valve (362). "A" shift solenoid pressure shifts the valve against spring pressure allowing 3rd fluid to enter the 4th and 4th clutch fluid passages. 4th fluid is directed to:

- the 4th clutch discrete pressure switch (218) to close the switch
- seats No. 10 checkball (372) to force the fluid through a feed orifice to stroke the 3-4 accumulator piston (428) while applying the 4th clutch

3-4 accumulator fluid on the spring side of the piston is forced through the 3-4 accumulator pin (426) into the 3-4 accumulator passage and exhausts at the 3-4 accumulator valve (350).

"A" Shift Solenoid (315): Energizes to prevent fluid from exhausting out of the line passage and "A" shift solenoid passage.

1-2 Shift Valve (318): Is held against "A" shift solenoid by spring pressure and "B" shift solenoid off fluid pressure at the end of the valve.

3-4 Shift Valve (362): When shifted against spring force, it allows 3rd fluid to enter the 4th and 4th clutch fluid passages. 4th fluid seats No. 10 checkball (372) and is forced through a feed orifice before stroking the 3-4 accumulator piston (428) and applying the fourth clutch.

4th Clutch Discrete Switch (218): A normally open switch that closes when 4th fluid pressure is fed from the 3-4 shift valve (362). When the switch is closed it completes the circuit from the PCM to ground and provides information to the PCM that the transaxle is in Overdrive Range – Fourth Gear.

No. 10 Checkball (372): Located in the valve body (300), it forces 4th fluid through a feed orifice into the 4th clutch passage to stroke the 3-4 accumulator piston (428).

3-4 Accumulator Piston (428): Cushions the apply of the 4th clutch by using 4th clutch fluid to stroke the piston against spring force and 3-4 accumulator fluid.

3-4 Accumulator Valve (350): Controls the rate of exhaust of 3-4 accumulator fluid during 4th clutch apply.

Figure 6-35 Oil circuit for a THM 4T60-E transaxle. (Courtesy of the Chevrolet Motor Division of General Motors Corp.) *(continued)*

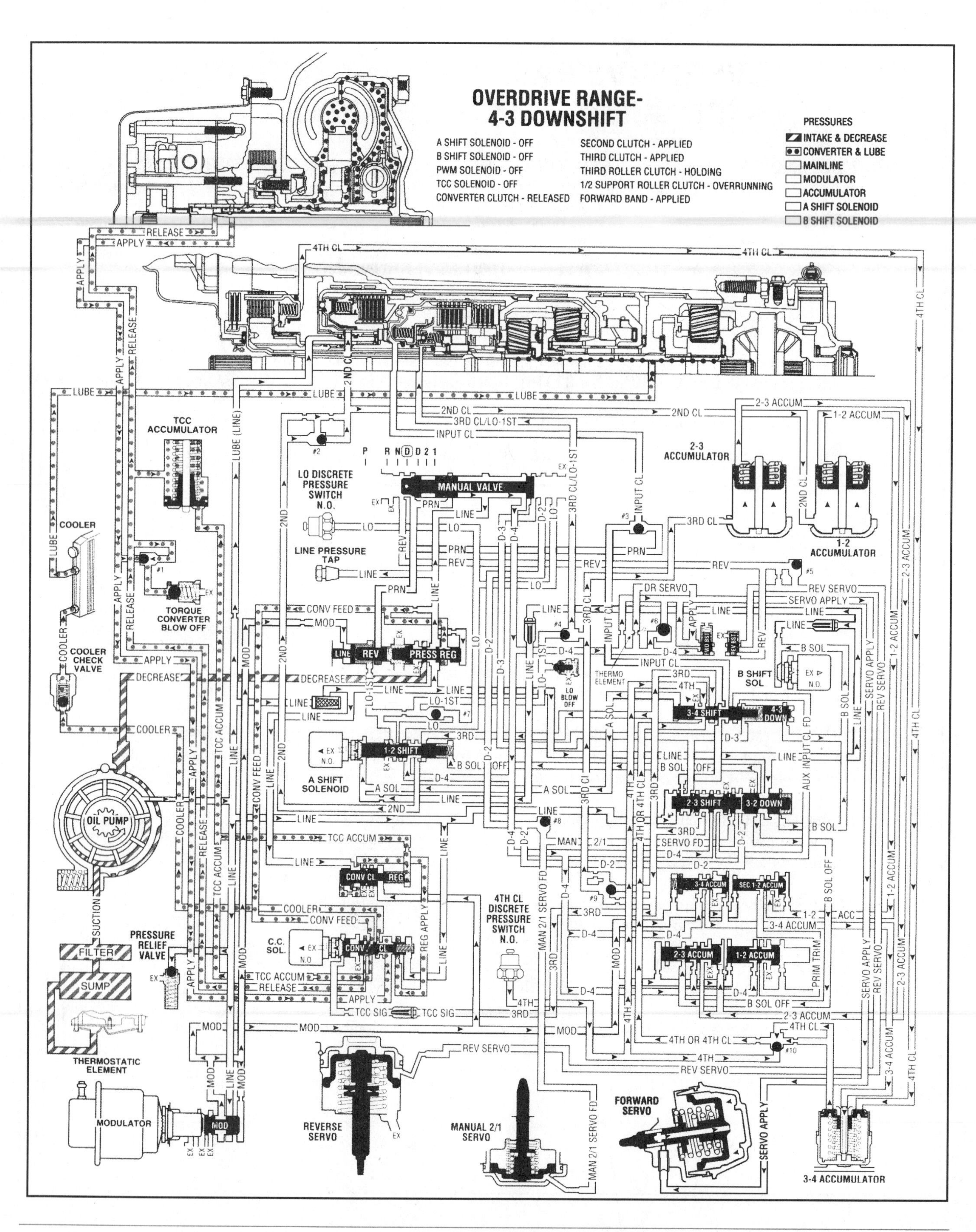

Figure 6-35 Oil circuit for a THM 4T60-E transaxle. (Courtesy of the Chevrolet Motor Division of General Motors Corp.) *(continued)*

OVERDRIVE RANGE - 4-3 DOWNSHIFT (CONVERTER CLUTCH - RELEASED)

"A" SHIFT SOLENOID - OFF

"B" SHIFT SOLENOID - OFF

TCC SOLENOID - OFF

SECOND CLUTCH - APPLIED

THIRD CLUTCH - APPLIED

THIRD ROLLER CLUTCH - HOLDING

1/2 SUPPORT ROLLER CLUTCH - OVERRUNNING

FORWARD BAND - APPLIED

Under light throttle acceleration, the torque converter clutch will release when the input signals sent to the PCM from the VS Sensor, TP Sensor and 4th clutch discrete switch (218) have commanded a 4-3 downshift. When the appropriate signals are received, the torque converter clutch solenoid (315) is "de-energized" or "turned off". Line pressure and spring force at the converter clutch valve (335) moves the valve allowing TCC signal fluid to exhaust through the converter clutch solenoid (315). Converter feed fluid enters the release passage, seats No. 1 checkball (372) against apply fluid and feeds to the release side of the torque converter clutch plate. Apply fluid from the torque converter (1) is directed through the apply passage to the converter clutch valve (335) where it enters the cooler passage. TCC accumulator fluid, from the TCC accumulator piston (416), is directed: (1) through the converter clutch valve (335) and enters the release passage, and (2) to the converter clutch regulator valve (330) to control the release feel of the converter clutch plate.

A 4-3 shift occurs when the PCM "de-energizes" or "turns off current" to "A" shift solenoid (315) allowing "A" shift solenoid fluid at the 3-4 shift valve (362) to exhaust through the solenoid. Spring force at the 3-4 shift valve (362) moves the valve allowing 4th clutch apply fluid (at the 4th clutch) to be directed through the 3-4 accumulator assembly (421-428), seat No. 10 checkball (372) against 4th clutch (apply) fluid and exhaust at the 3-4 shift valve (362).

Torque Converter Clutch Solenoid (315): De-energizes, allowing the converter clutch valve (335) to move and TCC signal fluid to exhaust through the solenoid.

Converter Clutch Valve (335): When the torque converter clutch solenoid (315) is "off", converter feed fluid passes through the valve into the converter release passage and converter apply fluid is directed into the cooler passage. TCC accumulator fluid also exhausts into the release passage and acts as a bias on the converter clutch regulator valve (330).

Converter Clutch Regulator Valve (330): Uses filtered line fluid to supply regulated apply fluid to the converter clutch valve (335). Regulation is controlled by TCC accumulator and modulator fluid pressures.

TCC Accumulator Assembly (413-416): Spring force and release fluid pressure returns the TCC accumulator piston (416) to TCC "off" position.

No. 1 Checkball (372): Located in the channel plate (400), converter release fluid seats the checkball against apply fluid to allow release of the torque conveter clutch plate.

"A" Shift Solenoid (315): De-energizes, allowing "A" shift solenoid fluid at the 3-4 shift valve (362) to exhaust through the solenoid.

3-4 Shift Valve (362): When "A" shift solenoid fluid exhausts, spring force moves the valve allowing 4th clutch fluid to exhaust through the valve.

3-4 Accumulator Assembly (421-428): Allows 4th clutch exhaust fluid to pass through the accumulator assembly to No. 10 checkball (372).

No. 10 Checkball (372): Located in the valve body (300), forces 4th clutch fluid through an exhaust orifice to the 3-4 shift valve (362) where it exhausts.

4th Clutch Discrete Switch (218): "Opens" a circuit to the PCM to signal that the transaxle is no longer operating in Overdrive Range – Fourth Gear.

NOTE: Under light accelerating conditions, normally the torque converter clutch will release prior to the transaxle making a 4-3 downshift. However, depending on throttle angle, vehicle load and road conditions, torque converter clutch release and a 4-3 shift may occur at the same time.

Figure 6-35 Oil circuit for a THM 4T60-E transaxle. (Courtesy of the Chevrolet Motor Division of General Motors Corp.) *(continued)*

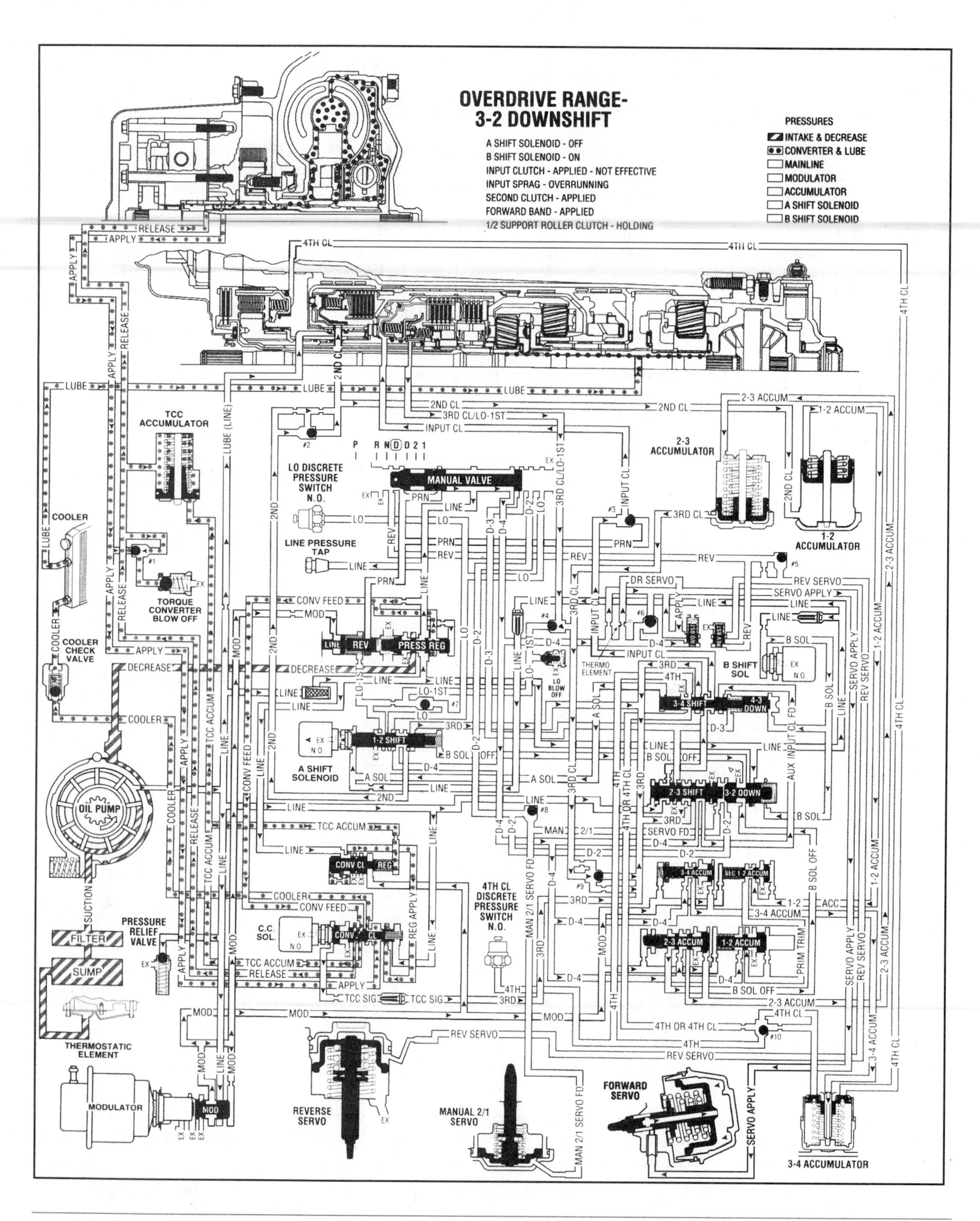

Figure 6-35 Oil circuit for a THM 4T60-E transaxle. (Courtesy of the Chevrolet Motor Division of General Motors Corp.) (*continued*)

OVERDRIVE RANGE - 3-2 DOWNSHIFT

"A" SHIFT SOLENOID - OFF

"B" SHIFT SOLENOID - ON

INPUT CLUTCH - APPLIED

INPUT SPRAG - OVERRUNNING

SECOND CLUTCH - APPLIED

1/2 SUPPORT ROLLER CLUTCH - HOLDING

FORWARD BAND - APPLIED

During all full throttle downshifts, engine manifold vacuum supply to the vacuum modulator (32) drops allowing the modulator valve (34) to move and increase modulator pressure to the line boost valve (304). At the same time, pump output increases through the higher engine speeds forcing a higher volume of fluid into the line circuit. The increased modulator fluid pressure and increased volume of fluid at the pressure regulator valve (313) creates higher line pressures for proper clutch and valve operation. A full throttle 3-2 downshift occurs when the PCM receives input signals from the VS sensor and TP sensor to "energize" or turn on "B" shift solenoid (315). Line pressure at the solenoid is forced into the "B" shift solenoid passage and is directed to the 4-3 manual downshift valve (360) and 3-2 manual downshift valve (356). "B" shift solenoid fluid forces the 3-2 manual downshift valve (356) to move also shifting the 2-3 shift valve (357). Line pressure at the 2-3 shift valve (357) is now directed into the input clutch feed passage, through the 3-4 shift valve (362) into the input clutch passage. Input clutch fluid seats #3 checkball (372) and applies the input clutch. The third clutch is released by allowing 3rd clutch/lo-lst fluid to exhaust to the #4 checkball (372) and into the 3rd clutch passage. 3rd clutch fluid is directed to #9 checkball (372) forcing it through the exhaust orifice and into the 3rd fluid passage to the 2-3 shift valve (357) where it exhausts.

Modulator Valve (34): Moves in response to engine manifold vacuum and increases modulator pressure to the line boost valve (304).

Line Boost Valve (304): Moves in response to modulator pressure to increase line pressure through the pressure regulator valve (313).

Pressure Regulator Valve (313): Increases line pressure in response to increased modulator pressure.

"B" Shift Solenoid (315): Energizes, allowing line pressure into the "B" shift solenoid passage to force a 3-2 downshift.

4-3 Manual Downshift Valve (360): Is shifted by "B" shift solenoid fluid and prevents the 3-4 shift valve (362) from moving.

3-2 Manual Downshift Valve (356): Is shifted by "B" shift solenoid fluid to downshift the 2-3 shift valve (357).

2-3 Shift Valve (357): When downshifted, it allows line pressure to enter into the input clutch feed passage and directs it to the 3-4 shift valve (362).

3-4 Shift Valve (362): Directs input clutch feed fluid into the input clutch passage and sends it to the #3 checkball (372).

#3 Checkball (372): Located in the channel plate (400), it is seated against the PRN passage to allow input clutch fluid to apply the input clutch.

#4 Checkball (372): Located in the channel plate (400), it directs 3rd clutch/lo-1st exhaust fluid into the 3rd clutch passage.

#9 Checkball (372): Located in the valve body (300), it directs 3rd clutch exhaust fluid into the 3rd fluid passage. 3rd fluid is sent to the 2-3 shift valve (357) where it exhausts.

Figure 6-35 Oil circuit for a THM 4T60-E transaxle. (Courtesy of the Chevrolet Motor Division of General Motors Corp.) *(continued)*

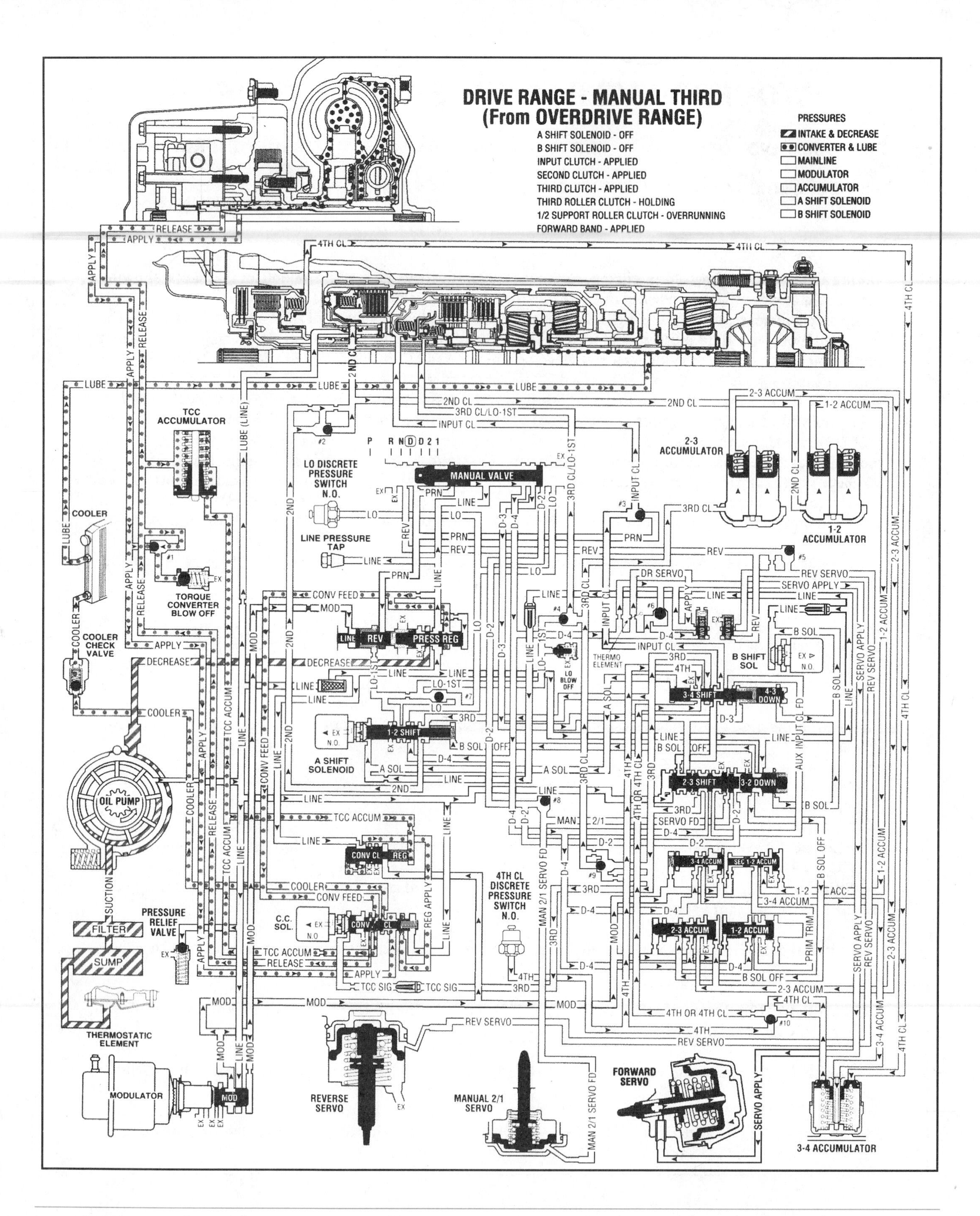

Figure 6-35 Oil circuit for a THM 4T60-E transaxle. (Courtesy of the Chevrolet Motor Division of General Motors Corp.) (*continued*)

DRIVE RANGE - MANUAL THIRD
(From OVERDRIVE RANGE FOURTH GEAR)

"A" SHIFT SOLENOID - OFF

"B" SHIFT SOLENOID - OFF

INPUT CLUTCH - APPLIED

SECOND CLUTCH - APPLIED

THIRD CLUTCH - APPLIED

THIRD ROLLER CLUTCH - HOLDING

1/2 SUPPORT ROLLER CLUTCH - OVERRUNNING

FORWARD BAND - APPLIED

When the gear selector lever is moved to the DRIVE RANGE (D), the manual valve (404) allows line pressure to enter the D-3 passage. D-3 fluid is then directed to the 3-4 shift valve (362) and moves the valve against "A" shift solenoid pressure. D-3 fluid is also sent to the 2-3 shift valve (357) where it enters the input clutch feed passage. Input clutch feed fluid passes through the 3-4 shift valve (362) into the input clutch passage, seats No. 3 checkball (372) against the PRN passage and applies the input clutch. 4th clutch apply fluid at the 4th clutch, is directed through the 3-4 accumulator assembly (421-428) to the No. 10 checkball (372). Exhausting 4th clutch fluid seats the No. 10 checkball (372) against the 4th passage directing fluid through an exhaust orifice up to the 3-4 shift valve (362) where it exhausts. 4th fluid pressure at the 4th clutch discrete switch (218) also exhausts at the 3-4 shift valve (362) allowing the switch to open and signal the PCM that the transaxle is no longer operating in 4th gear.

Since D-3 fluid overrides "A" shift solenoid fluid at the 3-4 shift valve (362), a manual 4-3 downshift will result even if "A" shift solenoid is "energized" or "turned on" (Drive Range Fourth Gear solenoid state is "A" shift solenoid on and "B" shift solenoid off). If the PCM receives the appropriate input signals from the VS Sensor and TP Sensor, "A" shift solenoid will "de-energize" or "turn off" and "A" shift solenoid fluid will then exhaust through the solenoid.

Manual Valve (404): Is moved by the gear selector lever and allows line pressure to enter the D-3 passage.

2-3 Shift Valve (357): Allows D-3 fluid to enter the input clutch feed passage and directs it to the 3-4 shift valve (362).

3-4 Shift Valve (362): Is downshifted when D-3 fluid is fed to the valve. In this position input clutch feed fluid enters the input clutch passage.

No. 3 Checkball (372): Located in the channel plate (400), it is seated against the PRN passage allowing input clutch fluid to apply the input clutch.

No. 10 Checkball (372): Located in the valve body (300), it is seated against 4th fluid allowing 4th clutch fluid to exhaust at the 3-4 shift valve).

4th Clutch Discrete Switch (218): "Opens" a circuit to the PCM to signal that the transaxle is no longer operating in Overdrive Range – Fourth Gear.

"A" Shift Solenoid (315): When de-energized, it allows "A" shift solenoid fluid from the 3-4 shift valve (362) to exhaust through the solenoid. However, this event does not have to occur in order to achieve a manual 4-3 downshift.

NOTE: When a Manual Third Gear Range is selected (from Overdrive Range Fourth Gear) 4th fluid pressure at the 4th clutch discrete switch (218) exhausts allowing the switch to open. The PCM senses this change in vehicle operation and releases the torque converter clutch to help eliminate harsh manual D4 to D3 downshifts. The torque converter clutch will re-apply after the PCM senses that normal Third Gear operating conditions (such as vehicle speed, throttle position, etc.) have once again been met.

Figure 6-35 Oil circuit for a THM 4T60-E transaxle. (Courtesy of the Chevrolet Motor Division of General Motors Corp.) *(continued)*

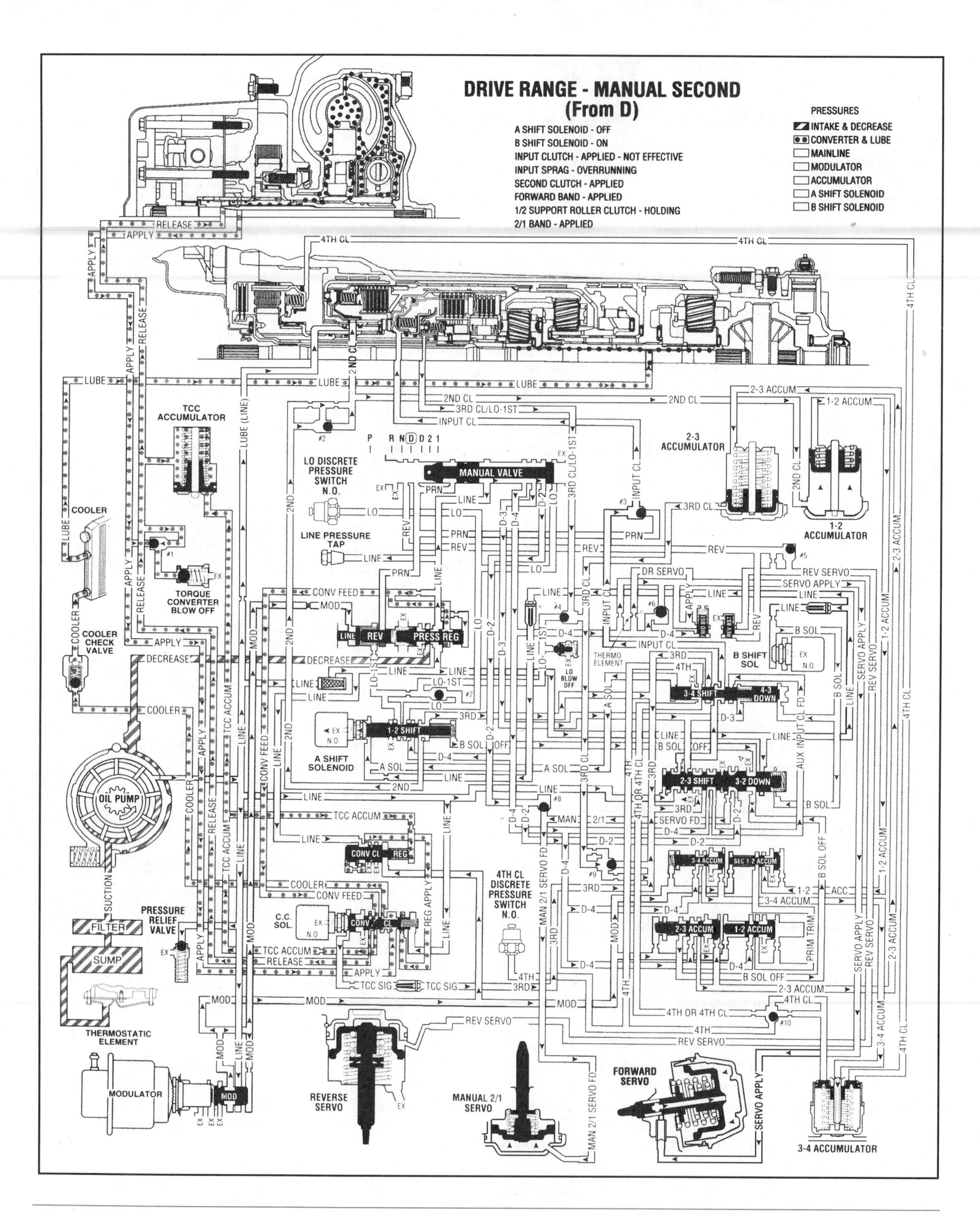

Figure 6-35 Oil circuit for a THM 4T60-E transaxle. (Courtesy of the Chevrolet Motor Division of General Motors Corp.) (*continued*)

DRIVE RANGE - MANUAL SECOND (From DRIVE RANGE THIRD GEAR)

"A" SHIFT SOLENOID - OFF

INPUT SPRAG - OVERRUNNING

FORWARD BAND - APPLIED

"B" SHIFT SOLENOID - ON

SECOND CLUTCH - APPLIED

2/1 BAND - APPLIED

INPUT CLUTCH - APPLIED/ NOT EFFECTIVE

1/2 SUPPORT ROLLER CLUTCH - HOLDING

When the gear selector lever is moved to the MANUAL SECOND RANGE (2), the manual valve (404) allows line pressure to enter the D-2 passage. D-2 fluid seats #8 checkball (372) and is directed to the 2-3 shift valve (357) to shift the valve. When shifted, D-2 fluid enters the manual 2-1 servo feed passage and strokes the manual 2/1 servo assembly (104-116) to apply the 2/1 band (680). The third clutch is released by allowing 3rd clutch/lo-lst apply fluid to exhaust by seating #4 checkball (372) against the lo-lst passage and into the 3rd clutch passage. 3rd clutch fluid is then directed to #9 checkball (372) and seats it against the 3rd fluid passage forcing the fluid through an exhaust orifice into the 3rd fluid passage. 3rd fluid is then sent to the 2-3 shift valve (357) where it exhausts. Since D-2 fluid overrides line pressure at the 2-3 shift valve (357), a manual 3-2 downshift will result even if "B" shift solenoid is "de-energized" or "turned off" (drive range 3rd gear solenoid state is: "A" shift solenoid off and "B" shift solenoid off). If the PCM receives the appropriate input signals from the VS sensor and TP sensor, "B" shift solenoid will "energize" or "turn on". "B" shift solenoid fluid is then directed to the 4-3 manual downshift valve (360) and the 3-2 manual downshift valve (356).

Manual Valve (404): Is moved by the gear selector lever allowing line pressure to enter the D-2 passage.

#8 Checkball (372): Located in the valve body (300), is fed D-2 fluid from the manual valve (404) and directs it to the 2-3 shift valve (357).

2-3 Shift Valve (357): When shifted by D-2 fluid, it allows D-2 fluid to enter the manual 2-1 servo feed passage to stroke the manual 2/1 servo assembly (104-116) and allows 3rd fluid to exhaust.

Manual 2/1 Servo Assembly (104-116): Applies the 2/1 band (680) during manual second and manual first gear ranges.

#4 Checkball (372): Located in the channel plate (400), it directs 3rd clutch/lo-lst exhaust fluid into the 3rd clutch passage.

"B" Shift Solenoid (315): When energized, it allows line pressure to enter the "B" shift solenoid passage and sends it to the 4-3 manual downshift valve (360) and 3-2 manual downshift valve (356). However, this event does not have to occur in order to achieve a manual 3-2 downshift.

3-2 Manual Downshift Valve (356): Is shifted by "B" shift solenoid fluid and prevents the 2-3 shift valve (357) from upshifting.

Figure 6-35 Oil circuit for a THM 4T60-E transaxle. (Courtesy of the Chevrolet Motor Division of General Motors Corp.) *(continued)*

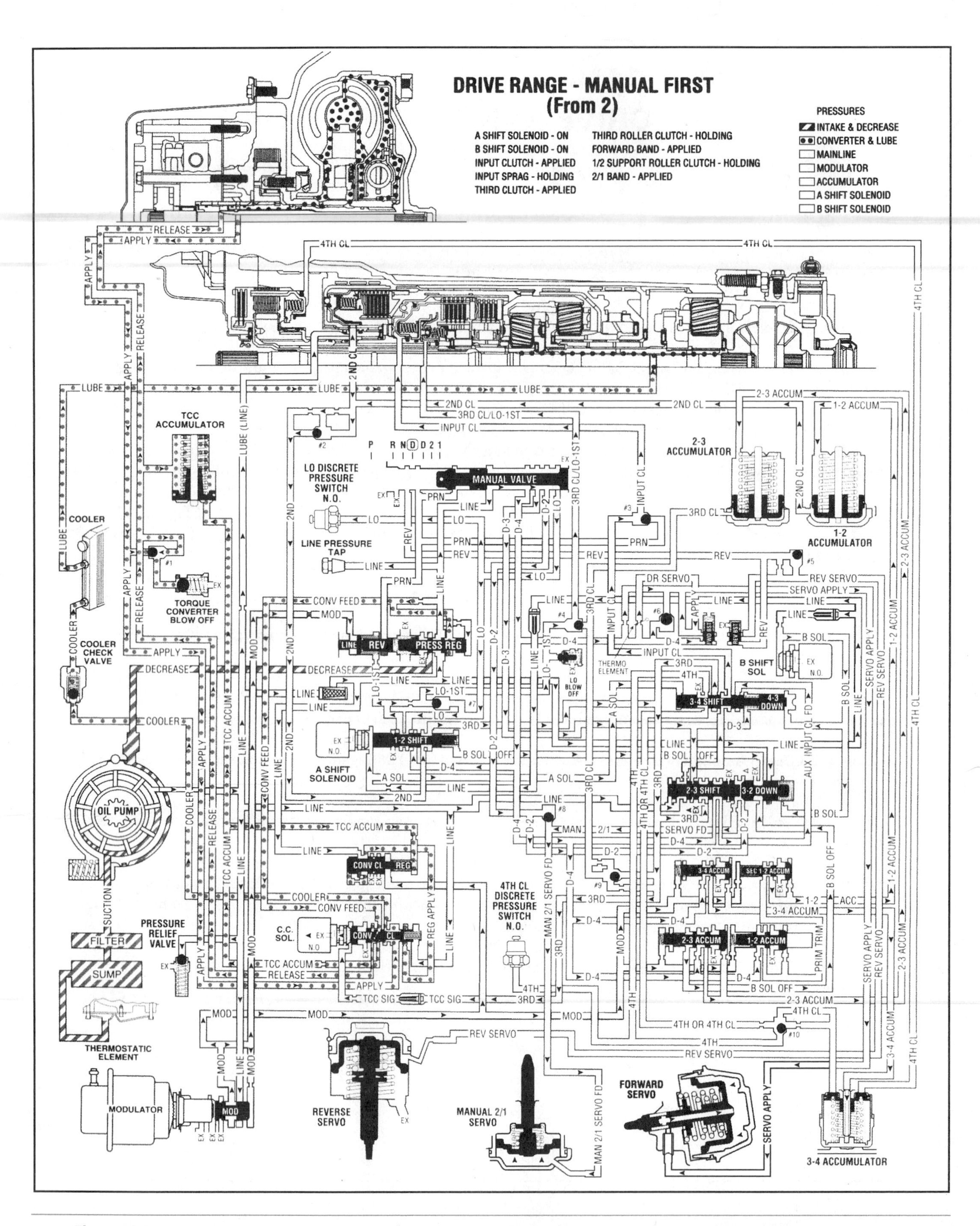

Figure 6-35 Oil circuit for a THM 4T60-E transaxle. (Courtesy of the Chevrolet Motor Division of General Motors Corp.) (*continued*)

DRIVE RANGE - MANUAL FIRST (From MANUAL SECOND)

"A" SHIFT SOLENOID - ON

"B" SHIFT SOLENOID - ON

THIRD CLUTCH - APPLIED

THIRD ROLLER CLUTCH - HOLDING

INPUT CLUTCH - APPLIED

INPUT SPRAG - HOLDING

1/2 SUPPORT ROLLER CLUTCH - HOLDING

FORWARD BAND - APPLIED

2/1 BAND - APPLIED

When the gear selector lever is moved to the MANUAL FIRST RANGE (1), the manual valve (404) allows line pressure to enter the lo passage. Lo fluid pressure is sent to the lo discrete switch (218), closes it thereby completing the circuit from the PCM to ground. Lo fluid pressure is also sent to No. 7 checkball (372) which seats against the lo-1st fluid passage. Based on PCM input signals from the VS Sensor, TP Sensor, lo discrete switch (218), and lo-1st lockout speed calibration, "A" shift solenoid is "energized" or "turned on" allowing line pressure to enter the "A" shift solenoid passage. "A" shift solenoid fluid then holds the 1-2 shift valve (318) and is also directed to the 3-4 shift valve (362).

The second clutch is released by 2nd clutch apply fluid exhausting to the No. 2 checkball (372), forcing it through the exhaust orifice into the 2nd fluid passage. 2nd fluid is then directed through the 1-2 shift valve (318) and into the 3rd fluid passage where it exhausts at the 2-3 shift valve (357). Simultaneously, lo fluid is directed through the 1-2 shift valve (318) into the lo-1st fluid passage. Lo-1st fluid is sent to the pressure regulator valve (313) to boost line pressure and at the same time to the lo blow off valve (407). Lo-1st fluid also seats No. 4 checkball (372) against the 3rd clutch passage and directs the fluid into the 3rd clutch/lo-lst passage to apply the 3rd clutch.

Manual Valve (404): Is moved by the gear selector lever and allows line pressure to enter the lo fluid passage.

Lo Discrete Switch (218): A normally open switch that closes when lo fluid pressure is fed from the manual valve (404). When the switch is closed it completes the circuit from the PCM to ground and provides information to the PCM that the gear selector lever has been moved to Manual First Gear Range.

No. 7 Checkball (372): Located in the valve body (300), directs lo fluid to the 1-2 shift valve (318).

1-2 Shift Valve (318): When shifted against spring force, allows lo fluid to enter the lo-lst passage to the pressure regulator valve (313). During release of the 2nd clutch, exhausting 2nd fluid passes through the valve and enters the 3rd fluid passage.

Pressure Regulator Valve (313): Boosts line pressure using lo-lst fluid pressure.

Lo Blow Off Valve (407): A pressure relief valve that will exhaust excess lo fluid pressures above 448 kPa (65 psi) in the 3rd clutch apply circuit.

No. 4 Checkball (372): Located in the channel plate (400), it seats against 3rd clutch fluid allowing lo-lst fluid to enter the 3rd clutch fluid passage and apply the 3rd clutch.

"A" Shift Solenoid (315): Energizes, allowing line pressure to feed into the "A" shift solenoid passage and to the 3-4 shift valve (357).

No. 2 Checkball (372): Located in the channel plate (400), forces exhausting 2nd clutch apply fluid through an orifice into the 2nd clutch passage and to the 1-2 shift valve (318).

NOTE: The transaxle will shift to Manual First Gear Range only when the following conditions have been met:

- Manual First Gear Range is selected
- Solenoid state is operating for First Gear ("A" Shift Solenoid and "B" Shift Solenoid – ON)
- Lo Discrete Switch provides ground to PCM
- Vehicle speed is below 56 kmh (35 mph)

Figure 6-35 Oil circuit for a THM 4T60-E transaxle. (Courtesy of the Chevrolet Motor Division of General Motors Corp.) *(continued)*

During light acceleration or slight load changes, the PCM will turn off the torque converter clutch. Often, the PCM will order a 4–3 downshift during these same conditions. When the TCC solenoid is turned off, line pressure and spring tension at the converter clutch valve move the valve to allow TCC signal fluid to exhaust through the converter clutch solenoid. Converter feed fluid enters the release passage, seats the #1 check ball against apply fluid and feeds the release side of the torque converter clutch plate. Apply fluid from the torque converter is directed through the apply passage to the converter clutch valve to the cooler passage. TCC accumulator fluid is directed through the converter clutch valve and enters the release passage. It is also directed to the converter clutch regulator valve to control the release of the converter clutch plate.

When the PCM directs a downshift, it turns off the "A" shift solenoid. This allows the "A" shift solenoid fluid at the 3–4 shift valve to exhaust through the solenoid. Spring pressure at the 3–4 shift valve moves the valve, which allows 4th clutch apply fluid to be directed through the 3–4 accumulator assembly, to seat the #10 check ball against the 4th clutch apply fluid, and to exhaust at the 3–4 shift valve. Third gear is engaged.

Other gear selector positions and gear ranges are possible. There is no need to cover them all in this discussion. However, there is a need for any student of automatic transmissions to understand the action of the valves and apply devices during gear shifting. Take some time to trace the oil circuits through any gear. It may also be helpful to play "what if" games.

Reaction Members

Reaction members are those parts of a planetary gear set that are held in order to produce an output motion. Other members of the planetary gear set react against the stationary or held member. Apply devices, such as multiple-disc clutches, friction bands, and one-way overrunning clutches, are used in automatic transmissions to hold or drive members of the planetary gear set in order to provide for the various gear ratios and directions. One-way overrunning clutches are purely mechanical devices, whereas clutches and bands are hydraulically controlled mechanical devices. All automatic transmissions use more than one type of these apply devices; some use all three.

Shop Manual
Chapter 6, page 187

Organic frictional materials are either paper pulp- or cellulose-based compounds.

Brake Bands

A band is an externally contracting brake assembly that is positioned around the outside of a **drum** (Figure 6-36). The drum is connected to a member of the planetary gear set. Bands are simply flexible metal strips lined with either a semimetallic or organic frictional material. When a band is applied, it wraps around a drum to hold a member of a planetary gear set and keeps it from rotating. A band is applied hydraulically by a servo assembly. Hydraulic pressure moves the servo piston, which compresses the servo return spring and applies the band through a mechanical linkage.

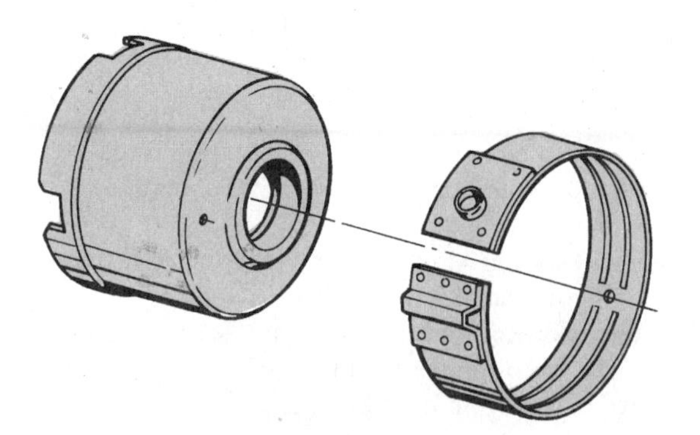

Figure 6-36 A clutch drum and band assembly. (Courtesy of the Hydra-Matic Division of General Motors Corp.)

To release the band, hydraulic pressure to the servo is diverted and the return spring moves the piston back into its bore.

A band and its servo are always used as holding devices and are never used to drive a member of the gear set. A band stops and holds the reaction member of the gear set. A band is anchored at one end and force is applied against the other end. As this force is applied, the band contracts around the rotating drum and squeezes it to a stop. The amount of pressure that stops the drum from rotating is determined by the length and width of the band and the amount of force applied against the band's unanchored end. Rods or struts are used to apply force against the band. They may be placed between the anchor and the band or be at both sides of the band.

When the band contracts in the same direction as drum rotation, the clamping pressure is increased. This is called the self-energizing effect.

A band can be positioned to allow the applying pressure to move against the direction of drum rotation or with it. If the band is mounted so that the force is applied in the same direction as drum rotation, the movement of the drum adds to the applying force and less hydraulic pressure is needed. When the band moves in the opposite direction to drum rotation, the drum opposes the band and more pressure is needed to stop the drum.

Rods are round metal bars, whereas struts are flat metal plates. Both are often referred to as operating links.

Although all transmission bands are made of flexible steel and their inside surfaces lined with frictional material, they differ in size and construction depending on the amount of work they are required to do. A band that is split with overlapping ends is called a **double-wrap band.** A one-piece band that is not split is called a **single-wrap band.**

Double-wrap bands are also called **split bands** or **dual bands.**

Two types of single-wrap bands (Figure 6-37) are commonly used in transmissions today. One type is made of light and flexible steel and the other type is made of heavy and rigid cast iron. The heavy bands are typically made with a metallic lining material that can withstand large gripping pressures. Light bands are lined with less-abrasive material that helps limit drum wear.

Double-wrap bands (Figure 6-38) have a smooth and uniform grip, and lend themselves to self-energizing. A double-wrap band readily conforms to the circular shape of a drum; therefore, it can provide greater holding power for a given application force. A double-wrap band also requires less hydraulic pressure than a single-wrap band to produce the same amount of holding power. All these features allow a double-wrap band to provide smooth gear changes.

To prevent harsh changing of gears, which would result from quickly stopping the movement of a gear set member, bands are designed to slip a little as they are being applied. The amount that a band slips increases as its lining wears. This wear increases the clearance between the band and

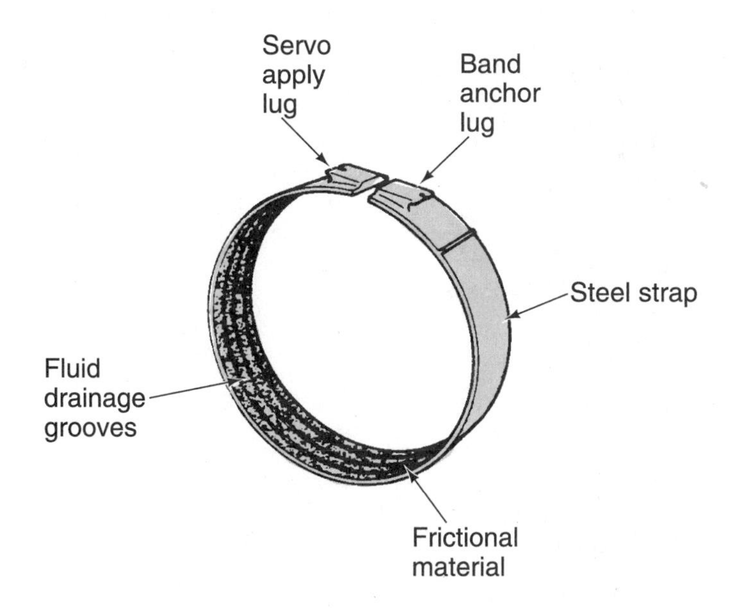

Figure 6-37 A single-wrap band. (Courtesy of Chrysler Corp.)

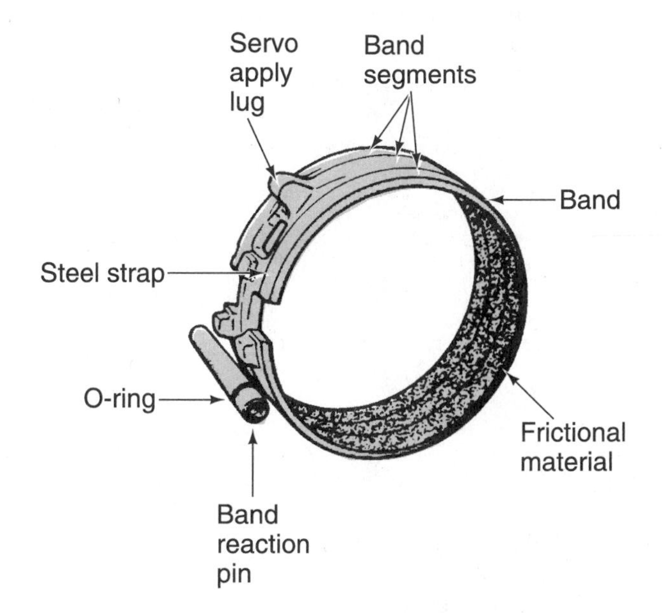

Figure 6-38 A double-wrap band. (Courtesy of Chrysler Corp.)

the drum, and reduces the holding force of the band. Because of this wear, the bands in most older automatic transmissions need to be adjusted periodically. However, as band designs have improved, periodic band adjustment is not needed on most newer transmissions. On units that require periodic adjustment of the bands, the clearance between the band and the drum is set with an adjustment screw that also serves as the anchor for the band. Excessive slippage will cause a band to burn or become glazed.

A servo assembly is a hydraulic piston and cylinder assembly that controls the contraction or application and release of a band.

Shop Manual
Chapter 6, page 191

Servos

A band is applied hydraulically by a servo unit (Figure 6-39). The servo contracts the band when hydraulic pressure pushes against the servo's piston and overcomes the tension of the servo's return spring. This action moves an operating rod toward the band, which squeezes the band around the drum. When hydraulic pressure to the servo apply port is stopped and exhausted, the band and servo are released and the return spring on the opposite side of the piston returns the piston to its original position. On some transmissions, the return of the piston is aided by hydraulic pressure sent to the release side of the servo. Once the piston has returned, this release pressure keeps the band from applying until apply pressure is again sent to the servo.

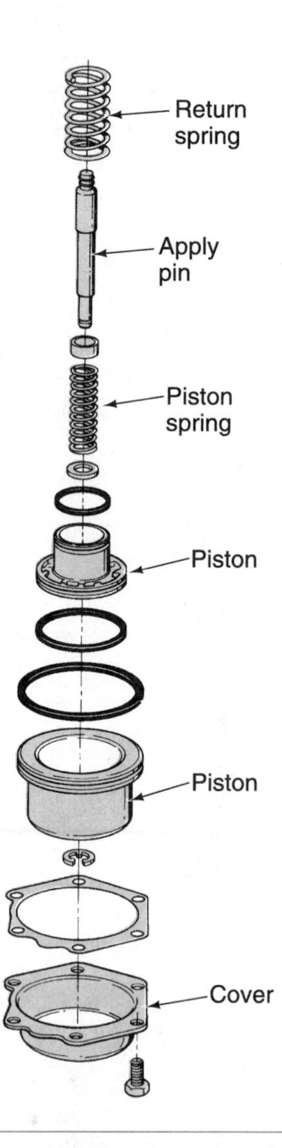

Figure 6-39 A typical dual-piston servo unit. (Courtesy of the Hydra-Matic Division of General Motors Corp.)

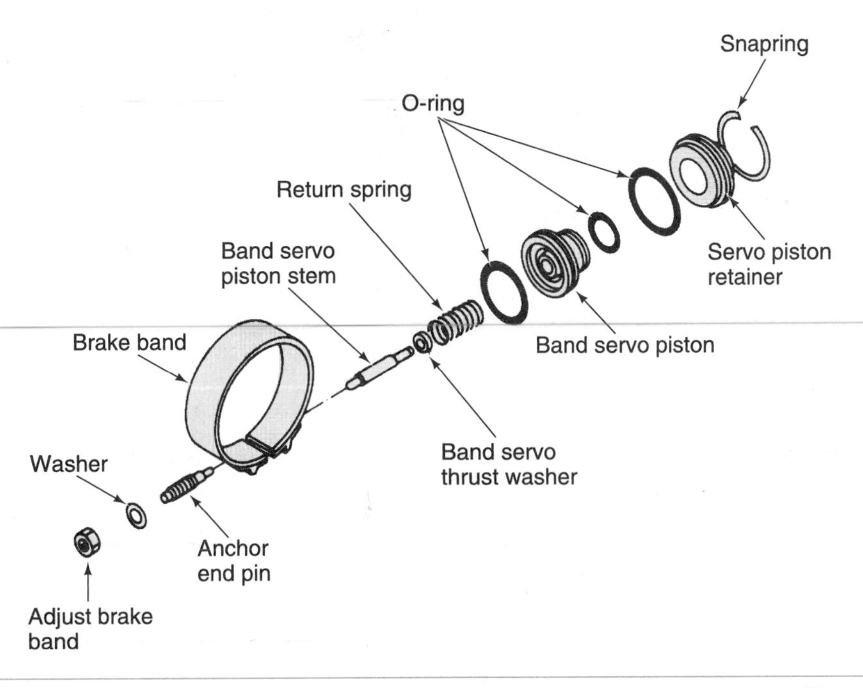

Figure 6-40 A typical band and servo assembly. (Courtesy of Nissan Motor Co., Ltd.)

A servo unit provides for the application and deenergizing of the transmission's bands (Figure 6-40). A band is energized by the action of the servo's piston on one end of the band while its other end is anchored to the transmission case. A servo unit consists of a piston in a cylinder and a piston return spring. The cylinder may be part of the transmission housing's casting or it may be a separate unit bolted to the case. The force from the servo's piston acts directly on the end of the band by an apply pin or through a lever arrangement that provides mechanical advantage or a multiplying force. The band squeezes a drum attached to a gear set member.

The servo unit must apply the band securely so that it can rigidly hold a member of the planetary gear set in order to cause forward or reverse gear reduction on every drive. To assist the hydraulic and mechanical apply forces, the servo and band anchor are positioned to take advantage of drum rotation. When the band is applied, it becomes self-energized and wraps itself around the drum in the same direction as drum rotation. This self-energizing effect reduces the force that the servo must produce to hold the band.

To release the servo apply action on the band, the servo apply fluid is exhausted from the circuit or a servo release fluid is introduced on the servo piston that opposes the apply fluid. When servo release fluid is introduced on a shift, the hydraulic mainline pressure acting on the top of the piston plus the servo return spring will overcome the servo apply fluid and move the piston down.

The clamping pressure of a band is determined by the application force exerted on it by the servo. Because the area of the servo's piston is relatively large, the application force of the servo is considerably greater than the line pressure delivered to the servo. This increase in force is necessary to clamp the band tightly enough around the spinning drum to bring it to a halt. The servos in most transmissions are designed to react to varying pressures and apply different amounts of application forces to meet the needs of the transmission. The total force of a piston is equal to the pressure applied to it multiplied by the surface area of the piston. Therefore, if the servo piston has a surface area of 2.5 square inches and the normal pressure sent to the servo is 50 psi, the application force of the servo on the band will be 125 psi. When the pressure sent to the servo increases to 100 psi, such as during heavy loads, the application force on the operating rod increases to 250 psi. This increase in force allows for increased clamping pressures during heavy load conditions and times when slipping will most likely take place.

The operating rod of a servo tightens a band through one of three basic linkage designs: straight, lever, or cantilever. The straight linkage design uses a straight rod or strut to transfer the piston's force to the free end of the band. This type linkage is only used when the servo is placed where it can act directly on a band and when the servo's piston is large enough to hold the band when maximum torque is applied to the drum. Some transmissions that use a straight linkage have specially designed rods, which are graduated. These rods are designed to minimize the need for periodic band adjustment. Band adjustment takes place during assembly by selecting the proper rod length. When the correct rod is installed in the servo, the piston will move a specific distance and apply a specific amount of force against a band.

The lever-type linkage uses a lever to move the rod or strut that actually applies the band. A lever-type linkage normally increases the application force of the piston because the **fulcrum** of the lever is closer to the band.

A fulcrum is the support and pivot point of a lever.

A **cantilever**-type linkage uses a lever and a cantilever to act on both ends of an unanchored band. As the servo piston applies force to the operating rod, the rod moves the lever and applies force to one end of the band. The movement of the piston also pulls the cantilever toward its pivot pin, thereby clamping the ends of the band together and tightening the band around the drum. A cantilever linkage increases band application force and allows for smooth band application because the band self-centers and contracts evenly around the drum.

A cantilever linkage acts on both ends of a band to tighten it around a drum. A cantilever is a lever that is anchored and supported at one end by its fulcrum or pivot pin and provides an opposing force at its opposite end.

Multiple-Disc Clutch Assembly

The purpose of a multiple-disc clutch is the same as a band; however, because there is much more surface area on the clutch, it has greater holding capabilities. The multiple-disc clutch assembly (Figure 6-41) can be used to drive or hold a member of the planetary gear set by connecting it to the transmission's case or to a clutch drum. A multiple-disc clutch pack consists of several clutch plates lined with frictional material and several **steel** separator **discs** that are placed alternately inside a clutch drum. The clutch plates have rough frictional material on their faces, whereas the steel discs have smooth faces without frictional material. Hydraulic pressure causes a piston inside the drum to compress the discs tightly together to apply the clutch. If this pressure is vented, return springs retract the piston, and the clutch disengages.

Shop Manual
Chapter 6, page 195

A clutch friction disc is basically a steel plate with its sides lined with frictional material. Metallic, semimetallic, and paper-based materials are used as this frictional lining. Paper cellulose is the most common frictional material because it offers good holding power without the high frictional wear of metallic materials. Friction plates often have grooves cut in them to help keep them cool, thereby increasing their effectiveness and durability. The grooves allow ATF to flow over the plates, thereby removing some of the heat from the plates.

The multiple-disc clutch assembly used in automatic transmissions is a "wet clutch" assembly.

A multiple-disc clutch assembly also contains one or more return springs, return spring retainers, seals, one or more pressure plates, and snaprings. The seals hold in the hydraulic pressure during application of the clutch. A **pressure plate** (Figure 6-42) is a heavy metal plate that provides the clamping surface for the clutch plates and is installed at one or both ends of the clutch pack. The snaprings are used to hold the parts in the clutch pack together. In a typical clutch, the apply piston at the rear of the drum is held in place by the return springs and by a spring retainer secured by a snapring. Hydraulic pressure moves the piston against return spring pressure and clamps the clutch plates against the pressure plate. The friction between the plates locks them together, causing them to turn as a unit.

Friction discs and steel discs are collectively called clutch packs.

Steel discs are sometimes also called apply plates or steels.

One set of plates has **splines** on its inner edges, while the other set is splined on its outer edges. The splines of each set fit into matching splines on either a shaft, a drum, a member of the gear set, or the transmission case. When the clutch is applied, the components meshed with the splines of the clutch plates are mechanically connected when the two sets of plates are locked together. Depending on what they are splined to, multiple-disc clutches can be either braking or driving devices.

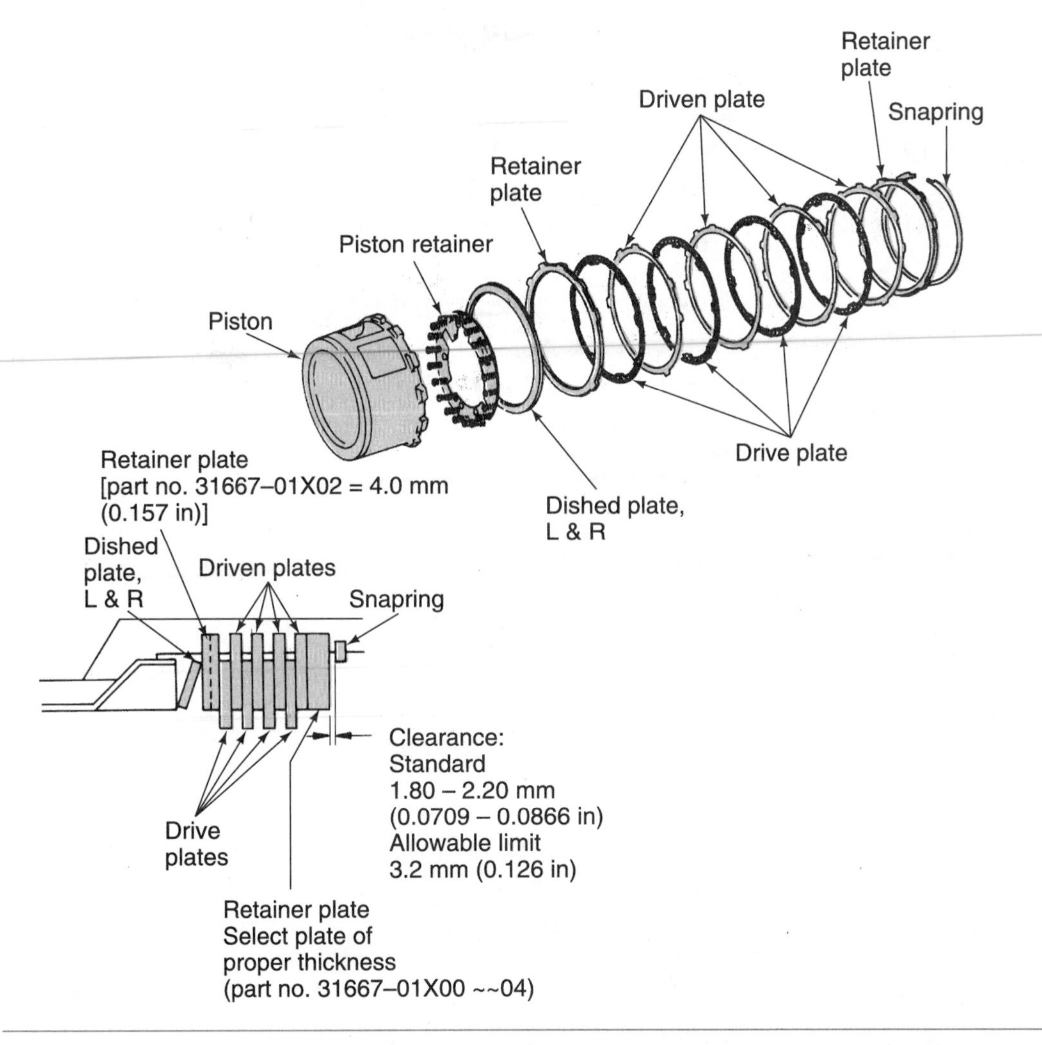

Figure 6-41 Typical multiple-disc clutch assembly. (Courtesy of Nissan Motor Co., Ltd.)

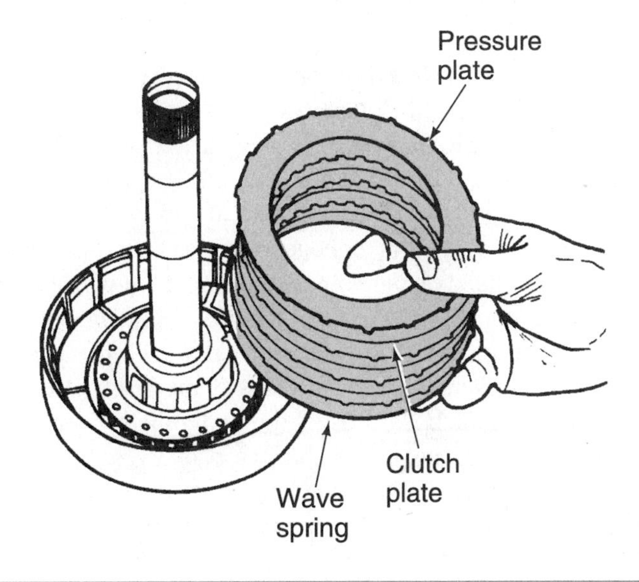

Figure 6-42 Location of the pressure plate in a multiple-disc clutch assembly. (Reprinted with the permission of Ford Motor Co.)

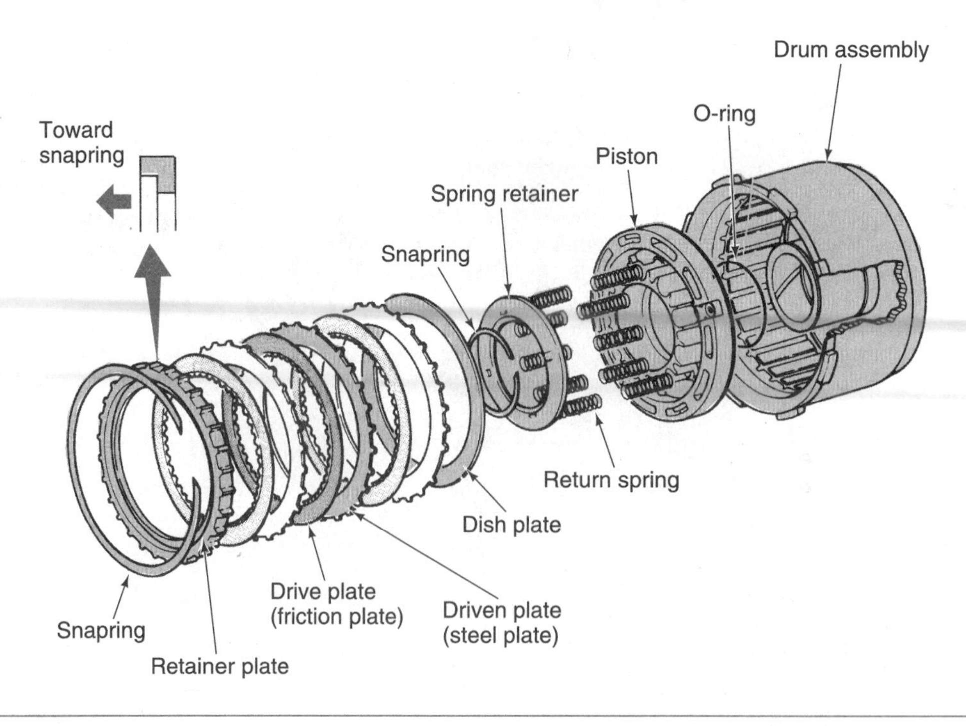

Figure 6-43 Clutch assembly used to lock two planetary gear set members together. (Courtesy of Nissan Motor Co., Ltd.)

Splines are used as a method for attaching a gear, hub, drum, or other component to a housing or shaft by aligning interior slots of one component with matching exterior slots of the other so that the two parts are locked together.

When multiple-disc clutch packs lock two members of a planetary gear set together, the clutch is a driving device (Figure 6-43). One set of discs is splined to one member of the planetary gear set and the other set of discs is splined to another member of the gear set. The two members of the gear set rotate together at the same speed when the clutch is applied.

Driving clutches can have a set of clutch friction plates splined to the transmission's input shaft. The set of steel plates is splined to the inside of the clutch drum, which is connected to a member of the planetary gear set. When the clutch is applied, the piston clamps the sets of the discs together. The drive discs then engage with the driven discs, which in turn rotate the drum and the attached gear set member at the same speed. Hydraulic pressure is applied to the clutch through a passage inside the input shaft. When the hydraulic pressure is exhausted, the drive discs rotate with the input shaft but are not engaged with the driven discs splined to the drum.

Driving clutches can also have a drum splined directly to the input shaft. Inside the drum and splined to it are the drive discs. The driven discs are splined to the outside of a clutch hub splined to the output shaft or attached to a member of the gear set. When the clutch is applied, the plates mechanically lock the drum and hub together.

A multiple-disc clutch is also used to hold one member of the planetary gear set (Figure 6-44). The friction discs are splined on their inner edges and are fit into matching splines on the outside of a drum. The steel discs are splined on their outer edges and fit into matching splines machined into the transmission case. When the clutch is applied, the gear set member cannot rotate, as it is locked to the transmission case. The apply piston for a holding clutch is in a bore located in the transmission case and is fed hydraulic pressure through a passage in the case.

When a multiple-disc clutch is released, the hydraulic pressure to the piston is stopped and exhausted. The piston return spring moves the piston, allowing for a clearance between the sets of plates, thereby disengaging them. To relieve any residual pressure when the clutch is released, the clutch has a vent port with a check ball (Figure 6-45). The check ball is forced against its seat when full hydraulic pressure is applied to the clutch. This holds all the pressure inside the drum. When the pressure supply to the clutch is stopped, only residual pressure remains on the check ball. Centrifugal force pulls the ball from its seat and allows the fluid to escape from the drum through the open vent port. If the residual pressure is not relieved, some pressure may remain on the pis-

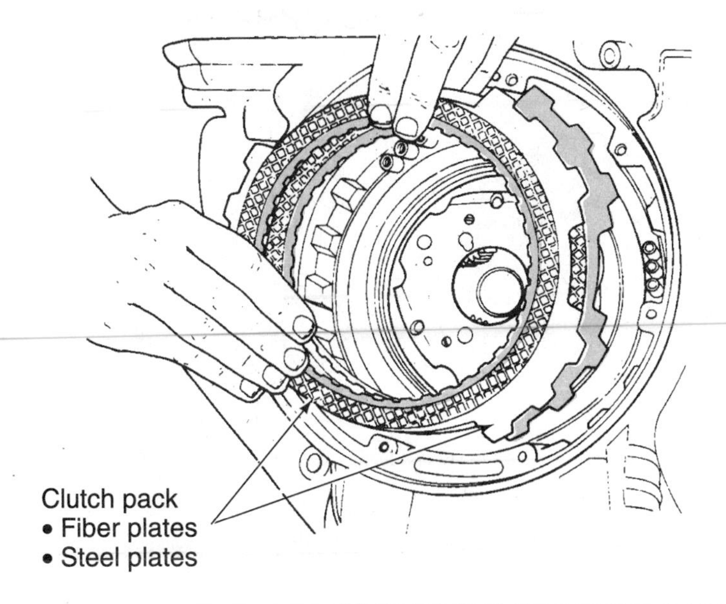

Figure 6-44 Clutch assembly used to lock a planetary gear set member to the transmission case. (Reprinted with the permission of Ford Motor Co.)

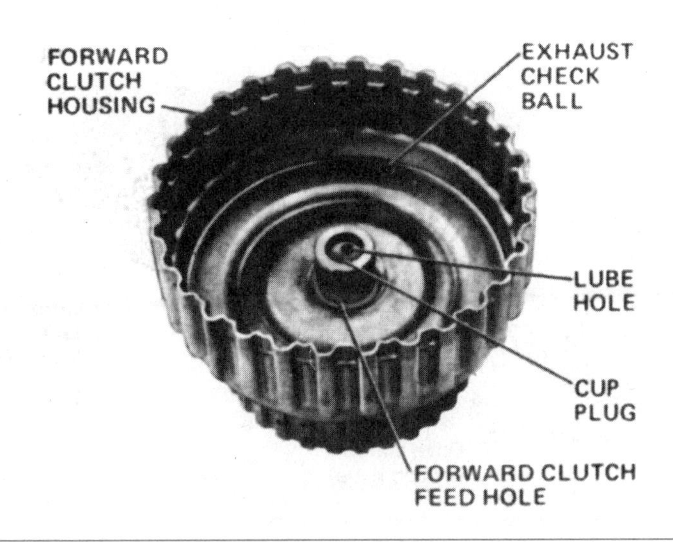

Figure 6-45 Location of exhaust check ball in a clutch housing. (Courtesy of the Oldsmobile Division of General Motors Corp.)

ton causing partial engagement of the clutch. Not only would this cause the discs to wear prematurely, bur also it would adversely affect shift quality.

The check ball and vent port are located in the drum or the piston. Some transmissions are fitted with a metered orifice in either the clutch drum or the piston. A metered orifice controls the release of the hydraulic pressure and therefore serves the same purpose as a vent port and check ball.

Shop Manual
Chapter 6, page 200

Clutch apply pistons are retracted by one large coil spring (Figure 6-46), several small springs, or a single belleville spring (Figure 6-47). The type and number of return springs used in a clutch pack are determined by the pressure needed to release the piston quickly enough to prevent clutch dragging. However, the amount of spring tension is limited to keep the resistance to piston application minimized. Pistons fitted with multiple springs have spring pockets machined into the assembly. There are often fewer springs used than there are pockets. This is an indication

A belleville spring is also called a diaphragm or **over-center spring.**

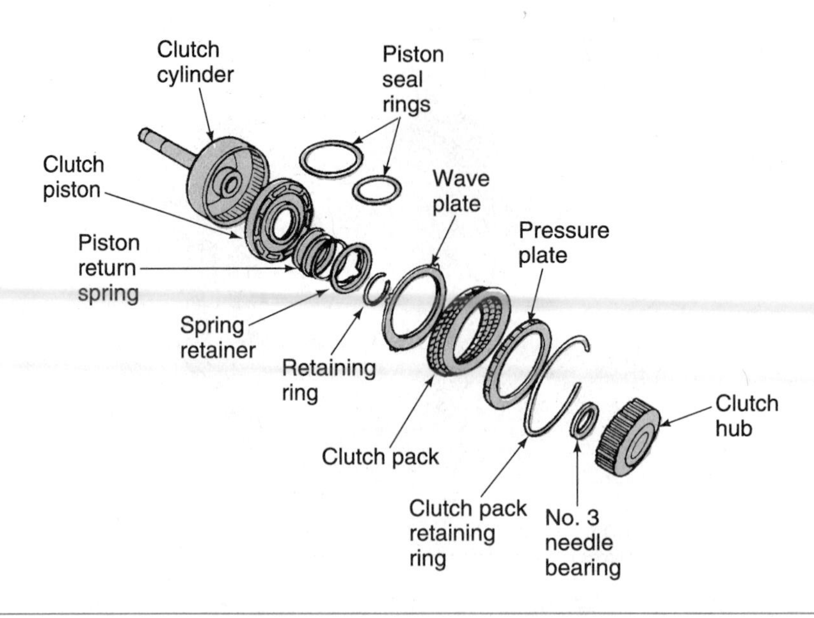

Figure 6-46 Clutch assembly using a single coil return spring for release. (Reprinted with the permission of Ford Motor Co.)

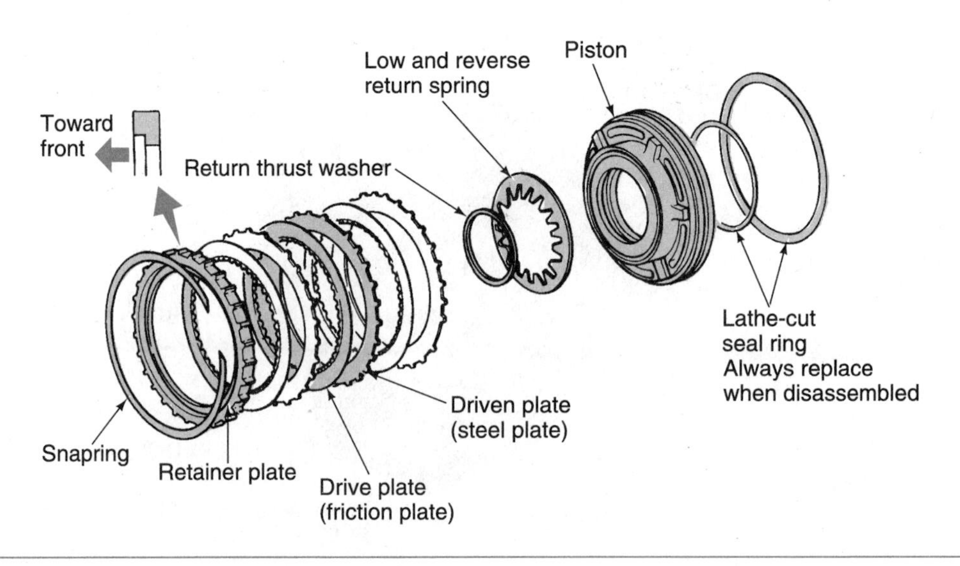

Figure 6-47 Clutch assembly using a single belleville-type return spring. (Courtesy of Nissan Motor Co., Ltd.)

that the manufacturer uses the same clutch assembly for different applications and springs are added or subtracted to meet the needs of those applications. The manufacturer may also make other changes to the basic assembly to change the load-carrying capacity of the clutch. One of these changes may be to use a different number of clutches in the assembly. This is done by using a different thickness of the piston or retainer plate, or by changing the location of the snapring groove in the clutch drum.

Shop Manual
Chapter 6, page 200

A belleville spring acts to improve the clamping force of the clutch assembly, and a piston return spring. The spring is locked into a groove inside the drum by a snapring. As the clutch piston moves to apply the clutch, it moves the inner ends of the belleville spring fingers into contact with the pressure plate to apply the clutch. The spring's fingers act as levers against the pressure plate and increase the application force of the clutch. When hydraulic pressure to the piston is stopped, the spring relaxes and returns to its original shape. The piston is forced back and the clutch is released.

One-Way Clutches

One-way or overrunning clutches are holding devices. These clutches operate mechanically, not hydraulically, and are considered apply devices. The main difference between a multiple-disc clutch or band and an overrunning clutch is that the one-way clutch allows rotation in only one direction and operates at all times, whereas a clutch or a band allows or stops rotation in either direction and operates only when hydraulic pressure is applied to it.

A one-way clutch is a mechanical holding device that prevents rotation in one direction but overruns to allow rotation in the other direction.

One-way overrunning clutches can be either roller or sprag type. A **roller clutch** utilizes roller bearings held in place by springs to separate the inner and outer race of the clutch assembly (Figure 6-48). Around the inside of the outer race are several cam-shaped indentations. The rollers and springs are located in these pockets. Rotation of one race in one direction locks the rollers between the two races, causing both to rotate together. When a race is rotated in the opposite direction, the roller bearings move into the pockets and are not locked between the races (Figure 6-49). This allows the races to turn independently.

Races are the parts in bearings or one-way overrunning clutches on which the rollers or ball bearings roll.

A one-way **sprag clutch** consists of a hub and a drum separated by figure-eight-shaped metal pieces called sprags. The sprags are shaped such that they lock between the races when a race is turned in one direction only. Between the inner and outer races of the clutch are the sprags, cages, and springs. The sprags are longer than the distance between the two races. The cages keep the sprags equally spaced around the diameter of the races. The springs hold the sprags at the correct angle and maintain contact of the sprags with the races for instantaneous engagement. When a race turns in one direction, the sprags tilt and allow the races to move independently (Figure 6-50). When a race is moved in the opposite direction, the sprags straighten up and lock the two races together.

When the two races are free to move independently of each other, an overrun condition exists.

Both types of one-way clutches apply and release quickly and evenly in response to the rotational direction of the races. This allows for smooth gear changes. Either type can be used to hold a member of the planetary gear set by locking the inner race to the outer race, which is held by the transmission housing. Both types are also effective as long as the engine powers the transmission. While the transmission is in a low gear and is coasting, the drive wheels rotate the transmission's output shaft with more power than received on the input shaft. This allows the sprags or rollers to unwedge and begin **freewheeling.** This puts the planetary gear set into neutral, disallowing

Freewheeling occurs when there is a power input but no transmission of power through the transmission.

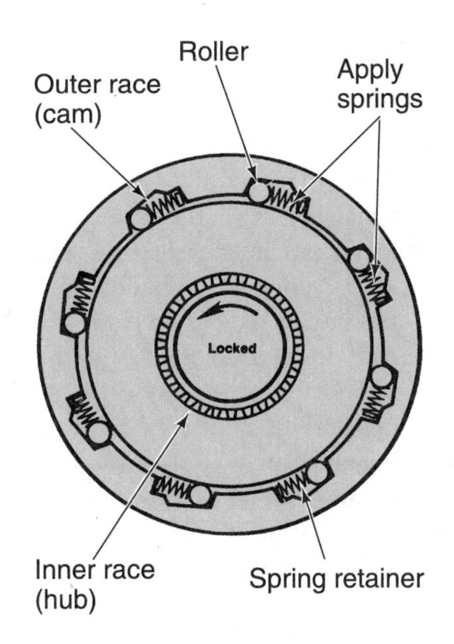

Figure 6-48 A roller-type one-way clutch. (Courtesy of Chrysler Corp.)

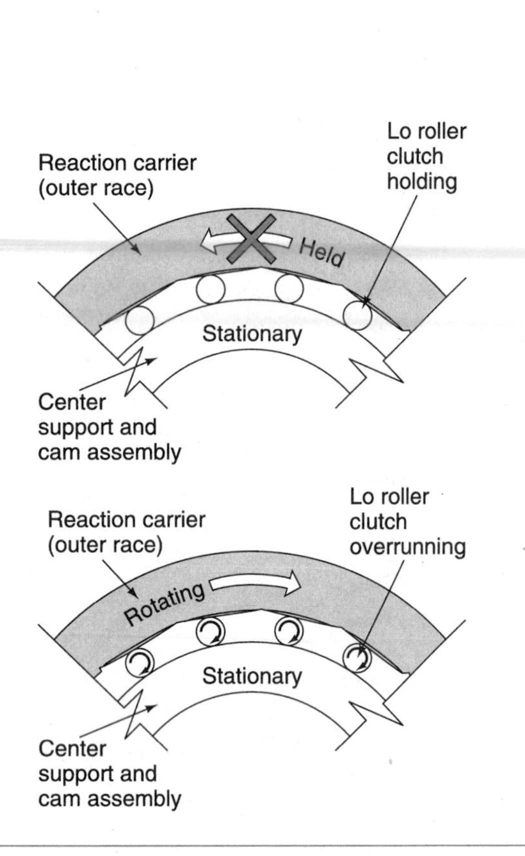

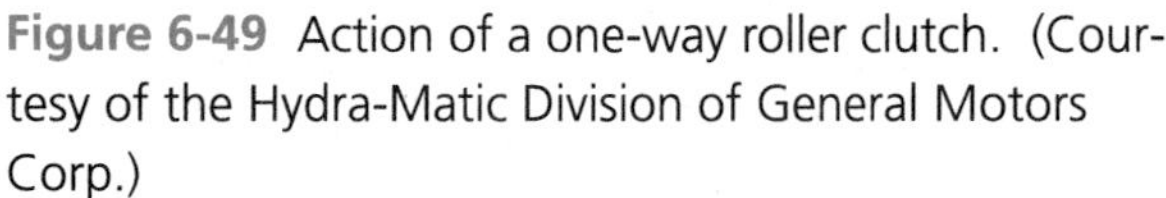

Figure 6-49 Action of a one-way roller clutch. (Courtesy of the Hydra-Matic Division of General Motors Corp.)

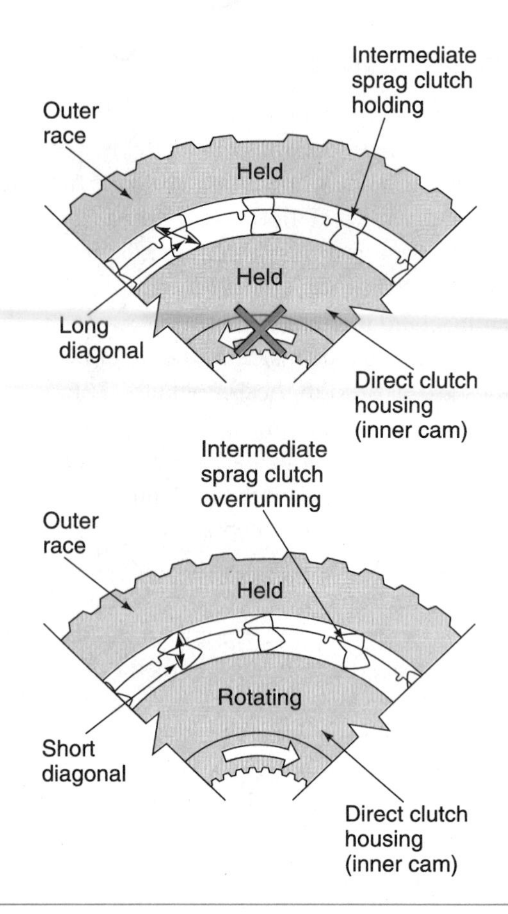

Figure 6-50 Action of a sprag-type one-way clutch. (Courtesy of the Hydra-Matic Division of General Motors Corp.)

engine compression to aid in the slowing down process or braking. Engine braking is provided in many different ways by the various transmission designs. It is normally provided by a band, which holds regardless of the power source.

Some late-model transmissions, such as the 4R70W, use a ratchet-type one-way clutch. These operate as a sprag clutch but have gears in place of the sprags.

Shift Feel

All transmissions are designed to change gears at the correct time, according to engine speed, load, and driver intent. However, transmissions are also designed to provide for positive change of gear ratios without jarring the driver or passengers. If a band or clutch is applied too quickly, a harsh shift will occur. "Shift feel" is controlled by the pressure at which each reaction member is applied or released, the rate at which each is pressurized or exhausted, and the relative timing of the apply and release of the members.

To improve shift feel during gear changes, a band is often released while a multiple-disc clutch is being applied. The timing of these two actions must be just right or both components will be released or applied at the same time, which would cause engine flare-up or drive line shudder. Several other methods are used to smooth gear changes and improve shift feel.

Clutch packs sometimes contain a wavy spring-steel separator plate that helps smooth the application of the clutch. Shift feel can also be smoothed out by using a restricting orifice or an accumulator piston in the band or clutch-apply circuit. A restricting orifice in the passage to the apply piston restricts fluid flow and slows the pressure increase at the piston by limiting the quantity of fluid that can pass in a given time. An accumulator piston slows pressure buildup at the apply piston by diverting a portion of the pressure to a second piston in the same hydraulic circuit. This delays and smooths the application of a clutch or band.

The use of hydraulic power to apply bands and clutches is confined to a single device, the hydraulic piston. The pistons are housed in cylinder units known as servo and clutch assemblies. The function of the piston is to convert the force in the fluid into a mechanical force capable of handling large loads. Hydraulic pressure applied to the piston strokes the piston in the cylinder and applies its load. During the power stroke, a mechanical spring (or springs) is compressed to provide a means of returning the piston to its original position. The springs also determine when the apply pressure buildup will stroke the piston. This is critical to clutch/band life and shift quality.

In rotating clutch units, a problem usually arises when the clutch is not engaged. With the clutch off, the clutch drum or housing still spins; at times this speed may be an overspeed condition. The high-speed rotation could create sufficient centrifugal force in the residual or remaining oil in the clutch-apply cylinder to partially engage the clutch. This creates an unwanted drag between the clutch plates. To prevent this problem, a centrifugal check ball relief valve is incorporated in the clutch drum or clutch piston.

The steel check ball operates in a cavity with a check ball seat. A small hole or orifice is tapped from the seat for pressure relief. When the clutch unit is applied, oil pressure holds the ball on its seat and blocks off the orifice. In the released position, centrifugal force created by the rotating clutch moves the ball off its seat, allowing the residual oil behind the clutch piston to be released.

Accumulators

Shift quality with a band or clutch is mostly dependent on how quickly the apply device is engaged by hydraulic pressure and the amount of pressure exerted on the piston. Some apply circuits utilize an accumulator to slow down application rates without decreasing the holding force of the apply device.

Shop Manual
Chapter 6, page 191

An accumulator is similar to a servo in that it consists of a piston and cylinder (Figure 6-51). However, its purpose is to cushion the application of an apply device. The pressure required to activate the piston in an accumulator is controlled by a spring or by fluid pressure. As the spring compresses, pressure at the servo or clutch pack increases. This increase in pressure activates the accumulator, which causes a delay in the servo or clutch pack receiving the high pressure. As a result, the shift takes slightly longer but is less harsh.

An accumulator (Figure 6-52) works like a shock absorber and cushions the application of servos and clutches. An accumulator cushions sudden increases in hydraulic pressure by temporarily diverting some of the apply fluid into a parallel circuit or chamber. This allows the pressure to gradually increase and provides for smooth engagement of a band or clutch.

Some transmission designs do not rely on an accumulator for shift quality, rather they have a restrictive orifice in the line to the servo or clutch piston. This restriction decreases the amount of initial apply pressure but will eventually allow for full pressure to act on the piston.

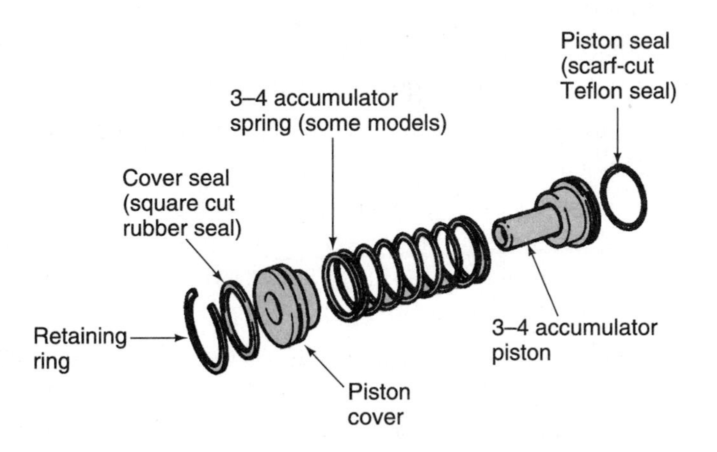

Figure 6-51 Typical accumulator assembly. (Reprinted with the permission of Ford Motor Co.)

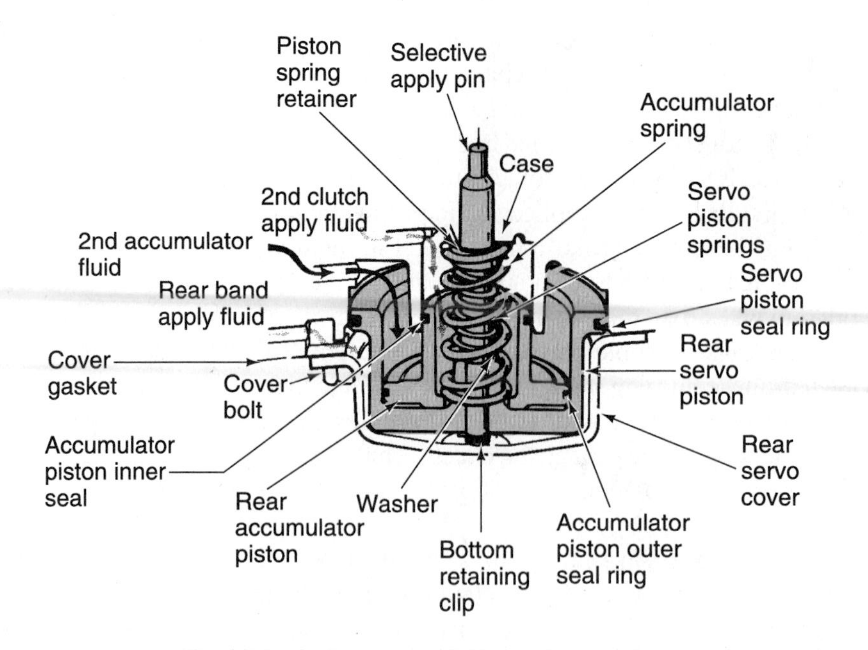

Figure 6-52 An accumulator works like a shock absorber to cushion the application of servos and clutches. (Courtesy of the Hydra-Matic Division of General Motors Corp.)

Summary

- ❑ An automatic transmission receives engine power through a torque converter, which is indirectly attached to the engine's crankshaft.
- ❑ The various gear ratios are selected by the transmission according to engine and vehicle speeds, engine load, and other operating conditions.
- ❑ Pressure-regulating valves change the pressure of the fluid to control the shift points of the transmission. Flow-directing valves direct the pressurized fluid to the appropriate reaction members, which cause a change in gear ratios.
- ❑ All fluids conform to the shape of their container.
- ❑ Pascal determined that force applied to liquid creates pressure, or the transmission of force through the liquid.
- ❑ Liquids have no shape of their own. They acquire the shape of the container they are put in.
- ❑ Liquids are basically incompressible.
- ❑ The pressure applied to a liquid in a sealed container is transmitted equally in all directions and to all areas of the system and acts with equal force on all areas.
- ❑ When a pressure is applied to a confined liquid, the pressure of the liquid is the same everywhere within the hydraulic system.
- ❑ ATF is a hydraulic fluid designed specifically for automatic transmissions. Its primary purpose is to transmit pressure to activate the transmission's bands and clutches. It also serves as a fluid coupling between the engine and the transmission, removes heat from the transmission, and lubricates the transmission's moving parts.
- ❑ Petroleum-based ATF typically has a clear, red color and will darken when it is burned or will become milky when contaminated by water.
- ❑ Nearly all manufacturers recommend the use of DEXRON-II ATF in their automatic transmissions.

Terms to Know

Accumulator
Ball-type valve
Belleville spring
Cantilever
Control valves
Crescent
Diaphragm spring
Double-wrap band
Drum
Dual band
Freewheeling
Fulcrum
Gear-type pumps
Hybrid valve
IX pump
Lands
Manual valve
MAP sensor
Mechanical advantage

- To trap dirt and metal particles from the circulating ATF, automatic transmissions have a fluid filter normally located inside the transmission case between the oil pump pickup and the bottom of the oil pan.
- Current automatic transmissions are fitted with one of three types of filter: a screen filter, paper filter, or felt filter.
- To remove the heat from the fluid, the ATF is directed through an oil cooler where it dissipates heat.
- Three types of pumps are commonly used: the gear-type, rotor-type, and vane-type. The gear-type is the most common.
- The pumps in all rear-wheel-drive and some front-wheel-drive transmissions are driven externally by the torque converter's drive hub. The hub has two slots or flats machined in it that fit into the oil pump drive.
- Many front-wheel-drive transaxles use a hex-shaped or splined shaft fitted in the center of the converter cover to drive the pump. This system is called an internal drive.
- While the pump is operating, the pump creates a low pressure area at its inlet port. The pan is vented to the atmosphere, and atmospheric pressure forces the fluid into the inlet port of the pump. The fluid is now moved by the pump and forced out, at a higher pressure, through the outlet port.
- Gear-type pumps consist of two gears in mesh to pressurize the fluid.
- A common type of gear pump is the crescent pump, which uses two gears. One gear of this pump has internal teeth and the other has external teeth. The smaller gear with external teeth is in mesh with one part of the larger gear.
- A rotor-type pump is a variation of the gear-type pump. However, instead of gears, an inner and an outer rotor turn inside the housing.
- In a vane-type pump, a vane ring is driven by the torque converter. The vane ring contains several sliding vanes that seal against a slide mounted in the pump housing.
- Vane-type pumps are variable-displacement pumps whose output can be reduced when high fluid volumes are not necessary.
- The valve body regulates and directs the fluid flow to provide for the gear changes.
- The basic types of valves used in automatic transmissions are ball, poppet, or needle check valves, and relief valves, orifices, and spool valves.
- A simple check valve is used to block a hydraulic passageway. It normally stops fluid flow in one direction, while allowing it in the opposite direction.
- A relief valve prevents or allows fluid flow until a particular pressure is reached. Then it either opens or closes the passageway. Relief valves are used to control maximum pressures in a hydraulic circuit.
- Orifices are used to regulate and control fluid volumes.
- Spool valves are normally used as flow-directing valves. They are the most commonly used type of valve in an automatic transmission.
- Fluid passages in the bore are closed and opened as the lands of the spool cover and uncover the passage openings.
- The movement of a spool valve is controlled by either mechanical or hydraulic forces.
- A control valve is used as a flow-directing valve.
- A relay valve is a spool valve with several spools, lands, and reaction areas. It is used to control the direction of fluid flow. A relay valve does not control pressure.
- The valve body is machined from aluminum or iron castings and has many precisely machined bores and fluid passages. Fluid flow to and from the valve body is routed through passages and bores.

Terms to Know (continued)

Modulator valve
Needle check valve
Orifice
Over-center spring
Poppet valve
Positive displacement pump
Pressure plate
Pressure-regulating valve
Reaction area
Relay valve
Relief valve
Roller clutch
Rotor-type pump
Separator plate
Shift valve
Single-wrap band
Splines
Split-band
Spool valve
Sprag clutch
Steel discs
Throttle valve
Transfer plate
Two-way check valve
Valley
Vane-type pump
Variable capacity pumps
Work
Worm tracks

- Valve bodies are normally fitted with three different types of valves: spool valves, check ball valves, and poppet valves.
- The valve body is composed of two or three main parts: a valve body, separator plate, and transfer plate. The separator and transfer plates are designed to seal off some of the passages and they contain some openings that help control and direct fluid flow through specific passages. The three parts are typically bolted together and mounted as a single unit to the transmission housing.
- All nonelectronically shifted automatic transmissions use three basic pressures to control their operation: mainline pressure, throttle pressure, and governor pressure.
- Mainline pressure is a regulated pump pressure that is the source of all other pressures in the transmission.
- Governor pressure is a regulated hydraulic pressure that varies with vehicle speed.
- Throttle pressure is a regulated pressure that varies with engine load or throttle position.
- Pressure-regulating valves use principles of both a pressure relief valve and a spool valve.
- The pressure regulator valve has three primary purposes and modes of operation: filling the torque converter, exhausting fluid pressure, and establishing a balanced condition by maintaining line pressure.
- A governor senses road speed and sends a fluid pressure signal to the valve body to either upshift or downshift.
- There are three common designs of governors: shaft-mounted, gear-driven with check balls, and gear-driven with a spool valve. All of these designs have primary and secondary weights, springs, and valves connected to the output shaft by a valve shaft.
- The shaft-mounted governor has a spring-loaded spool valve attached to the output shaft of the transmission.
- Gear-driven governors with check balls mount to the transmission at a right angle to the output shaft.
- The gear-driven governor with a spool valve is driven by and mounted at a right angle to the transmission output shaft.
- A PM generator is an electronic device that utilizes a magnetic pickup sensor located on the transmission/transaxle housing and a trigger wheel mounted on the output shaft.
- Throttle pressure causes an increase in the fluid pressure applied to the apply devices of the planetary units to hold them tightly to reduce the chance of slipping while the vehicle is operating under heavy load.
- Transmissions that are not equipped with a vacuum modulator use a throttle cable or electronic devices to sense engine load and change fluid pressures.
- A vacuum modulator measures engine vacuum to sense the load placed on the engine and the drive train. A vacuum modulator is a load-sensing device that increases or decreases fluid pressure in response to a vacuum signal, which varies with throttle opening and vehicle load.
- Throttle pedal movement is used to control the operation of the throttle valve.
- The throttle valve converts mainline pressure into a variable throttle pressure based on the position of the throttle plate.
- Engine load can be monitored electronically through the use of various electronic sensors that send information to an electronic control unit, which in turn controls the pressure at the valve body. The most commonly used sensor is the manifold absolute pressure (MAP) sensor. The MAP sensor senses air pressure or vacuum in the intake manifold.
- Shift timing is determined by throttle pressure and governor pressure acting on opposite ends of the shift valve.

- ❑ Shift valves control the upshifting and downshifting of the transmission by controlling the flow of pressurized fluid to the apply devices that engage the different gears.
- ❑ The manual valve is a spool valve operated manually by the gear selector linkage.
- ❑ A valve body is fitted with a kickdown circuit to provide a downshift when additional torque is needed.
- ❑ Shifting electronically allows for shifting at better times than does shifting by hydraulics, and is based on inputs to a control computer from various sensors, such as engine temperature, engine speed, engine load, vehicle speed, throttle position, and gear selector position.
- ❑ The valve body is the master flow control for an automatic transmission. It contains the worm tracks and numerous bores fitted with many spool-type and other type valves. Each passage, bore, and valve forms a specific oil circuit with a specific function.
- ❑ All transmissions are designed to change gears at the correct time, according to engine speed, load, and driver intent.
- ❑ A band is an externally contracting brake assembly that is positioned around the outside of a drum. The drum is connected to a member of the planetary gear set.
- ❑ A servo is a hydraulically operated piston assembly used to apply a band.
- ❑ A servo contracts the band when hydraulic pressure pushes against the servo's piston and overcomes the tension of the servo's return spring. This action moves an operating rod toward the band, which squeezes the band around the drum. When hydraulic pressure to the servo apply port is stopped and exhausted, the band and servo are released and the return spring on the opposite side of the piston returns the piston to its original position.
- ❑ The operating rod of a servo tightens a band through one of three basic linkage designs: straight, lever, or cantilever.
- ❑ The purpose of a multiple-disc clutch is the same as a band; however, it can also be used to drive a planetary member.
- ❑ One-way overrunning clutches are purely mechanical devices, whereas clutches and bands are hydraulically controlled mechanical devices. All automatic transmissions use more than one type of these apply devices; some use all three.
- ❑ The multiple-disc clutch assembly can be used to drive or hold a member of the planetary gear set by connecting it to the transmission's case or to a clutch drum.
- ❑ When multiple-disc clutch packs lock two members of a planetary gear set together, the clutch is a driving device.
- ❑ A multiple-disc clutch is used to hold one member of the planetary gear set.
- ❑ Clutch-apply pistons are retracted by one large coil spring, several small springs, or a single belleville spring.
- ❑ The main difference between a multiple-disc clutch or band and an overrunning clutch is that the one-way clutch allows rotation in only one direction and operates at all times, whereas a clutch or a band allows or stops rotation in each direction and operates only when hydraulic pressure is applied to it.
- ❑ One-way overrunning clutches can be either roller or sprag type. A roller clutch utilizes roller bearings held in place by springs to separate the inner and outer race of the clutch assembly.
- ❑ A sprag clutch is a one-way overrunning clutch that uses a series of sprags placed between two races.
- ❑ Shift feel is controlled by the pressure at which each apply device is applied or released, the rate at which each is pressurized or exhausted, and the relative timing of the apply and release of the members.

Review Questions

Short Answer Essays

1. Describe the path of line pressure when a GM THM 4T60-E is operating in first gear with the gear selector in overdrive.
2. What are the purposes of ATF?
3. What drives a transmission oil pump?
4. How can a multiple-disc clutch assembly be used to drive or hold a member of the planetary gear set?
5. What is the purpose of the transfer and separator plates in a valve body assembly?
6. Explain each of the different operating pressures in a typical transmission.
7. What is the purpose of a vacuum modulator?
8. What determines the timing of the shifts in an automatic transmission?
9. What are the benefits of controlling shift timing electronically?
10. What controls the feel of the shifts?

Fill in the Blanks

1. Power flow through the gears is controlled by ________________ - ________________ clutches, ________________ - ________________ clutches, or ________________ ________________.
2. ________________ - ________________ valves change the pressure of the oil to control the shift points of the transmission. ________________ - ________________ valves direct the pressurized oil to the appropriate reaction members, which cause a change in gear ratios.
3. Current automatic transmissions are fitted with one of three types of filters: a ________________ filter, ________________ filter, or ________________ filter.
4. Three types of oil pumps are commonly used: the ________________ -type, ________________ -type, and ________________ -type. The ________________ -type is the most common.
5. Clutch-apply pistons are retracted by one large ________________ spring, several ________________ ________________ springs, or a single ________________ spring.
6. The operating rod of a servo tightens a band through one of three basic linkage designs: ________________ , ________________ , or ________________ .
7. When multiple-disc clutch packs lock two members of a planetary gear set together, the clutch is a ________________ device.

8. The valve body is composed of two or three main parts: a ________________ ________________, ________________ ________________, and ________________ ________________.

9. All automatic transmissions use three basic pressures to control their operation: ________________ pressure, ________________ pressure, and ________________ pressure.

10. There are three common designs of governors: ________________-mounted, ________________-driven with ________________, and ________________-driven with a ________________.

ASE-Style Review Questions

1. *Technician A* says a simple check valve can be used to block a hydraulic passageway.
   Technician B says a check valve stops fluid flow in one direction, while allowing it in the opposite direction.
   Who is correct?
   A. A only **C.** Both A and B
   B. B only **D.** Neither A nor B

2. While discussing brake band operation:
   Technician A says a servo is a hydraulically operated piston assembly used to apply the band.
   Technician B says an accumulator is a hydraulic piston assembly that helps a servo quickly apply the band.
   Who is correct?
   A. A only **C.** Both A and B
   B. B only **D.** Neither A nor B

3. *Technician A* says one-way overrunning clutches can be either roller or sprag type.
   Technician B says a roller clutch utilizes roller bearings held in place by sprags, which separate the inner and outer race of the clutch assembly.
   Who is correct?
   A. A only **C.** Both A and B
   B. B only **D.** Neither A nor B

4. *Technician A* says a multiple-disc clutch can be used to hold one member of the planetary gear set.
   Technician B says a multiple-disc clutch can be used to drive one member of the planetary gear set.
   Who is correct?
   A. A only **C.** Both A and B
   B. B only **D.** Neither A nor B

5. *Technician A* says that one of the primary purposes of the pressure regulator valve is to fill the torque converter with fluid.
   Technician B says the pressure regulator valve directly controls throttle pressure.
   Who is correct?
   A. A only **C.** Both A and B
   B. B only **D.** Neither A nor B

6. *Technician A* says spool valves are the most commonly used type of valve in an automatic transmission.
   Technician B says spool valves are normally used as relief valves.
   Who is correct?
   A. A only **C.** Both A and B
   B. B only **D.** Neither A nor B

7. *Technician A* says engine load can be monitored electronically through the use of various electronic sensors, which send information to an electronic control unit.
   Technician B says the most commonly used load sensor is the manifold absolute pressure (MAP) sensor.
   Who is correct?
   A. A only **C.** Both A and B
   B. B only **D.** Neither A nor B

8. *Technician A* says all types of ATF become lighter when the friction modifiers are depleted.
   Technician B says petroleum-based ATF typically has a clear, red color and will darken when it is burned.
   Who is correct?
   A. A only **C.** Both A and B
   B. B only **D.** Neither A nor B

9. *Technician A* says in a vane-type pump, the vane ring contains several sliding vanes that seal against a slide mounted in the pump housing. *Technician B* says vane-type pumps are variable-displacement pumps whose output can be reduced when high fluid pressures are not necessary. Who is correct?

   A. A only
   B. B only
   C. Both A and B
   D. Neither A nor B

10. *Technician A* says the movement of a spool valve is controlled by mechanical forces. *Technician B* says the movement of a spool valve is controlled by hydraulic forces. Who is correct?

   A. A only
   B. B only
   C. Both A and B
   D. Neither A nor B

Common Automatic Transmissions

Upon completion and review of this chapter, you should be able to:

- ❑ Describe the construction and operation of typical Chrysler Corporation Simpson gear train–based transmissions.
- ❑ Describe the construction and operation of typical Ford Motor Company Simpson gear train–based transmissions.
- ❑ Describe the construction and operation of typical General Motors Corporation Simpson gear train–based transmissions.
- ❑ Describe the construction and operation of other typical Simpson gear train–based transmissions.
- ❑ Describe the construction and operation of typical Simpson gear train–based transmissions with add-on overdrive units.
- ❑ Describe the construction and operation of typical Ford Motor Company Ravigneaux gear train–based transmissions.
- ❑ Describe the construction and operation of typical General Motors Corporation Ravigneaux gear train–based transmissions.
- ❑ Describe the construction and operation of other typical Ravigneaux gear train–based transmissions.
- ❑ Describe the construction and operation of typical Ravigneaux gear train–based transmissions with overdrive.
- ❑ Describe the construction and operation of Ford AXOD and General Motors' 4T60 transaxles that use planetary gear sets in tandem.
- ❑ Describe the construction and operation of Honda automatic transmissions that use helical gears in constant mesh.

Introduction

Most automatic transmissions use planetary gear sets to provide for the different gear ratios. The gear ratios are selected manually by the driver or automatically by the hydraulic control system, which engages and disengages the clutches and bands used to shift gears. A single planetary gear set consists of a sun gear, a planet carrier with two or more planet pinion gears, and a ring gear. Planet gears are always in constant mesh; therefore, they allow quick, smooth, and precise gear changes without the worry of clashing or partial engagement.

Most common automatic transmissions use compound planetary gear sets to achieve the different forward and reverse gear ratios. The Simpson planetary gear set is the most commonly used, although some transmissions use a Ravigneaux gear set, tandem planetary gear sets, or helical gear sets. Because there is some similarity between models of transmissions according to the type of gear sets they use, a discussion of the operation and construction of commonly used transmissions is best presented in groupings of gear sets.

Simpson Gear Train

The Simpson gear train is an arrangement of two separate planetary gear sets with a common sun gear, two ring gears, and two planet pinion carriers. One half of the compound set or one planetary unit is referred to as the front planetary and the other planetary unit is the rear planetary. For the most part, each automobile manufacturer uses different parts of the planetary assemblies as input, output, and reaction members. This also varies with the different transmission models from the same manufacturer. There are also many different apply devices used in the various transmission designs.

Chrysler Torqueflite Transmissions

Shop Manual
Chapter 7, page 238

Chrysler Corporation introduced the Torqueflite transmission in 1956. This transmission was the first modern three-speed automatic transmission with a torque converter and the first to use the Simpson compound planetary gear train. Nearly all Torqueflite-based transmissions and transaxles use a rotor-type oil pump and all use a Simpson gear train.

The original Torqueflite transmission was called the A-466 and had a cast iron case with separate aluminum castings for the bell housing and the extension housing. Late-model Torqueflites have a one-piece aluminum housing that incorporates the bell housing and a bolt-on extension.

Chrysler transmissions have been used by AMC and International Harvester.

There are two basic versions of the three-speed Torqueflite transmission: the A-904 and the A-727. Introduced in 1960, the 904 is the light-duty version, whereas the 727, which was introduced in 1962, is the heavy-duty version. The A-904 Torqueflite is typically used behind small displacement engines on RWD vehicles. Since 1978, some vehicles equipped with this transmission have been fitted with a lockup torque converter.

The A-998 and A-999 transmissions are newer versions of the A-904, which were introduced in the early 1970s. The A-998 was designed specifically for the 318 cid (5.2 L) V-8, whereas the A-999 was used behind the 360 cid (5.8 L) V-8. All three designs look very similar and share the same basic components. Through the years, the A-904 has been modified by changing the size of internal parts and the reinforcement of the case. These changes allowed the use of the basic transmission in many vehicle applications and with many different engines.

The A-727 Torqueflite has a larger case than the A-904 and was originally designed to be used behind large displacement engines. Through the years, it has been modified so that it can be used in a variety of vehicles and behind a variety of engines, and the basic design is still used today.

Torqueflite transmissions are now also referred to as Loadflite transmissions.

In 1978, the basic Torqueflite transmission was modified for use as a FWD transaxle. Torqueflite transaxles contain the same basic parts as the A-904 transmission, except a **transfer shaft,** final drive gears, and a differential unit have been added to the assembly.

Shop Manual
Chapter 7, page 233

Many different transaxle models have been used since then, most designed for use with a particular engine (Figure 7-1). The A-404 was used in 1978 and later Omni/Horizon models that were equipped with the 1.7-L engine, built by Volkswagen. This transmission continued to be used until 1983. The A-413 was introduced in 1982 and was used with Chrysler-built 2.2-L and 2.5-L engines. Also in 1982, the A-470 was introduced and used with the Mitsubishi-built 2.6-L engine. In 1985, the A-415 was released and used with the Peugeot-built 1.6-L engine. The A-415 was discontinued in 1986. In 1987, the A-670 was introduced for use in Plymouth and Dodge minivans equipped with the 3.0-L V-6 engine. In 1987, the A-670 was added for use with the Mitsubishi-built 3.0-L V-6 engine.

Both the FWD and RWD model transmissions have been available in both standard and wide-ratio gear sets. The range of gear ratios for low gear is 2.45:1 to 2.69:1, and for second gear is 1.55:1 to 1.45:1. Third gear is always direct drive, and reverse gears range from 2.10:1 to 2.21:1.

All the Torqueflite-based transmissions and transaxles used today have two multiple-disc clutches, an **overrunning clutch,** two servos and bands, and two planetary gear sets to provide three forward gear ratios and a reverse ratio. The two multiple-disc clutches are called the **front clutch** and **rear clutch** packs. The servos and bands are also referred to by their location, front and rear, or by their function, kickdown and low/reverse.

Input Devices

On some transmissions, the front clutch hub and rear clutch drum are built as a single assembly.

The front and rear multiple-disc clutches serve as the input devices. The turbine of the torque converter rotates the transmission's input shaft clockwise. Since the front clutch hub and rear clutch drum are splined together at one end of the input shaft, they also rotate clockwise (Figure 7-2). The rear ring gear carries the output of the transmission to the final drive unit.

The inner edges of the front clutch friction discs are splined to the outer edges of the front clutch hub (Figure 7-3) and therefore turn with the hub. The outer edges of the front clutch steel

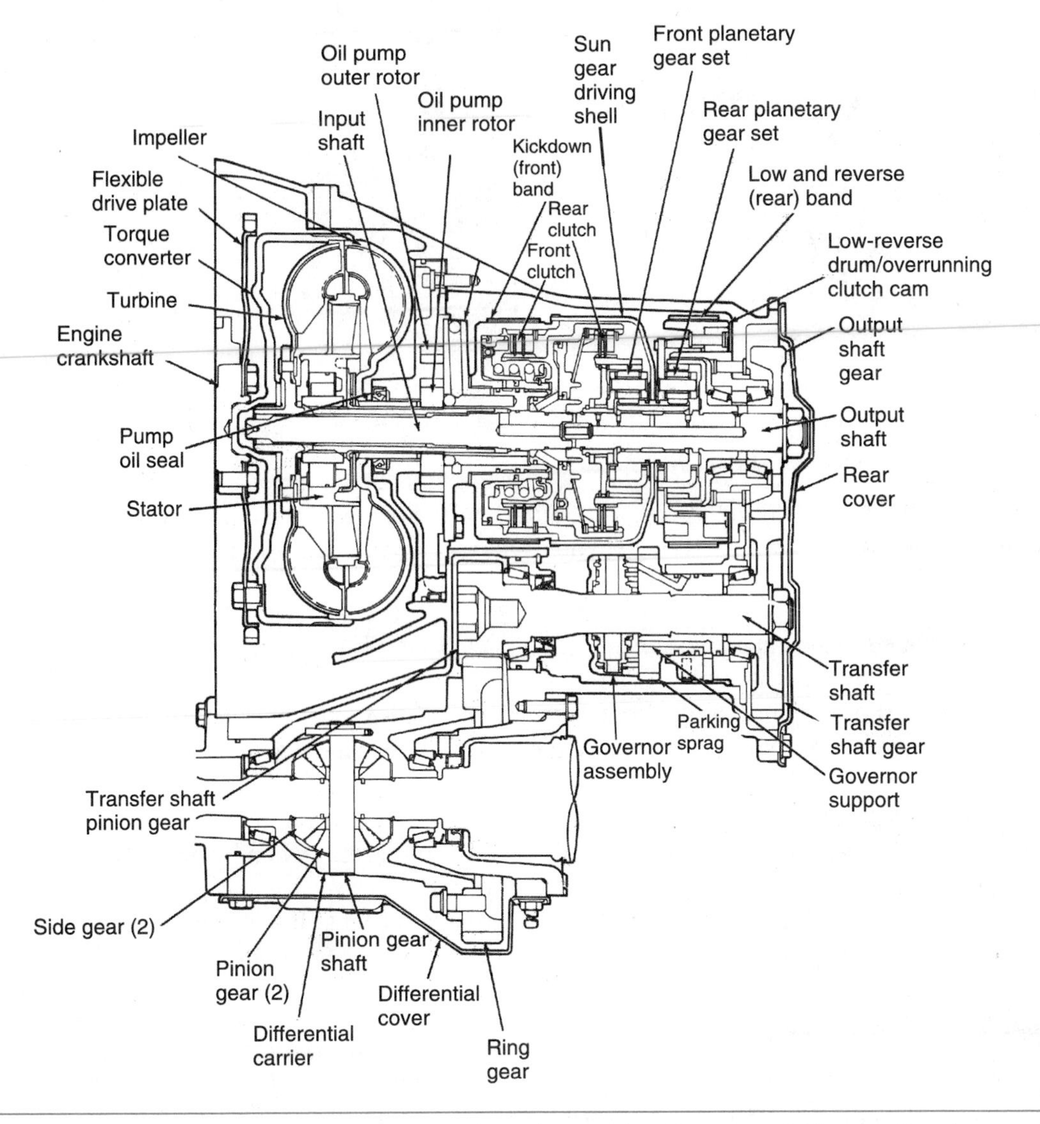

Figure 7-1 A Chrysler Simpson gear set–based transaxle. (Courtesy of Chrysler Corp.)

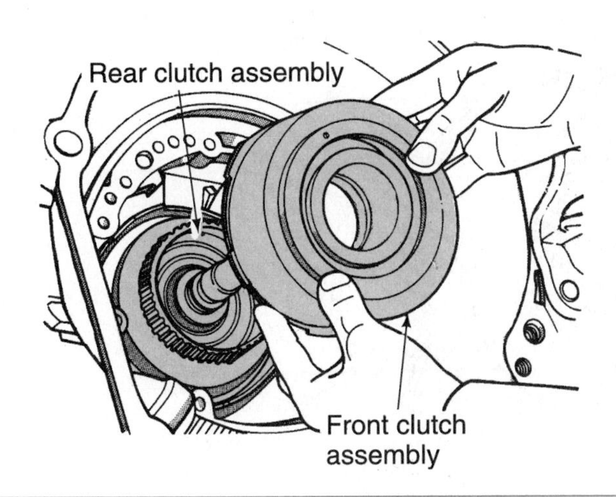

Figure 7-2 The front clutch hub and the rear clutch drum are splined together. (Courtesy of Chrysler Corp.)

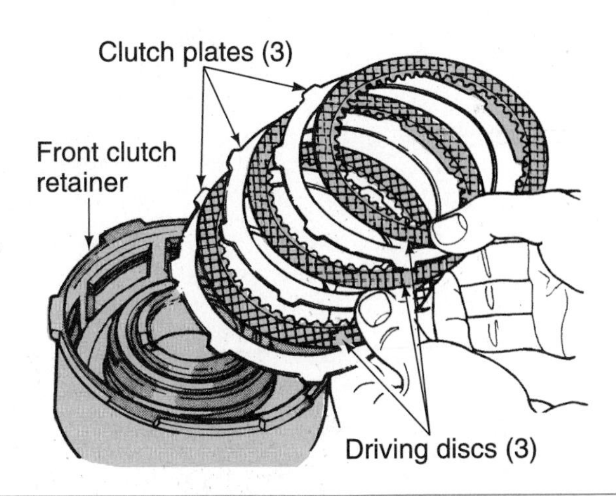

Figure 7-3 The friction discs are splined to the front clutch hub and the steel plates are splined to the front clutch drum. (Courtesy of Chrysler Corp.)

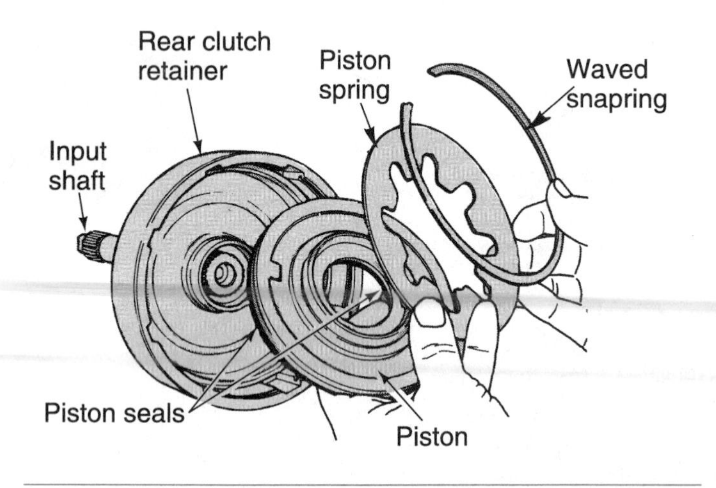

Figure 7-4 The rear clutch uses a belleville return spring. (Courtesy of Chrysler Corp.)

Output shaft
Overrunning clutch cam assembly

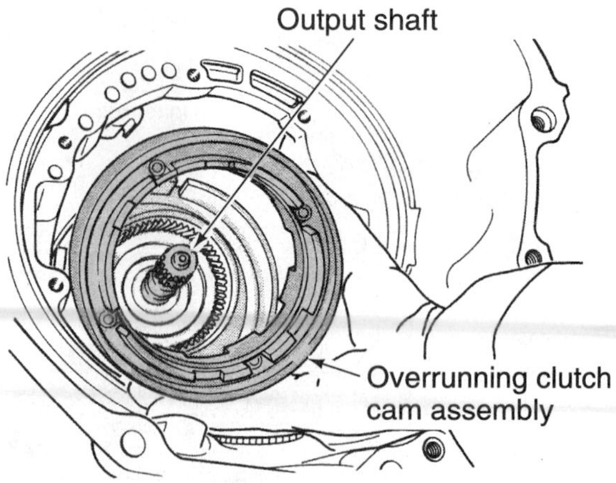

Figure 7-5 The one-way overrunning clutch is used to hold the rear carrier. (Courtesy of Chrysler Corp.)

plates are splined to the inner edges of the front clutch drum. The drum rotates with the input shaft whenever the front clutch is applied.

An input shell is splined to the front clutch drum and the common sun gear, which rotates on the output shaft but is not splined to it. Since the front and rear pinion gears mesh with the sun gear, the drum, the input shell, and the sun gear rotate with the input shaft when the clutch is applied.

The front clutch is applied in third and reverse gears to drive the sun gear. The front clutch is released by either one large coil spring or several small coil springs when hydraulic pressure at the clutch is released.

The rear clutch is applied in all forward gears. It uses a belleville spring to multiply the applying force of the piston and to help the piston retract into its bore (Figure 7-4).

Holding Devices

The front and rear bands and the one-way overrunning clutch (Figure 7-5) are the holding devices for the transmission. The **front** or kickdown **band** is used only in second gear and holds the input shell and the sun gear stationary (Figure 7-6).

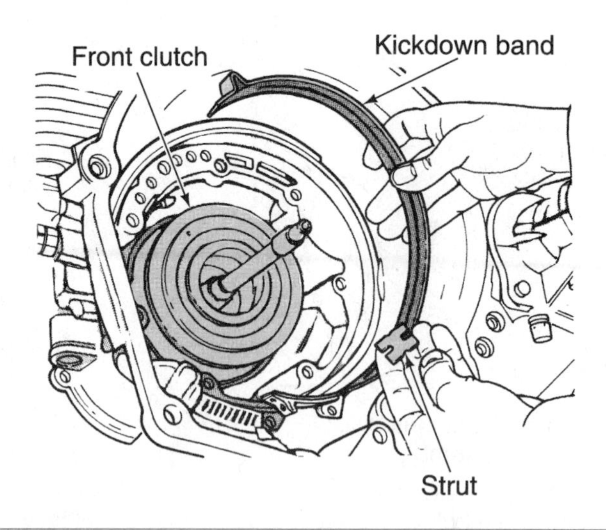

Figure 7-6 The kickdown band holds the input shell and sun gear stationary. (Courtesy of Chrysler Corp.)

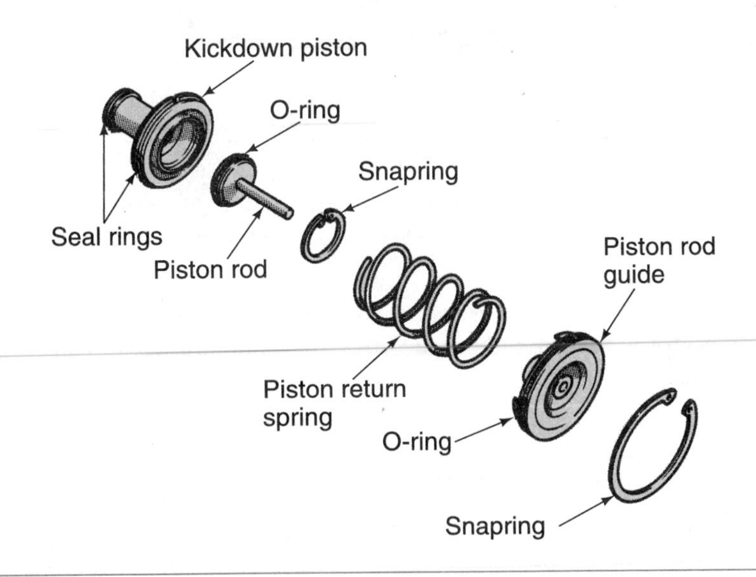

Figure 7-7 An exploded view of a Chrysler controlled load servo unit. (Courtesy of Chrysler Corp.)

The **rear band** is applied in reverse and manual low and holds the rear planet carrier. In manual low, the band is ineffective because the one-way overrunning clutch is holding. The rear band is effective during coasting or deceleration. When the one-way clutch begins to freewheel, the rear band holds the drum to allow for engine braking. If the rear band slips or cannot hold the drum, the one-way clutch will hold but will not provide engine braking.

Apply Devices

Most Torqueflite transmissions are equipped with a **controlled load** front **servo** (Figure 7-7). This type of servo has two pistons and allows for the quick release of the band during shifts from second to third gear and second to first gear. The servo also allows for smooth engagement of the band during first to second and third to second gear shifts.

Some A-727 Torqueflites are not equipped with a controlled load front servo. Rather, these use a conventional servo, which firmly applies and releases the band. Most Torqueflites have an external adjustment screw for the front band.

Power flow through Torqueflite transmissions occurs by the engagement and disengagement of the clutches and bands. Refer to the clutch and band application chart (Figure 7-8) while reading through the power flow.

GEAR SELECTOR POSITION	OPERATING GEAR	REAR CLUTCH	REAR BAND	FRONT BAND	FRONT CLUTCH	ONE-WAY CLUTCH
P– Park	None					
R– Reverse	Reverse		X		X	
N– Neutral	None					
D– Drive	1st gear	X				X
	2nd gear	X		X		
	3rd gear	X			X	
2– Man. 2nd	1st gear	X				X
	2nd gear	X		X		
1– Man. 1st	1st gear	X	X			X

Figure 7-8 Clutch and band application chart for typical Chrysler transmissions.

Power Flow in PARK or NEUTRAL

The power inputs the transmission through the turbine of the torque converter. When the gear selector is in PARK and NEUTRAL, the power flow ends at the front and rear clutches because neither is applied; therefore, no power is applied to the output shaft. When the selector is placed in PARK, a parking pawl is mechanically moved by the linkage and locks the parking gear to the transmission case. The parking gear is on the outer circumference of the governor support.

Power Flow in REVERSE

When the gear selector is placed in REVERSE, the front clutch is applied and engages the input shell and the sun gear. The rear band is also applied and holds the rear planet carrier. The front clutch drives the sun gear in a clockwise direction. Because the rear planet carrier is held, the sun gear drives the rear planet gears in a counterclockwise direction. The rotation of the planet gears drives the rear ring gear and the output shaft in a counterclockwise direction, resulting in reverse gear with gear reduction.

Power Flow in First Gear

When the gear selector is moved to the DRIVE position, fully automatic shifting with three gear ranges is available. The shift points are determined by vehicle speed and load. When the transmission is operating in first gear, the rear clutch is applied and serves as the input member for the planetary gear set. Because the rear clutch is applied, input torque flows from the turbine of the torque converter through the input shaft to the rear clutch, which drives the front ring gear in a clockwise direction.

The front planet carrier is splined to the output shaft and is therefore held by the weight of the vehicle and the drive wheels. This causes the ring gear to drive the front planet gears in a clockwise direction. The pinion gears, which are in mesh with the sun gear, rotate it counterclockwise. The rotation of the common sun gear causes the rear planet pinion gears to rotate in a clockwise direction, due to the carrier being held by the overrunning clutch. The planet gears cause the rear ring gear and the output shaft to turn in a clockwise direction.

The rear planet carrier is held by the overrunning clutch whenever the engine's torque is driving the planetary unit. During coast or deceleration, the weight of the vehicle and its momentum drive the planetary units through the output shaft. This causes the rear carrier to rotate clockwise and release the one-way clutch. This results in a neutral condition and doesn't allow for engine braking.

When the gear selector is placed in Manual Low (1), the rear band and clutch are applied. The overrunning clutch holds the rear planet carrier, as does the rear band. The band holds the carrier during deceleration when the one-way clutch freewheels. This allows for engine braking during coasting.

Power Flow in Second Gear

When the gear selector is in DRIVE, the transmission will automatically shift from first to second gear when the vehicle speed has reached a particular point and some load is overcome. The rear clutch remains applied and the front band engages to hold the sun gear stationary. The input shaft causes the front ring gear to rotate clockwise. Because the sun gear is held, the planet gears walk around the sun gear and drive the front planet carrier and output shaft in a clockwise, but speed reduced, direction.

The rear planetary set is effective during this gear because the rear planet carrier is rotating clockwise. This allows the one-way clutch to freewheel, thereby providing a neutral condition in the rear planetary gear set.

Power Flow in Third Gear

Third gear is a direct drive gear. When the transmission shifts into third, both the front and rear clutches become inputs for the planetary gear sets. This locks the gear sets into direct drive.

Whenever two members of a gear set are locked together, the pinion gears are unable to rotate on their individual shafts and the planetary gear set is locked. In this transmission, the two members that are locked together are the front ring gear and the sun gear.

Add-On Overdrive

Chrysler introduced the A-500 and A-518, which are four-speed automatic transmissions, in 1989. These transmissions are used exclusively in the mid-size Dodge Dakota and full-size Dodge RAM pickups and vans. The first three forward gear ratios are the same as the A-904 and A-727 Loadflite units. Fourth gear is provided by a separate planetary gear set and controlled by an overdrive clutch, direct clutch, and overrunning clutch in overdrive assembly attached to the rear of the transmission. In overdrive, the output ratio is 0.69 to 1. The A-500 is actually a modified A-999 three-speed transmission. The A-518 is a modified A-727 and is designed for heavy-duty use.

Overdrive or fourth-gear operation is controlled by a manually operated overdrive switch on the instrument panel. The electrical overdrive switch and the **Single Board Engine Controller (SBEC)** control the overdrive solenoid on the valve body.

The SBEC receives information pertaining to coolant temperature, engine speed, vehicle speed, throttle position, shift lever position, and engine load to determine if the transmission should be shifted into overdrive. The SBEC is bypassed when the overdrive switch is open, as the transmission will not shift into overdrive if the switch is not closed, regardless of the conditions.

These transmissions have extra-long extension housings, which hold the additional planetary gear set. An additional shaft was also added to the basic three-speed models. This shaft serves as the output shaft. The three-speed output shaft became an intermediate shaft linking the output from the Simpson gear train to the overdrive assembly.

Power flow through the first three gears is the same as other Torqueflite transmissions. See the clutch band application chart (Figure 7-9). However, to control the operation of the overdrive planetary, two multiple-disc clutches (direct and **overdrive clutches**) and a one-way overrunning clutch are used. The intermediate shaft is locked to the output shaft whenever the one-way clutch is locked. This locking results in bypassing the overdrive planetary and provides for direct drive.

The direct clutch locks the sun and ring gears together to prevent freewheeling of the overrunning clutch during coasting and deceleration. This provides for engine braking. The spring used in the direct clutch assembly is a heavy-tension single-coil spring that applies great holding force onto the clutch discs.

The intermediate shaft also drives the planet carrier of the overdrive gear set. When the transmission shifts from third to fourth gear, the overdrive clutch piston moves the clutch's hub to relieve the spring tension on the direct clutch assembly. It also applies pressure to the overdrive

GEAR SELECTOR POSITION	OPERATING GEAR	FRONT CLUTCH	REAR CLUTCH	OVERRUNNING CLUTCH	FRONT BAND	REAR BAND	OVERDRIVE CLUTCH	ONE-WAY CLUTCH	DIRECT CLUTCH
D	1st gear		X	X				X	X
	2nd gear		X		X			X	X
	3rd gear	X	X					X	X
	Overdrive	X	X				X		X
2	1st gear		X	X				X	X
	2nd gear		X		X			X	X
1	1st gear		X	X		X		X	X
R	Reverse	X				X			X

Figure 7-9 Clutch and band application chart for typical Chrysler transmissions with an add-on fourth gear.

clutch, which locks the sun gear to the transmission case. With the sun gear held, the planet carrier forces the ring gear and output shaft to rotate in an overdrive condition.

Ford Motor Company Simpson Gear Train Transmissions

Shop Manual
Chapter 7, page 246

Ford Motor Company began to use the Simpson gear train with the introduction of the C-4 transmission in 1964. Previous transmissions were based on the Ravigneaux design, as are some of their current models. Ford Motor Company has five basic transmissions that use the Simpson gear train and they are quite similar to each other: the C-3, C-4, C-5, C-6, and A4LD. The A4LD is a four-speed transmission, whereas the others are three-speed transmissions. All these units use a gear and crescent-type oil pump as the source of all hydraulic pressures.

The C-4 continued to be used until 1982, when it was replaced by the C-5, which was an updated C-4 with a lockup torque converter. The C-5 was used until 1986. The C-6 (Figure 7-10), which was introduced in 1966, is the heavy-duty version of Ford transmissions. The C-6 has been continually updated and is still being used in light trucks.

The C-3 was introduced in 1974 and was used until 1986. The A4LD is a modified version of the C-3. Its housing was enlarged and an additional planetary gear set was added to provide for an overdrive fourth gear. The A4LD was the first electronically controlled Ford automatic transmission and revised models of it are still being used.

The gear train for the first three speeds of these transmissions is identical except for the C-6, which uses a **low/reverse clutch** instead of the **low/reverse band** used by the other models.

The gear ratios of these transmissions are also similar. The typical ratio for first gear is 2.46:1, second gear is 1.46:1, and reverse gear is 2.19:1.

In appearance, these transmissions are similar and all have a separate and removable extension housing. The major external difference between the models is the size and the reinforcements

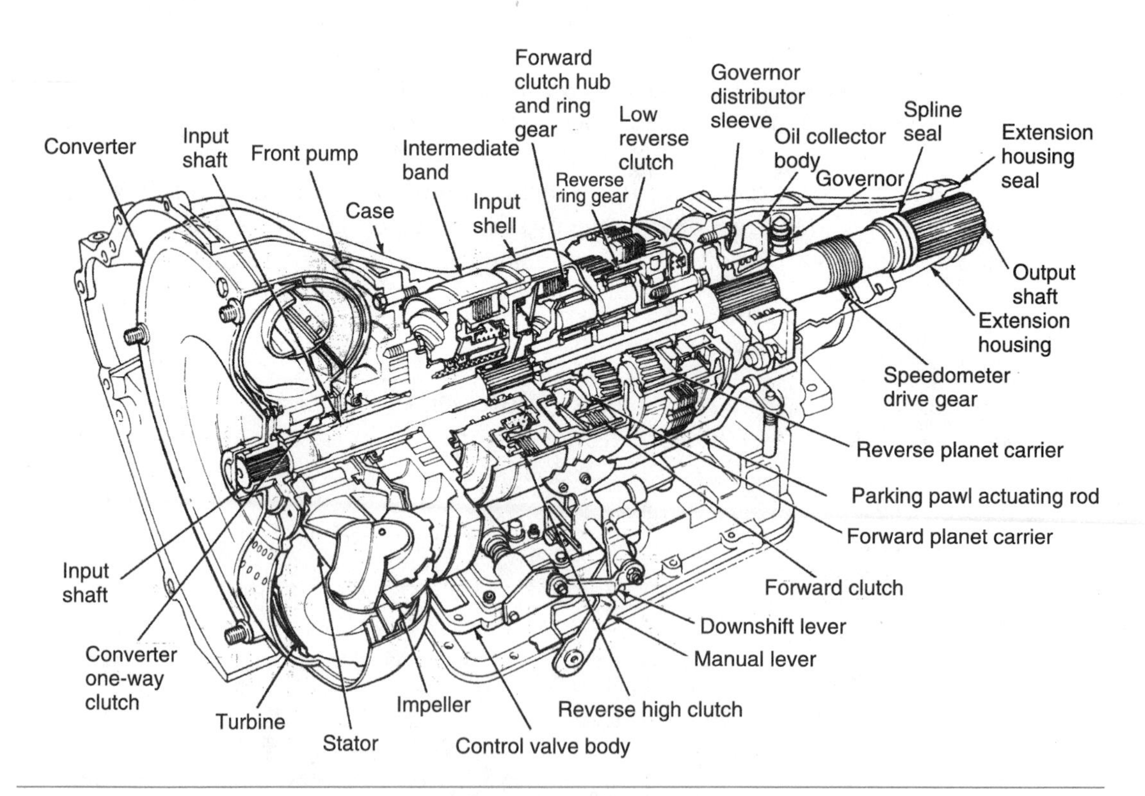

Figure 7-10 Simpson gear set–based Ford C-6 transmission. (Reprinted with the permission of Ford Motor Co.)

of the case. The bell housings of the C-3, C-4, and C-5 are removable from the transmission case, whereas the bell housing of the C-6 is part of the case's casting.

To control planetary gear action, these transmissions use two multiple-disc clutches, two bands, and one overrunning clutch. The two clutches are referred to as the high, front, or reverse/high and the rear or forward clutches. The two bands are the intermediate, front, or kick-down and low/reverse or rear units. Again the exception to this is the C-6, which uses a clutch in place of a band for low and reverse.

Input Devices

The front planet carrier and the rear ring gear are splined to the output shaft; therefore, either of these planetary gear set members will always serve as the output member. Both the front carrier and the rear ring gear can either drive the output shaft or be driven by it, depending on which bands and clutches are applied. The input shaft is splined to the **forward clutch** drum (Figure 7-11) and the **high/reverse clutch** hub assembly (Figure 7-12). The inner edges of the high/reverse clutch friction discs are splined to the outer edge of the high/reverse clutch hub and rotate with it. The outer edges of the high/reverse clutch steel discs are splined to the inner edge of the high/reverse clutch drum. When the high/reverse clutch is applied, the drum is rotated clockwise by the input shaft. The sun gear is splined to the input shell and becomes the input member when the clutch is applied.

The high/reverse clutch is applied in third and reverse gears to hold or drive the sun gear. These clutch packs typically use several small coil springs to return the clutch piston; however, some C-4 transmissions were fitted with a single, large coil spring.

The outer edges of the forward clutch steel discs are splined to the inner edges of the forward clutch drum, which is also splined to the input shaft. The inner edges of the forward clutch friction discs are splined to the outer edges of the forward clutch hub. When the forward clutch is applied, the front ring gear becomes the input member of the gear set. The outside of the front ring gear is the forward clutch hub in C-3 and C-6 transmissions, whereas the forward clutch hub is splined to the ring gear in C-4 and C-5 models.

The forward clutch is applied in all forward gears. The number of friction discs and steel plates in a forward clutch will vary with the different transmission models. The forward clutch in a C-3 or A4LD transmission has several small coil springs to return the clutch piston, whereas the clutches in C-4, C-5, and C-6 transmissions are equipped with belleville springs.

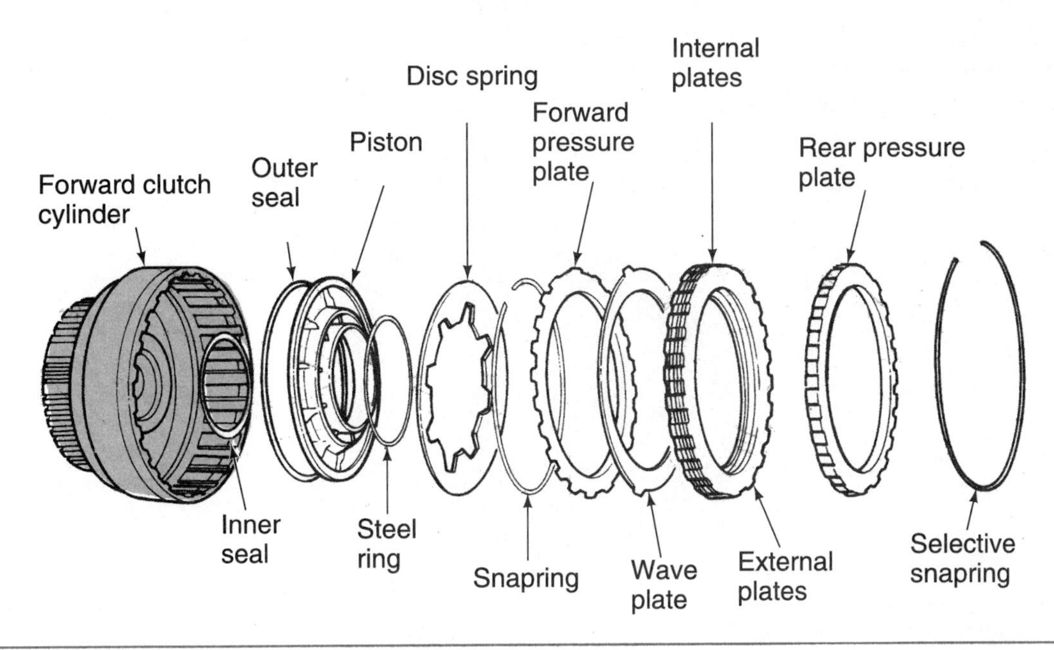

Figure 7-11 The input shaft is splined to the forward clutch cylinder. (Reprinted with the permission of Ford Motor Co.)

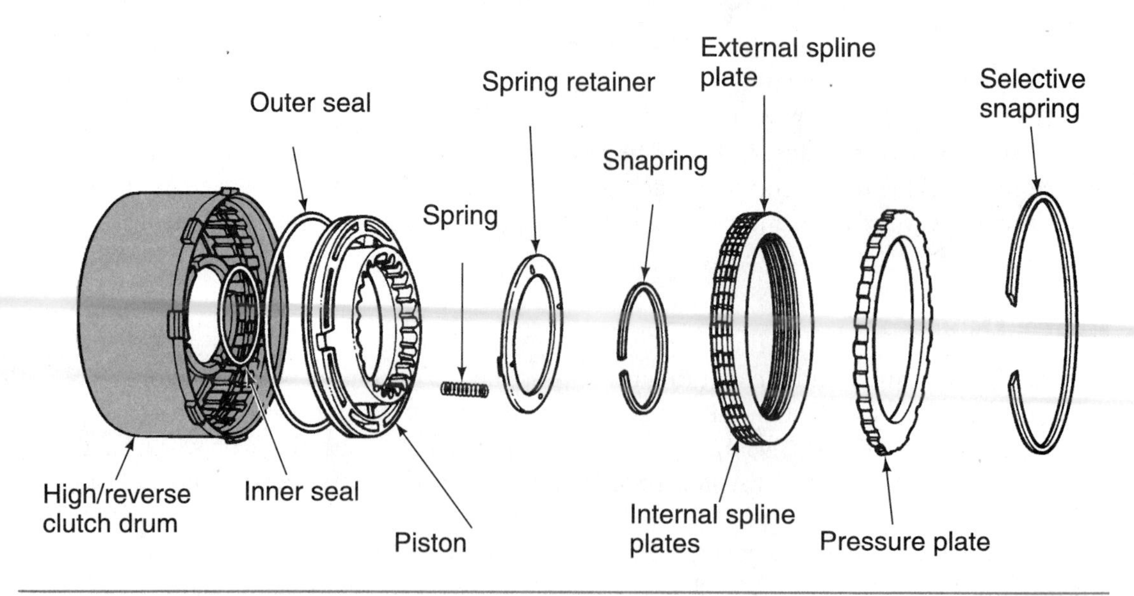

Figure 7-12 The input shaft is also splined to the high/reverse clutch drum. (Reprinted with the permission of Ford Motor Co.)

Holding Devices

As input is received on the front ring gear, the front pinion gears rotate in the front carrier and drive the sun gear. Torque is sent to the rear planetary unit through the common sun gear. The rear planet carrier is splined to the inside of the low/reverse hub and a one-way roller clutch is attached to the rear of the hub (Figure 7-13). The one-way clutch locks and holds the rear carrier when the low/reverse hub tries to rotate counterclockwise. This happens when the gear selector is placed in DRIVE and the transmission is operating in LOW.

The rear planet carrier can also be held by the low/reverse band or clutch. In all models but the C-6, a low/reverse band wraps around the low/reverse drum to hold the rear carrier. In the C-6, the friction discs of a low/reverse clutch are splined to the outside of the low/reverse drum and the steel discs are splined to the transmission case. The low/reverse band or clutch is applied in reverse and manual low. Several small coil springs are used to return the clutch piston.

The **intermediate band** is wrapped around the outside of the high/reverse clutch drum. When the transmission is operating in second gear, the intermediate band is applied and the clutch drum, the input shell, and the sun gear are held stationary.

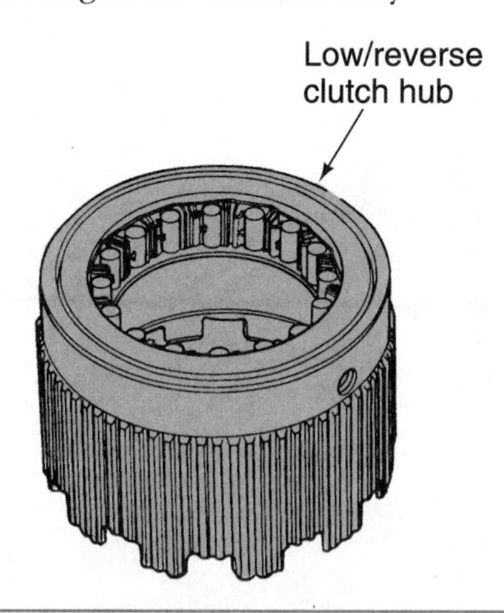

Figure 7-13 The rear planet carrier is splined to the inside of the low/reverse clutch hub and the one-way clutch is attached to the rear of the hub to hold the rear carrier. (Reprinted with the permission of Ford Motor Co.)

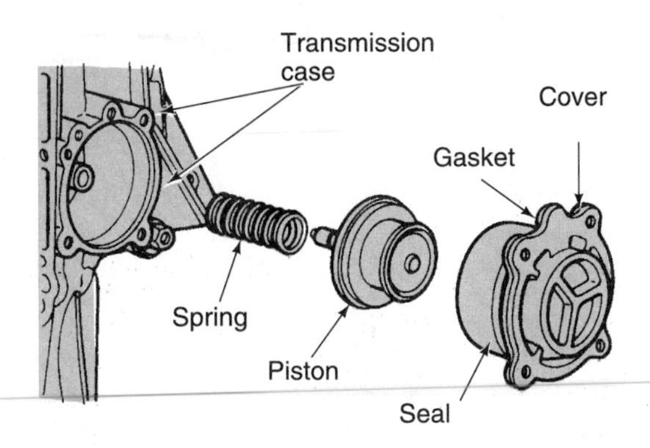

Figure 7-14 External mounting of the intermediate servo. (Reprinted with the permission of Ford Motor Co.)

Apply Devices

The low/reverse servo applies the band to hold the low/reverse drum and rear carrier. The servos in the C-4 and C-5 are accessible from the outside of the case and the bands are adjustable. The servo in the C-3 is located inside the case and the oil pan must be removed for access. The band in a C-3 is not adjustable and is controlled by selective piston rod lengths during assembly of the transmission.

The intermediate servo applies the intermediate band in second gear and is accessible from the outside of the case (Figure 7-14). The intermediate band on these transmissions is adjustable.

When the gear selector is placed in PARK, a mechanical lever, rod, or pawl engages with a large gear on the transmission's output shaft and locks it to the transmission case (Figure 7-15).

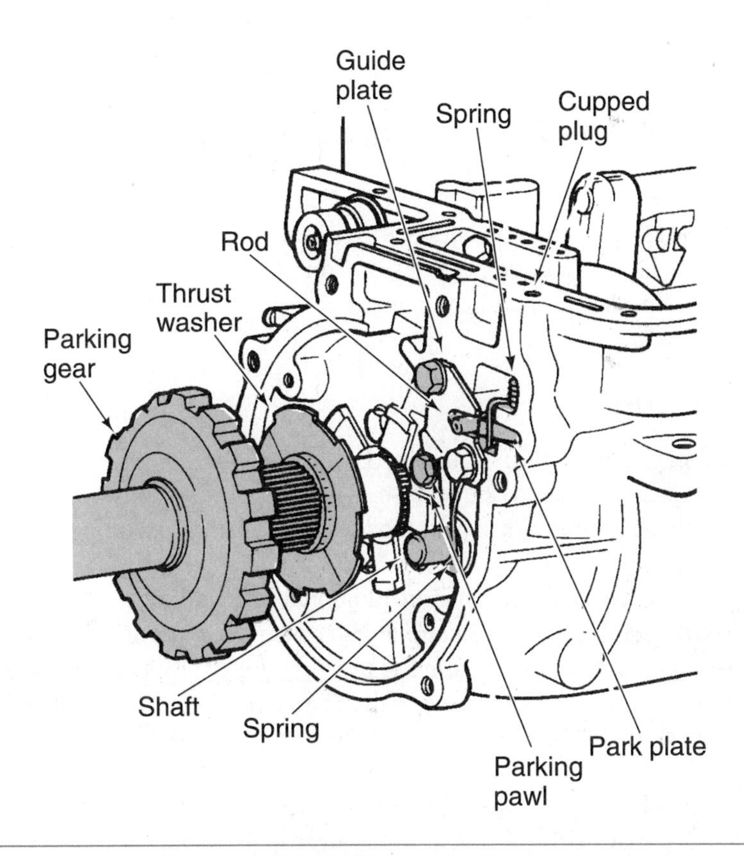

Figure 7-15 Parking gear and pawl assembly in a C-6 transmission. (Reprinted with the permission of Ford Motor Co.)

GEAR SELECTOR POSITION	OPERATING GEAR	FORWARD CLUTCH	LOW/REVERSE BAND	INTERMEDIATE BAND	HIGH/REVERSE CLUTCH	ONE-WAY CLUTCH
P– Park	None					
R– Reverse	Reverse		X		X	
N– Neutral	None					
D– Drive	1st gear	X				X
	2nd gear	X		X		
	3rd gear	X			X	
2–Man. 2nd	2nd gear	X		X		
1– Man. 1st	1st gear	X	X			X

Figure 7-16 Clutch and band application chart for Ford C-3, C-4, and C-5 transmissions.

GEAR SELECTOR POSITION	OPERATING GEAR	FORWARD CLUTCH	LOW/REVERSE CLUTCH	INTERMEDIATE BAND	HIGH/REVERSE CLUTCH	ONE-WAY CLUTCH
P– Park	None					
R– Reverse	Reverse		X		X	
N– Neutral	None					
D– Drive	1st gear	X				X
	2nd gear	X		X		
	3rd gear	X			X	
L2– Man. 2nd	2nd gear	X		X		
L1– Man. 1st	1st gear	X	X			X

Figure 7-17 Clutch and band application chart for a Ford C-6 transmission.

Power Flow

The power flow through Ford transmissions with Simpson gear trains is similar to that of the power flow through a Chrysler Torqueflite transmission. Refer to the clutch and band application chart (Figure 7-16) for the C-3, C-4, and C-5 transmissions and the chart (Figure 7-17) for the C-6 transmission.

Add-On Overdrive

Shop Manual
Chapter 7, page XXX

The A4LD transmission (Figure 7-18) is a modified version of the C-3. It is a fully automatic transmission with four forward speeds and one reverse. The transmission consists of a torque converter, planetary gear train, three multiple-disc clutch packs, three bands, two one-way clutches, and a hydraulic control system.

The housing of a C-3 was enlarged and an additional planetary gear set was added to provide for an overdrive fourth gear. The additional planetary gear set was added in front of the normal three-speed Simpson gear train. Input from the torque converter passes through the overdrive planetary gear set (Figure 7-19) before it reaches the Simpson gear set. To control planetary action of the overdrive gear set, the A4LD uses an **overdrive band,** overdrive clutch, and an **overdrive one-way clutch.**

The overdrive band holds the overdrive sun gear to provide for fourth gear overdrive. The overdrive clutch locks the overdrive sun gear to the overdrive planet carrier to provide for engine braking while the transmission is operating in DRIVE. The overdrive one-way roller clutch locks the

A4LD
Automatic overdrive transmission

Band and clutch application/gear ratio

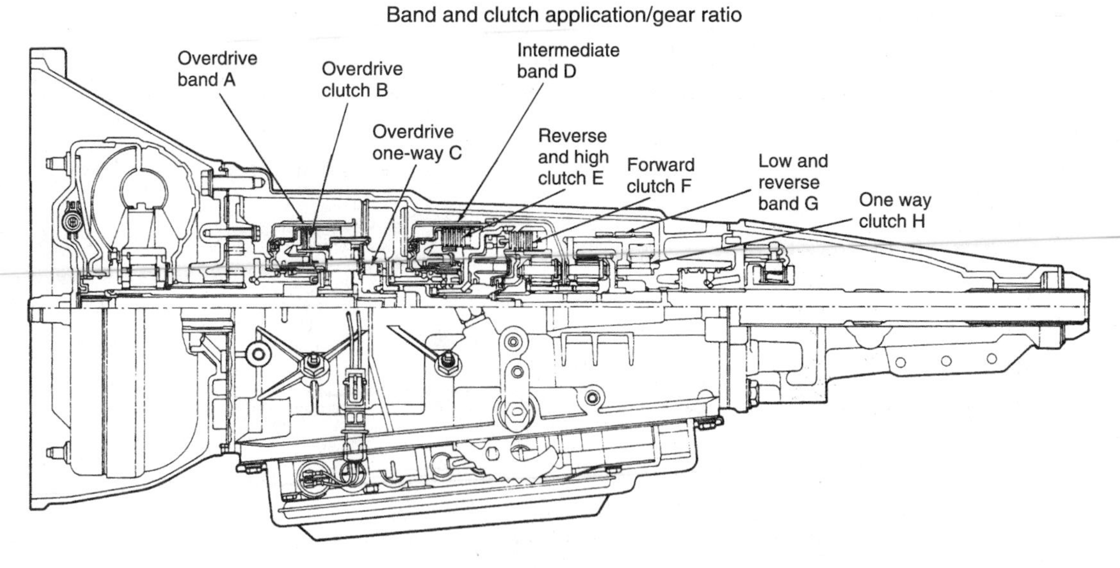

GEAR	OVERDRIVE BAND A	OVERDRIVE CLUTCH B	OVERDRIVE ONE WAY CLUTCH C	INTERMEDIATE BAND D	REVERSE AND HIGH CLUTCH E	FORWARD CLUTCH F	LOW AND REVERSE BAND G	ONE WAY CLUTCH H	GEAR RATIO
1 — MANUAL FIRST GEAR (LOW)		APPLIED	HOLDING			APPLIED	APPLIED	HOLDING	2.47:1
2 — MANUAL SECOND GEAR		APPLIED	HOLDING	APPLIED		APPLIED			1.47:1
D — DRIVE AUTO — 1ST GEAR		APPLIED	HOLDING			APPLIED		HOLDING	2.47:1
D — O/D AUTO — 1ST GEAR			HOLDING			APPLIED		HOLDING	2.47:1
D — DRIVE AUTO — 2ND GEAR		APPLIED	HOLDING	APPLIED		APPLIED			1:47:1
D — O/D AUTO — 2ND GEAR			HOLDING	APPLIED		APPLIED			1.47:1
D — DRIVE AUTO — 3RD GEAR		APPLIED	HOLDING		APPLIED	APPLIED			1.0:1
D — O/D AUTO — 3RD GEAR			HOLDING		APPLIED	APPLIED			1.0:1
D — OVERDRIVE AUTOMATIC FOURTH GEAR	APPLIED				APPLIED	APPLIED			0.75:1
REVERSE		APPLIED	HOLDING		APPLIED		APPLIED		2.1:1

Figure 7-18 A Ford A4LD transmission. (Reprinted with the permission of Ford Motor Co.)

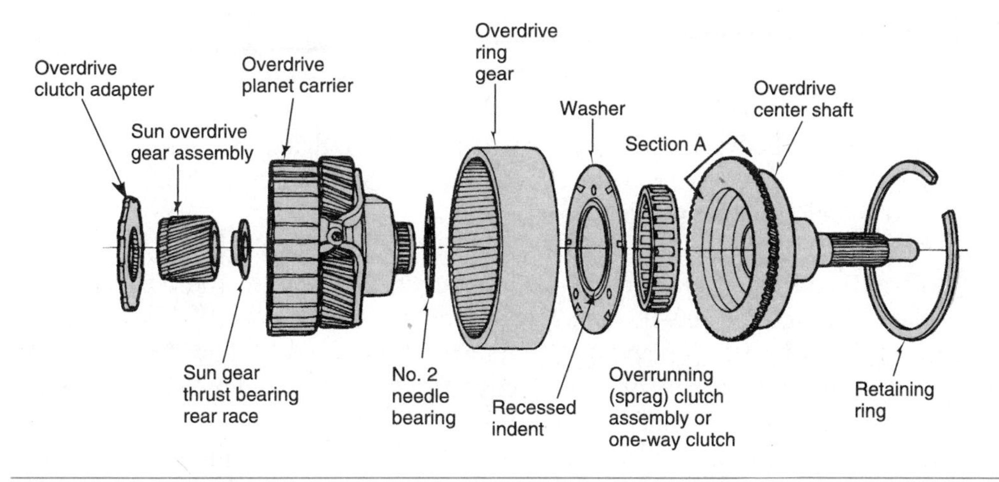

Figure 7-19 Add-on planetary gear set, which provides fourth gear. (Reprinted with the permission of Ford Motor Co.)

input from the torque converter's turbine shaft directly to the input shaft of the Simpson gear train and provides direct drive through the planetary gear set by locking the carrier to the ring gear.

Direct drive through the overdrive planetary unit is present whenever the transmission is operating in first, second, third, and reverse gears. During these conditions, all input power flows through the overdrive one-way clutch. The overdrive clutch is also applied to provide for engine braking.

The prefix THM stands for Turbo-Hydramatic, which is the division of General Motors that is responsible for automatic transmission development and production.

The Buick version of the 400 is sometimes called the Super Turbine 400.

Some parts catalogs refer to the 400 as the M-40 transmission.

Some parts catalogs refer to the 425 as the M-42 transmission.

In recent years, General Motors has redesignated its transmissions. The 400/475 transmissions are now labeled as 3L80 transmissions.

The suffix C, which follows some transmission model numbers, indicates that the transmission is equipped with a lockup torque converter.

The 375 is a medium-duty version of the 400 and should not be confused with the 375B, which is the heavy-duty version of the 350.

Some parts catalogs may refer to the THM 200 as an M-29 transmission.

When overdrive is selected, the overdrive band is applied and holds the sun gear. The overdrive clutch is released and the planet carrier becomes the input member driving the ring gear at an overdrive ratio of approximately 0.75:1.

Other than the addition of the overdrive planetary gear set, minor changes were made to accommodate the fourth gear and to make it more suitable for its different applications. Five of the major changes are:

1. The front clutch is applied in third, fourth, and reverse gears to hold or drive the sun gear.
2. The outer race of the one-way clutch is locked to the low/reverse drum and the inner race is bolted to the case.
3. A low/reverse band is used to hold the low/reverse drum and rear planet carrier.
4. The rear clutch uses several coil springs to return the clutch piston.
5. The kickdown and overdrive bands are adjustable from outside the case.

General Motors Transmissions

Most General Motors transmissions and transaxles are based on the Simpson gear train. The development of three-speed units through the years had led to the heavy use of the compound planetary gear set on light-, medium-, and heavy-duty vehicles.

The THM 375, 400, and 475 transmissions are heavy-duty transmissions and are generally used in full-size RWD cars and trucks. The 425 is based on the same design as the others, but it was modified for use in FWD vehicles with longitudinally placed engines. The 400 was introduced in 1964 and has been used by all divisions of GM, as well as by many other manufacturers, such as Jaguar and Rolls-Royce. The 425 was introduced in 1966 and was used until 1978, when it was replaced by the 325.

The THM 250, 250C, 350, 350C, and 375B transmissions are three-speed units. The 250 and 250C are light-duty units that were used primarily in 1974 to 1984 compact and intermediate models. The 350 (Figure 7-20) and 350C are medium-duty transmissions used in intermediate and full-size cars and light trucks from 1969 to 1986. The 375B is a heavy-duty version of the 350 that has an additional direct clutch for increased torque capacity.

The primary difference between a 250 and a 350 transmission is that a 250 uses a band as the second gear holding device, whereas a 350 uses a band plus a multiple-disc clutch and a one-way roller clutch as holding devices during second gear operation.

The 200 (Figure 7-21) was introduced in 1976 and was replaced by the 200C in 1979. The 200, 200C, 200-4R, 325, and 325-4L are light- to medium-duty transmissions. The 200, 200C, and 325 are three-speed units and the 200-4R and 325-4L have four forward speeds. The 200 series transmissions are used in RWD vehicles and the 325 transmissions are used in FWD models. The 325 (Figure 7-22) was introduced in 1979. It replaced the 425 in FWD Buick Riviera, Cadillac Eldorado, and Oldsmobile Toronado models.

The 325 and 425 transmissions are not classified as transaxles because the differential and final drive gears are not built into the transmission case. Rather, a separate final drive unit is bolted to the transmission to drive the front wheels.

The turbine shaft of the torque converter in 325 and 425 transmissions is splined to a drive sprocket. This drive sprocket rotates with the turbine shaft and drives a multiple-link chain that connects the drive sprocket to a driven sprocket. The driven sprocket is connected to the transmission's input shaft. This arrangement allows engine torque to take a 90-degree turn before it reaches the gear train. After the input shaft, the internal components are identical to a 200 or 400 transmission.

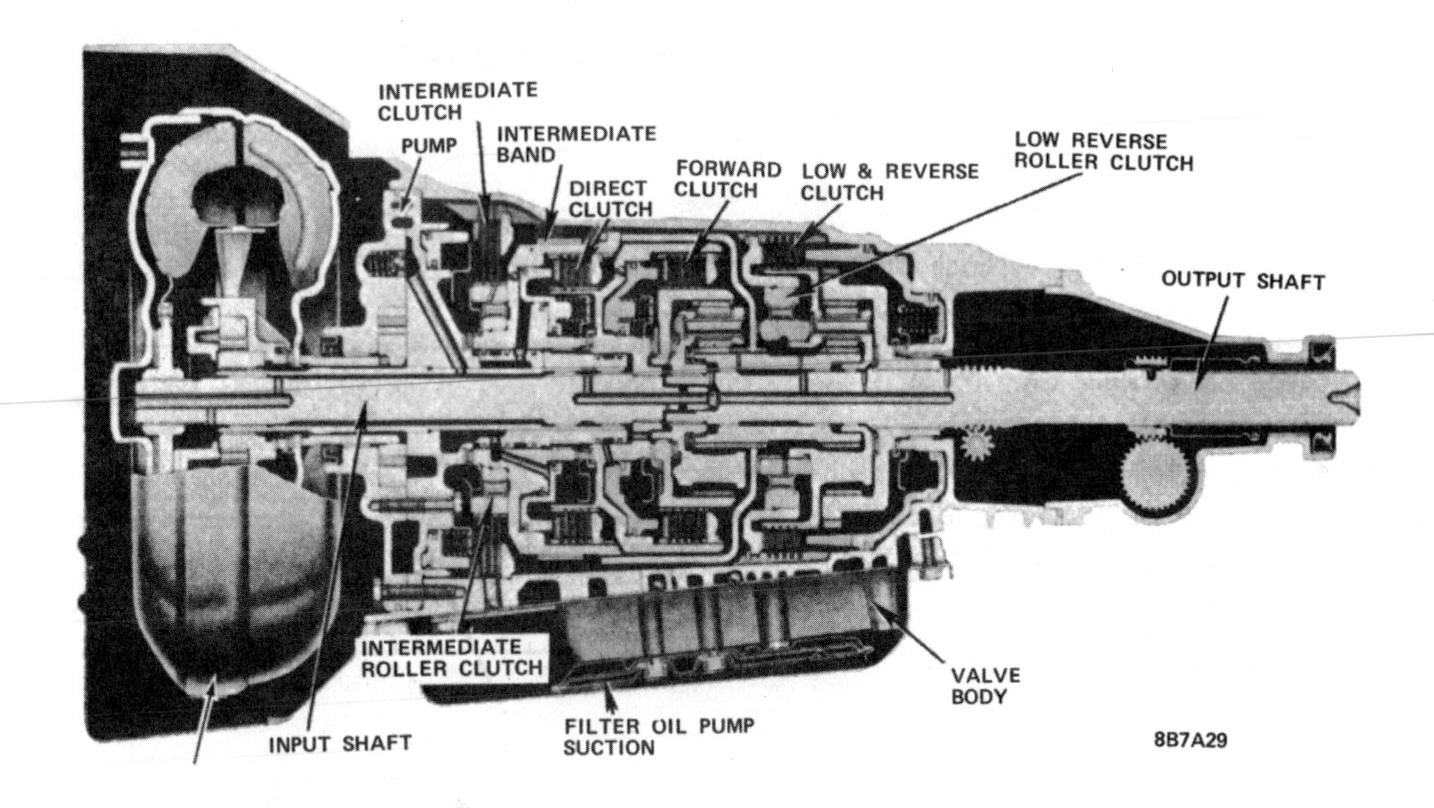

Figure 7-20 A THM 350 transmission. (Courtesy of the Oldsmobile Motor Division of General Motors Corp.)

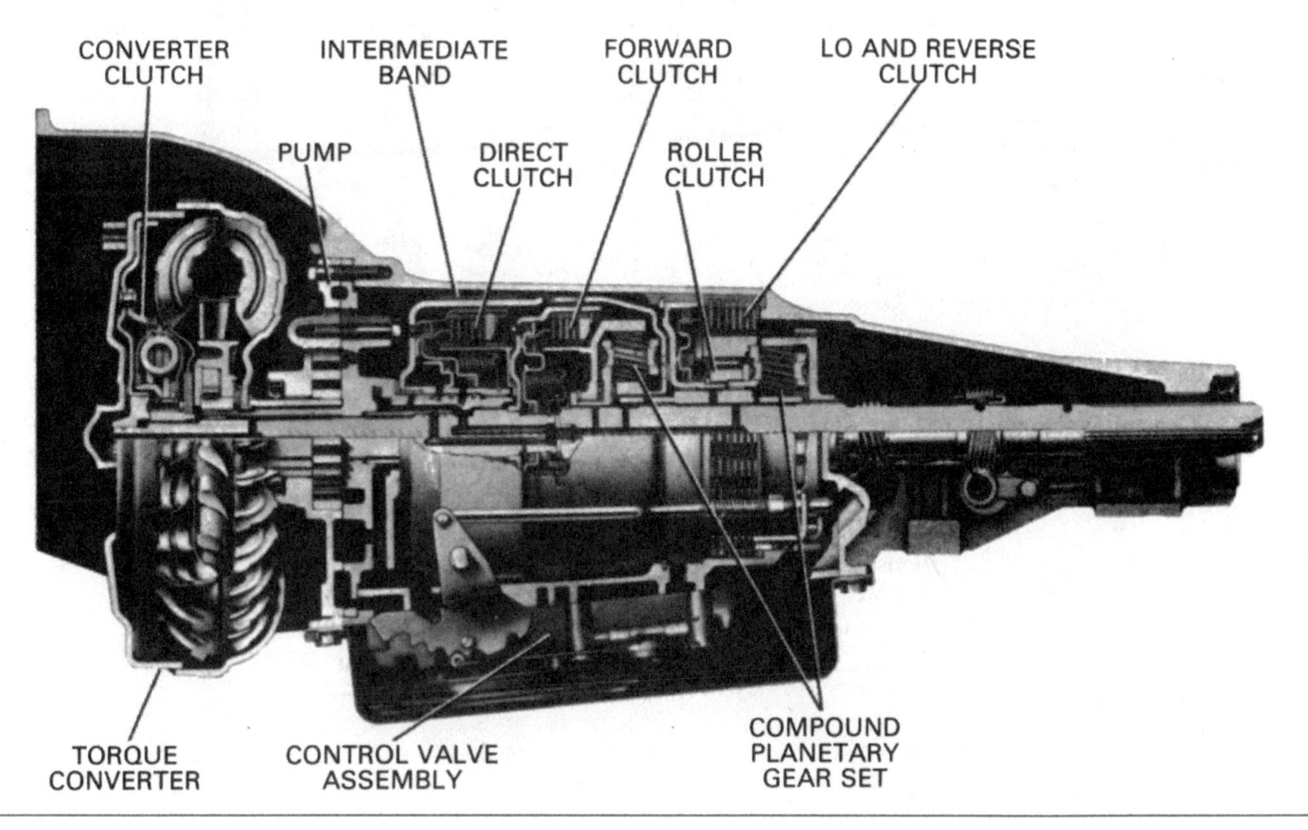

Figure 7-21 A THM 200 transmission. (Courtesy of the Oldsmobile Motor Division of General Motors Corp.)

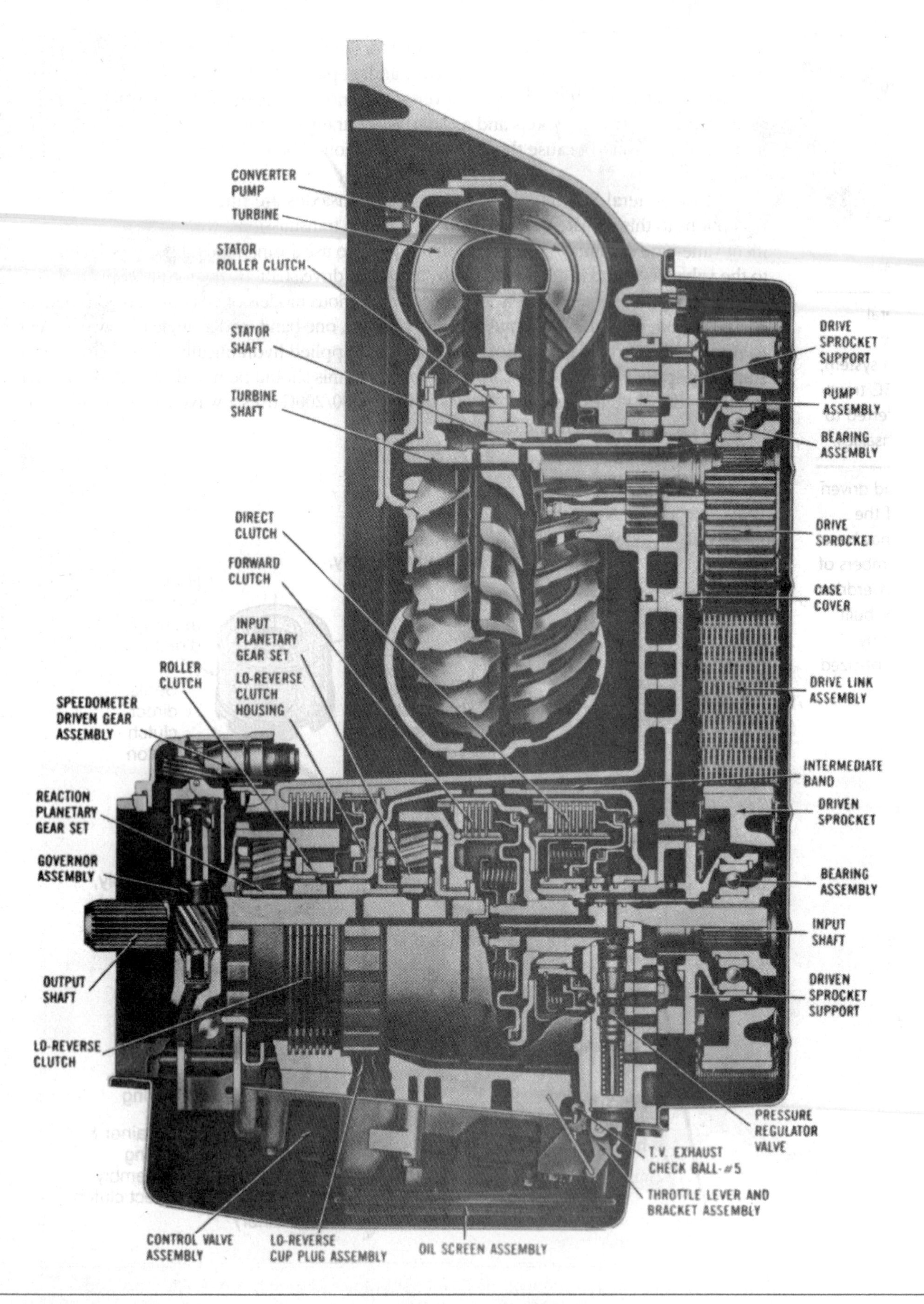

Figure 7-22 A THM 325 transmission. (Courtesy of the Oldsmobile Motor Division of General Motors Corp.)

The THM 125 was introduced in 1980 and the 125C was added in 1981. The 125 and 125C are three-speed automatic transaxles designed for light-duty use. The 125/125C (3T40) housing is a one-piece aluminum casting with the case cover and oil pan bolted to it. The bell housing is integral with the transaxle case. The torque converter does not directly drive the transaxle; rather, the turbine shaft drives two sprockets and a chain, which transfer engine torque to the gear train. This transfer is necessary because the gear train is positioned behind and below the centerline of the crankshaft.

Most General Motors transmissions and transaxles are fitted with gear-type oil pumps. Exceptions to this are late-model 200-4R and 700-R4 transmissions, which use a variable displacement vane-type oil pump. The 125/125C models also use a vane-type oil pump, which is mounted to the valve body and is driven indirectly, through a drive shaft, by the torque converter.

Although there are differences between the various models of GM transmissions with a Simpson gear train, most use three multiple-disc clutches, one band, and two single one-way roller clutches to provide the various gear ratios. Each clutch is applied hydraulically and released by several small coil springs (Figure 7-23). Two exceptions to this should be noted. The 350/350C has four multiple-disc clutches instead of three. And the 200/200C uses a waved spring (Figure 7-24) to cushion the application of the clutch.

Shop Manual
Chapter 7, page 268

Some parts catalogs refer to the 125 as the M-34 transaxle and the 125C as the MD-9 transaxle.

Under General Motors new classification system, the 125/125C transaxles are referred to as 3T40 transaxles.

The drive and driven sprockets of the 125/125C may have different numbers of teeth. An overdrive is sometimes built into the unit by using different-size sprockets.

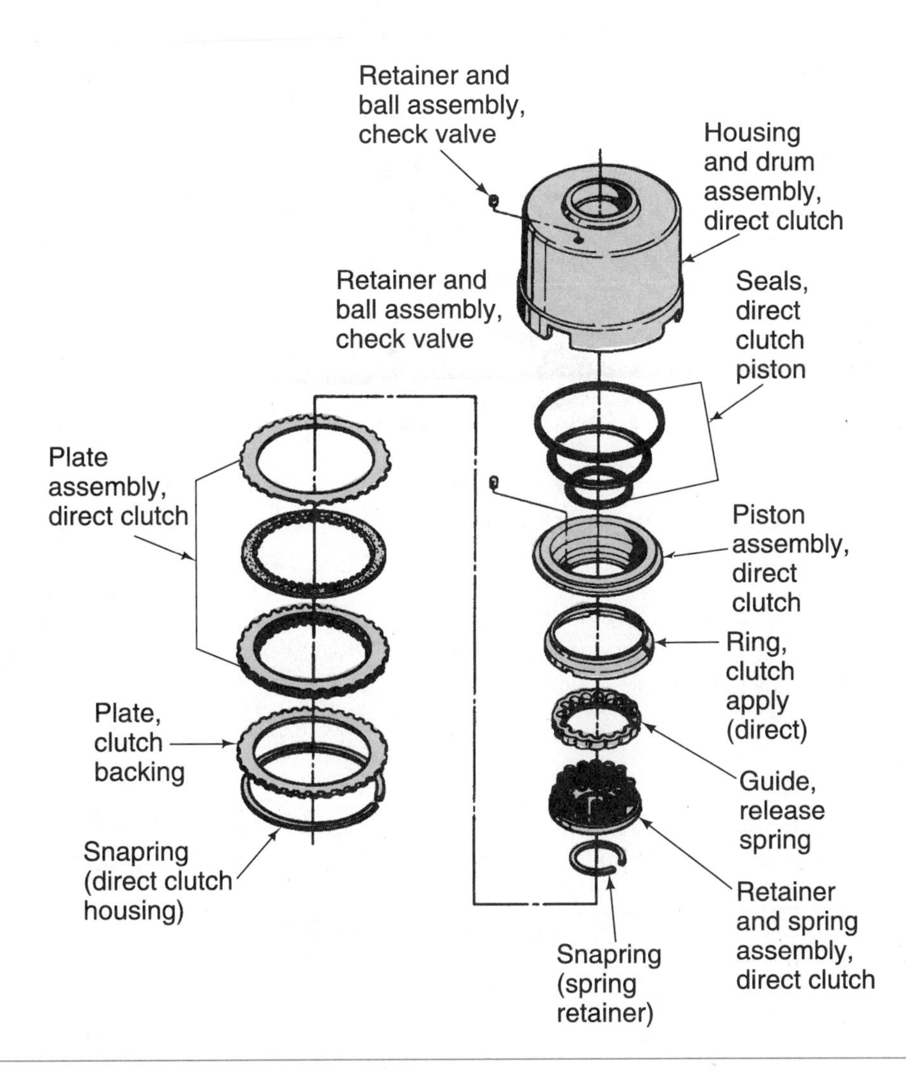

Figure 7-23 Most multiple-disc clutch assemblies in GM transmissions are released by several small coil springs. (Courtesy of the Hydra-Matic Division of General Motors Corp.)

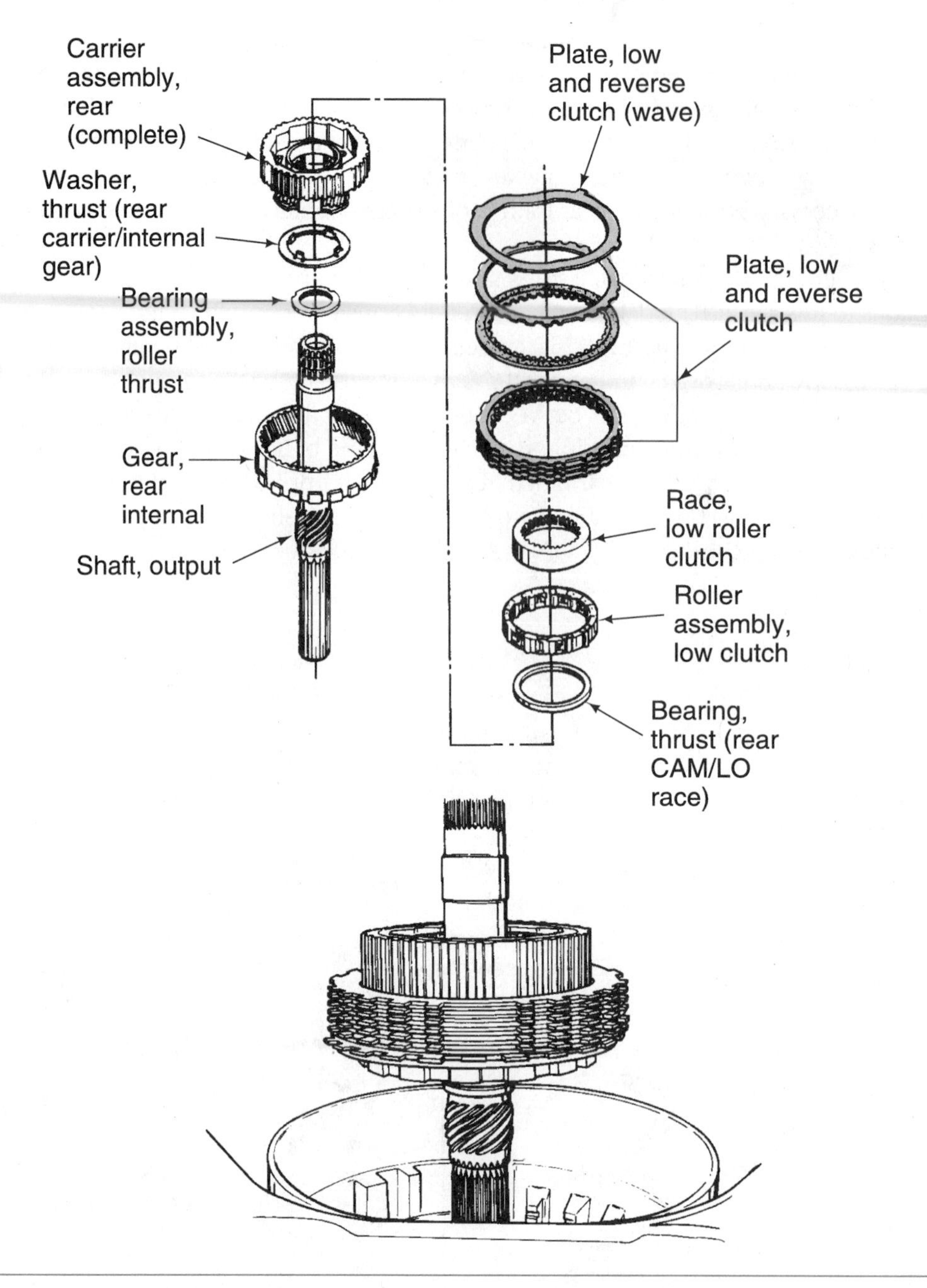

Figure 7-24 The THM 200 uses a waved spring in the low/reverse clutch assembly. (Courtesy of the Hydra-Matic Division of General Motors Corp.)

Input Devices

None of the members of the Simpson gear train is directly attached to the input shaft. However, two members are splined to the output shaft (Figure 7-25). The front carrier and the rear ring gear will always be the output members of the gear set. The outside of the rear ring gear serves as the parking gear, which is locked to the transmission/transaxle case by the parking pawl when the gear selector is placed in the PARK position.

The forward clutch is the input device and the low/reverse clutch is the reaction member for all these transmission models (Figure 7-26). The overrunning clutch is applied in drive, manual second, and manual low gear selector positions. The forward clutch of all these Simpson-based transmissions is applied in all forward gears. The **direct clutch** of all these transmissions is applied in third and reverse gears.

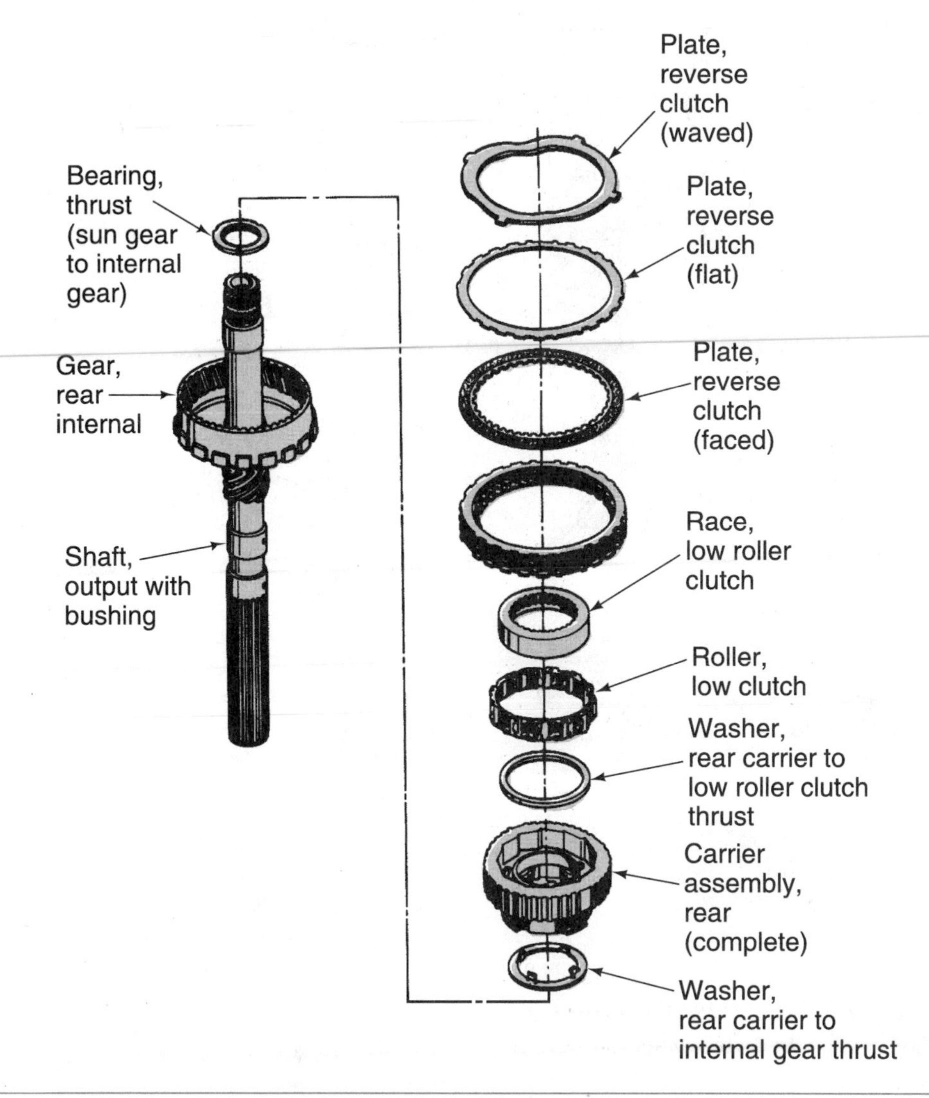

Figure 7-25 The front planet carrier and the rear ring gear are splined to the output shaft. (Courtesy of the Hydra-Matic Division of General Motors Corp.)

The input shaft is splined to the direct clutch hub and the forward clutch drum; therefore, the hub, drum, and input shaft rotate as a unit (Figure 7-27). The inside edges of the direct clutch friction discs are splined to the outside of the direct clutch hub. The outside edges of the steel discs are splined to the inside of the direct clutch drum. When the direct clutch is applied, the drum is rotated clockwise by the input shaft.

The drum is splined to the input shell, which is splined to the common sun gear. The sun gear, therefore, rotates with the input shell. The sun gear is the input member for the gear set when the direct clutch is applied.

The direct clutch is designed to allow for greater clamping pressures when the transmission is operated in reverse gear. The clutch piston is designed with two separate surfaces on which hydraulic pressure may act. When the transmission is operating in high gear, normal loads are low and high clamping pressures are not needed. Hydraulic pressure is applied only to the smaller, inner area of the clutch piston. While in reverse, hydraulic pressure is routed to the inner area and to the larger outer area of the piston. This allows the clutch to develop greater holding force in reverse.

The inside edges of the forward clutch's friction discs are splined to the outside edge of the front planetary ring gear. The outside edges of the steel discs are splined to the inside of the forward clutch drum, which is splined to the input shaft. When the forward clutch is applied, the front ring gear rotates with the input shaft.

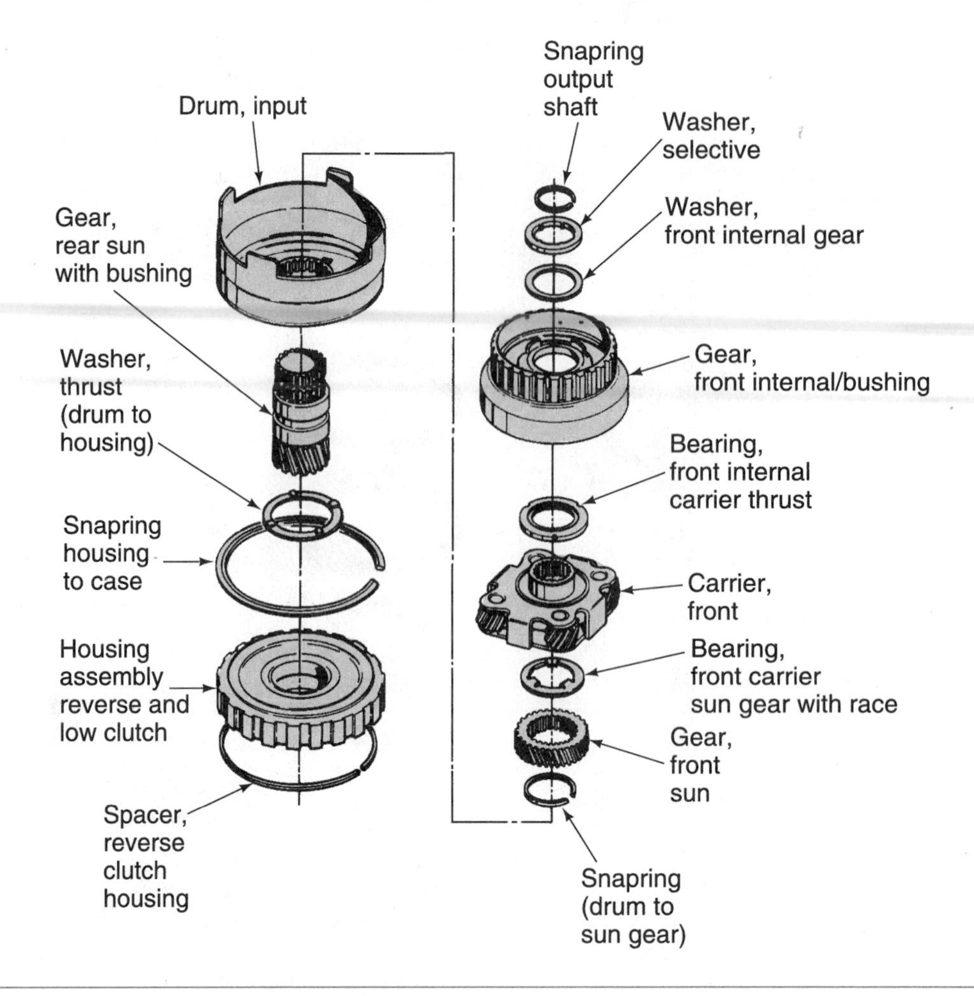

Figure 7-26 The forward and direct clutches are the input devices for all THM transmissions. (Courtesy of the Hydra-Matic Division of General Motors Corp.)

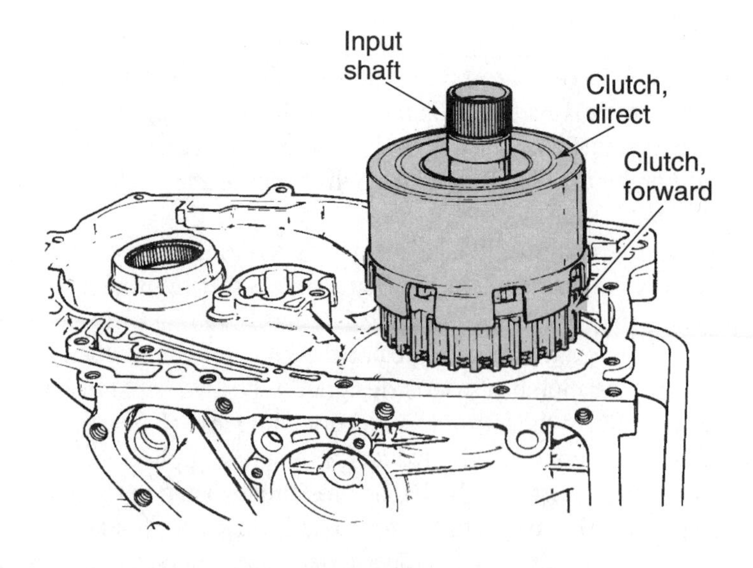

Figure 7-27 The input shaft is splined to the direct clutch hub and the forward clutch drum. (Courtesy of the Hydra-Matic Division of General Motors Corp.)

Holding Devices

The main difference between the different models of THM transmissions is based on which members of the planetary gear set are held and how they are held. These differences also cause the power flows through the various transmissions to be slightly different from each other.

The 400 transmissions use front and rear bands plus a **low/reverse roller clutch,** an intermediate one-way clutch, and an intermediate clutch as the holding devices. The 125, 200, 250, 350, and 325 transmissions use a low/reverse clutch, low roller clutch, and an intermediate band as the holding devices (Figure 7-28). The 350 transmissions use an intermediate roller clutch in addition to the other holding devices.

The rear band, used in the 400 series (Figure 7-29), is applied in manual low and reverse gears. The rear band cannot be adjusted from the outside of the transmission. Rather, band adjustment is controlled by the length of the servo apply pin or rod, which is selected during the assembly of the transmission.

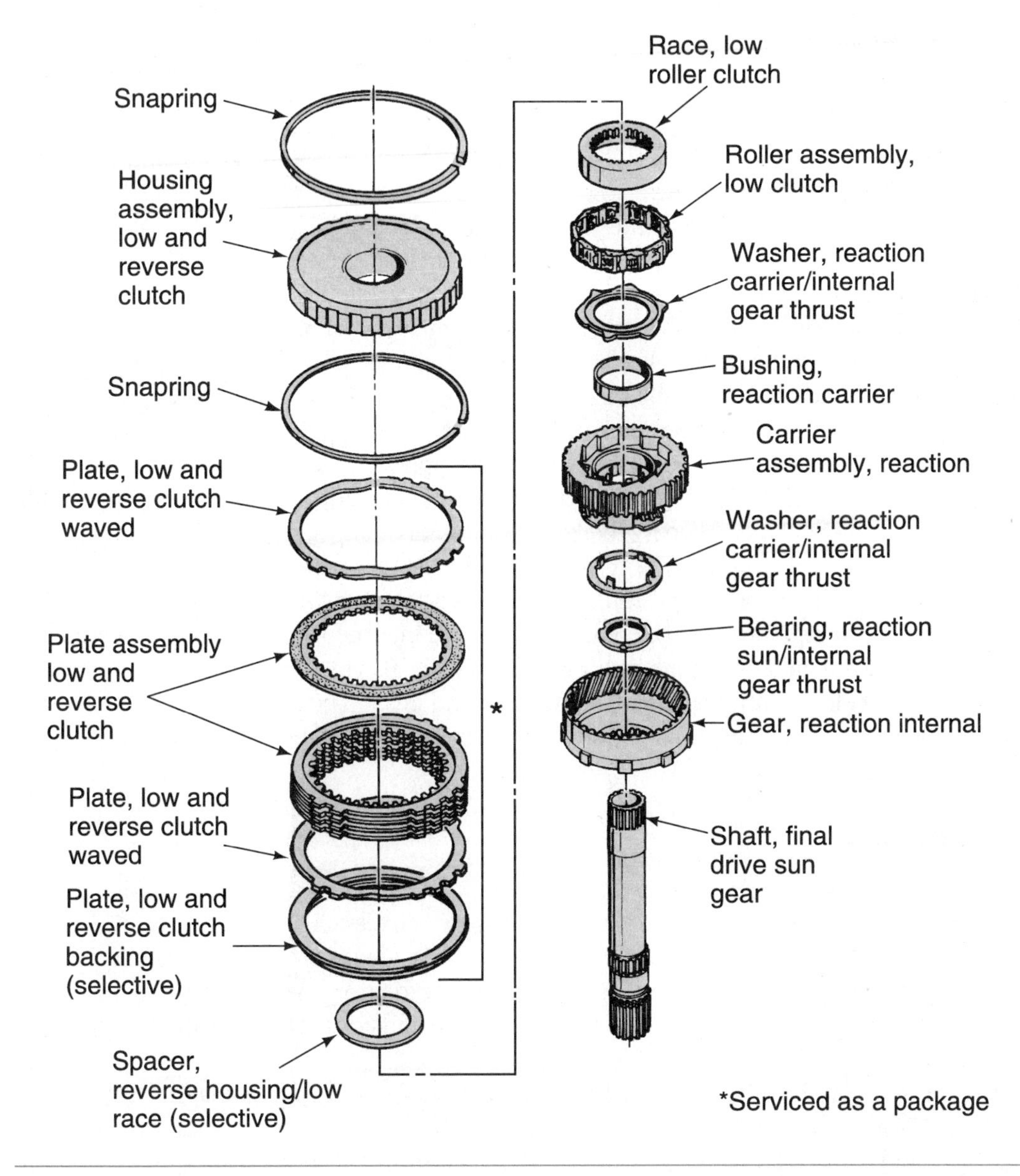

Figure 7-28 The low/reverse clutch, low one-way clutch, and intermediate band are used as the holding devices in most THM transmissions. (Courtesy of the Hydra-Matic Division of General Motors Corp.)

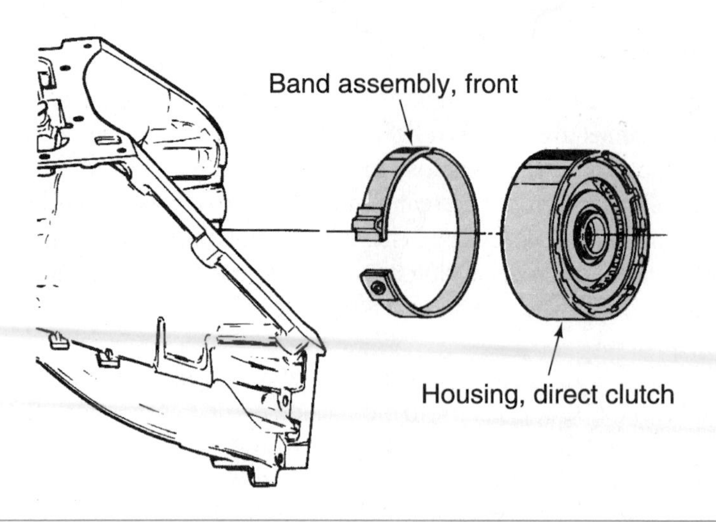

Figure 7-29 The rear band holds the direct clutch drum in manual low and reverse gears in a THM 3L80 transmission. (Courtesy of the Hydra-Matic Division of General Motors Corp.)

The other transmission models use a low/reverse clutch, which is applied in manual low and reverse. The inside edges of the low/reverse clutch friction discs are splined to the outside edges of the rear carrier and the outside edges of the steel discs are splined to the inside of the case. When the low/reverse clutch is applied, it holds the rear carrier from turning in either direction. In the 250 and 350 series, the low/reverse clutch is designed to react with greater clamping pressure when the transmission is operating in reverse. The method used is similar to that used in the direct clutch.

The 350 and 400 transmissions utilize a front band in manual second gear. The front band in the 350 series is used and applied only in manual second gear. The other models apply the front band whenever the transmission is operating in second gear. The front band is wrapped around the outside of the direct clutch drum. When the front band is applied, the clutch drum, input shell, and sun gear assembly are all held. Typically, the front band cannot be adjusted from the outside of the transmission. Adjustment is controlled by the length of the servo apply pin, which is installed during the assembly of the transmission. The front band of the 250 models is externally adjustable.

The intermediate clutch of the 350 and 400 models is applied in second and third gears; however, it is ineffective during third-gear operation because the intermediate roller clutch overruns.

In the 350 and 400 models, the intermediate roller clutch also freewheels during deceleration but locks during acceleration in drive and manual second gear. The low one-way clutch of all transmission models holds during acceleration in drive low, but freewheels during deceleration or coasting.

Apply Devices

The 400 series has two servos: the rear servo, which applies the rear band and also contains the 1–2 accumulator piston that cushions application of the intermediate clutch, and the front servo, which applies the front band.

The other transmission models use one servo assembly to apply the front band or intermediate overrun band. The servo bore is in the transmission case and consists of a piston, an apply pin, a cushion spring, and a cover.

The front servo in the 125, 200, and 325 series also serves as an accumulator for the direct clutch (Figure 7-30). During first to second gear shifts, hydraulic pressure causes the band to tighten. The band is released when pressure is exhausted from the servo. During upshifts from second to third gear, the apply pressure for the direct clutch is routed to the spring side of the servo piston and the servo acts as an accumulator to smooth the application of the clutch.

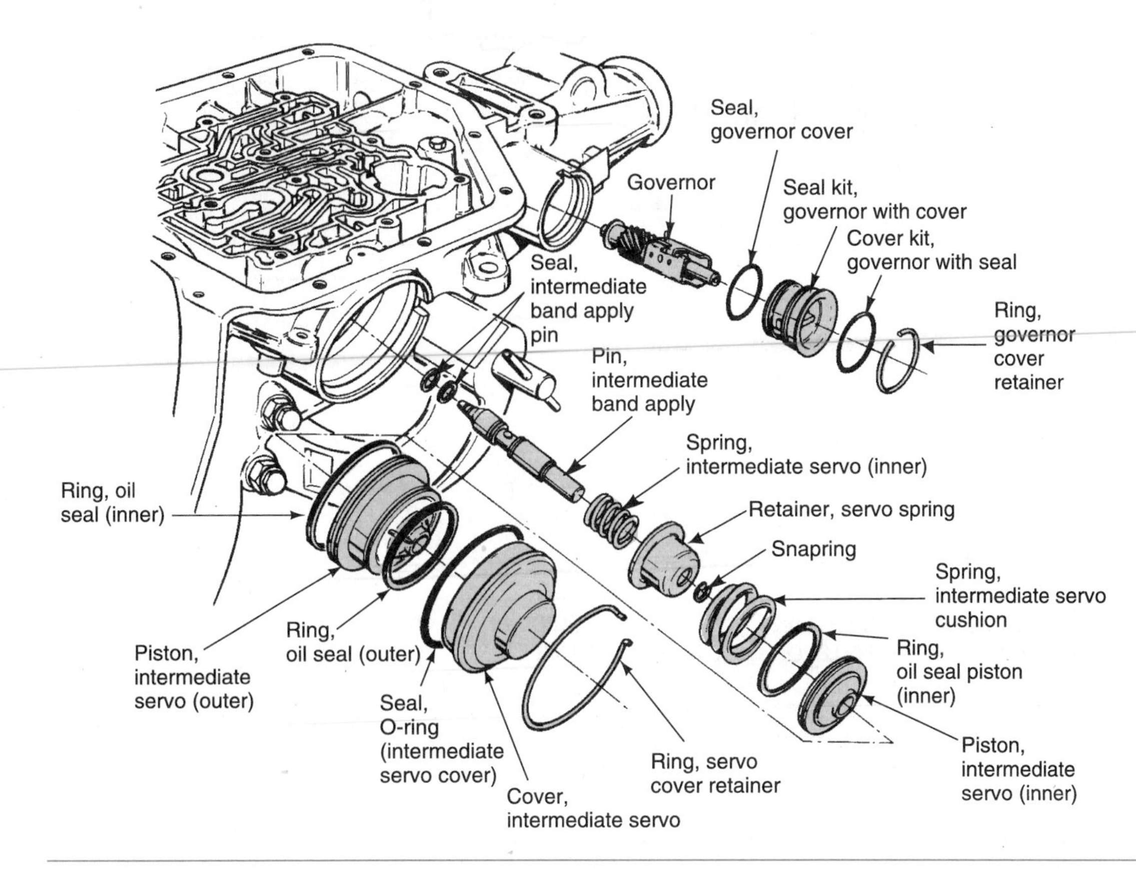

Figure 7-30 The intermediate servo in most THM transmissions also serves as an accumulator for the direct clutch. (Courtesy of the Hydra-Matic Division of General Motors Corp.)

Power Flow

The power flow through General Motors transmissions with Simpson gear trains is similar to that of the power flow through a Chrysler Torqueflite transmission. There are some slight variations within the different model types, but these depend on the operating characteristics of each model (Figures 7-31, 7-32, 7-33, and 7-34).

Add-On Overdrive

In 1982, General Motors replaced the 200C and 325 with the 200-4R and 325-4L. These transmissions were actually the same transmissions with an overdrive planetary gear set added to the front of the Simpson gear train.

GEAR SELECTOR POSITION	OPERATING GEAR	FORWARD CLUTCH	DIRECT CLUTCH	FRONT BAND	INTERMEDIATE CLUTCH	INTERMEDIATE ROLLER CLUTCH	LOW ROLLER CLUTCH	REAR BAND
D	1st gear	X					X	
	2nd gear	X			X	X		
	3rd gear	X	X		X			
2	1st gear	X					X	
	2nd gear	X		X	X	X		
1	1st gear	X					X	X
R	Reverse		X					X

Figure 7-31 Clutch and band application chart for a THM 375/400/425 transmission.

GEAR SELECTOR POSITION	OPERATING GEAR	INTERMEDIATE CLUTCH	DIRECT CLUTCH	FORWARD CLUTCH	LOW/REVERSE CLUTCH	INTERMEDIATE BAND	LOW/ROLLER CLUTCH	INTERMEDIATE ROLLER CLUTCH
D	1st gear			X			X	
	2nd gear	X		X				X
	3rd gear	X	X	X				
2	1st gear			X			X	
	2nd gear	X		X		X		X
1	1st gear			X	X		X	
R	Reverse		X		X			

Figure 7-32 Clutch and band application chart for a THM 250/350 transmission.

GEAR SELECTOR POSITION	OPERATING GEAR	OVERRUN CLUTCH	INTERMEDIATE BAND	OVERDRIVE ROLLER CL	DIRECT CLUTCH	LOW ROLLER CLUTCH	4TH CLUTCH	FORWARD CLUTCH	LOW/REVERSE CLUTCH
O/D	1st gear			X		X		X	
	2nd gear		X	X				X	
	3rd gear			X	X			X	
	Overdrive				X		X	X	
3	1st gear	X						X	
	2nd gear	X	X			X		X	
	3rd gear	X			X			X	
L2	1st gear	X						X	
	2nd gear	X	X			X		X	
L1	1st gear	X						X	X
R	Reverse			X	X				X
N/P	Neutral			X					

Figure 7-33 Clutch and band application chart for a THM 4L30 transmission.

GEAR SELECTOR POSITION	OPERATING GEAR	DIRECT CLUTCH	FORWARD CLUTCH	LOW & REVERSE CLUTCH	INTERMEDIATE BAND	LOW ROLLER CLUTCH
D	1st gear		X			X
	2nd gear		X		X	
	3rd gear	X	X			
2	1st gear		X			X
	2nd gear		X		X	
1	1st gear		X	X		X
R	Reverse	X		X		

Figure 7-34 Clutch and band application chart for a THM 125 (3T40) transmission.

GEAR SELECTOR POSITION	OPERATING GEAR	2–4 BAND	REVERSE INPUT CLUTCH	OVERRUN CLUTCH	FORWARD CLUTCH	FORWARD SPRAG CLUTCH	3–4 CLUTCH	LOW ROLLER CLUTCH	LOW/REVERSE CLUTCH
OD	1st gear				X	X		X	
	2nd gear	X			X	X			
	3rd gear				X	X	X		
	Overdrive	X			X		X		
D	1st gear			X	X	X		X	
	2nd gear	X		X	X	X			
	3rd gear			X	X	X	X		
2	1st gear			X	X	X		X	
	2nd gear	X		X	X	X			
R	Reverse		X						X

Figure 7-35 Clutch and band application chart for a THM 700-R4 (4L60) transmission.

The 200-4R and 325-4L use the same multiple-disc clutches as the 200, but they also have two additional clutches: the drive clutch and **fourth clutch.** Each of these clutches is applied hydraulically. The forward, direct, and fourth clutches are released by several small coil return springs and the drive and low/reverse clutches are released by wave springs.

With GM's new designation, the 200-4R became the 4L30 transmission.

Also in 1982, GM released the 700-R4, which is used behind medium and large displacement engines. This transmission has been used in most full-size GM cars and trucks and in some heavy-duty and high-performance vehicles.

Some parts catalogs list the 700-R4 as the MD-8 transmission.

The THM 700-R4 is a fully automatic transmission consisting of two planetary gear sets, five multiple-disc clutches, one sprag clutch, one roller clutch, and a band. Refer to the clutch and band application chart (Figure 7-35) for more details. These provide four forward gears, including an overdrive. The five multiple-disc clutches are applied hydraulically and released by several small coil springs. The 700-R4 has a one-piece case casting that incorporates the bell housing. The extension housing is a separate casting bolted to the rear of the case.

GM's new designation for the 700-R4 is the 4L60 transmission.

Input Devices

The forward clutch, **forward sprag, 3–4 clutch,** and **reverse input clutches** are the possible input devices for the 700-R4 transmission (Figure 7-36). The forward clutch is applied in all forward gears and drives the front sun gear. The forward sprag clutch also drives the front sun gear in first, second, and third gears, but freewheels in fourth gear.

The 3–4 (high/overdrive) clutch is applied in third and fourth gears and drives the front ring gear and rear planetary carrier. The reverse input clutch is applied only in reverse gear and it locks the rear sun gear to the input shaft.

Holding Devices

The low/reverse clutch, low roller clutch, and **2–4 band (intermediate/overdrive band)** are the holding devices for this transmission. The low/reverse clutch is applied in reverse and manual low. This clutch locks the rear planet carrier to the transmission case. Like other THM transmissions, the low/reverse clutch of the 700-R4 is designed to allow greater holding force when the transmission is operating in reverse. The clutch's piston has two separate surfaces on which the hydraulic pressure may act. When operating in manual low, pressure is routed to the smaller, inner area of the piston. When in reverse, the hydraulic pressure is applied to the inner area of the piston and to a larger outer area. This allows the clutch to have a greater clamping force while operating in reverse gear.

The low roller clutch holds the rear carrier and front ring gear in low gear. The drive overrunning clutch is applied in manual third, manual second, and manual low gears. It locks the front

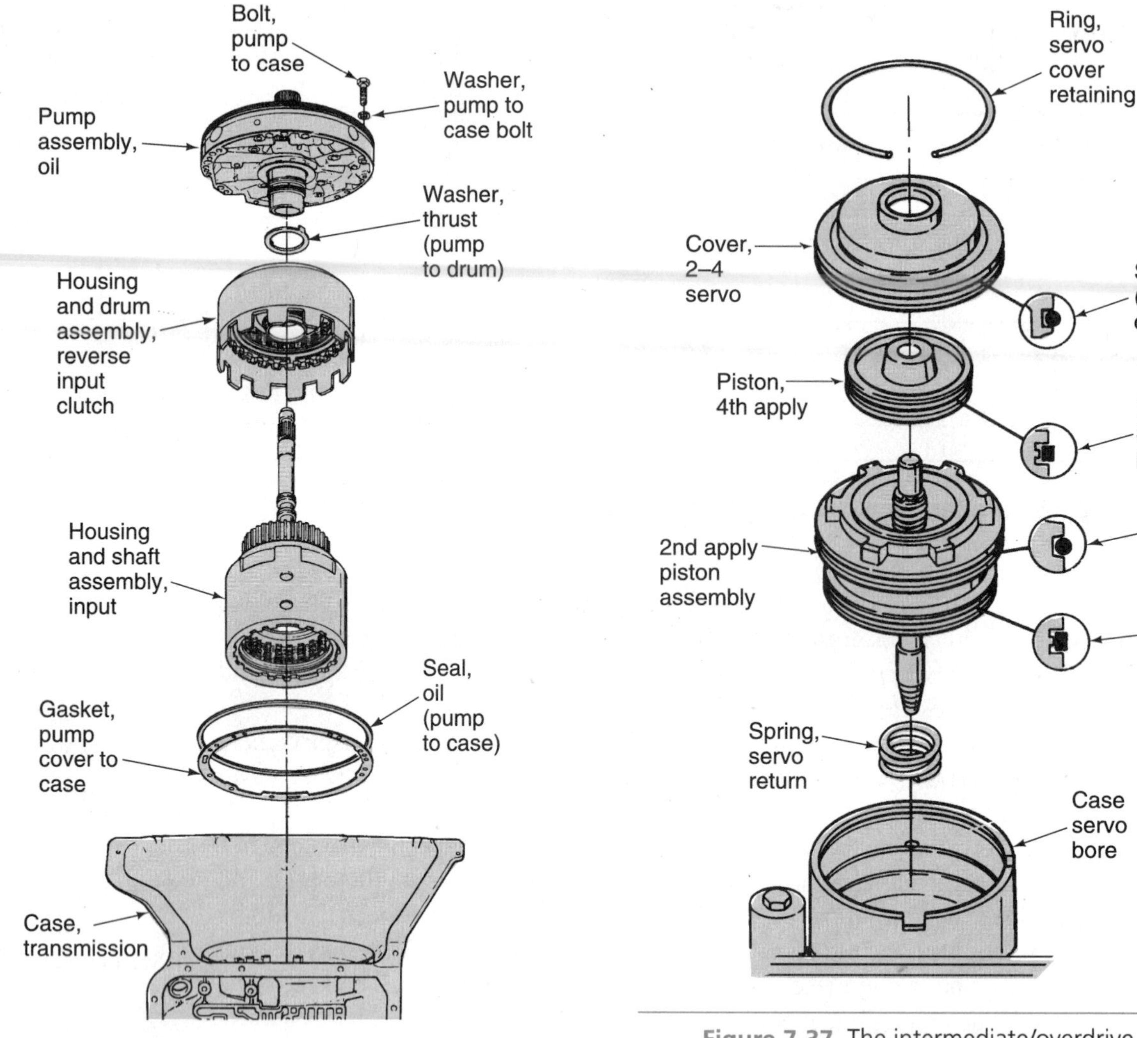

Figure 7-36 In a THM 700-R4 transmission, the forward, forward one-way, high/overdrive, and reverse clutches are the possible input devices. (Courtesy of the Hydra-Matic Division of General Motors Corp.)

Figure 7-37 The intermediate/overdrive servo unit for a THM 700-R4 transmission. (Courtesy of the Hydra-Matic Division of General Motors Corp.)

sun gear to the input shaft. The overrunning clutch prevents the front sun gear from freewheeling during deceleration in the manual low, manual second, and manual third gear ranges. The 2–4 band is applied only in second and fourth gears.

Apply Devices

The 700-R4 has a single servo that is used to apply the 2–4 band (Figure 7-37). The band cannot be adjusted from the outside of the transmission. Adjustment is made during the assembly of the transmission through the selection of different length servo apply pins or servo rods.

Other Simpson-Based Transmissions

There are many different automatic transmissions that are based on the Simpson gear train. Most have similar power flows and major components. The primary differences between most of the transmissions is in the nomenclature and holding devices used by the different manufacturers. The following transmissions are presented to illustrate the similarities of all Simpson-based transmissions and to expose you to some of the nomenclature used by other manufacturers.

Nissan Motor Company Transmissions

A widely used Nissan RWD transmission is the L4N71B/E4N71B series. These transmissions provide four forward gears through the use of a Simpson gear set and an additional planetary unit mounted in front of the Simpson gear set (Figure 7-38). The primary difference between the "L" and the "E" series is that the E-model provides converter lockup in third and fourth gear.

Shop Manual
Chapter 7, page 302

These transmissions use four multiple-disc clutches, two servos and bands, and two one-way clutches to provide for the different ranges of gears. The direct clutch is applied in all gears except overdrive. Its purpose is to bypass the overdrive planetary gear set and transfer engine torque to the Simpson gear set. The forward or rear clutch is applied in all forward gears. The high/reverse

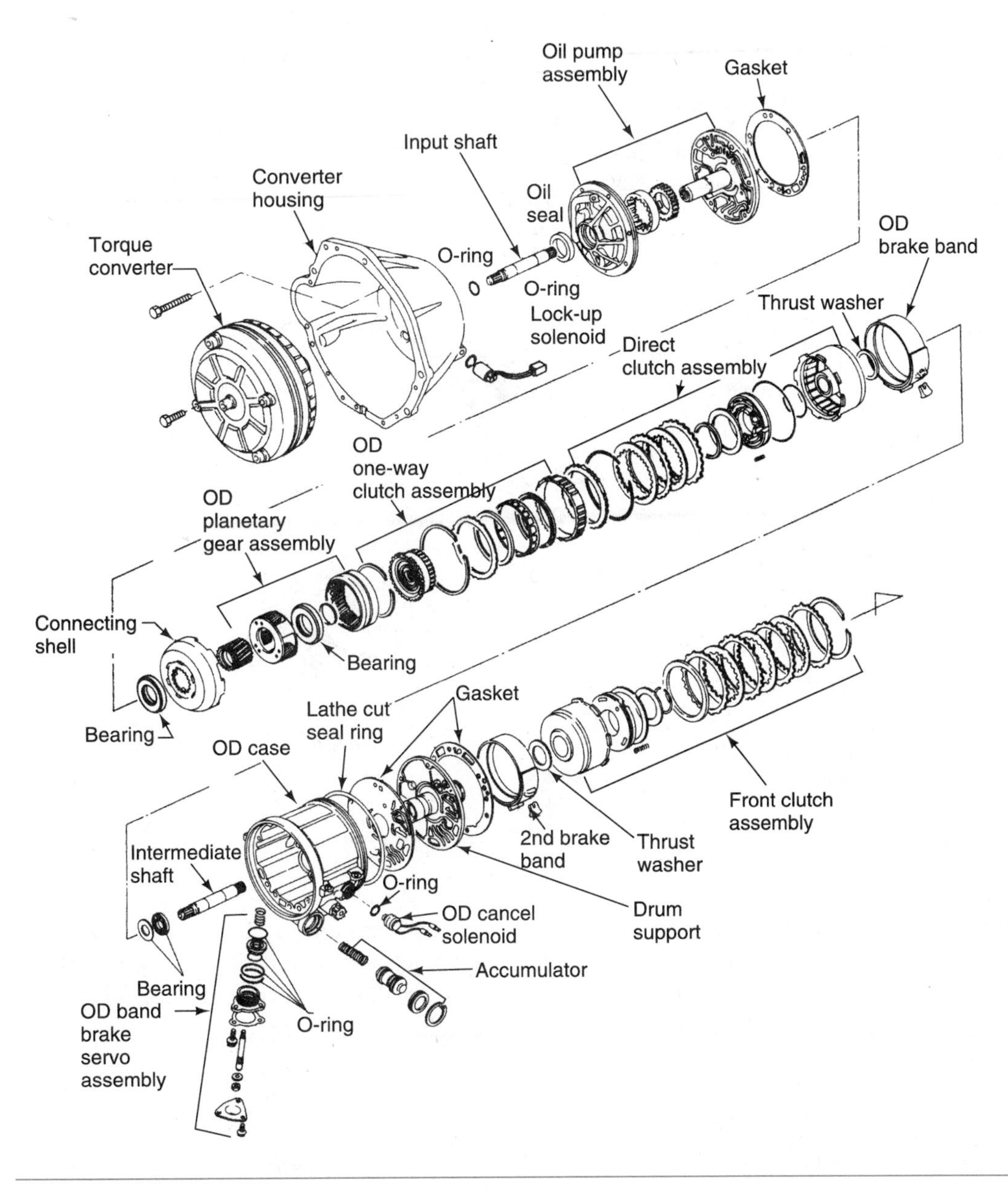

Figure 7-38 Exploded view of a Nissan L4N71B Simpson-based transmission. (Courtesy of Nissan Motor Co., Ltd.)

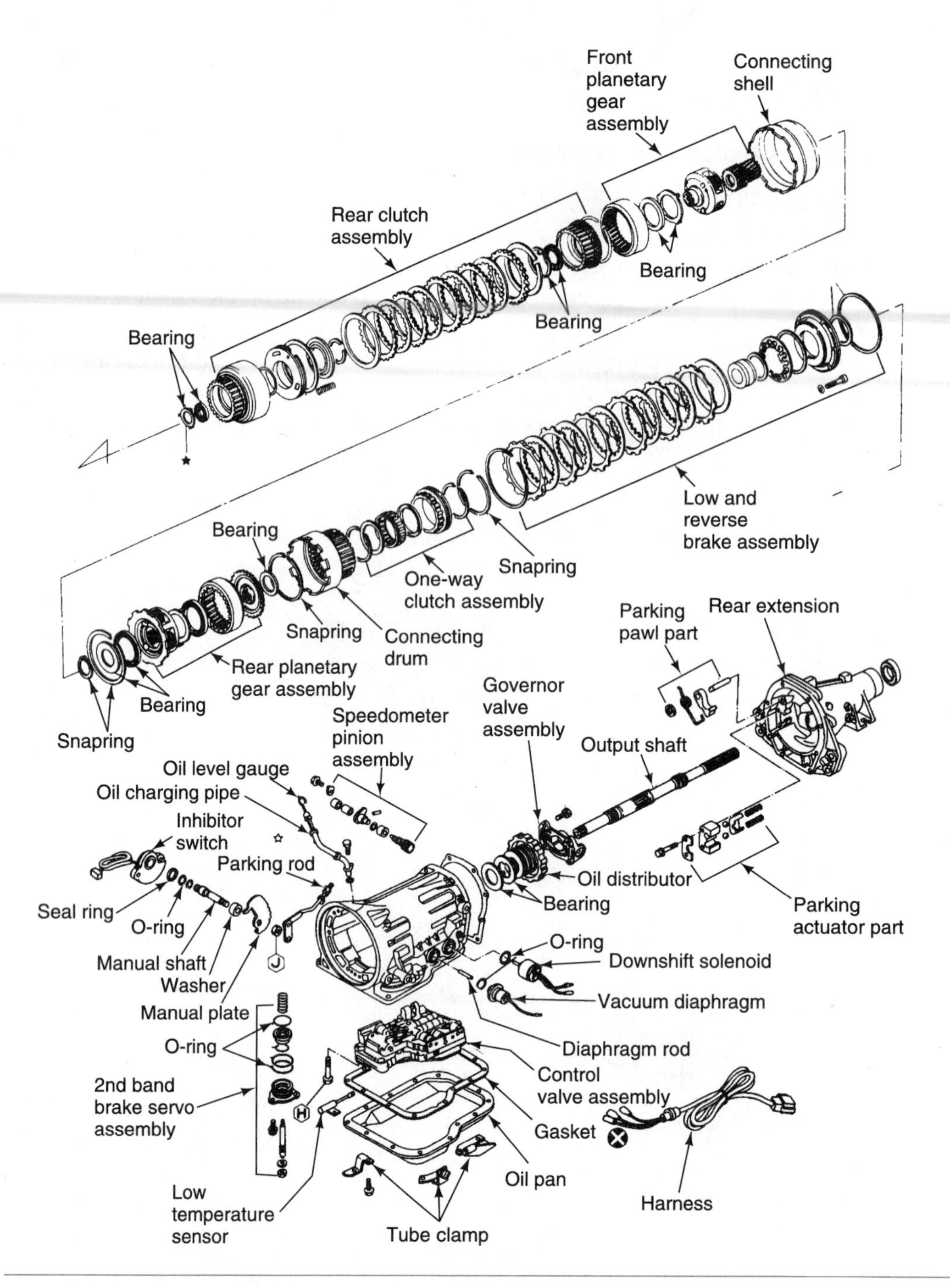

Figure 7-38 Continued

clutch (front clutch) is applied in reverse gear and third and fourth gears. The low/reverse clutch is applied in reverse and in manual low. This clutch serves as a brake and allows for engine braking during manual low operation. All of these clutches except the low/reverse unit are released by several small coil springs. The low/reverse unit utilizes a belleville-type spring for greater clamping pressures.

The bands and the one-way clutch are used as holding members. The **second band** is applied only in second gear. The overdrive band (Figure 7-39) is applied only when the transmission is operating in fourth gear. The one-way clutch is locked during first-gear operation but freewheels and is ineffective during deceleration or coasting. The bands are adjustable. Refer to the clutch and band application chart (Figure 7-40) for more details on power flow.

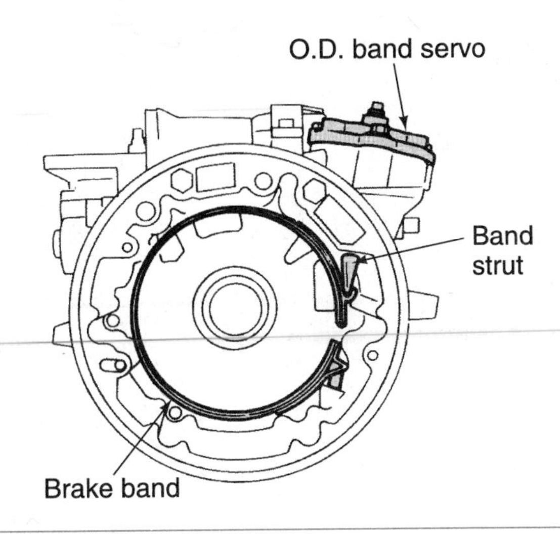

Figure 7-39 The overdrive band is externally adjustable and the servo is contained in its own bore in the transmission case. (Courtesy of Nissan Motor Co., Ltd.)

GEAR SELECTOR POSITION	OPERATING GEAR	REAR CLUTCH	ONE-WAY CLUTCH	BAND	FRONT CLUTCH	LOW & REVERSE BAND
D	1st gear	X	X			
	2nd gear	X		X		
	3rd gear	X			X	
2	2nd gear	X		X		
1	1st gear	X	X			X
R	Reverse				X	X

Figure 7-40 Clutch and band application chart for a Nissan L4N71B transmission.

Toyota Motors Transmissions

Many different transmissions and transaxles have been used by Toyota. A common early design is the A-140E/A-140L, used in the Camry and Celica models. This transaxle combines a three-speed transmission with an overdrive assembly. The A140E (Figure 7-41) is used on Camry and Celica models, whereas the A140L is used on the Celica GT-S. The primary differences between the L- and E-type transaxles are the main valve body, operating mechanism, and electronic control. The E refers to an **electronic controlled transaxle (ECT).** The ECT is different from the oil pressure control transaxle (A140L) in that it is controlled by a microprocessor located behind the glove box. The A140L uses an electronic overdrive solenoid system.

Although these transaxles are electronically controlled, the basis for operation is a Simpson gear train in line with a single overdrive planetary gear set. The transaxles use four multiple-disc clutches, two band and servo assemblies, and three one-way clutches to provide the various gear ranges.

The multiple-disc clutches are an overdrive clutch, forward clutch, direct clutch, and a second coast drum. The latter clutch provides engine braking during second gear deceleration and coasting. The two band assemblies are the **second coast band** and the first/reverse brake. The one-way overrunning clutches are called the #1, #2, and overdrive one-way clutches. Refer to the clutch and band application chart (Figure 7-42) for details on the power flow through this transmission.

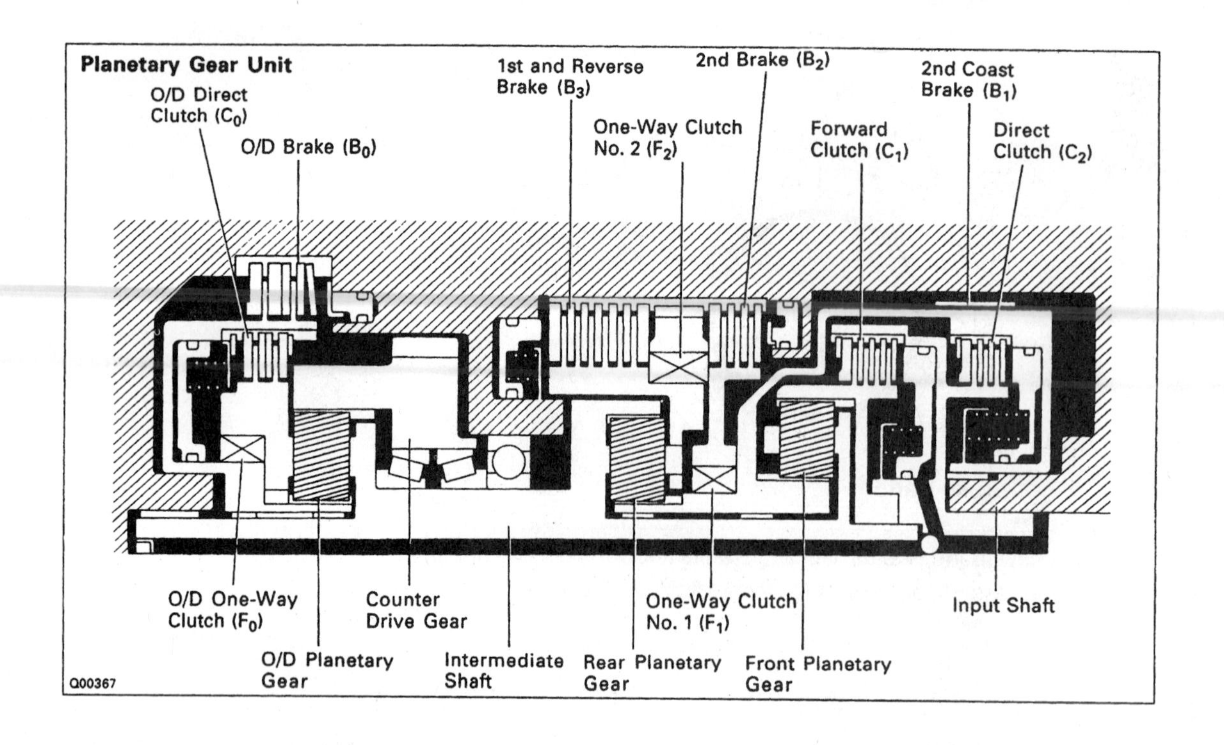

Figure 7-41 The apply devices for the planetary gear set in an A140E transaxle. (Reprinted with permission.)

GEAR SELECTOR POSITION	OPERATING GEAR	OD CLUTCH	FORWARD CLUTCH	DIRECT CLUTCH	2ND COAST BAND	2ND COAST DRUM	1ST & REVERSE BRAKE	OD ONE-WAY CLUTCH	#1 ONE-WAY CLUTCH	#2 ONE-WAY CLUTCH
D-Drive	1st gear	X	X					X		X
	2nd gear	X	X			X		X	X	
	3rd gear	X	X	X		X		X		
	Overdrive		X	X		X				
2-Second	1st gear	X	X					X		X
	2nd gear	X	X		X	X		X	X	
	3rd gear	X	X	X		X		X		
L-Low	1st gear	X	X				X	X		X
	2nd gear	X	X		X	X		X	X	
R	Reverse	X		X			X			
N/P	Neutral	X								

Figure 7-42 Clutch and band application chart for a Toyota A140 transmission.

In the mid-1990s Toyota introduced the A541E transaxle (Figure 7-43), which was a revised copy of the A140E. The biggest change was in the electronic controls, which now had adaptive learning. This transaxle has been used in many different models of Toyotas, especially the Camry and Avalon. The band and clutch location (Figure 7-44), as well as the application of each (Figure 7-45), is the same as the previous models.

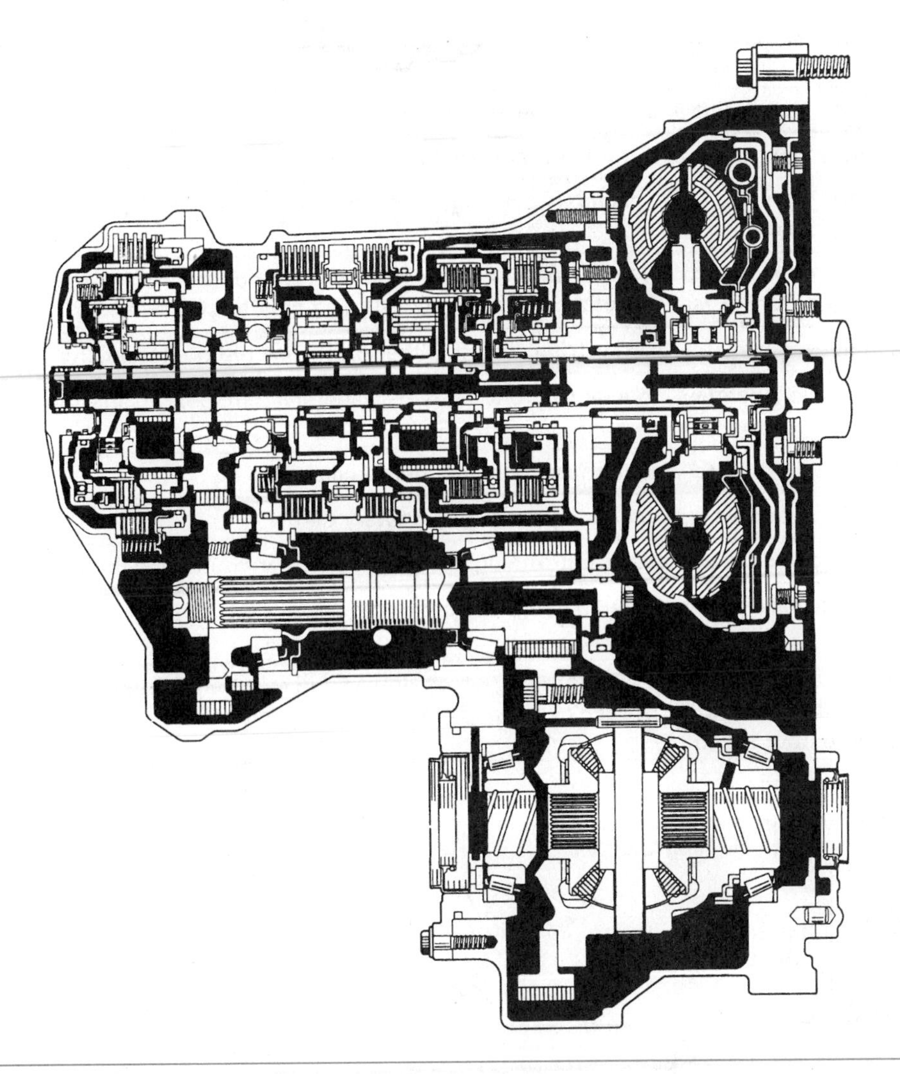

Figure 7-43 Toyota's A541E transaxle. (Reprinted with permission.)

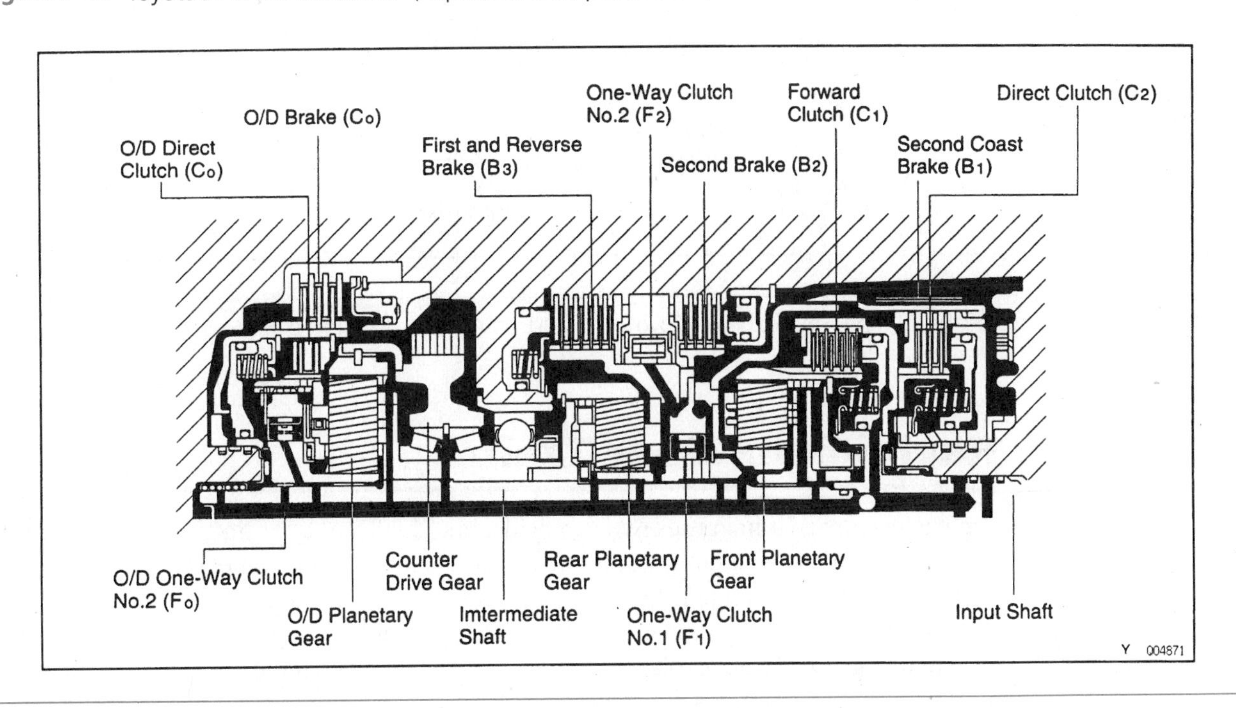

Figure 7-44 Band and clutch location for Toyota's A541E transaxle. (Reprinted with permission.)

COMPONENT		FUNCTION
Forward Clutch	C_1	Connects input shaft and front planetary ring gear
Direct Clutch	C_2	Connects input shaft and front & rear planetary sun gear
2nd Coast Brake	B_1	Prevents front & rear planetary sun gear from turning either clockwise or counterclockwise
2nd Brake	B_2	Prevents outer race of F_1 from turning either clockwise or counterclockwise, thus preventing front & rear planetary sun gear from turning counterclockwise
1st & Reverse Brake	B_3	Prevents rear planetary carrier from turning either clockwise or counterclockwise
No.1 One-Way Clutch	F_1	When B_2 is operating, prevents front & rear planetary sun gear from turning counterclockwise
No.2 One-Way Clutch	F_2	Prevents rear planetary carrier from turning counterclockwise
O/D Direct Clutch	C_0	Connects overdrive sun gear and overdrive planetary carrier
O/D Brake	B_0	Prevents overdrive sun gear from turning either clockwise or counterclockwise
O/D One-Way Clutch	F_0	When transaxle is being driven by engine, connects overdrive sun gear and overdrive carrier
Planetary Gears		These gears change the route through which driving force is transmitted in accordance with the operation of each clutch and brake in order to increase or reduce the input and output speed

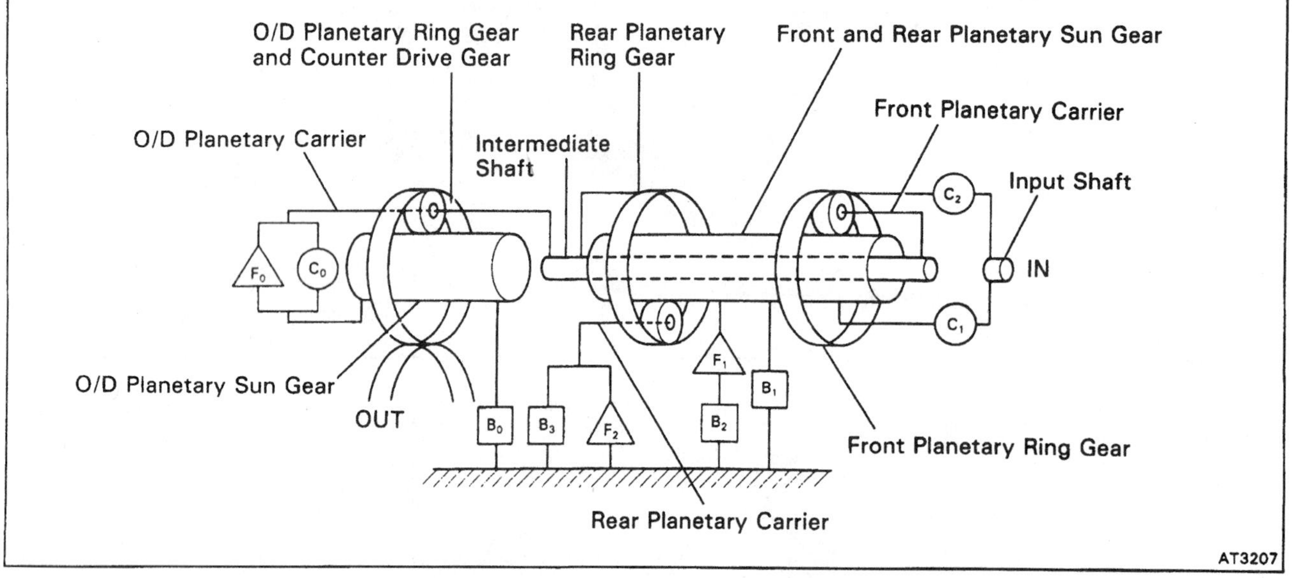

Figure 7-45 Purpose of the clutches and bands in Toyota's A541E transaxle. (Reprinted with permission.)

Ravigneaux Gear Train

The Ravigneaux gear train uses two sun gears, one small and one large, and two sets of planetary pinion gears, three long pinions, and three short pinions. The planetary pinion gears rotate on their own shafts, which are fastened to a common planet carrier. The small sun gear is meshed with the short planetary pinion gears. These short pinions act as idler gears to drive the long planetary pinion gears. The long planetary pinion gears mesh with the large sun gear and the ring gear. A single ring gear surrounds the complete assembly.

Ford ATX Ravigneaux-Based Transaxle

The Ford ATX (Figure 7-46), a three-speed automatic transaxle, was introduced in 1980 in the Escort and Lynx model cars. In 1984, the Tempo and Topaz models were also equipped with the ATX. In 1986, the ATX was modified for use in the Taurus and Sable. The ATX transaxle uses a Ravigneaux planetary gear train. Its one-piece aluminum housing is one of the most compact units installed in domestic automobiles.

Shop Manual
Chapter 7, page 255

ATX transaxles use a gear-type oil pump that is driven indirectly by the torque converter. Because the oil pump is located at the opposite end of the transaxle, as is the torque converter, a drive shaft is used to rotate the oil pump.

The transaxle has been continually updated and modified to fit different applications. The most obvious change is the torque converter. The ATX has been available with three different designs of torque converters: the conventional three-element converter, a converter with centrifugal lockup, and a **split-torque converter.**

The original ATX transaxles were equipped with a splitter gear torque converter. This unit contains a planetary gear set that divides the delivery of input torque. Part of the torque is transmitted mechanically and the remainder by fluid force on the torque converter's turbine. In second gear, 62 percent of the torque is transmitted mechanically, while in third gear 93 percent is

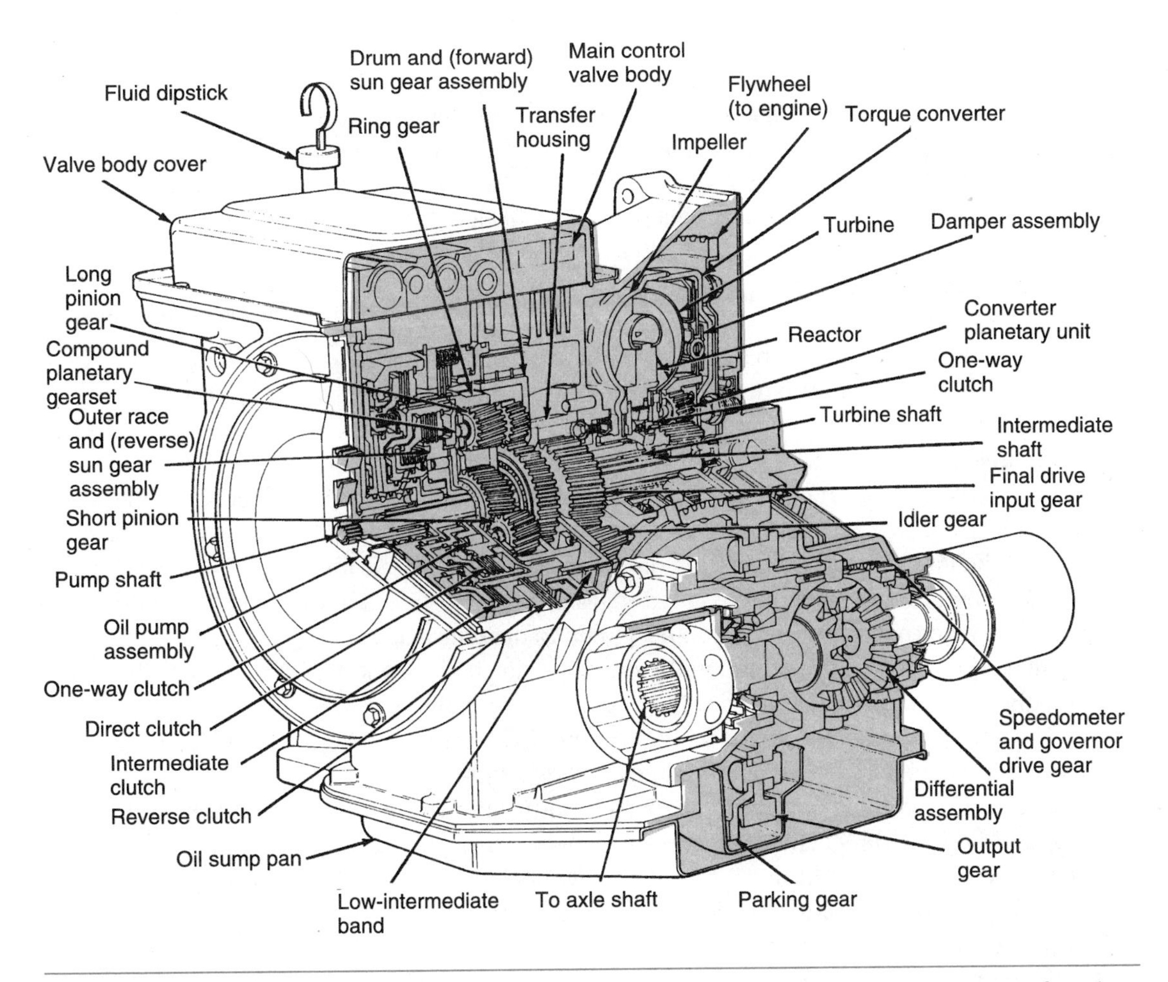

Figure 7-46 A Ford ATX Ravigneaux-based transaxle. (Reprinted with the permission of Ford Motor Co.)

mechanically transmitted. In low gear and reverse, full engine torque is transmitted by normal converter fluid flow.

The split-torque converter is used in Escort and Lynx ATX transaxles and in the transaxles of 1984–1986 Tempo and Topaz models. The ATX in the Taurus and Sable models used a centrifugally locking torque converter. By 1988, all ATX transaxles used in the Tempo, Topaz, Taurus, and Sable models were equipped with a conventional fluid-linked torque converter.

The small sun gear of the ATX gear train is called the reverse sun gear and is meshed with the short pinion gears. The short pinion gears serve as idler gears between the reverse sun gear and the long pinion gears. The large sun gear is called the forward sun gear and it meshes with the long planetary pinion gears. The short pinion gears are meshed with the long gears and both sets of pinions can rotate on their own shafts held by the carrier. The planet carrier serves as the output for the gear train. The ring gear surrounds the other members of the planetary gear set. The long pinions are in constant mesh with the short pinions, the forward sun gear, and the ring gear.

The ATX has a wide-ratio gear train with a 2.8:1 first gear, a 1.6:1 second gear, and direct drive for third gear. The ATX is equipped with three multiple-disc clutches, one band, and a one-way roller clutch. The three multiple-disc clutches are called the top gear (direct), second gear (intermediate/high), and reverse clutches. The band is referred to as the front or **low/intermediate band.** The parking gear is located on the output gear of the final drive assembly.

Input Devices

The one-way clutch, second gear clutch, and **top gear clutch** serve as the input devices for the planetary gear set. The reverse sun gear is attached to the outer race of the one-way clutch and to the direct clutch hub. The reverse sun gear can be driven by the torque converter's turbine shaft if the one-way clutch is holding or by the application of the top gear clutch.

The input shaft is attached to the inner race of the one-way clutch on the top gear clutch drum. The outer race of the one-way clutch is located inside the reverse sun gear. The outer edges of the top gear clutch steel discs are splined to the top gear clutch drum and the inner edges of the friction discs are splined to the outer race of the one-way clutch and to the outside of the forward sun gear.

The forward sun gear is part of the input shell that is surrounded and held by the **front brake band.** The forward sun gear meshes with the long pinion gears and the short pinions mesh with the reverse sun gear. Both the long and short pinions turn in the carrier, which is splined to the final drive input gear and is always the output member of the gear set.

The additional input shaft is called the direct shaft.

An intermediate shaft passes through the inside of the input shaft and attaches to the second gear clutch drum. The outer edges of the clutch's steel discs are splined to the inside of the clutch drum, and the inner edges of the friction discs are splined to the outside of the ring gear. The outside of the ring gear is splined to the inner edges of the reverse clutch friction discs, and the outer edges of the reverse clutch steel plates are splined to the transaxle case.

All engine torque passes through the one-way clutch. The one-way clutch locks the turbine shaft to the reverse sun gear as long as the engine is driving the gear set.

The **second gear clutch** allows the intermediate shaft to engage and drive the ring gear. This clutch is applied in second and third gears.

The torque converter turbine shaft passes through the **input one-way clutch** and, if engaged, drives the direct clutch. Either of these will drive the reverse sun gear in a clockwise direction. When both are applied, the gear set works in a 1:1 ratio. To allow for engine braking while in manual low, the direct clutch is applied and locks the turbine shaft to the reverse sun gear. The direct clutch is applied in manual first gear, third gear drive, and reverse.

The top gear (Figure 7-47) and second gear clutches rely on several small coil springs to move their pistons back into their bores when hydraulic pressure is released from the clutch.

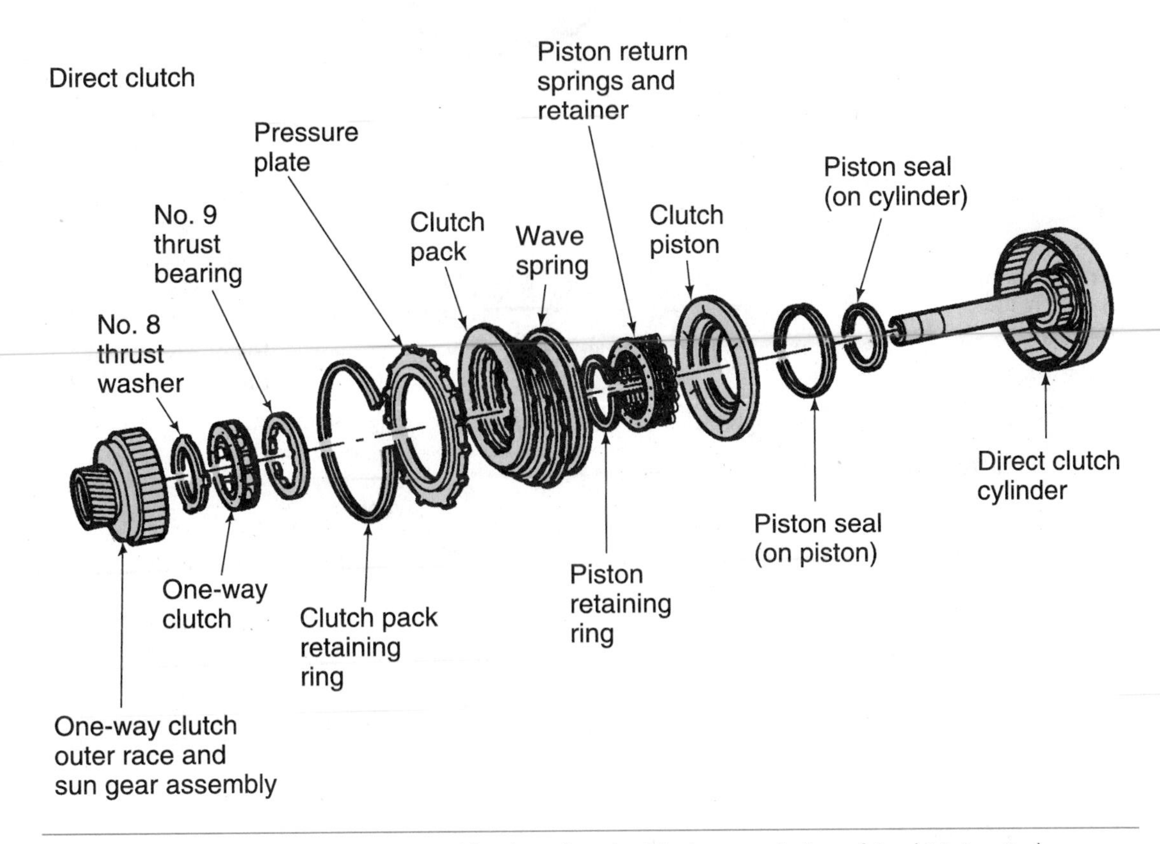

Figure 7-47 Direct clutch assembly. (Reprinted with the permission of Ford Motor Co.)

Holding Devices

The front brake band and the reverse clutch serve as the holding devices for the gear set. The band holds the forward sun gear stationary and is applied in first and second gears.

The reverse clutch holds the ring gear stationary by linking it to the case and is applied only in reverse. The reverse clutch is housed in a recess of the case. The clutch's piston is forced back into its bore when hydraulic pressure is exhausted by several small coil springs (Figure 7-48) and a waved spring.

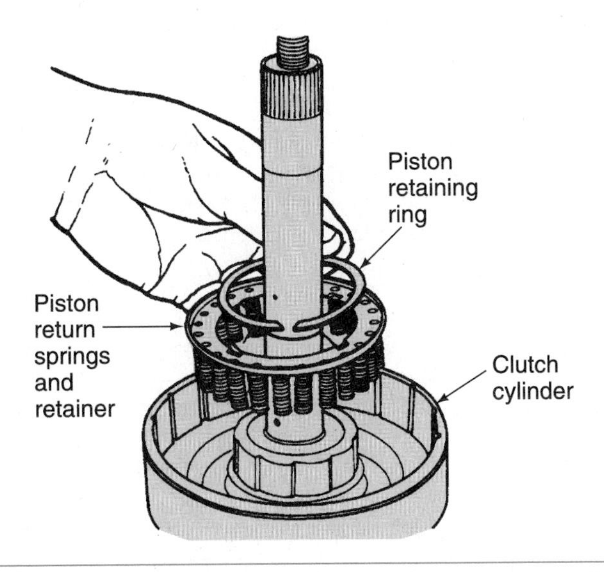

Figure 7-48 The multiple-disc clutches rely on several small coil springs for release. (Reprinted with the permission of Ford Motor Co.)

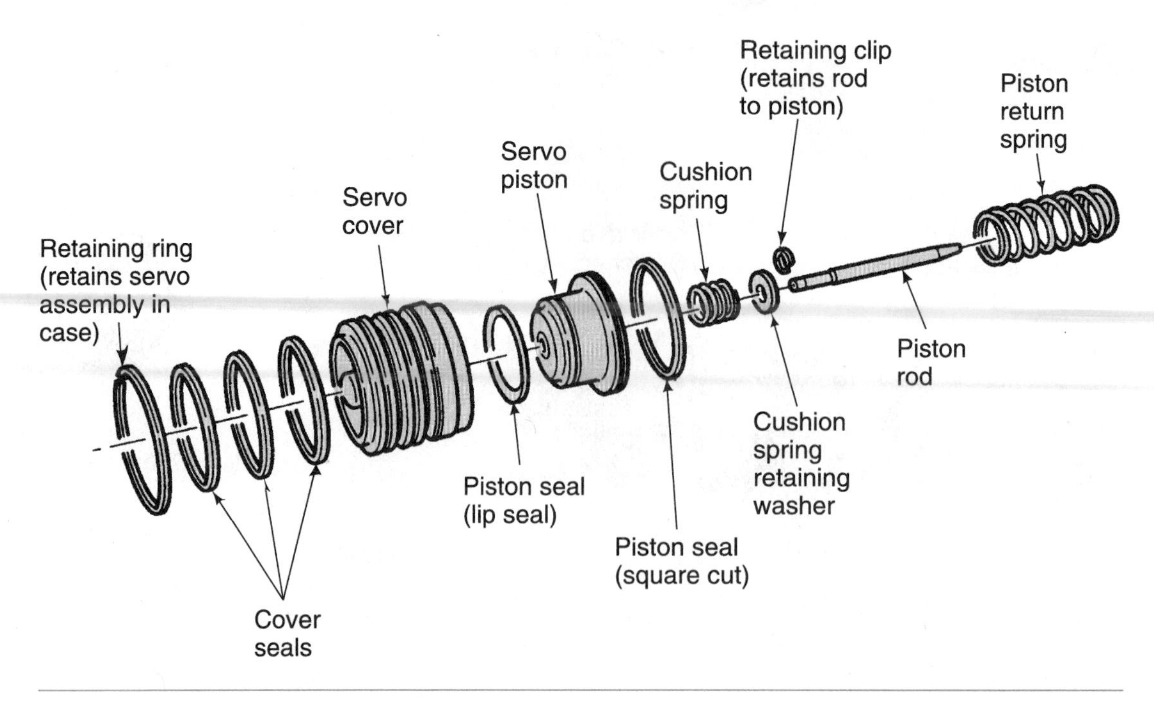

Figure 7-49 An exploded view of the band apply servo in an ATX transaxle. (Reprinted with the permission of Ford Motor Co.)

Apply Devices

The front brake servo (Figure 7-49) applies the front brake band and is installed in a machined bore in the transmission case. The front brake band holds the forward sun gear input shell in all forward gear ranges except drive third. The band is not externally or internally adjustable.

Refer to the clutch and band application chart (Figure 7-50) while reading through the power flow through this transaxle.

Power Flow in NEUTRAL or PARK

When the gear selector is placed in NEUTRAL or PARK, the one-way clutch locks the input from the torque converter's turbine shaft to the sun gear. Since no other clutch or band is applied, power

GEAR SELECTOR POSITION	OPERATING GEAR	FRONT BAND	TOP-GEAR CLUTCH	2ND GEAR CLUTCH	REVERSE CLUTCH	ONE-WAY CLUTCH
P– Park	None					x
R– Reverse	Reverse		x		x	x
N– Neutral	None					x
D– Drive	1st gear	x				x
	2nd gear	x		x		
	3rd gear		x	x		
2– Man. 2nd	1st gear	x				x
	2nd gear	x		x		
1– Man. 1st	1st gear	x	x			x

Figure 7-50 Clutch and band application chart for a Ford ATX transaxle.

does not flow through the gear train. When the gear selector is in PARK, the parking gear is locked to the transaxle case by the parking pawl.

Power Flow in First Gear

The one-way roller clutch locks and allows the reverse sun gear to be driven by the turbine shaft. The reverse sun gear drives the short pinion gears in a counterclockwise direction. The short pinions, in turn, drive the long pinion gears in a clockwise direction. The long pinions walk around the forward sun gear, which is held stationary by the front brake band, and drive the planet carrier and final drive clockwise. The top gear clutch is applied in manual low to provide for engine braking.

Power Flow in Second Gear

Engine torque is transmitted through the planet carrier to the applied second gear clutch, which locks the intermediate shaft to the ring gear. The front brake band is also applied and holds the forward sun gear. The turning of the ring gear causes the long pinion gears to rotate clockwise and walk around the forward sun gear. This action drives the planet carrier and final drive clockwise.

Power Flow in Third Gear

When operating in third gear, the second gear clutch of the ATX remains applied and locks the intermediate shaft to the ring gear. The top gear clutch is also applied and locks the reverse sun gear to the turbine shaft. Since the planetary gear train has two inputs, direct drive results. By allowing input on both the reverse sun gear and the ring gear, the long and short pinion gears are trapped between the ring gear and reverse sun gear. The planet carrier is the output.

Power Flow in Reverse

While operating in reverse gear, the ATX applies the reverse clutch, which holds the ring gear. The applied top gear clutch and one-way clutch lock the reverse sun gear to the turbine shaft. The clockwise rotation of the sun gear causes the short pinion gears to rotate in a counterclockwise direction. The short pinion gears drive the long pinion gears in a clockwise direction; however, since the ring gear is held, the planet carrier is forced to rotate in a counterclockwise direction. The carrier drives the final drive and the vehicle moves in reverse.

Ford's AOD Transmission

The AOD (Figure 7-51) was introduced in 1980 to replace the C-6 in cars. With the introduction of this transmission, Ford became the first U.S. manufacturer to offer an automatic transmission with a built-in overdrive. The AOD is a four-speed transmission with a 0.67:1 overdrive that uses the Ravigneaux gear train.

Another feature of the AOD is the split-torque-type torque converter. This arrangement consists of two input shafts driven by the converter. One of these shafts relays torque into the transmission through the normal hydraulic action of a converter and the other shaft provides a mechanical link from the engine to the gear train when the transmission is operating in third and fourth gears.

The AOD uses the common ring gear as the output member (Figure 7-52). Two sun gears are used, a small one called the forward sun gear and a larger one called the reverse sun gear. The planetary gear set has two sets of planetary pinion gears, three long ones and three short ones. The planetary pinion gears are able to rotate on their own shafts while sharing the common planet carrier.

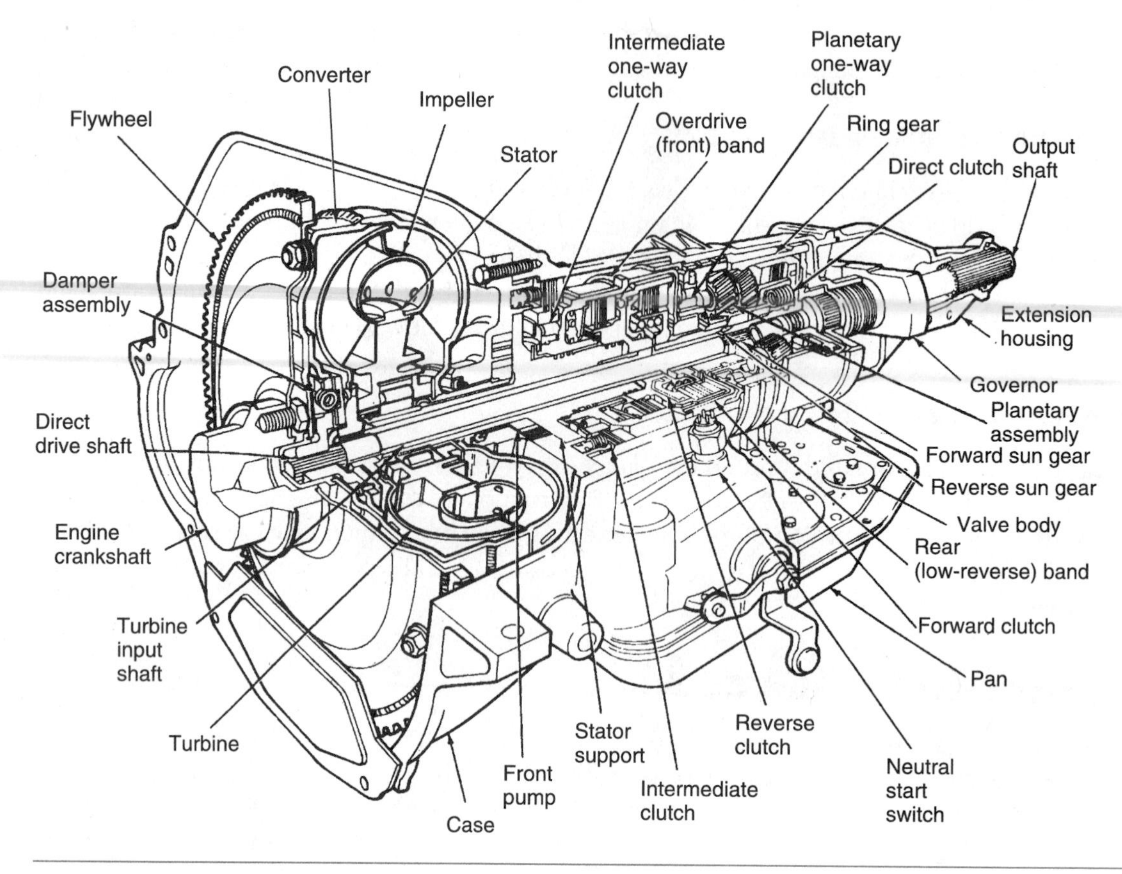

Figure 7-51 A Ford Automatic Overdrive Transmission (AOD). (Reprinted with the permission of Ford Motor Co.)

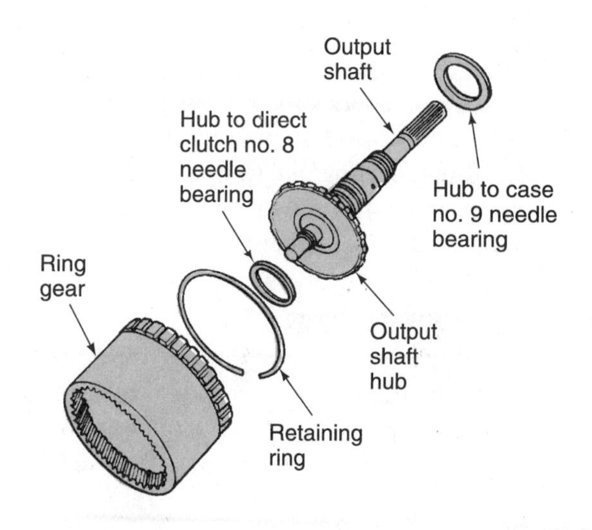

Figure 7-52 The AOD uses the common ring gear as the output member. (Reprinted with the permission of Ford Motor Co.)

The AOD uses four multiple-disc clutches, two one-way clutches, and two bands to obtain the various gear ranges. See the clutch and band application chart (Figure 7-53). The forward sun gear is driven by the turbine shaft when the forward clutch is applied. The reverse sun gear is driven by the turbine input shaft when the reverse clutch is applied and is held, during overdrive, by the overdrive band. The reverse sun gear also can be held by applying the intermediate clutch in second, third, and fourth gears.

GEAR SELECTOR POSITION	OPERATING GEAR	INTER-MEDIATE CLUTCH	INTER-MEDIATE ONE-WAY CLUTCH	OVERDRIVE BAND	REVERSE CLUTCH	FORWARD CLUTCH	PLANETARY ONE-WAY	LOW/REVERSE BAND	DIRECT CLUTCH
OD	1st gear					X	X		
	2nd gear	X	X			X			
	3rd gear	X				X			X
	4th gear	X		X					X
3	1st gear					X	X		
	2nd gear	X	X			X			
	3rd gear	X				X			X
1	1st gear					X	X	X	
	2nd gear	X	X	X		X			
R	Reverse				X			X	

Figure 7-53 Clutch and band application chart for a Ford AOD transmission.

The single planetary carrier assembly can be held by either the low/reverse band or the **low one-way clutch**. The long pinion gears are in constant mesh with the ring gear, the reverse sun gear, and the short pinion gears. The short pinion gears are in constant mesh with the forward sun gear and the long pinions. The short pinions do not mesh with the ring gear but can drive it through the long pinion gears. The ring gear is splined to a flange on the output shaft and the ring gear is always the output member.

Input Devices

The AOD uses a torque converter that splits power flow from the engine. The converter drives two input shafts to provide both a hydraulic and mechanical input into the transmission. A typical input shaft from the converter's turbine hydraulically drives the forward clutch drum. An additional input shaft linked to the converter cover mechanically links the output of the engine to the direct clutch (Figure 7-54). The amount of input from either or both sources varies according to the operating gear. When the transmission is operating in first, second, and reverse gears, 100 percent of the input is delivered hydraulically by the turbine shaft. When it is operating in third gear, the gear

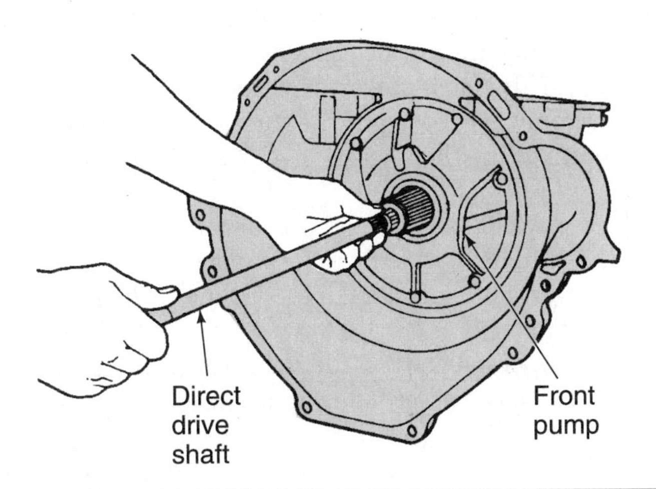

Figure 7-54 Because the AOD is equipped with a split-torque converter, it has an additional input shaft that is attached to the converter cover and passes through the center of the pump. (Reprinted with the permission of Ford Motor Co.)

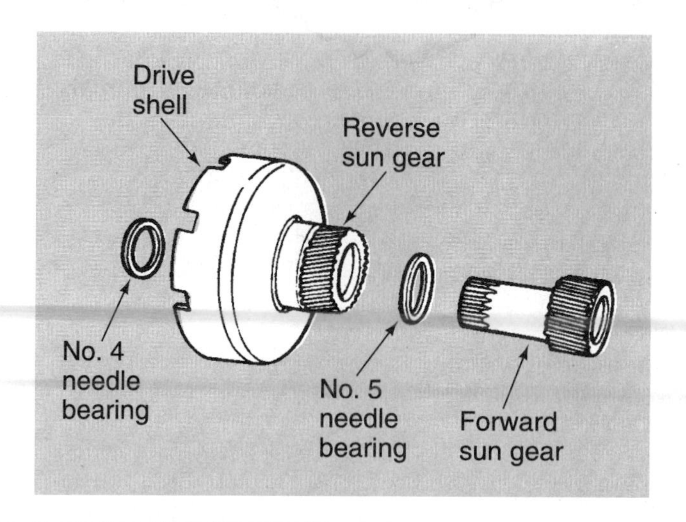

Figure 7-55 The Ravigneaux gear train has two sun gears. (Reprinted with the permission of Ford Motor Co.)

train receives approximately 40 percent of its input from the turbine shaft and 60 percent through the mechanical linkage. In fourth gear, all the input to the gear train is through the mechanical linkage.

Regardless of the input into the transmission, the gear train receives power through the application of the forward, direct, or reverse clutches, which serve as the input devices for the gear train. The forward clutch drives the forward sun gear and is applied in first, second, and third gears. The clutch is applied hydraulically and is released by the tension of a single, large coil spring when hydraulic pressure at the clutch is exhausted.

The direct clutch is applied in third and fourth gears. This clutch connects the input from the converter cover to the planet carrier. The direct clutch is released by several small coil springs when hydraulic pressure to the clutch is relieved.

The reverse clutch is applied only in reverse and drives the reverse sun gear (Figure 7-55). The reverse clutch is released by a belleville spring, which also increases the clamping force of the clutch.

Holding Devices

The gear train relies on the **intermediate clutch,** both one-way clutches, the low/reverse band, and the overdrive band for holding gear set members. The intermediate clutch is applied in second, third, and fourth gears. This clutch prevents the reverse sun gear from turning counterclockwise, but it is effective only in second gear because the **intermediate one-way clutch** freewheels in third and fourth gear. The intermediate clutch is released by several small coil springs.

The intermediate one-way clutch locks and prevents the reverse sun gear from turning counterclockwise during acceleration in second gear. During deceleration, the intermediate one-way clutch freewheels and prevents engine braking. The intermediate one-way clutch is also locked during first gear operation.

The low one-way clutch locks and prevents the planet carrier from turning counterclockwise while operating in the drive range and accelerating in first gear. The low one-way clutch is also locked during manual low operation, but does not provide engine braking.

The low/reverse band is applied during manual low and reverse gear operation. It is also applied in second gear to provide engine braking when a downshift to manual low is made at high speeds. The low/reverse band holds the planet carrier stationary. The overdrive band holds the reverse sun gear to provide for an overdrive condition in fourth gear.

Apply Devices

The low/reverse and the overdrive servos are located in bores in the bottom of the transmission case. Neither band is adjustable. Each is only adjusted during transmission assembly through the selection of different length apply stems.

GM's THM 180/180C Transmission

The 180/180C transmission is a three-speed automatic that uses the Ravigneaux compound planetary gear set. The THM 180 was used from 1969–1975 in the imported Opel, which was sold by Buick. The 180 was also used in the Chevrolet Chevette and Pontiac T1000 from 1977–1981. The 180C replaced the 180 for the 1982–1986 models of these same cars. The 180 and 180C are identical except the 180C is equipped with a lockup torque converter. The 180C is currently used in the Geo Tracker.

Some parts catalogs refer to the THM 180 as the MD-3 transmission and the 180C as the MD-2.

These transmissions are built in Strasbourg, France, and have a separate casting for the bell housing and the rear extension housing. They have three multiple-disc clutches, referred to as the second, third, and reverse, or the intermediate/high, direct, and reverse clutches; one band and a single one-way (sprag) clutch provide for the various gear ratios. Each clutch is hydraulically applied and released by several small coil springs when hydraulic pressure is routed away from the bore of the clutch's piston. This General Motors transmission uses a gear-type oil pump.

The new designation for the 180/180C is the 3L30 transmission.

The planet carrier is the output member of the gear set. The power flow through the gears is identical to that of the Ford ATX, except the location of the planetaries is reversed. See the figure for clutch and band applications throughout the operating gear ranges (Figure 7-56).

Input Devices

The one-way, intermediate/high, and direct clutches serve as the input devices for the gear set. The direct clutch drum and intermediate/high clutch hub are welded to the end of the input shaft. The entire assembly rotates clockwise with the turbine of the torque converter.

The intermediate/high clutch is applied in second and third gears. The inside edges of the intermediate/high friction discs are splined to the outside of the intermediate/high clutch hub. The outside edges of the intermediate/high clutch steel discs are splined to the inside of the clutch drum (Figure 7-57). The ring gear is splined to the inside of the intermediate/high clutch drum; therefore, the ring gear turns at input speed whenever the intermediate/high clutch is applied.

The direct clutch is called the third clutch.

The direct clutch is applied in third gear, manual low, and reverse gear. The outside of the direct clutch steel discs are splined to the inside of the direct clutch drum and the inside edges of the friction discs are splined to the direct clutch hub (Figure 7-58). The primary sun gear is driven clockwise at turbine speed whenever the direct clutch is applied.

GEAR SELECTOR POSITION	OPERATING GEAR	ONE-WAY CLUTCH	INTERMEDIATE/ HIGH CLUTCH	DIRECT CLUTCH	REVERSE CLUTCH	LOW BAND
Drive	1st gear	X				X
	2nd gear		X			X
	3rd gear	X	X	X		
2	2nd gear		X			X
Low	1st gear	X		X		X
R	Reverse	X		X	X	

Figure 7-56 Clutch and band application chart for a THM 180 transmission.

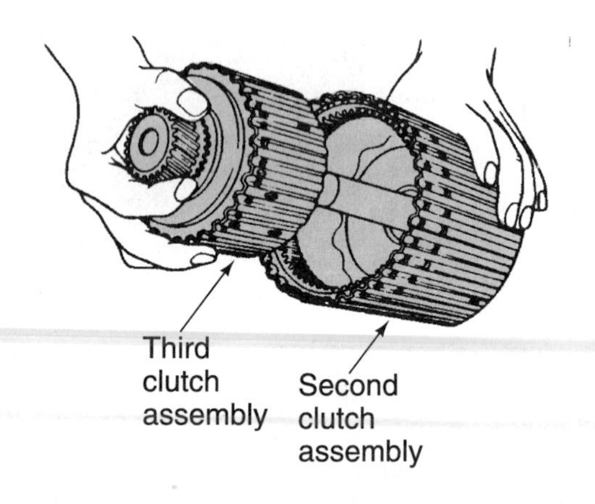

Figure 7-57 The third (direct) clutch assembly is splined to the second (intermediate/high) clutch assembly. (Courtesy of the Hydra-Matic Division of General Motors Corp.)

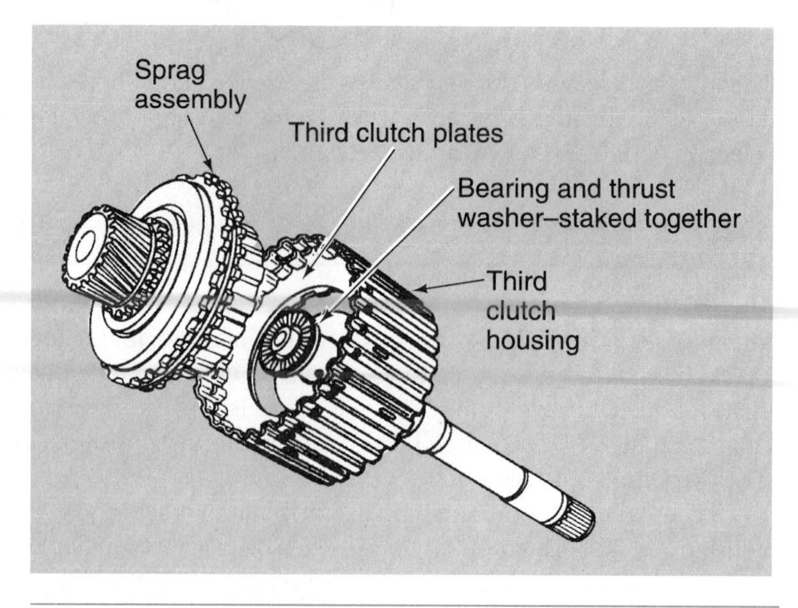

Figure 7-58 The third clutch assembly. (Courtesy of the Hydra-Matic Division of General Motors Corp.)

The one-way clutch is locked during first-gear operation. The outer race of the one-way clutch is splined to the inside of the direct clutch drum, and its inner race is part of the primary sun gear assembly. The primary sun gear is driven clockwise and at turbine speed whenever the direct clutch drum rotates. The one-way clutch also provides input during manual low and reverse gear operation if the direct clutch fails, but it cannot provide for engine braking during deceleration or coasting.

Holding Devices

The low/intermediate band and reverse clutch serve as the holding devices for the gear train. The low/intermediate band is applied in first and second gears, whereas the reverse clutch is applied in reverse only.

Apply Devices

The low/intermediate servo used in the 180/180C is similar to the servos used in the 125, 200, and 325 series transmissions. It serves two purposes: it applies the low/intermediate band and acts as an accumulator for the direct clutch during upshifts from second to third gear. During first to second gear shifts, hydraulic pressure causes the band to tighten. The band is released when pressure is exhausted from the servo. During upshifts from second to third gear, the apply pressure for the direct clutch is routed to the spring side of the servo piston and the servo acts as an accumulator to smooth the application of the clutch.

The low/intermediate band is not adjustable; rather, band adjustment is controlled by selecting a servo apply pin of the correct length during the assembly of the transmission.

Other Ravigneaux-Based Transmissions

Although the Simpson gear train is the most common compound planetary gear train, the Ravigneaux design has been and is being used by some manufacturers. Perhaps the most unusual of the Ravigneaux-based transmissions are the Mitsubishi KM 170/171/172 transaxles.

These transaxles are used in a variety of Mitsubishi vehicles, as well as some Chrysler and Hyundai models. These transaxles are three-speed units, and some are equipped with a lockup torque converter.

GEAR SELECTOR POSITION	OPERATING GEAR	DIRECT/ REVERSE CLUTCH	FRONT CLUTCH	SECOND GEAR CLUTCH	LOW/ REVERSE BAND	ONE-WAY CLUTCH
P– Park	None					
R– Reverse	Reverse		X		X	
N– Neutral	None					
D– Drive	1st gear	X				X
	2nd gear	X		X		
	3rd gear	X	X			
2– Man. 2nd	1st gear	X				X
	2nd gear	X		X		
1– Man. 1st	1st gear	X			X	X

Figure 7-59 Clutch and band application chart for KM 170/171/172 transmissions.

The unique feature of these transaxles does not lie in the arrangement or control of the gear train, rather in its physical location in the vehicle. These transaxles mount on the passenger side with the engine mounted on the driver's side of the vehicle. This arrangement requires the output to be reversed before it reaches the drive wheels. This is accomplished by an idler gear placed between the output gear and the final drive input gear. This arrangement of gears is referred to as the **transfer assembly** of the transaxle.

The complete transaxle assembly is contained in a single housing and consists of a torque converter, transmission, transfer assembly, and differential. The transaxle uses three multiple-disc clutches: a front clutch, **direct/reverse clutch,** and low/reverse clutch; a second gear band; and a one-way clutch to provide the various gear ranges. Refer to the clutch and band application chart (Figure 7-59) for details.

Planetary Gear Sets in Tandem

Rather than relying on the use of a compound gear set, some automatic transmissions use two simple planetary units in series. In this type of arrangement, gear set members are not shared. Instead, the holding devices are used to lock different members of the planetary units together. The front planet carrier is normally locked to the rear ring gear and the front ring gear is locked to the rear planet carrier. Two common transaxles use this planetary gear arrangement: the THM 440-T4 and the Ford AXOD. Both of these will be covered in detail, although they are similar in design and power flow.

Shop Manual
Chapter 7, page 283

Some parts catalogs may refer to the 440-T4 as the ME-9 transaxle.

The new GM designation for the 440-T4 is the 4T60 transaxle.

There was a special heavy-duty version of the 440-T4 made for the Cadillac Allanté; some parts catalogs refer to this as the F-7.

THM 440-T4 Transaxle

The THM 440-T4 was the first domestically produced four-speed automatic transaxle built for FWD vehicles. The 440-T4 was introduced in 1984 and has been used in nearly all of GM's intermediate and full-size cars. Overdrive fourth gear is obtained by using the same planetary units as are used for the first three speeds, instead of adding a planetary unit just for overdrive.

Although the gear train is based on two simple planetary gear sets operating in **tandem,** the combination of the two planetary units does function much like a compound unit. The two tandem units do not share a common member; rather, certain members are locked together or are integral with each other. The front planet carrier is locked to the rear ring gear and the front ring gear is locked to the rear planet carrier. The transaxle houses a third planetary unit, which is used only as the final drive unit and not for overdrive.

The 440-T4 uses a variable displacement vane-type oil pump and four multiple-disc clutches, two bands, and two one-way clutches to provide the various gear ranges. One of the one-way clutches is a roller clutch, the other is a sprag. The four multiple-disc clutches are released by several small coil springs when hydraulic pressure is diverted from the clutch's piston.

Input Devices

Engine torque is transferred to the transaxle through the torque converter and a sprocket-and-drive link assembly (Figure 7-60). The torque converter does not directly drive the gear train of the transaxle. The converter drives a sprocket that is linked by a drive chain to an input sprocket at the gear train. This arrangement allows the gear train to be positioned away from the centerline of the engine's crankshaft.

Input through the gear train is controlled by the **input clutch,** input one-way clutch, and second clutch (Figure 7-61). The input clutch is applied in all gears except third and fourth. The input one-way clutch locks in first, manual third, and reverse gears and overruns in second gear. The input clutch and the input one-way clutch drive the front sun gear.

The second clutch is applied in second, third, and fourth gears and drives the front planet carrier and the rear ring gear, which is locked to the front carrier.

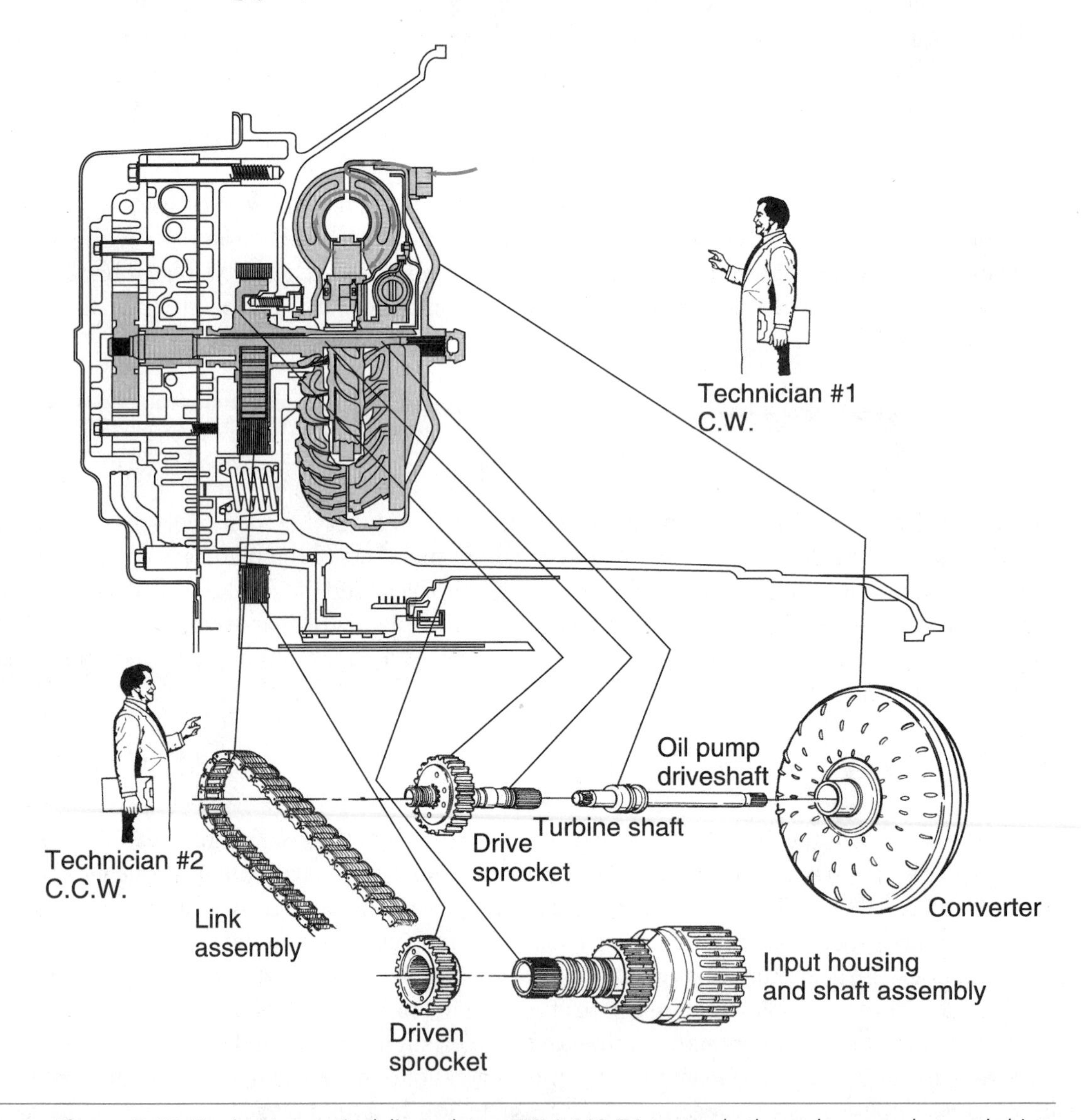

Figure 7-60 Engine torque is delivered to a THM 440-T4 transaxle through a sprocket-and-drive link assembly. (Courtesy of the Hydra-Matic Division of General Motors Corp.)

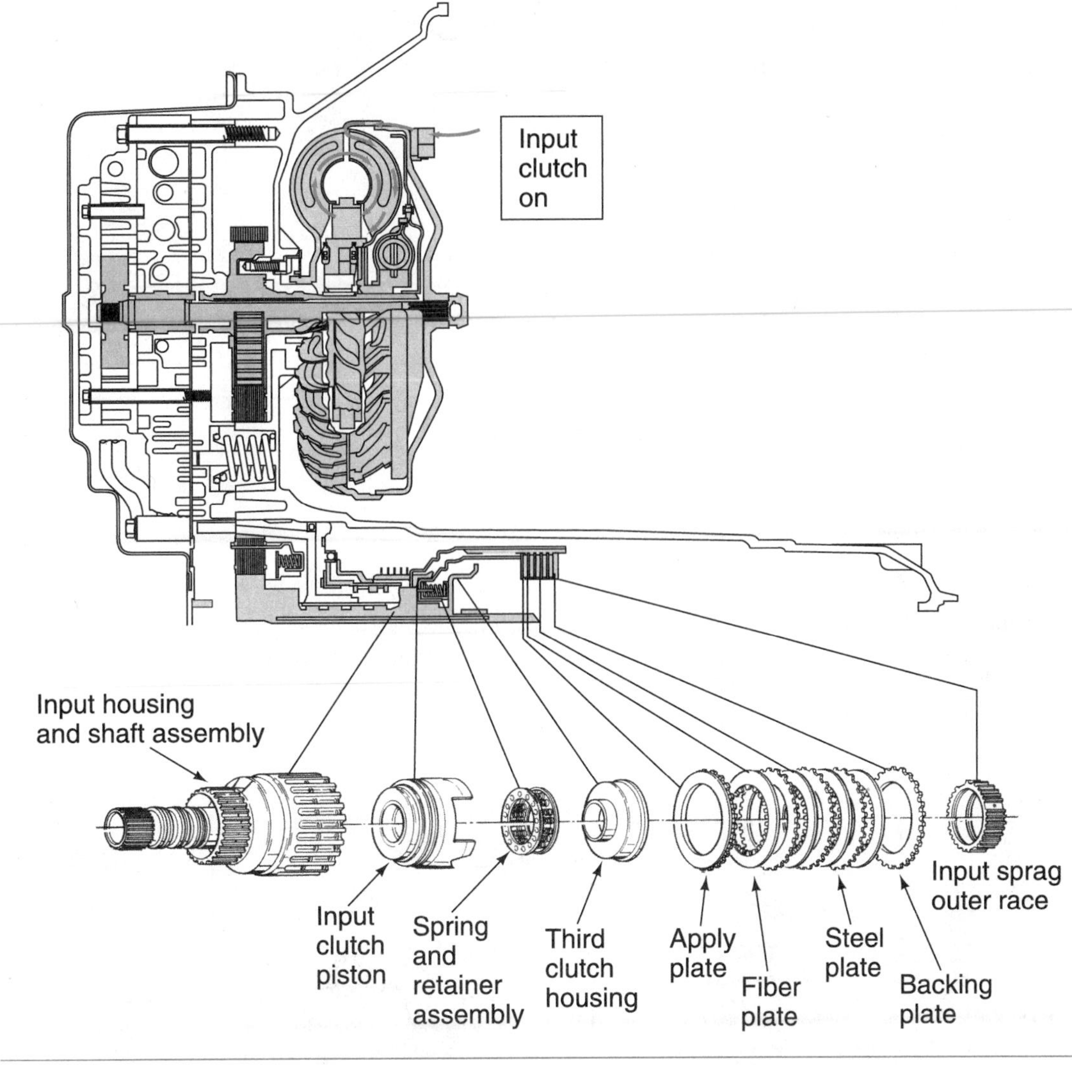

Figure 7-61 Input through the gear train is controlled by the direct clutch, direct one-way clutch, and intermediate/high clutch. (Courtesy of the Hydra-Matic Division of General Motors Corp.)

Holding Devices

The fourth clutch, 1–2 band, and reverse band are the holding devices for the gear train. The third clutch is applied in third, fourth, and manual low gears. The third roller clutch is locked in third gear and in the manual low position. The third roller clutch overruns in fourth gear. Both the third clutch and third roller clutch prevent the drive wheels from rotating the front sun gear at a speed faster than input shaft speed.

The fourth clutch holds the front sun gear and is applied in fourth gear. The 1–2 band is applied in first and second gears, whereas the reverse band is applied only in reverse.

Apply Devices

Like other GM transmissions, the 440-T4 uses servos fitted with two pistons. These servos not only apply the bands but also act as an accumulator for the application of the clutch involved in upshifting. Band adjustment is controlled by the length of the servo rod or apply pin, which is selected during transaxle assembly. External adjustment is possible after assembly, but it requires the use of a special tool that helps select the proper length rod or apply pin.

Refer to the clutch and band application chart (Figure 7-62) while reading through the power flow of this transmission.

GEAR SELECTOR POSITION	OPERATING GEAR	4TH CLUTCH	REVERSE BAND	2ND CLUTCH	3RD CLUTCH	3RD ROLLER CLUTCH	INPUT CLUTCH	INPUT SPRAG CLUTCH	1–2 BAND
D	1st gear						X	X	X
	2nd gear			X			X		X
	3rd gear			X	X	X			
	Overdrive	X		X	X				
3	3rd gear			X	X	X	X	X	
2	2nd gear			X			X		X
Low	1st gear				X	X	X	X	X
R	Reverse		X				X	X	

Figure 7-62 Clutch and band application chart for a THM 440-T4 (4T60) transaxle.

Power Flow

Refer to the band and clutch application chart while going through the power flow.

While operating in first gear, the input is received through the front sun gear, which is locked to and drives the rear ring gear. The 1–2 band is applied and holds the rear sun gear. This forces the rear carrier to be driven by the rear ring gear at a speed reduction and a ratio of 2.92:1 (Figure 7-63).

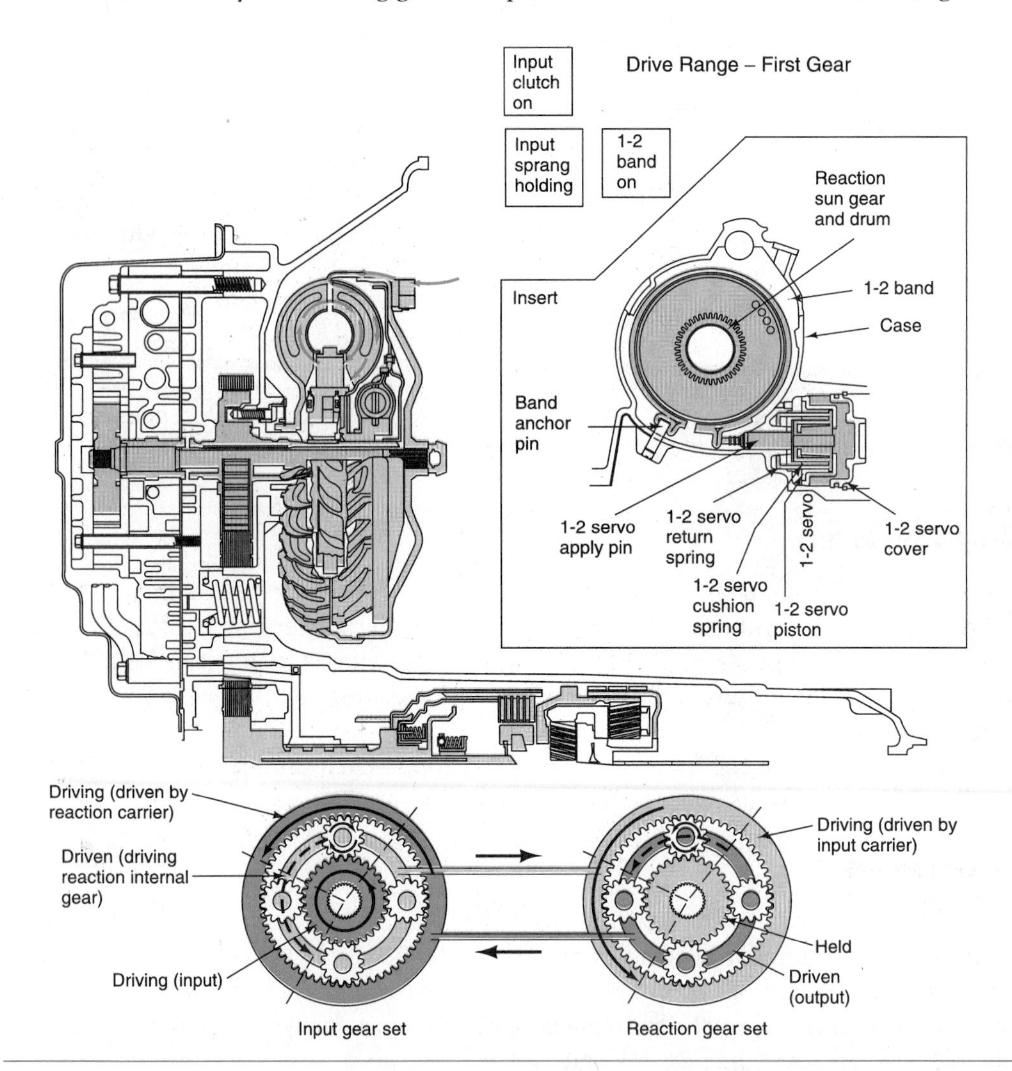

Figure 7-63 Power flow through first gear. (Courtesy of the Hydra-Matic Division of General Motors Corp.)

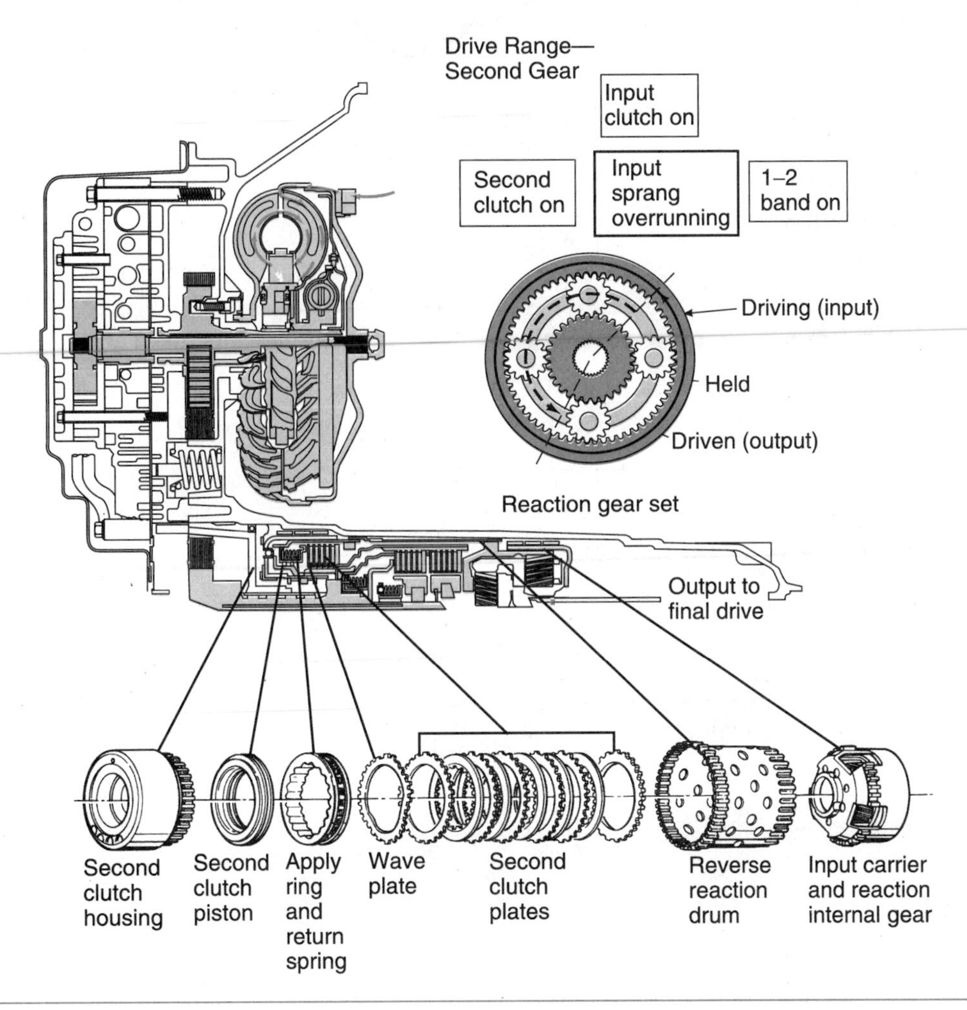

Figure 7-64 Power flow through second gear. (Courtesy of the Hydra-Matic Division of General Motors Corp.)

Input for second-gear operation is received by the front planet carrier when the second clutch is applied. Since the rear ring gear is an integral part of the front planet carrier, the two rotate at the same speed and in the same direction. The rear sun gear is held stationary by the one-way clutch; therefore, the rear planet carrier rotates around it at a gear reduction ratio of 1.57:1 (Figure 7-64).

Third gear offers direct drive. Input is received by the gear train through the front sun gear by the application of the third clutch. The second clutch is still applied and directs input through the front planet carrier. Since two of the three planetary members are rotating at the same speed, the ring gear is forced to rotate with them. The ring gear drives the rear planet carrier at the same speed and direct drive results (Figure 7-65).

In fourth gear, the second clutch remains applied and input is received through the front planet carrier. The fourth clutch is also applied and it holds the front sun gear. This causes the front ring gear to rotate around the carrier at a speed greater than input speed. Since the rear planet carrier is connected to the front ring gear, the output from the gear sets is an overdrive with a ratio of 0.70:1 (Figure 7-66).

Reverse operation is achieved by holding the front planet carrier with the reverse band. With the carrier held, the input on the front sun gear causes the front ring gear to rotate in the opposite direction. Because the front ring gear is locked to the rear carrier, the output of this gear combination is in the opposite direction to the input at a ratio of 2.38:1 (Figure 7-67).

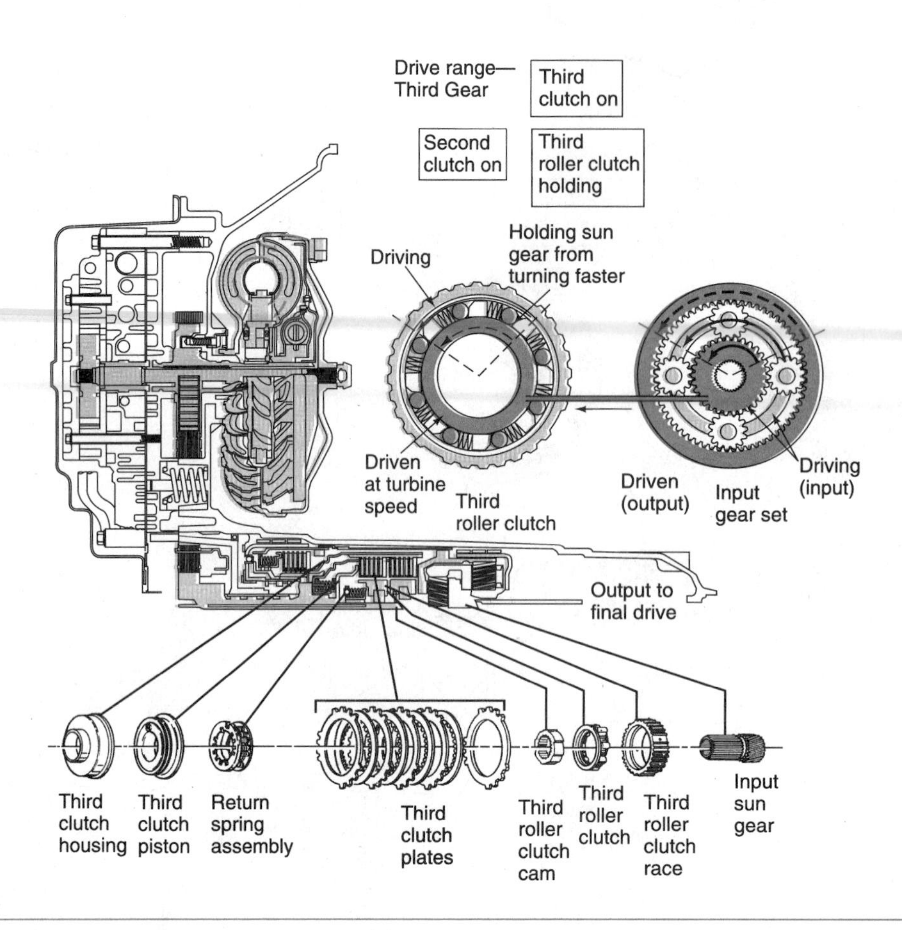

Figure 7-65 Power flow through third gear. (Courtesy of the Hydra-Matic Division of General Motors Corp.)

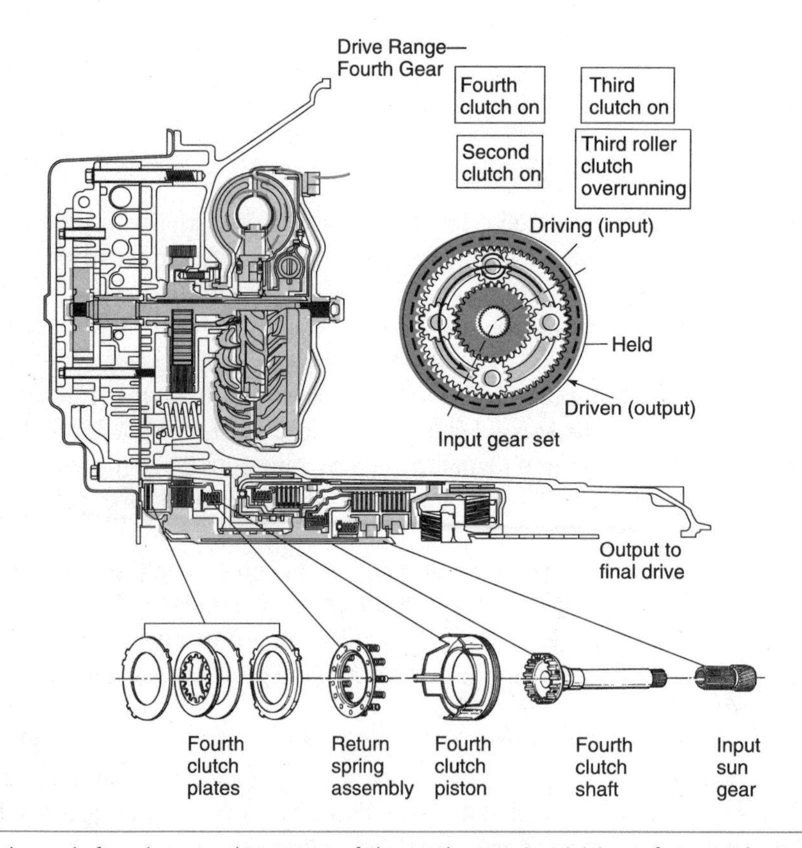

Figure 7-66 Power flow through fourth gear. (Courtesy of the Hydra-Matic Division of General Motors Corp.)

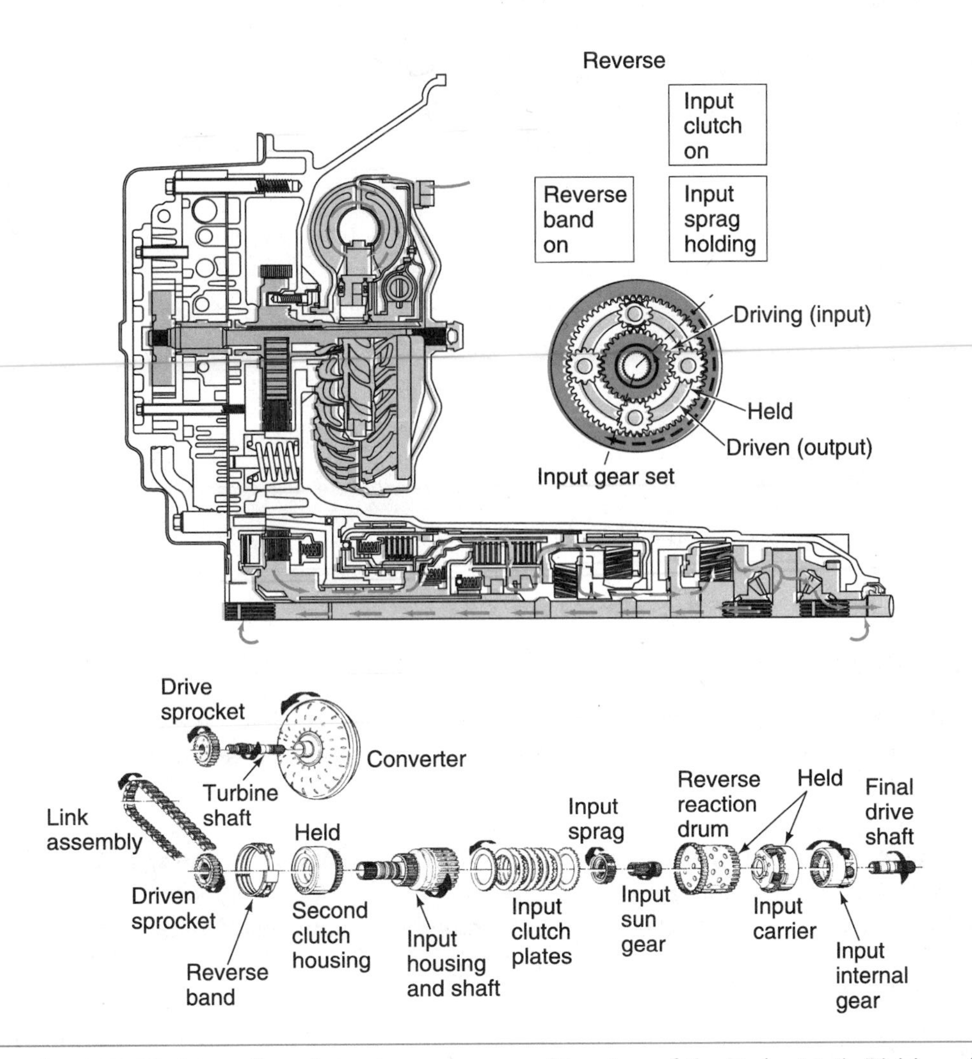

Figure 7-67 Power flow through reverse gear. (Courtesy of the Hydra-Matic Division of General Motors Corp.)

Ford AXOD Transaxle

Shop Manual
Chapter 7, page 260

The AXOD four-speed transaxle was introduced in 1986 on Taurus/Sable models with the V-6 engine. Since then, its application has expanded. This transaxle doesn't use a Ravigneaux or a Simpson gear set. It relies on two simple planetary units that operate in tandem. The planet carriers of each planetary unit are locked to the other planetary unit's ring gear. Each planetary set has its own sun gear and set of planetary pinion gears. There are no transmissions used by Ford that have any similarity to the AXOD. The similarity lies with the construction and operation of the THM 440-T4.

The AXOD combines an automatic transmission and differential into a single unit designed for FWD vehicles. The transmission is a fully automatic unit with four forward speeds, reverse, neutral, and park. The AXOD uses four multiple-disc clutches, two band assemblies, and two one-way clutches to control the operation of the planetary gear set. The planetary units provide a 2.77:1 first gear, a 1.54:1 second gear, a direct drive third gear, and a 0.69:1 fourth gear. Output from the planetary units is through the rear planet carrier and front ring gear.

All the multiple-disc clutches are hydraulically applied and released by several small coil springs when hydraulic pressure is diverted from the clutch's piston. For more insight into the operation of the AXOD, refer to the clutch and band application chart (Figure 7-68).

GEAR SELECTOR POSITION	OPERATING GEAR	LOW/ INTERMEDIATE BAND	OVERDRIVE BAND	FORWARD CLUTCH	INTER-MEDIATE CLUTCH	DIRECT CLUTCH	REVERSE CLUTCH	LOW ONE-WAY CLUTCH	DIRECT ONE-WAY CLUTCH
OD	1st gear	x		x				x	
	2nd gear	x		x	x				
	3rd gear			x	x	x			x
	Overdrive		x		x	x			x
D	1st gear	x		x				x	
	2nd gear	x		x	x			x	
	3rd gear			x	x	x			
Low	1st gear	x		x		x		x	x
R	Reverse			x			x	x	

Figure 7-68 Clutch and band application chart for a Ford AXOD transaxle.

Input Devices

Like the 440-T4, the AXOD torque converter does not directly drive the gear train of the transaxle. The converter drives a sprocket that is connected by a chain to a sprocket on the input shaft of the gear train.

Input to the gear train is received by the front sun gear, the front planet carrier, or the rear ring gear. The forward clutch, low one-way clutch, direct clutch, direct one-way clutch, and intermediate clutch serve as the input devices.

The low one-way clutch is locked by the action of the forward clutch in low, second, third, and reverse gears. The one-way clutch is effective only in reverse and low, because it freewheels in second and third gears.

The **direct one-way clutch** is locked by the action of the direct clutch in third and fourth gears and when the transmission is operating in manual low. The one-way clutch is effective only in third gear and manual low. It freewheels in fourth gear. In third gear, the direct one-way clutch transmits input torque from the direct clutch to the front sun gear. Also in third gear, the direct clutch drives the front sun gear through the direct one-way clutch and locks the gear train to provide for engine braking.

The intermediate clutch is applied in second, third, and fourth gears and drives the front planet carrier and rear ring gear in second and third gears.

Holding Devices

The low/intermediate band holds the rear sun gear stationary in first and second gears. The overdrive band is applied in fourth gear and holds the front sun gear. The reverse clutch holds the front planet carrier and rear ring gear for reverse gear.

Apply Devices

The AXOD has two servos: one to operate the low/intermediate band and the other to operate the overdrive band. Both servo units consist of a cover, a piston, an apply rod, and a return spring. The low/intermediate servo also acts as an accumulator for second to third gear shifting. The hydraulic pressure that is sent to apply the direct clutch is routed into the release side of the low/intermediate servo. This causes a delay in the application of the direct clutch and smooths the upshift. Both AXOD bands can be adjusted internally by substituting different length apply rods.

Ford's CD4E Transaxle

The CD4E transaxle is used primarily in Ford Contour and Mercury Mystique model cars. This transaxle is similar in design to the AXOD-E. It is a four-speed unit with electronic control. It uses two simple planetary gear sets and a planetary final drive gear set. A chain drive is used to connect the output of the planetary gears to the final drive.

The transaxle uses five multiple-disc clutches, two one-way clutches, and a band. They may be equipped with a transmission control switch or a manual hold switch. Depending on how the transaxle is equipped, the labels on the gear selector may vary from those shown in Figure 7-69. The type of switch also determines the availability of certain gears during manual shift operations, the function of holding members during coast, and the electronic control system. This transaxle is fitted with a lockup torque converter controlled by modulated pressure.

Unique features of this transaxle include an oil filter that can be accessed only when the transaxle is disassembled. It also has a thermostatic valve that allows for increased fluid storage in the main control cover when the fluid is at normal operating temperature. This means that fluid checks must be taken when the transaxle is hot. There is also a two-piece chain cover that prevents the rotating chain from aerating the oil in the sump.

The main control assembly has no check balls. However it does contain all the transaxle's valves, accumulators, and solenoids. The solenoid body contains a built-in filtering screen. There are five solenoids and one sensor connected to the main valve body. The electronic pressure control (EPC) solenoid varies hydraulic pressure in response to directions from the PCM. There are two shift solenoids and a 3–2 T/CCS (3–2 timing/coast clutch solenoid). This solenoid varies pressure to two valves. One valve controls the apply of the coast clutch while the other regulates the synchronous release of the direct clutch and apply of the intermediate/overdrive band during 3–2 downshifts. The TCC solenoid is also mounted to the main valve body assembly. The engagement of the TCC is controlled by the PCM, which varies the pulse width of the solenoid.

Three accumulators are used to cushion engagements into reverse and drive, as well as the application of the intermediate/overdrive band during 1–2 and 3–4 upshifts. The intermediate/overdrive band accumulator is not used during 3–2 downshifts since the PCM controls the timing of the shift.

These transaxles have a fail-safe mode that is used only when the PCM detects an electronic problem. During this mode, the transaxle will have maximum pressure in all gear selector positions, coast braking in third gear, and the TCC will be released.

GEAR SELECTOR POSITION	OPERATING GEAR	INTERMEDIATE OVERDRIVE BAND	REVERSE CLUTCH	DIRECT CLUTCH	FORWARD CLUTCH	FORWARD ONE-WAY CLUTCH		COAST CLUTCH	LOW/ REVERSE CLUTCH	LOW ONE-WAY CLUTCH	
						DRIVE	COAST			DRIVE	COAST
D	1st gear		X		X		Overrunning			X	
	2nd gear	X			X		Overrunning			Overrunning	Overrunning
	3rd gear			X	X		Overrunning			Overrunning	Overrunning
	4th gear	X		X	X	Overrunning	Overrunning			Overrunning	Overrunning
S	1st gear				X			X		X	Overrunning
	2nd gear	X			X			X		Overrunning	Overrunning
	3rd gear			X	X			X		Overrunning	Overrunning
L	1st gear				X			X	X	X	
R	Reverse								X		

Figure 7-69 Band and clutch application chart for a Ford CD4E transaxle.

Honda's Nonplanetary-Based Transmission

Shop Manual
Chapter 7, page 289

The Honda CA, F4, and G4 transaxles are used in many Honda and Acura cars. These transmissions are unique in that they do not use a planetary gear set to provide for the different gear ranges. Constant-mesh helical and square-cut gears (Figure 7-70) are used in a manner similar to that of a manual transmission.

These transaxles have a mainshaft and countershaft on which the gears ride. To provide the four forward and one reverse gears, different pairs of gears are locked to the shafts by hydraulically controlled clutches (Figure 7-71). Reverse gear is obtained through the use of a shift fork, which slides the reverse gear into position. The power flow through these transaxles is similar to that of a manual transaxle.

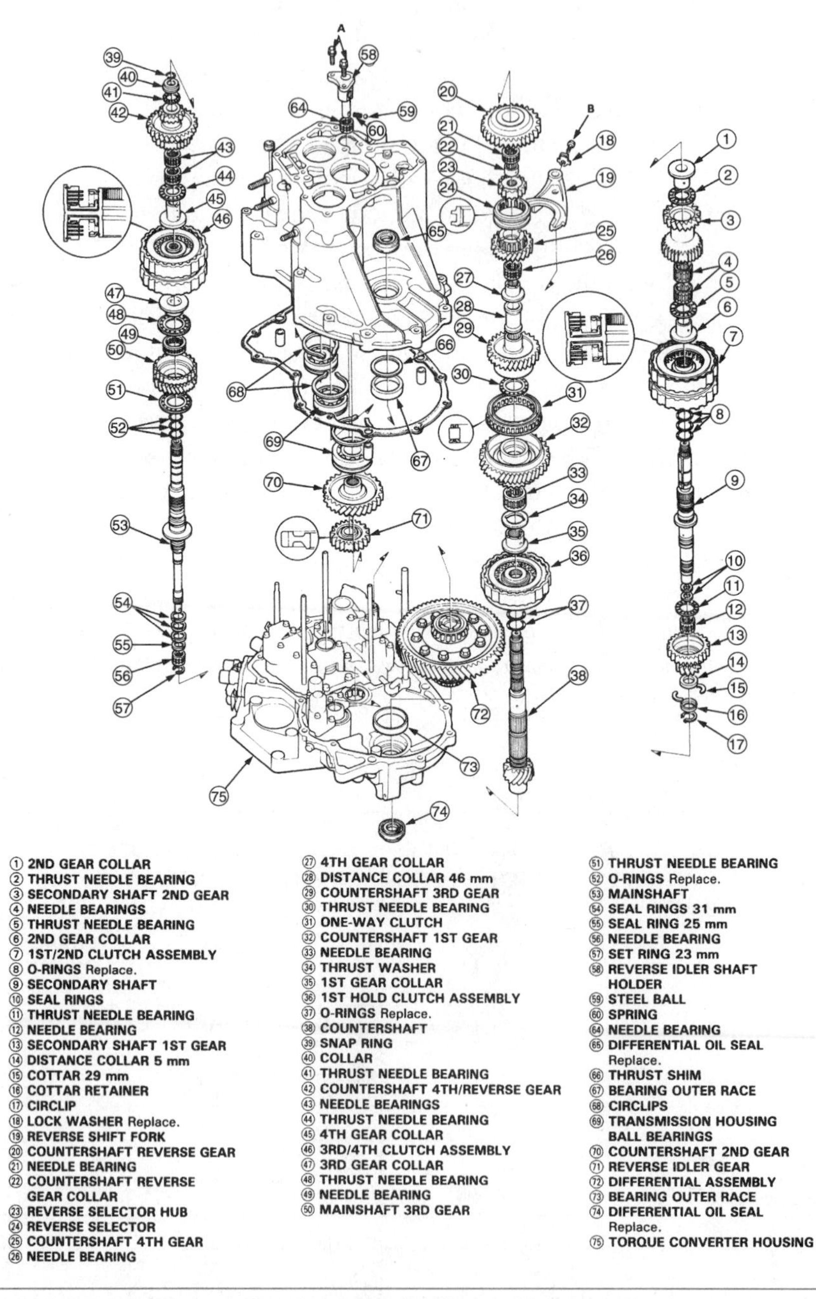

Figure 7-70 An exploded view of Honda's automatic transaxle, which uses constant-mesh helical gears instead of planetary gear sets. (Courtesy of Honda Motor Company, Ltd.)

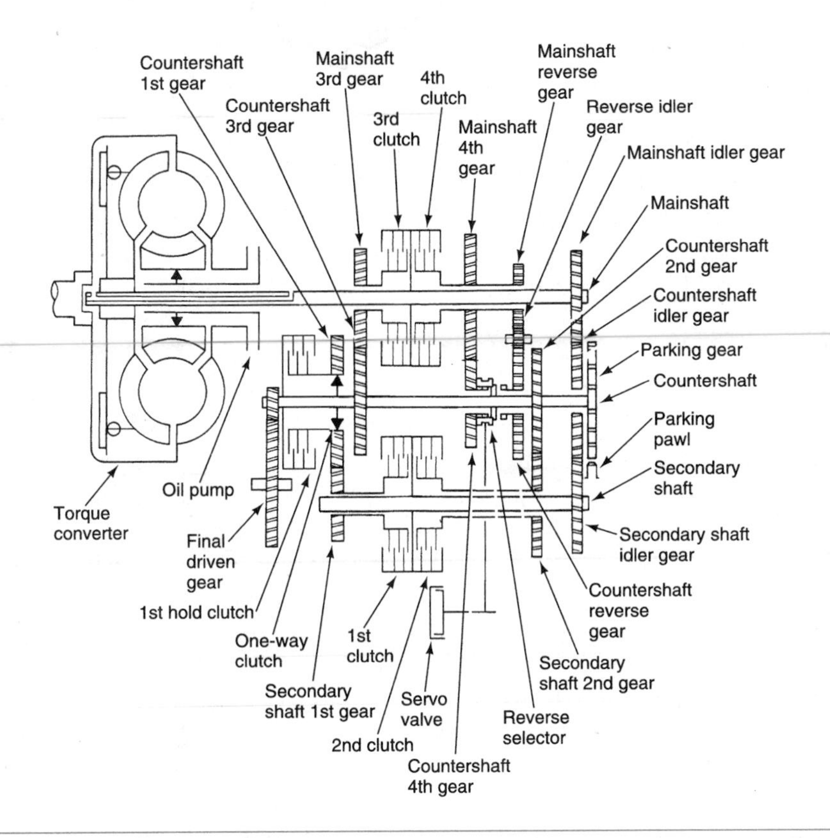

Figure 7-71 Arrangement of gears and reaction devices in a typical Honda transaxle. (Courtesy of the Honda Motor Company, Ltd.)

Four multiple-disc clutches, the sliding reverse gear, and a one-way clutch are used to control the gears. These are designated by the gear they activate: **first gear clutch, first gear one-way clutch,** second gear clutch, **third gear clutch, fourth gear clutch,** and reverse gear. Refer to the clutch and band application chart (Figure 7-72) for details on the power flow through this transaxle.

Power flow through Honda's nonplanetary transaxle is similar to that through a manual transaxle. The action of the clutches is much the same as the action of the synchronizer assemblies in a manual transmission. When Honda's transaxle is in PARK or NEUTRAL, none of the clutches is applied and no power is transmitted to the countershaft.

This transmission is typically referred to as a **dual shaft** automatic **transmission.**

GEAR SELECTOR POSITION	OPERATING GEAR	1ST CLUTCH	1ST ONE-WAY CLUTCH	2ND CLUTCH	3RD CLUTCH	4TH CLUTCH	REVERSE GEAR
D4	1st gear	X	X				
	2nd gear	X		X			
	3rd gear	X			X		
	4th gear	X				X	
D3	1st gear	X	X				
	2nd gear	X		X			
	3rd gear	X			X		
2	2nd gear			X			
R	Reverse					X	X

Figure 7-72 Clutch and band application chart for a Honda CA/F4/G4 transaxle.

When DRIVE is selected and the transaxle is in first gear, hydraulic pressure is applied to the first clutch, which rotates with the mainshaft (Figure 7-73). This causes first gear to rotate. Power is transmitted from first gear on the mainshaft to first gear on the countershaft. From there, power moves through the final drive unit to the drive wheels.

When the correct conditions are present, the transaxle shifts into second gear. To do this, hydraulic pressure is applied to the second clutch on the mainshaft (Figure 7-74). This allows for power on the second gear, which is meshed with second gear on the countershaft. Power is transmitted across the gears, driving the countershaft connected to the final drive unit.

Third gear engagement is made after the right operating conditions are met. To engage third gear, hydraulic pressure is applied to the third clutch. Power from the mainshaft is transmitted to the countershaft third gear. Hydraulic pressure is also sent to the first clutch, but since the rotational speed of third gear exceeds that of first gear, power from first gear is cut off at the one-way clutch. From the third gear on the countershaft, power is then transmitted to the final drive gear (Figure 7-75).

When engine load and vehicle speeds permit, the transmission upshifts into fourth gear. Hydraulic pressure is applied to the fourth clutch, which rotates together with the mainshaft causing the mainshaft fourth gear to rotate. Hydraulic pressure is also sent to the first clutch, but since the rotational speed of fourth gear exceeds that of first gear, power from first gear is cut off at the one-way clutch. Power is transmitted to the fourth countershaft gear, which drives the countershaft connected to the final drive unit (Figure 7-76).

When the driver selects reverse gear, hydraulic pressure is switched by the manual valve to the servo valve, which moves the reverse shift fork to the reverse position. The reverse shift fork engages with the reverse selector, reverse selector hub, and the countershaft reverse gear. Hydraulic pressure is also applied to the fourth clutch. Power is transmitted from the mainshaft reverse gear via the reverse idler gear to the countershaft reverse gear. The rotational direction of the countershaft reverse gear is changed by the reverse idler gear (Figure 7-77).

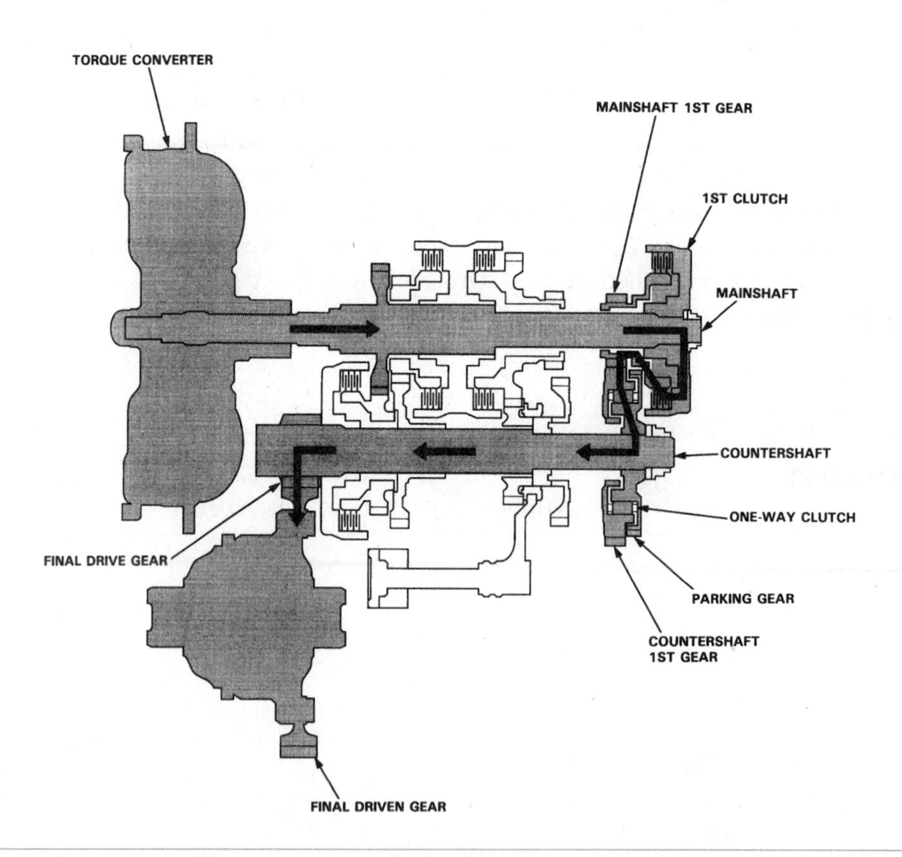

Figure 7-73 Power flow for a Honda automatic transmission in first gear. (Courtesy of Honda Motor Company, Ltd.)

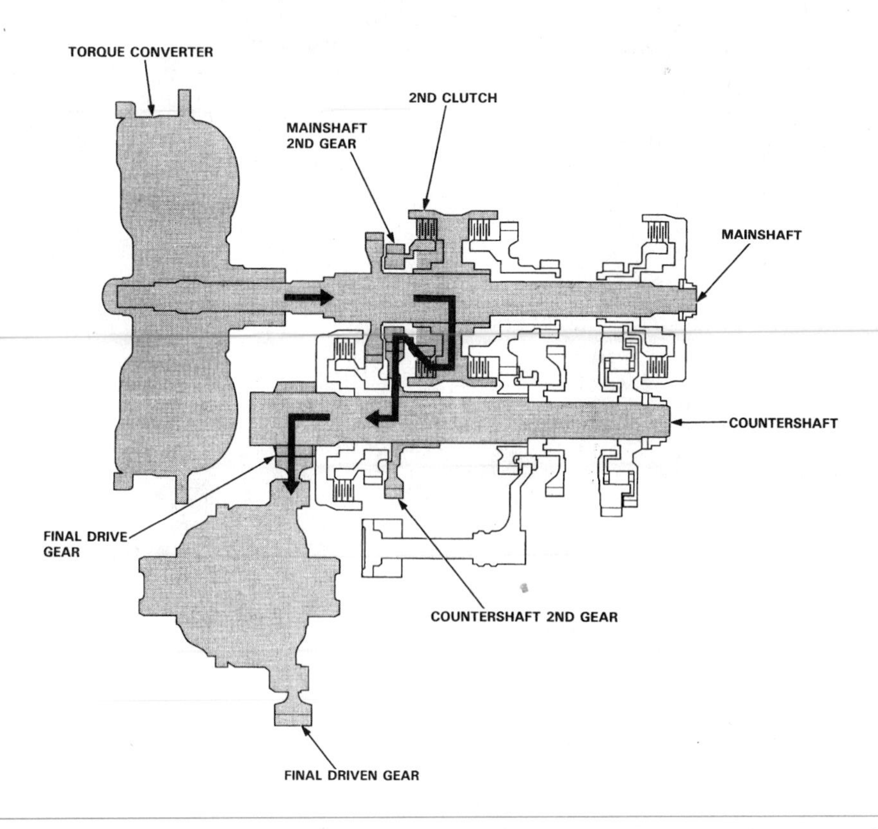

Figure 7-74 Power flow for a Honda automatic transmission in second gear. (Courtesy of Honda Motor Company, Ltd.)

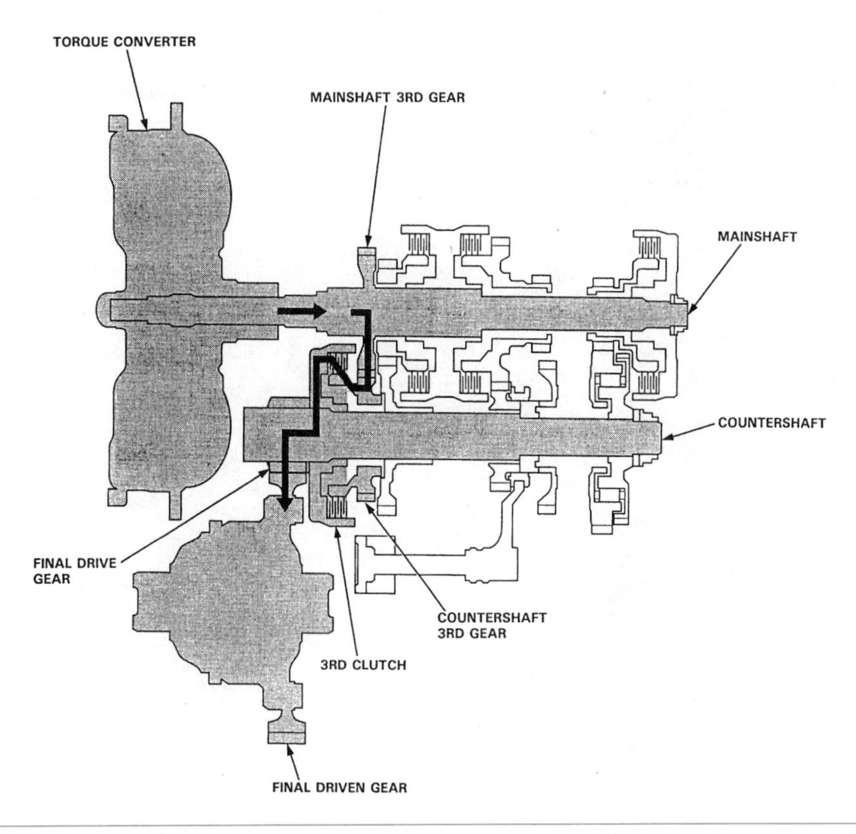

Figure 7-75 Power flow for a Honda automatic transmission in third gear. (Courtesy of Honda Motor Company, Ltd.)

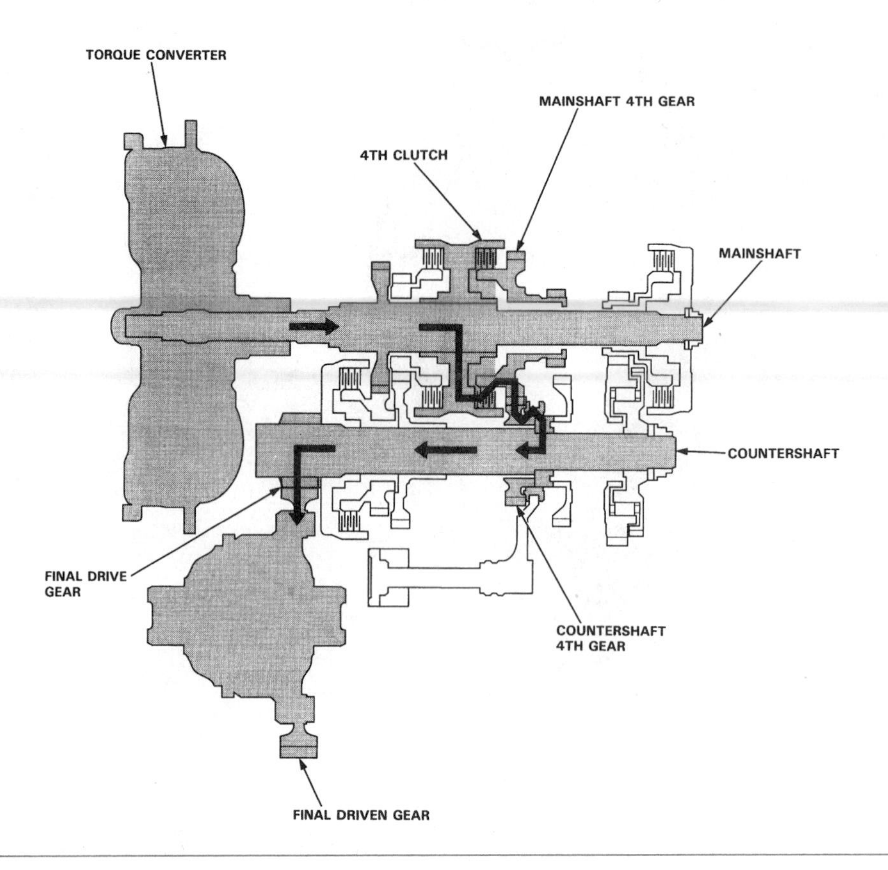

Figure 7-76 Power flow for a Honda automatic transmission in fourth gear. (Courtesy of Honda Motor Company, Ltd.)

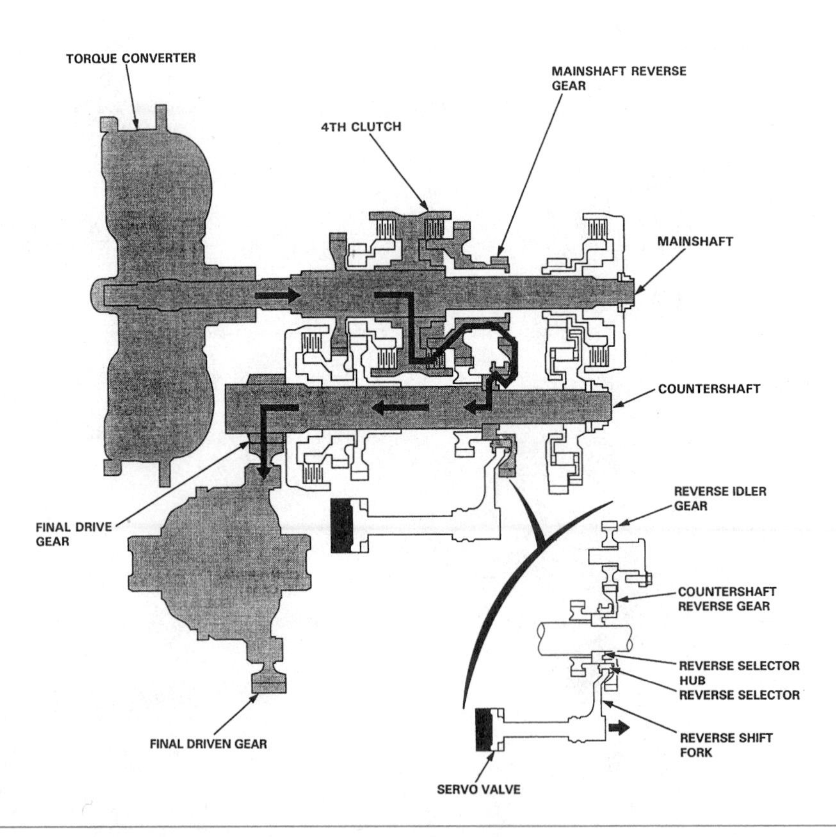

Figure 7-77 Power flow for a Honda automatic transmission in reverse. (Courtesy of Honda Motor Company, Ltd.)

Honda's Continuously Variable Transmission (CVT)

Another unconventional automatic transmission found in some late-model Honda Civics is the continuously variable transmission (CVT). A few other manufacturers, such as Subaru, have also used this transmission design. Basically, a CVT is a transmission with no fixed forward speeds. The gear ratio varies with engine speed and temperature. These transmissions are equipped with a one-speed reverse gear. The transaxles do not have a torque converter; instead, a manual transmission-type flywheel is used with a start clutch. Rather than rely on planetary or helical gear sets to provide drive ratios, CVTs are equipped with two pulleys and a steel belt. CVT transaxles do have a planetary gear set that provides neutral, forward, and reverse gear ranges (Figure 7-78).

One pulley is the driven member and the other is the drive. Each pulley has a moveable face and a fixed face. When the moveable face moves, the effective diameter of the pulley changes. The change in effective diameter changes the effective pulley (gear) ratio. The driven and drive pulleys are linked by a steel belt.

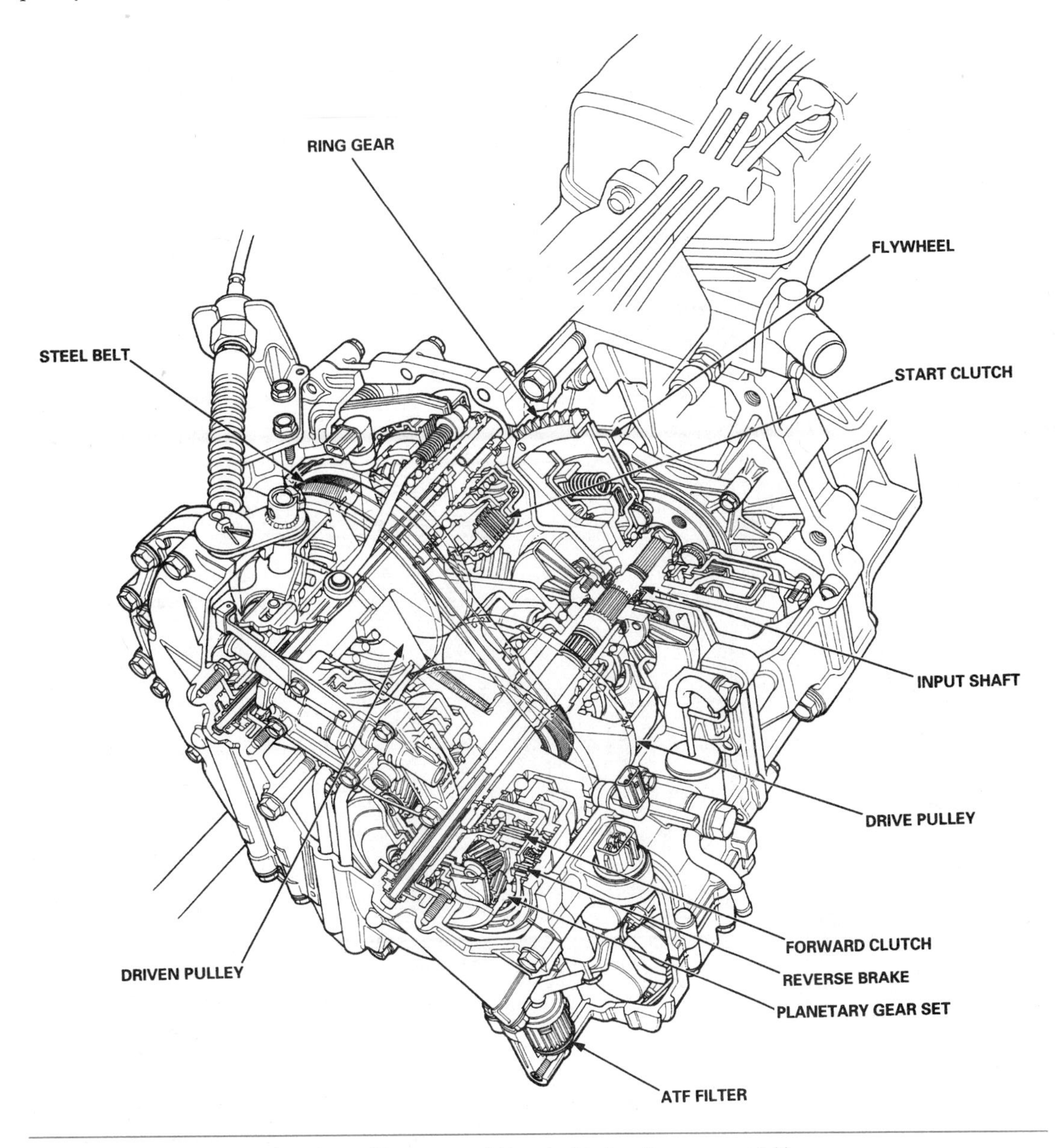

Figure 7-78 Honda's CVT. (Courtesy of Honda Motor Company, Ltd.)

To achieve a low pulley ratio, high hydraulic pressure works on the moveable face of the driven pulley to make it larger. Since the belt connects the two and proper tension of the belt is critical, a reduction of pressure at the drive pulley allows it to decrease in diameter. If the change in the drive pulley is the equivalent to the change of the driven pulley, the belt will have the correct stretch or tension. This means the increase of hydraulic pressure at the driven pulley is proportional to the decrease of pressure at the drive pulley. The opposite is true for high pulley ratios. Low hydraulic pressure causes the driven pulley to decrease in size whereas high pressure increases the size of the drive pulley.

This transaxle has four parallel shafts: the input shaft, drive pulley shaft, driven pulley shaft, and secondary shaft (Figure 7-79). The input shaft is in line with the engine's crankshaft and one end of it is the sun gear for a planetary gear set. The drive pulley shaft holds the drive pulley and

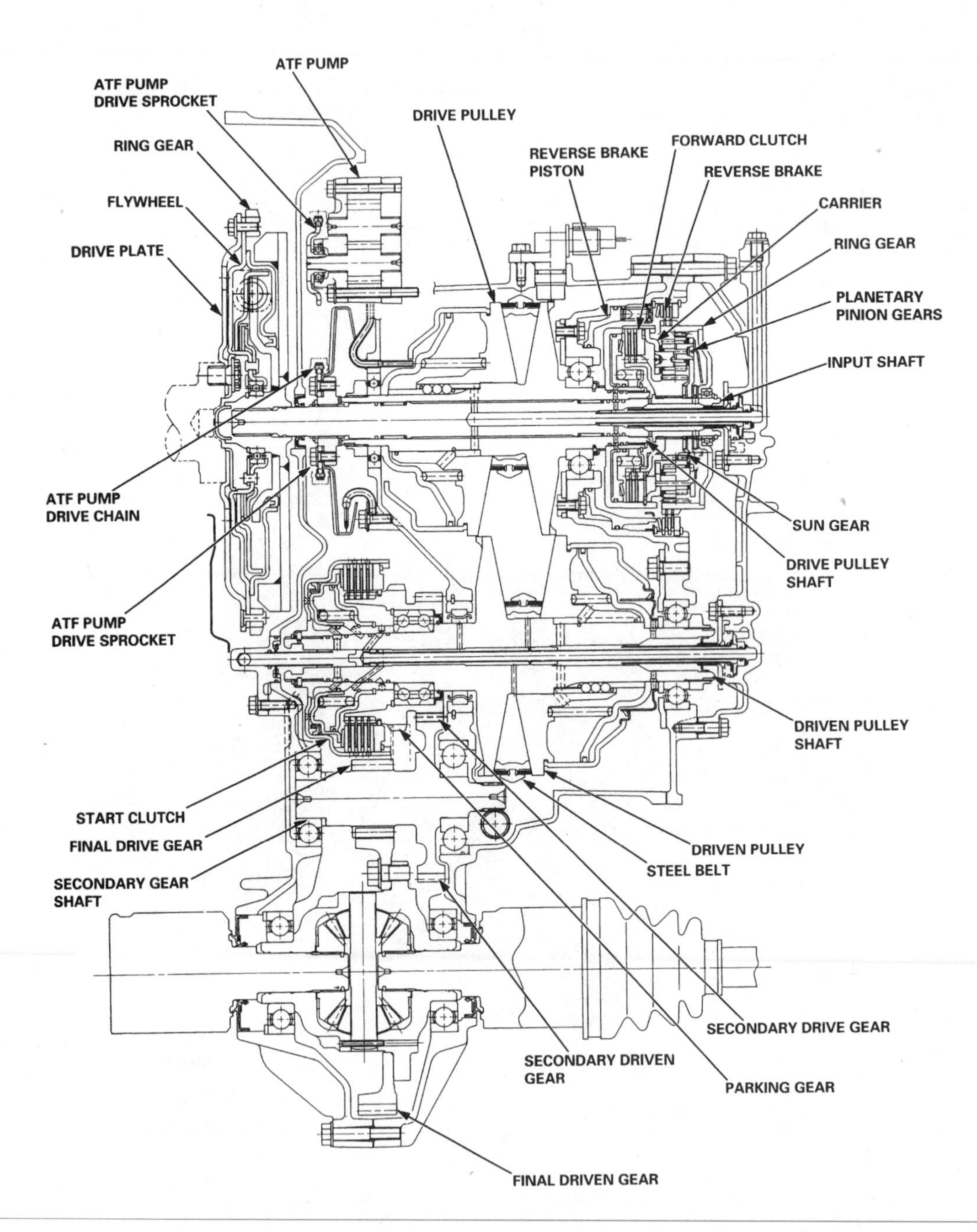

Figure 7-79 Shaft and pulley arrangement for Honda's CVT. (Courtesy of Honda Motor Company, Ltd.)

the forward clutch. The driven pulley shaft includes the start clutch and the secondary drive gear. The secondary gear shaft is located between the secondary drive gear and the final driven gear. The secondary gear shaft includes the secondary driven gear, which provides reverse gear.

The sun gear of the planetary gear set is part of the input shaft. The forward clutch on the drive pulley shaft is mounted to the carrier on the forward clutch drum. The carrier assembly includes pinion gears that mesh with the sun gear and the ring gear. The ring gear has a hub-mounted reverse brake multiple-disc assembly and is used only for switching the rotational direction of the pulley shafts. The parking gear is an integral part of the secondary drive gear.

The gear selector has six positions: PARK, REVERSE, NEUTRAL, DRIVE, SECOND, and LOW. Although these positions seem to indicate fixed gear ratios, this is not true of the forward ranges on this type of transmission. When LOW is selected, the transaxle shifts into the lowest range of available pulley ratios. This gear is intended for engine braking and power for climbing steep grades or for very heavy loads. Second gear is designed for quicker acceleration while the car is moving at highway speeds and for moderate engine braking. When this range is selected, the transaxle shifts into a lower range of pulley ratios. When DRIVE is selected, the transaxle automatically adjusts to the pulley ratios to keep the engine at the best speed for the current driving conditions. During these forward ranges, the start and forward clutches are engaged.

When the selector is in PARK, the parking pawl is engaged to the parking gear on the driven pulley shaft. This locks the front wheels. While the transaxle is in PARK and NEUTRAL, the start clutch and the forward clutch are released. When the selector is placed in REVERSE, the reverse brake clutch is engaged, as is the start clutch.

In the forward range the pinion gears do not rotate on their shafts. They revolve with the sun gear. This causes the carrier and forward clutch drum to rotate. The sun gear is driven by the input shaft. The carrier outputs the power to the drive pulley shaft. In REVERSE, the reverse clutch locks the ring gear to the housing. The pinion gears revolve around the sun gear causing the carrier to rotate in the opposite direction to the rotation of the sun gear. The start clutch engages and disengages the secondary drive gear to transfer power to the final drive unit.

Hydraulic Circuits

Honda's CVT is an electronically controlled transaxle. Shifting (or pulley ratio changes) is totally controlled by electronics. Three solenoids control the flow of hydraulic pressure, which in turn controls the action of the transaxle. The transaxle's oil pump is connected to the input shaft by sprockets and a chain. There are two valve body assemblies within the transaxle. The primary valve body, called the lower valve body, is mounted to the lower part of the transaxle case. The manual shift valve body is bolted to the intermediate housing of the transaxle.

The lower valve body includes the main valve body, the secondary valve body, the low pressure (PL) regulator valve body, the start clutch control valve body, and the shift valve body. The main valve body consists of the high pressure (PH) control valve, the lubrication valve, and the pilot regulator valve. The PH control valve supplies high control pressure to the PH regulator valve, which also regulates high pressure. The lubrication valve controls and maintains the pressure of the ATF going to the shafts. The pilot regulator valve controls the start clutch pressure in relationship to the engine speed when a fault is detected in the electronic system.

The secondary valve body contains the PH regulator valve, clutch reducing valve, start clutch valve accumulator, and shift inhibitor valve. The PH regulator valve maintains the pressure of the fluid from the oil pump and supplies fluid to the shafts and the rest of the hydraulic control circuit. The clutch-reducing valve receives high pressure from the PH regulator valve and controls the clutch-reducing pressure. The clutch-reducing valve supplies fluid to the manual valve and the start clutch control valve. It also supplies signal pressure to the PH-PL pressure control valve, shift control valve, and inhibitor solenoid valve. The start clutch valve accumulator stabilizes the hydraulic pressure applied to the start clutch. The switching of the hydraulic passages to the start clutch control when there is an electronic failure is the duty of the shift inhibitor valve. This valve also supplies clutch-reducing pressure to the pilot regulator valve and the pilot lubrication pipe.

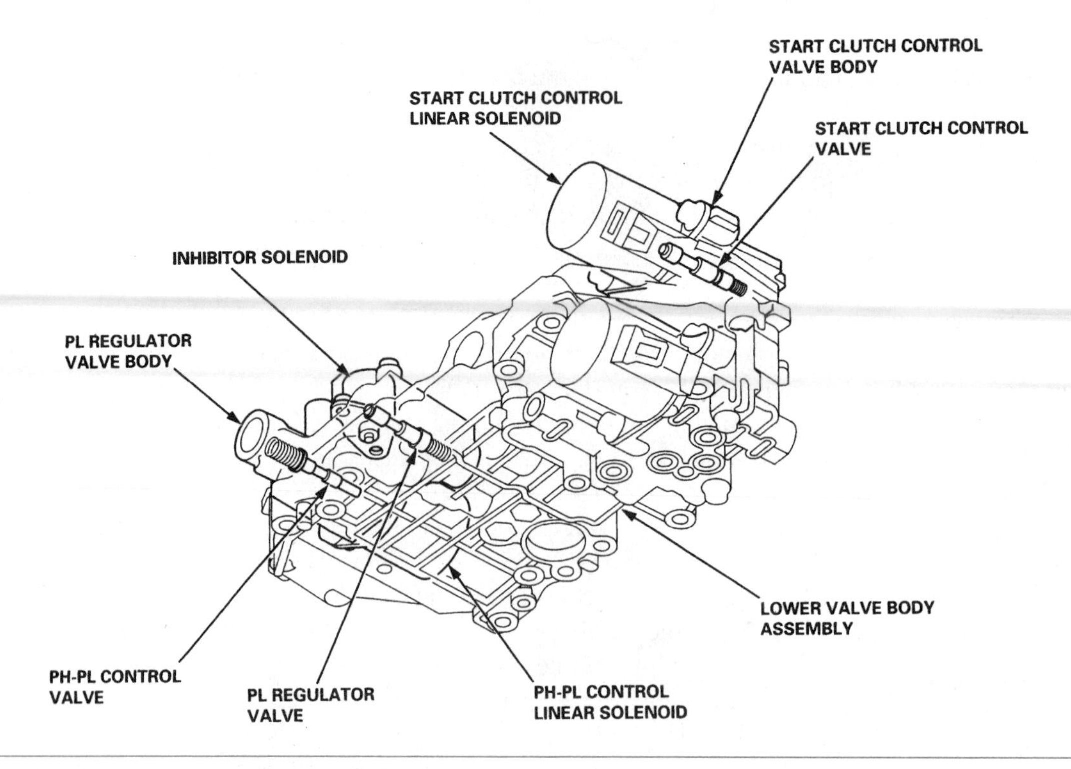

Figure 7-80 Valve body for Honda's CVT. (Courtesy of Honda Motor Company, Ltd.)

The low pressure valve body contains the PL regulator valve and the PH-PL control valve (Figure 7-80). Two solenoids are mounted to this portion of the valve body: the PH-PL control solenoid and the inhibitor solenoid. Both of these solenoids are controlled by the TCM. The PL regulator valve supplies low pressure to the pulleys when necessary. The PH-PL control valve controls the PL regulator valve according to engine torque. The PH-PL control valve supplies control pressure to the PH control valve to regulate PH higher than PL. The action of the PH-PL control valve is controlled by the PH-PL solenoid. The inhibitor solenoid controls the reverse inhibitor valve by cycling on and off. The inhibitor solenoid also controls high control pressure by applying reverse inhibitor pressure to the PH control valve.

The start clutch control valve regulates start clutch engagement according to throttle opening. The start clutch control valve is controlled by the start clutch solenoid, which is controlled by TCM. If there is an electronic failure, the start clutch control circuit switches to hydraulic controls based on engine speed.

The shift valve body contains the shift valve and the shift control valve. The shift valve is controlled by shift valve pressure from the shift control valve. The shift valve distributes high and low pressure to pulleys to change the pulley ratios. The shift control valve controls the shift valve according to vehicle speed and throttle opening. The action of the shift control valve is controlled by the shift control solenoid. When there is an electronic failure, the shift control valve switches the shift inhibitor valve to uncover the port leading to the pilot regulator pressure to the start clutch. This allows the transaxle to work in spite of electrical problems.

Power Flow

When PARK is selected, the start, forward, and reverse clutches are released. Therefore power is not transmitted to the secondary drive gear. The secondary drive gear is locked to the case by the parking pawl. Gear selector position N has the same operating conditions as P except the secondary drive gear is not locked.

When a forward range (D, S, or L) is selected, the start and forward clutch are engaged. The forward clutch drives the drive pulley shaft, which drives the driven pulley shaft via the steel belt. The pulley ratios are controlled by the hydraulic pressures applied to the moveable pulley faces according to the throttle opening. The driven pulley shaft drives the secondary drive gear through the start clutch. Power is then transferred to the final drive unit by the secondary driven gear (Figure 7-81).

When REVERSE is selected, the start and reverse clutches are engaged. Hydraulic pressure is applied to the reverse clutch, which locks the ring gear. This causes the carrier to revolve around the sun gear in an opposite direction. The carrier rotates the drive pulley shaft through the forward clutch drum. The driven pulley shaft drives the secondary gear shaft, which transmits the power to the final drive unit through the start clutch.

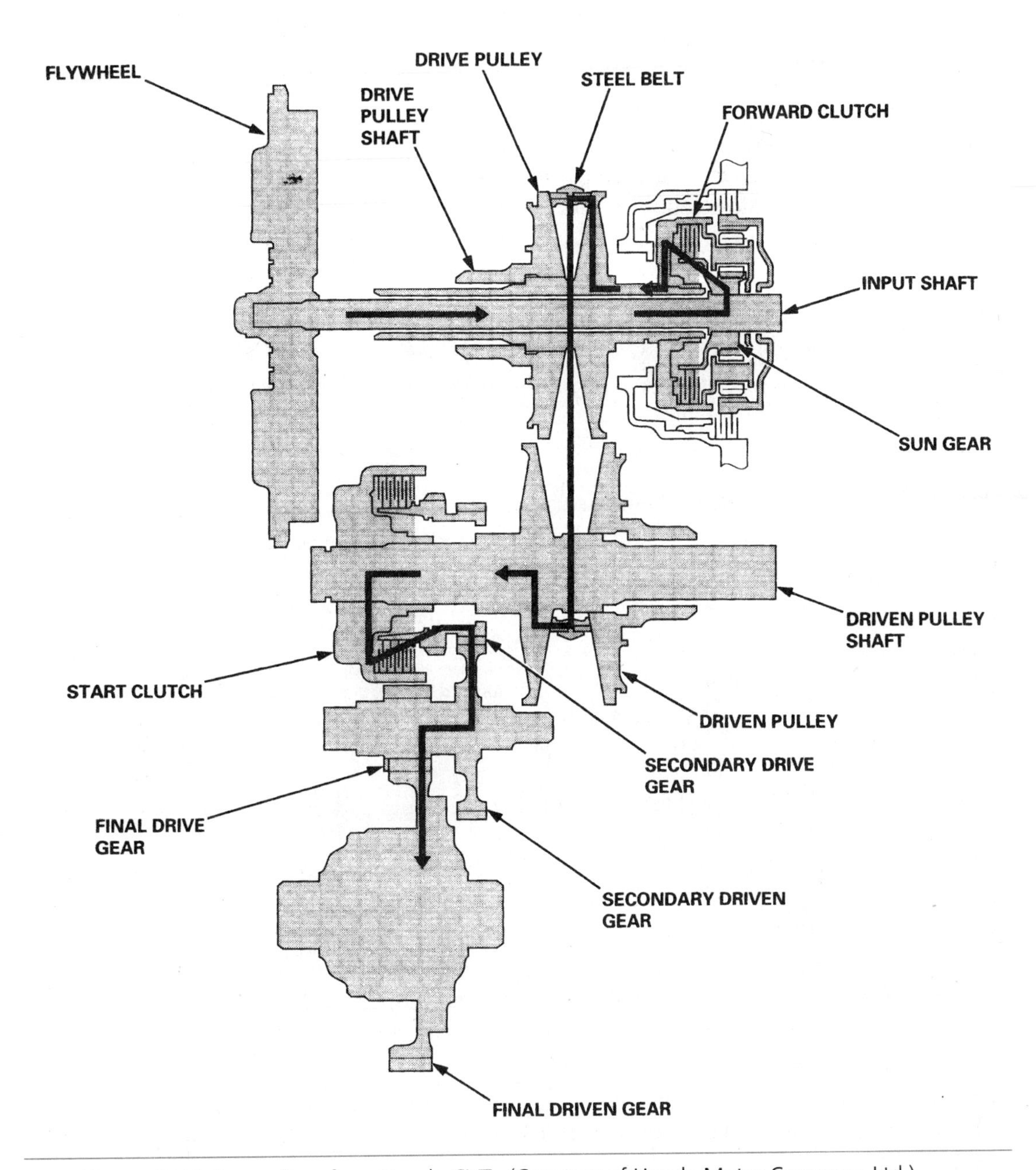

Figure 7-81 Power flow for a Honda CVT. (Courtesy of Honda Motor Company, Ltd.)

Summary

- There are two common designs of compound gear sets: the Simpson gear set in which two planetary gear sets share a common sun gear; and the Ravigneaux gear set, which has two sun gears, two sets of planet gears, and a common ring gear.
- A Simpson gear train is the most commonly used compound planetary gear set. It is used to provide three forward gears.
- The Torqueflite transmission was the first modern three-speed automatic transmission with a torque converter and the first to use the Simpson compound planetary gear train.
- There are two basic versions of the three-speed Torqueflite transmission: the A-904 and the A-727.
- The basic Torqueflite transmission was modified for use as a FWD transaxle and contains the same basic parts as the A-904 transmission, with the addition of a transfer shaft, final drive gears, and a differential unit.
- All the Torqueflite-based transmissions and transaxles have two multiple-disc clutches, an overrunning clutch, two servos and bands, and two planetary gear sets to provide three forward gear ratios and a reverse ratio. The two multiple-disc clutches are called the front and rear clutch packs. The servos and bands are also referred to by their location, front and rear, or by their function, kickdown and low/reverse.
- The front and rear multiple-disc clutches serve as the input devices in a Torqueflite transmission.
- The front and rear bands and the one-way overrunning clutch are the holding devices for a Torqueflite transmission.
- Most Torqueflite transmissions are equipped with a controlled load front servo that has two pistons and allows for the quick release of the band during shifts from second to third gear and second to first gear.
- The A-500 and A-518 are four-speed automatic transmissions with the first three forward gear ratios the same as the A-904 and A-727 Loadflite units. Fourth gear is provided for by a separate planetary gear set and controlled by an overdrive clutch, direct clutch, and overrunning clutch in an overdrive assembly.
- Ford Motor Company began to use the Simpson gear train with the introduction of the C-4 transmission.
- Ford Motor Company has five basic transmissions that use the Simpson gear train; they are quite similar to each other: the C-3, C-4, C-5, C-6, and A4LD.
- The A4LD is a modified version of the C-3.
- The A4LD was the first electronically controlled Ford automatic transmission and variations of it are still being used.
- The gear trains for the first three speeds of Ford transmissions are identical except for the C-6, which uses a low/reverse clutch instead of the low/reverse band used by the other models.
- To control planetary gear action, Ford transmissions use two multiple-disc clutches, two bands, and one overrunning clutch. The two clutches are referred to as the high, front, or reverse/high and the rear or forward clutches. The two bands are the intermediate, front, or kickdown and low/reverse or rear units. Again, the exception to this is the C-6, which uses a clutch in place of a band for low and reverse.
- The front planet carrier and the rear ring gear of Ford transmissions are splined to the output shaft; therefore, either of these planetary gear set members will always serve as the output member.

Terms to Know

2–4 band
3–4 clutch
Controlled load servo
Direct clutch
Direct one-way clutch
Direct/reverse clutch
Dual shaft transmission
ECT
First gear clutch
First gear one-way clutch
Forward clutch
Forward sprag
Fourth clutch
Fourth gear clutch
Front band
Front brake band
Front clutch
High/reverse clutch
Input clutch
Input one-way clutch
Intermediate band
Intermediate clutch
Intermediate one-way clutch
Intermediate/overdrive band
Low/intermediate band
Low one-way clutch
Low/reverse band
Low/reverse clutch
Low/reverse roller clutch
Overdrive band
Overdrive clutch

- ❑ The input shaft of most Ford transmissions is splined to the forward clutch drum and high/reverse clutch hub assembly.
- ❑ The rear planet carrier of Ford transmissions can be held by the one-way clutch or low/reverse band or clutch.
- ❑ The intermediate band in Ford transmissions is wrapped around the outside of the high/reverse clutch drum and is applied when the transmission is operating in second gear.
- ❑ The servos in the C-4 and C-5 are accessible from the outside of the case and the bands are adjustable.
- ❑ The bands in a C-3 are not adjustable and are controlled by selective piston rod lengths during assembly of the transmission.
- ❑ The intermediate servo applies the intermediate band in second gear and is accessible from the outside of the case.
- ❑ To control planetary action of the overdrive gear set, the A4LD uses an overdrive band, overdrive clutch, and an overdrive one-way clutch.
- ❑ Most General Motors transmissions and transaxles are based on the Simpson gear train.
- ❑ The THM 375, 400, and 475 transmissions are heavy-duty transmissions that are generally used in full-size RWD cars and trucks.
- ❑ The THM 250, 250C, 350, 350C, and 375B transmissions are three-speed units. The 250 and 250C are light-duty units, the 350 and 350C are medium-duty units, and the 375B is a heavy-duty version of the 350 with an additional direct clutch disc for increased torque capacity.
- ❑ The primary difference between a 250 and a 350 transmission is that a 250 uses a band as the second-gear holding device, and a 350 uses a band plus a multiple-disc clutch and a one-way roller clutch as holding devices during second-gear operation.
- ❑ The 200, 200C, 200-4R, 325, and 325-4L are light- to medium-duty transmissions.
- ❑ The 200, 200C, and 325 are three-speed units and the 200-4R and 325-4L have four forward speeds.
- ❑ The 325 and 425 transmissions are not classified as transaxles because the differential and final drive gears are not built into the transmission case; rather, a separate final drive unit is bolted to the transmission to drive the front wheels.
- ❑ The THM 125 and 125C are three-speed automatic transaxles designed for light-duty use.
- ❑ The 200-4R and 700-R4 transmissions use a variable displacement vane-type oil pump.
- ❑ The 125/125C models use a vane-type oil pump that is mounted to the valve body and is driven indirectly through a drive shaft by the torque converter.
- ❑ Most GM transmissions with a Simpson gear train use three multiple-disc clutches, one band, and a single one-way roller clutch to provide the various gear ratios.
- ❑ The front carrier and the rear ring gear of GM transmissions will always be the output member of the gear set.
- ❑ The forward clutch is the input device for all GM transmissions.
- ❑ The 400 series uses front and rear bands plus a low/reverse roller clutch, intermediate one-way clutch, and intermediate clutch as the holding devices.
- ❑ The 125, 200, 250, 350, and 325 series use a low/reverse clutch, low roller clutch, and an intermediate band as the holding devices.
- ❑ The 350 series uses an intermediate roller clutch in addition to the other holding devices.
- ❑ The 200-4R and 325-4L use the same multiple-disc clutches as the 200, but they also have two additional clutches—the overrunning clutch and fourth clutch.

Terms to Know (continued)

Overdrive one-way clutch
Overrunning clutch
Rear band
Rear clutch
Reverse input clutch
SBEC
Second band
Second coast band
Second gear clutch
Split-torque converter
Tandem
Third gear clutch
Top gear clutch
Transfer assembly
Transfer shaft

- The THM 700-R4 is a fully automatic transmission consisting of two planetary gear sets, five multiple-disc clutches, one sprag clutch, one roller clutch, and a band.
- The forward, forward sprag drive clutch, 3–4 clutch, and reverse input clutches are the possible input devices for this transmission.
- The low/reverse clutch, low roller clutch, and 2–4 band (intermediate/overdrive band) are the holding devices for a 700-R4 transmission.
- The power flow through General Motors transmissions with a Simpson gear train is similar to the power flow through a Chrysler Torqueflite transmission.
- The Nissan L4N71B/E4N71B series has four forward gears through a Simpson gear set and an additional planetary unit mounted in front of the Simpson gear set.
- The Nissan L4N71B/E4N71B transmissions use four multiple-disc clutches, two servos and bands, and a one-way clutch.
- The bands and the one-way clutch of the L4N71B/E4N71B are used as holding members.
- Toyota's A-140E/A-140L model transaxle combines a three-speed transmission with an overdrive assembly.
- Late-model Toyota transmissions are designated A540 or A541.
- The Ford ATX is a three-speed automatic transaxle that uses a Ravigneaux planetary gear train.
- The original ATX transaxles were equipped with a splitter gear torque converter.
- The one-way clutch, second gear clutch, and top gear clutch serve as the input devices for the planetary gear set in an ATX.
- The front brake band and the reverse clutch serve as the holding devices for the ATX.
- The AOD replaced the C-6 in cars. With the introduction of this transmission, Ford became the first manufacturer to offer an automatic transmission with a built-in overdrive.
- The AOD is a four-speed transmission with a split-torque-type torque converter.
- The AOD uses four multiple-disc clutches, two one-way clutches, and two bands to obtain the various gear ranges.
- Regardless of the input into an AOD transmission, the gear train receives power through the application of the forward, direct, or reverse clutches, which serve as the input devices for the gear train.
- The AOD gear train relies on the intermediate clutch, both one-way clutches, the low/reverse band, and the overdrive band for holding gear set members.
- GM's 180/180C is a three-speed automatic that uses the Ravigneaux compound planetary gearset.
- The THM 180 uses three multiple-disc clutches, which are referred to as the second, third, and reverse, or the intermediate/high, direct, and reverse clutches; one band, and a single one-way (sprag) clutch provide for the various gear ratios.
- The one-way, intermediate/high, and direct clutches serve as the input devices for the 180 gear set.
- The low/intermediate band and reverse clutch serve as the holding devices for the 180 gear train.
- A unique Ravigneaux-based transmission is the Mitsubishi KM 170/171/172 transaxle, which is a three-speed unit that mounts on the passenger side of the vehicle.
- The KM 170/171/172 uses three multiple-disc clutches—a front, direct/reverse, and low/reverse clutch; a second gear band; and a one-way clutch—to provide the various gear ranges.
- The THM 440-T4 was the first domestically produced four-speed automatic transaxle built for FWD vehicles.

- ❑ In the 440-T4, the two tandem planetary units do not share a common member. The front planet carrier is locked to the rear ring gear and the front ring gear is locked to the rear planet carrier.
- ❑ Input through the 440-T4 gear train is controlled by the input clutch, input one-way clutch, and second clutch.
- ❑ The third clutch, third roller clutch, fourth clutch, 1–2 band, and reverse band are the holding devices for the 440-T4 gear train.
- ❑ The AXOD four-speed transaxle has the planet carriers of each planetary unit locked to the other planetary unit's ring gear. Each planetary set has its own sun gear and set of planetary pinion gears.
- ❑ The AXOD uses four multiple-disc clutches, two band assemblies, and two one-way clutches to control the operation of the planetary gear set.
- ❑ The forward clutch, low one-way clutch, direct clutch, direct one-way clutch, and intermediate clutch serve as the input devices for the AXOD.
- ❑ In an AXOD, the low/intermediate band holds the rear sun gear stationary in first and second gears, the overdrive band is applied in fourth gear and holds the front sun gear, and the reverse clutch holds the front planet carrier and rear ring gear for reverse gear.
- ❑ The Honda CA, F4, and G4 transaxles use constant-mesh helical and square-cut gears in much the same manner as a manual transmission.
- ❑ Honda transaxles use four multiple-disc clutches, a sliding reverse gear, and a one-way clutch to control the gears. These are designated by the gear they activate: first gear clutch, first gear one-way clutch, second gear clutch, third gear clutch, fourth gear clutch, and reverse gear.
- ❑ Honda's CVT is based on pulleys and a belt.

Review Questions

Short Answer Essays

1. What basic Torqueflite transmissions were the basis for designing the A-500 and A-518? What major changes were made?
2. The A4LD has an overdrive fourth gear. What controls the action of the overdrive planetary gear set?
3. The THM 325 and 425 were used on FWD model cars. Why are these transmissions not classified as transaxles?
4. What drives the oil pump in THM 125/125C transaxles?
5. The 200-4R shares many of the same components as other THM 200 transmissions. What is different about the 200-4R?
6. Ford's AOD has a unique power input setup. Describe it.
7. What is unique about Honda automatic transmissions?
8. What are the two common designs of compound planetary gear sets? How do they differ?
9. When a Simpson-based transmission is modified to provide for a fourth (overdrive) gear, what is the major modification?
10. When a transmission is described as having two planetary gear sets in tandem, what does this mean?

Fill in the Blanks

1. In recent years, General Motors has redesignated their transmissions. The 400/475 series transmissions are now labeled as ________________ transmissions.
2. Under General Motors' new classification system, the 125/125C transaxles are referred to as ________________ transaxles.
3. With GM's new designation, the 200-4R became the ________________ or ________________ transmission.
4. GM's new designation for the 700-R4 is the ________________ transmission.
5. The new designation for the 180/180C is the ________________ transmission.
6. The new GM designation for the 440-T4 is the ________________ transaxle.
7. All the Torqueflite-based transmissions and transaxles have two multiple-disc clutches, called the ________________ and ________________ clutch packs. The servos and bands are referred to as ________________ and ________________, or ________________ and ________________.
8. The A4LD is a modified version of the ________________.
9. To control planetary gear action, Ford transmissions use two multiple-disc clutches, referred to as the ________________, ________________, or ________________ and the ________________ or ________________ clutches, and two bands called the ________________, ________________, or ________________ and ________________ or ________________ units.
10. The ________________ clutch, ________________ ________________ clutch, and ________________ ________________ clutch serve as the input devices for the planetary gear set in an ATX.

ASE-Style Review Questions

1. *Technician A* says a Simpson gear set is two planetary gear sets that share a common sun gear. *Technician B* says a Ravigneaux gear set has two sun gears, two sets of planet gears, and a common ring gear. Who is correct?
   A. A only
   B. B only
   C. Both A and B
   D. Neither A nor B
2. While discussing the THM 440-T4: *Technician A* says the THM 440-T4 was the first domestically produced four-speed automatic transaxle built for FWD vehicles. *Technician B* says the 440-T4 uses two tandem planetary units that do not share a common gear set member. Who is correct?
   A. A only
   B. B only
   C. Both A and B
   D. Neither A nor B
3. *Technician A* says the front and rear bands are holding devices in a Torqueflite transmission. *Technician B* says the one-way overrunning clutch is an apply device in a Torqueflite transmission. Who is correct?
   A. A only
   B. B only
   C. Both A and B
   D. Neither A nor B
4. *Technician A* says the 200-4R transmission uses a variable displacement vane-type oil pump. *Technician B* says the 700-R4 transmission uses a variable displacement rotor-type oil pump. Who is correct?
   A. A only
   B. B only
   C. Both A and B
   D. Neither A nor B

5. While discussing Simpson-based Ford transmissions:
Technician A says the front planet carrier may serve as the output member of the gear set.
Technician B says the rear ring gear may serve as the output member of the gear set.
Who is correct?
A. A only **C.** Both A and B
B. B only **D.** Neither A nor B

6. *Technician A* says the AOD replaced the C-6 in cars and trucks.
Technician B says with the introduction of the AOD, Ford became the first manufacturer to offer an automatic transmission with a built-in overdrive.
Who is correct?
A. A only **C.** Both A and B
B. B only **D.** Neither A nor B

7. While discussing Torqueflite-based transaxles:
Technician A says the transaxles were developed for FWD cars and have little in common with the other Torqueflite transmissions.
Technician B says the transaxles contain the same basic parts as the A-904 transmission, with the addition of a transfer shaft, final drive gears, and a differential unit.
Who is correct?
A. A only **C.** Both A and B
B. B only **D.** Neither A nor B

8. While discussing the servos used in Torqueflite transmissions:
Technician A says most Torqueflite transmissions are equipped with a controlled load front servo.
Technician B says most Torqueflite transmissions provide for a delayed release of the front band during shifts from second to third gear and second to first gear.
Who is correct?
A. A only **C.** Both A and B
B. B only **D.** Neither A nor B

9. *Technician A* says the planet carriers of each planetary unit are locked to the other planetary unit's ring gear in an AXOD transaxle.
Technician B says in an AXOD, each planetary set has its own sun gear and set of planetary pinion gears.
Who is correct?
A. A only **C.** Both A and B
B. B only **D.** Neither A nor B

10. While discussing Torqueflite transmissions:
Technician A says they were the first automatic transmissions with a torque converter.
Technician B says they were the first transmission to use the Simpson compound planetary gear train.
Who is correct?
A. A only **C.** Both A and B
B. B only **D.** Neither A nor B

Electronic Automatic Transmissions

Upon completion and review of this chapter, you should be able to:

- ❑ Identify the three conditions required for converter lockup in a system using electrical controls.
- ❑ Explain the advantages of using electronic controls for transmission shifting.
- ❑ Explain an elementary shift logic chart.
- ❑ Describe what determines the shift characteristics of each selector lever position.
- ❑ Discuss the operation of the hydraulic control system.
- ❑ Identify the input and output devices in a typical electronic control system and briefly describe the function of each.

A BIT OF HISTORY

In the mid-1980s, Toyota introduced the A140E transaxle. This was the first automatic transmission with electronic shift controls. This transmission began the trend in which, by 1990, all domestic manufacturers offered at least one electronic automatic transmission **(EAT).**

Introduction

Conventional automatic transmissions rely on hydraulic and mechanical devices and have been refined through the years. Much of the development has been centered on increasing their efficiency and prolonging their useful life. These automatic transmissions waste a good amount of the torque produced by the engine through the heat generated by the moving fluid. Also, because gear changes depend on the movement of fluid, upshifts and downshifts are somewhat lazy. Manufacturers could not get the transmissions to respond immediately to the needs of the vehicle without jarring the driver and the vehicle. By using electronic controls for transmission operation, the amount of wasted power can be reduced and the overall operation of the transmission can be more responsive with better reliability.

Electronic controls also allow for the production of less complicated transmissions, thereby reducing manufacturing and repair costs. The use of electronics eliminates many of the complex circuits and valves in a typical valve body (Figure 8-1). With this simplicity comes increased reliability, in addition to reduced costs.

Basics of Electronic Controls

The computers used in electronic control systems are called controllers, microprocessors, microcomputers, or control modules.

The basic part of all electronic control systems is a computer. A **computer** is an electronic device that receives information, stores information, processes information, and communicates information. All the information it works with is really nothing more than electricity. To a computer, certain voltage and current values mean something, and based on these values, the computer becomes informed.

Computers receive information from a variety of input devices that send voltage signals to the computer. These signals tell the computer the current condition of a particular part or the conditions that a particular part is operating in. After the computer receives these signals, it stores them

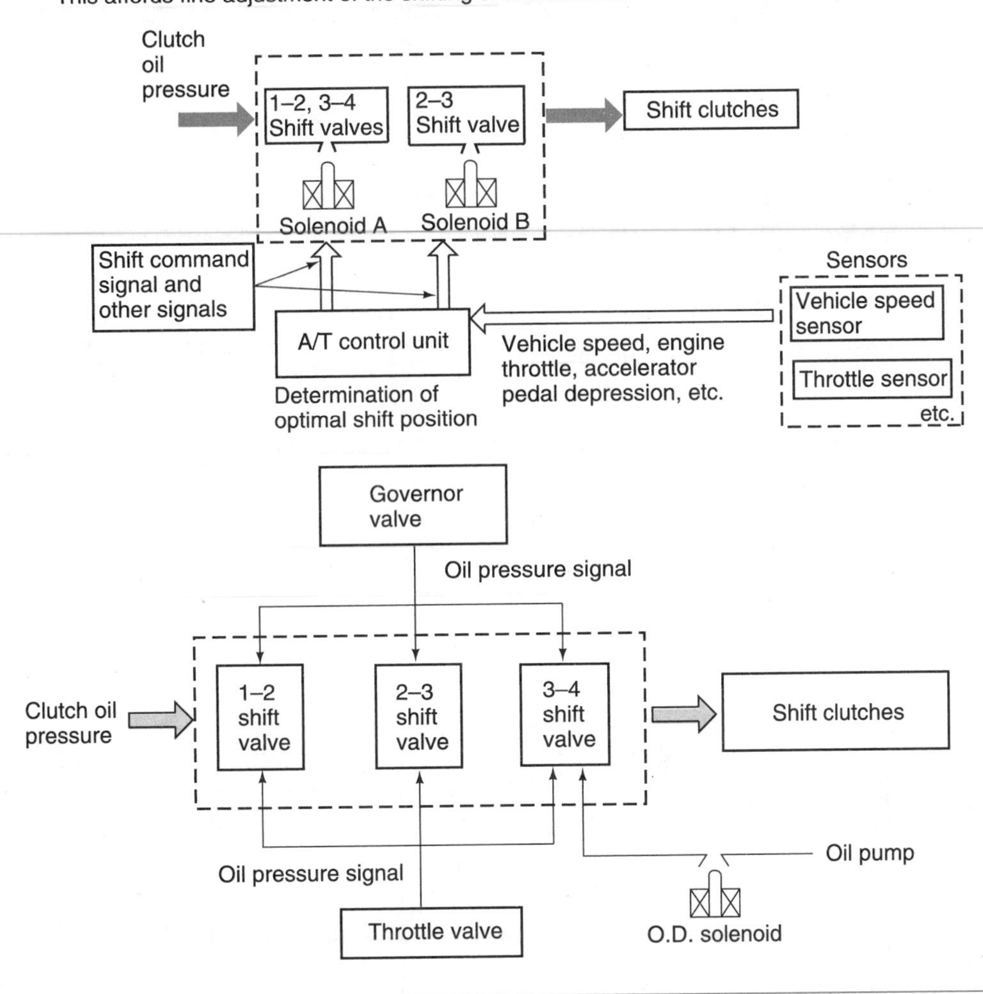

Figure 8-1 A comparison of electronic (top) and hydraulic (bottom) transmission controls. (Courtesy of Nissan Motor Co., Ltd.)

and interprets the signals by comparing the values to data it has in its memory. By processing these data, the computer knows what conditions the input data represent. It also can search its memory to identify any actions it should take in response to those conditions. If an action is required, the computer will send out a voltage signal to the device that should take the action, causing it to respond and correct the situation.

This entire process describes the operation of an electronic control system. Information is received by a microprocessor from some input sensors, the computer processes the information, then sends commands to output devices (Figure 8-2). Most control systems are also designed to monitor their own work and will check to see if their commands produced the expected results. If the result is not what was desired, the computer will alter its command until the desired outcome is achieved.

Inputs

The input devices (Figure 8-3) used in electronic control systems vary with each system; however, they can be grouped into distinct categories: **reference voltage sensors** and **voltage generators.** Voltage generation devices are typically used to monitor rotational speeds. The most common of these is the PM generator used as a vehicle speed sensor. A speed sensor is a magnetic pickup that

Shop Manual
Chapter 8, page 324

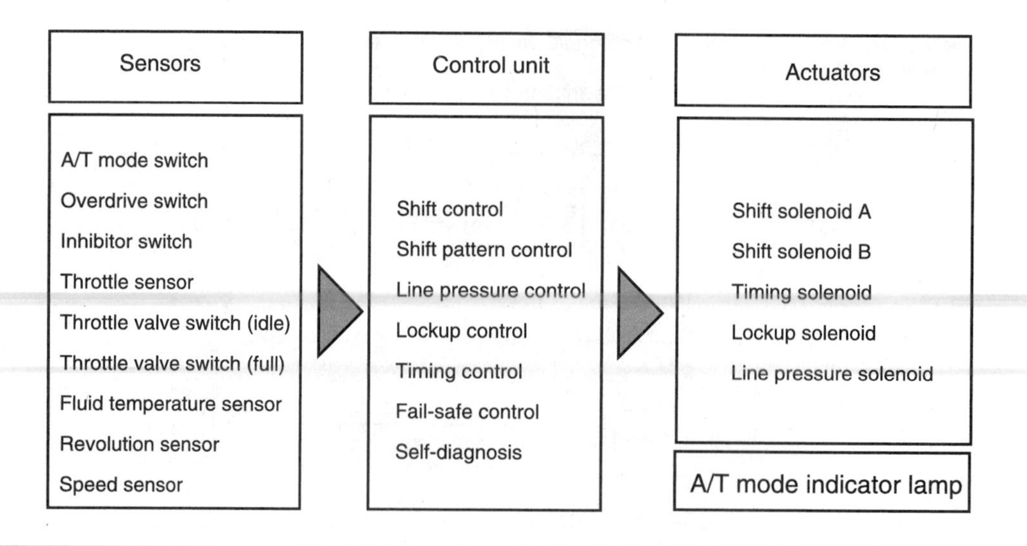

Figure 8-2 Summary of a typical electronic transmission control circuit. (Courtesy of Nissan Motor Co., Ltd.)

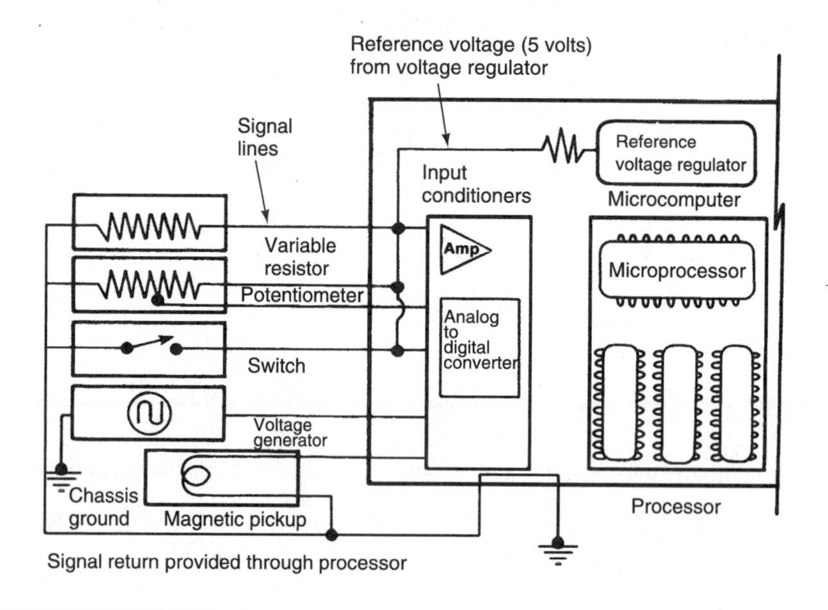

Figure 8-3 Common input circuits to a transmission control unit. (Reprinted with the permission of Ford Motor Co.)

senses and transmits low-voltage pulses (Figure 8-4). These pulses are generated each time a reluctor or toothed wheel, normally located on the circumference of a shaft or rotating component, passes by the wire coil of the pickup unit. The pulses of voltage are the signal sent to the **computer.** The computer then compares the signal to its clock or counter, which converts the signal into speed.

Common voltage reference sensors include on/off switches, **potentiometers, thermistors,** and **pressure sensors.** These sensors normally complete a circuit to and from the computer. The computer sends a reference voltage to them and reads the voltage sent back to it by them. The change in voltage represents a condition that the computer can again identify by looking at its program and comparing values. Potentiometers, thermistors, and pressure sensors are designed to change their electrical resistance in response to something else changing. Normally, a potentiometer is linked to devices such as the throttle linkage. As the throttle is moved, the resistance of the potentiometer changes. This causes a varied return signal to the computer as the throttle opening changes.

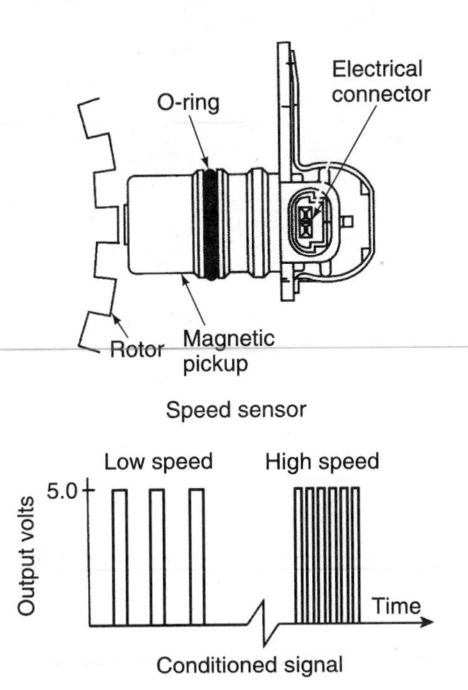

Figure 8-4 A typical speed sensor (top). The signal a speed sensor sends to the computer (bottom). (Courtesy of the Hydra-Matic Division of General Motors Corp.)

Thermistors also change their resistance values in response to conditions (Figure 8-5). However, they respond to changes in heat (Figure 8-6). Pressure sensors respond to pressure applied to a movable diaphragm in the switch. As pressure increases, so does the movement of the diaphragm and the amount of resistance in the sensor (Figure 8-7).

On/off switches are simple in operation. When a particular condition exists, the switch is forced either closed or open. When the switch is closed, the computer receives a return signal from the switch. When it is open, there is no return signal.

The return signal from a switch represents circuit on and off times. If a switch is cycled on/off very rapidly, the return signal is a rapid on/off one. This describes a digital signal—a series of on/off pulses. A computer is a digital device that processes information through a series of on/off signals. In order for a computer to process information, it must receive its data digitally.

Shop Manual
Chapter 8, page 254

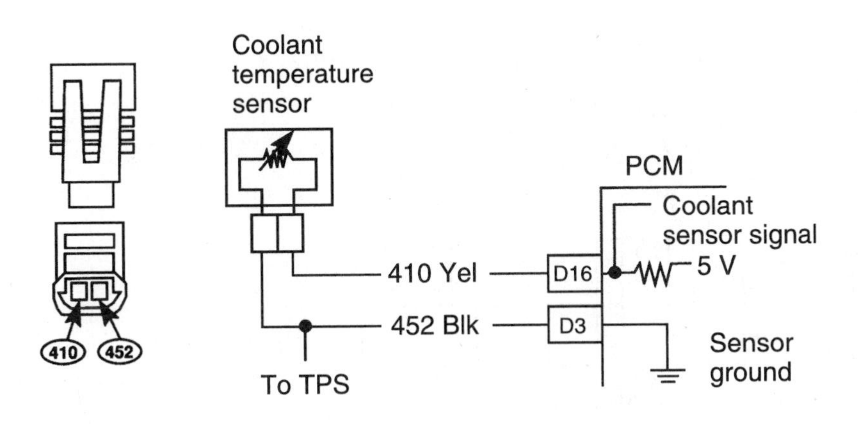

Figure 8-5 The resistance of a thermistor changes with changes in temperature; therefore, the return voltage from an ECT sensor to the computer changes with changes in temperature.

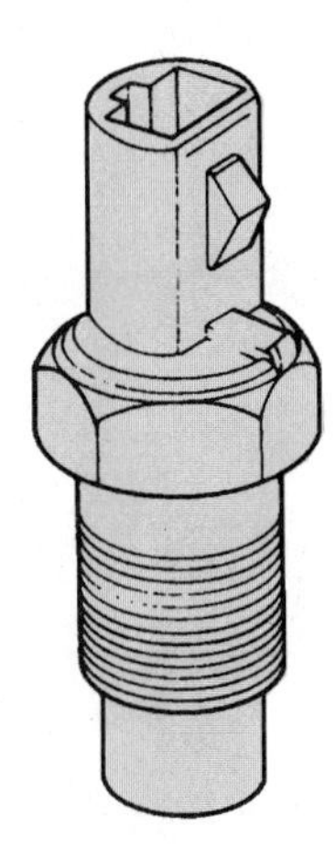

Figure 8-6 A typical ECT sensor. (Reprinted with the permission of Ford Motor Co.)

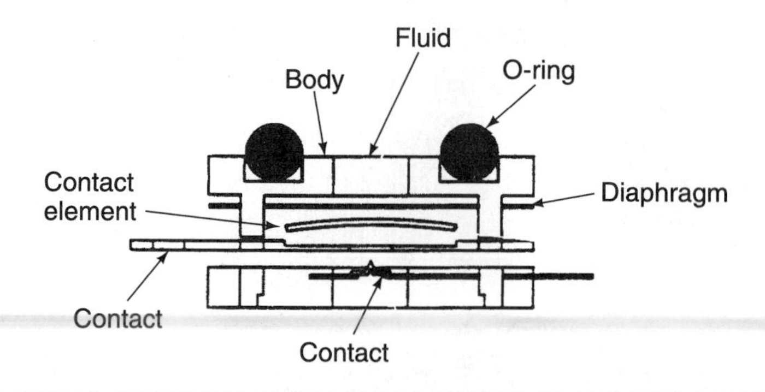

Figure 8-7 The working parts of a pressure switch. (Courtesy of the Hydra-Matic Division of General Motors Corp.)

The inputs from voltage reference sensors arrive at the computer as a change in voltage, not as a digitized signal. Therefore, in order for the computer to analyze the data, the data must first be changed to a digital signal (Figure 8-8). This is the first processing task of a computer—to translate the analog signals into a digital signal.

Input sensors commonly found in electronic transmission control systems include the following:

mass airflow (MAF) sensor
intake air temperature (IAT) sensor
manifold absolute pressure (MAP) sensor
engine coolant temperature (ECT) sensor
crankshaft position (CKP) sensor
transmission range (TR) sensor
throttle position (TP) sensor
brake switch
transmission fluid temperature (TFT) sensor
transmission pressure switches
vehicle speed sensor (VSS)
output shaft speed (OSS) sensor

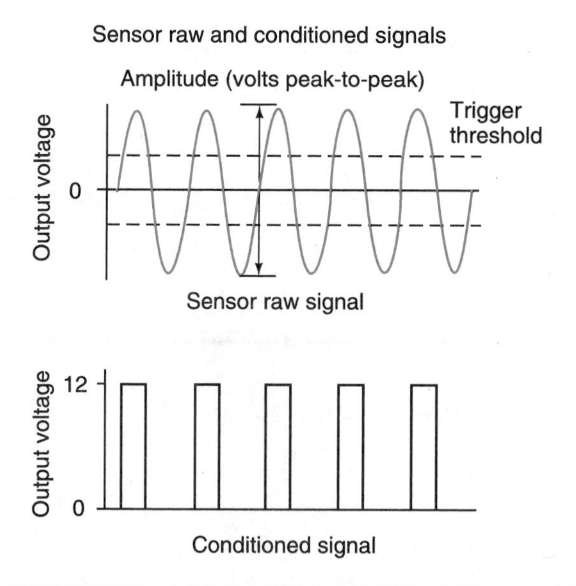

Figure 8-8 The ac signals (top) from many sensors must be translated or conditioned into a digital dc signal (bottom) before the computer can process the information. (Courtesy of the Hydra-Matic Division of General Motors Corp.)

The MAF, IAT, MAP, ECT, and CKP sensors provide the TCM with information on the operating condition of the engine. Through these sensors, the TCM is able to control shifting and TCC operation according to the temperature, speed, and load of the engine.

The TR sensor informs the transmission control module **(TCM)** of the gear selected by the driver. This sensor normally also contains the neutral safety switch and the reverse light switch. The TP sensor sends a voltage signal to the TCM in response to throttle position. This signal not only is used to inform the TCM of the driver's intent, but also is used in place of the hydraulic throttle pressure linkage. The brake switch informs the TCM when the brake pedal has been depressed. In response to the signal from the brake switch, the TCM will cancel TCC lockup and provide gear shifts that allow for engine braking.

The TFT sensor monitors the temperature of the transmission's fluid. When the signal from this sensor is normal, the transmission will operate within its normal range. However, when the signal indicates that the fluid is overly hot, the TCM will allow the transmission to operate only in a way that will allow the transmission to cool down. This prevents damage to the transmission. When the TFT signal indicates that the fluid is cooler than normal, the TCM will alter the shift schedule.

Various transmission pressure switches can be used to tell the TCM which hydraulic circuits are pressurized and which clutches and bands are applied. These input signals can serve as verification to other inputs and as self-monitoring or feedback signals.

The **VSS** and **OSS** sensors are used to monitor output of the transmission and vehicle speed. In some electronic control systems, only one of these sensors is used. When a vehicle has both sensors, the OSS signal is used as a verification signal for the VSS by the engine control system. The VSS is used as a verification signal for the OSS by the TCM. Some transmissions use these speed-related inputs in place of a governor. These signals are used to regulate hydraulic pressure and shift points and to control TCC operation.

Outputs

After the information has been processed, the computer sends commands to its output devices. The typical output devices are solenoids and motors, which cause something mechanical or hydraulic to move. The movement of these outputs is controlled by the commands of the computer. Sometimes, the command is merely the application of voltage to operate or energize the device. Other times, it is a variable signal that causes the device to cycle in response to the signal.

Processing

The primary purpose of a computer is to process information. It does so by comparing all data it receives against data it has stored. All stored data remain in the computer's memory and are used as needed to analyze and correct operating conditions. The computers used to control transmission functions rely on programming stored in their memory to provide gearshifting at the optimal time. The decision to shift or not to shift is based on shift schedules and logic programmed into the memory of the computer.

In order for a computer in any electronic control system to determine when to initiate a gearshift change, it must be able to refer to **shift schedules** that it has stored in its memory. A shift schedule contains the actual shift points to be used by the computer according to the input data it receives from the sensors. Shift schedule logic chooses the proper shift schedule for the current conditions of the transmission. It uses the shift schedule to select the appropriate gear, then determines the correct shift schedule or pattern that should be followed.

Current operating conditions are typically referred to as "real-time" conditions.

The first input a computer looks at to determine the correct shift logic is the position of the gearshift lever. All shift schedules are based on the gear selected by the driver. The choices of shift schedules are limited by the type and size of engine that is coupled to the automatic transmission. Each engine/transmission combination has a different set of shift schedules. These schedules are coded by selector lever position and current gear range and use throttle angle and vehicle speed as primary determining factors. The computer also looks at different temperature, load, and engine operation inputs for more information.

The shift schedules set the conditions that need to be met for a change in gears. Since the computer frequently reviews the input information, it can make quick adjustments to the schedule if needed and as needed. The result of the computer's processing of this information and commanding outcomes according to a logical program is optimum shifting of the automatic transmission. This results in improved fuel economy and overall performance.

The electronic control systems used by manufacturers differ with the various transmission models and the engines to which they are attached. The components in each system and the overall operation of the system also vary with the different transmissions. However, all operate in a similar fashion and use basically the same parts.

Electronically Controlled Transmissions

Electronic transmission control has become increasingly common on today's cars. These controls provide automatic gear changes when certain operating conditions are met. Through the use of electronics, transmissions have better shift timing and quality. As a result, the transmissions contribute to improved fuel economy, lower exhaust emission levels, and improved driver comfort. Although these transmissions function the same way as earlier hydraulically based transmissions, their shift points are determined by a computer. The computer uses inputs from several different sensors and matches this information to a predetermined schedule.

Hydraulically controlled transmissions rely on signals from a governor and throttle pressure device to force a shift in gears. Electronically controlled transmissions typically have neither governors nor throttle pressure devices. Hydraulically controlled transmissions rely on pressure differentials at the sides of a shift valve to hold or change a gear. Electronic transmissions do, too. However, the pressure differential is caused by the action of shift solenoids that allow for changes in pressure on the side of a shift valve. These solenoids are controlled by the computer. The solenoids do not directly control the transmission's clutches and bands. These are engaged or disengaged the same way as hydraulically controlled units. The solenoids simply control the fluid pressures in the transmission; they do not perform a mechanical function.

Most electronically controlled systems are complete computer systems. There is a central processing unit, inputs, and outputs. Often the central processing unit is a separate computer designated for transmission control. This computer is often called the transmission control module. Other transmission control systems use the **PCM** (power train control module) or the BCM (body control module) to control shifting. When transmission control is not handled by the PCM, the controlling unit communicates with the PCM.

The inputs include transmission operation monitors plus the some of the sensors used by the PCM. Input sensors, such as the throttle position (TP) sensor, supply information for many different systems. This information is shared by the control modules. The system outputs are solenoids. The different transmission control systems used by the various vehicle manufacturers differ mostly by the number and type of solenoids used.

Adaptive Controls

Many late-model transmissions have systems that allow the TCM to change transmission behavior in response to the operating conditions and the habits of the driver. The system monitors the condition of the engine and compensates for any changes in the engine's performance. It also monitors and memorizes the typical driving style of the driver and the operating conditions of the vehicle. With this information, the TCM adjusts the timing of shifts and converter lockup to provide good shifting at the appropriate time.

These systems are constantly learning about the vehicle and driver. The TCM adapts its normal operating procedures to best meet the needs of the vehicle and the driver. When systems are capable of doing this, they are said to have **adaptive learning** capabilities. To store this information, the TCM includes a long-term adaptive memory.

Manual Shifting

One of the most publicized features of electronically controlled transmissions is the availability of manual shift controls. Although not all electronic transmissions have this feature, they all could. Basically, these systems allow the driver to manually upshift and downshift the transmission at will, much like a manual transmission. Unlike a manual transmission, the driver does not need to depress a clutch pedal. All the driver does is move a gear selector or hit a button and the transmission changes gears. If the driver doesn't change gears and engine speed is high, the transmission shifts on its own. If the driver elects to let the transmission shift automatically, a switch disconnects the manual control and the transmission operates automatically.

Marketed as a sport option and a combination of a manual and automatic transmission, these transmissions are based on an automatic transmission with a torque converter. Therefore performance numbers are not quite as good as if the vehicle were equipped with a manual transmission. In fact, manual control of an automatic transmission often results in slower 0–60 mph times than when the transmission shifts by itself.

All manually shifted automatics do not behave in the same way, nor do they control the same things. Actually the behavior of the transmission depends on the car it sits in. Some of these cars are pure high-performance cars, whereas others are moderate-performance family sedans. The following are basic descriptions of the manually shifted automatic transmissions available in some common cars. More exotic cars also have this option but are not included here.

BMW's Steptronic—Available in the BMW 840Ci, this is a five-speed automatic transmission with a manual control option. Manual shifting is performed by moving a control on the console. Moving the selector forward provides for an upshift. A movement down allows for a downshift.

Chrysler AutoStick—This is the most familiar system and is currently available in some models of the Dodge Intrepid and Stratus, Chrysler Sebring, and Eagle Vision. Manual shifting is performed by moving a control on the console. Moving the selector to the right provides for an upshift. A movement to left allows for a downshift.

Honda's Sequential SportShift—Available on the Acura NSX and Honda Prelude, manual shifting is performed by moving a control on the console. Moving the selector forward provides for an upshift. A movement down allows for a downshift. This transmission is unique in that it will not automatically upshift if the driver brings the engine's speed too high. All other transmissions of this design will upshift automatically at a predetermined engine speed to prevent damage to the engine.

Tiptronic—This is a five-speed transmission available in the Porsche Boxster. Audi also has a Tiptronic transmission, but it is different from Porsche's. Manual shifting on the Audi is performed by moving a control on the console. Moving the selector forward provides for an upshift. A movement down allows for a downshift. To shift Porsche's transmission, the driver moves the gear selector into the manual gate, next to the automatic ranges, and depresses buttons on the steering wheel.

Chrysler Transmissions

Late-model Chrysler 42RE and 46RE transmissions and the transaxles used in most Chrysler/Mitsubishi vehicles rely on electronics to control the shifting into overdrive gear. FWD models are equipped with an A-604/41TE transaxle, which is a very advanced electronically controlled transmission. Electronics are involved in the control of the converter clutch in all other Chrysler transmissions.

Shop Manual
Chapter 8, page 337

The 42RE and 46RE transmissions are fully automatic units with an overdrive attached to the rear of the transmission. Most of the transaxles have been modified to include a fourth gear. The first through third gearshifts are provided and controlled hydraulically through apply devices in the main body of the transmission.

Fourth gear operation is controlled by a manually operated overdrive switch on the instrument panel, center console, or shift lever. The overdrive switch is wired into a circuit with an overdrive solenoid, which is located on the valve body, and the single board engine controller **(SBEC).** The torque converter clutch lockup solenoid is also wired into this circuit. The overdrive switch

The overdrive solenoid is also referred to as the overdrive lock-out solenoid.

Shop Manual
Chapter 8, page 338

Adaptive controls are those that perform their functions based on real-time feedback sensor information.

By putting information on the CCD bus bar, the information is made available to a number of circuits; therefore, this is a multiplexed system.

prevents a shift into fourth gear when the switch is off. Therefore, the operation of the torque converter clutch is also inhibited when the overdrive switch is off.

The shift from third to fourth gears is electronically controlled and hydraulically activated by the SBEC. The SBEC controls the overdrive solenoid using information received from several sensors.

Chrysler A-604/41TE Controls

The A-604/41TE transaxle is an electronically controlled four-speed transaxle. The transaxle uses hydraulically operated clutches controlled by the transaxle controller. The controller receives information from various inputs and controls a solenoid assembly through the electronic automatic transaxle **(EATX)** relay. The solenoid assembly consists of four solenoids (Figure 8-9) that control hydraulic pressure to four of the five clutches in the transaxle and to the lockup clutch of the torque converter (Figure 8-10).

The controller has an adaptive learn characteristic that learns the release rate and application rate of various transaxle components during various operating conditions. Adaptive learning allows the controller to compensate for wear and other events that might occur and cause the normal shift programming to be inefficient.

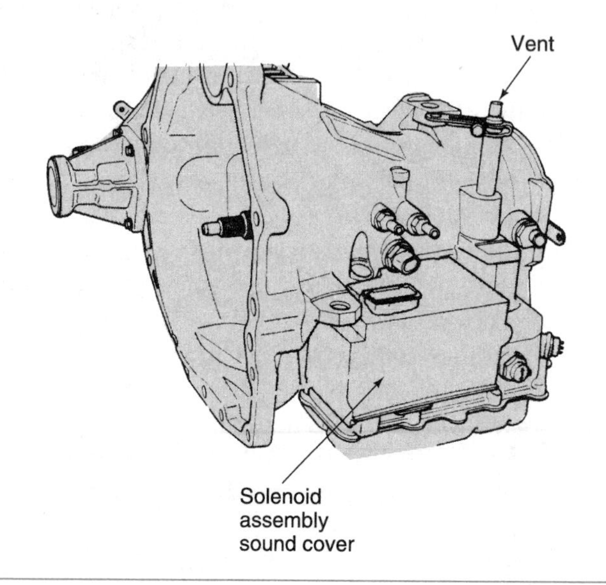

Figure 8-9 Location of solenoid assembly on a Chrysler A-604 transaxle. (Courtesy of Chrysler Corp.)

GEAR SELECTOR POSITION	OPERATING GEAR	UNDERDRIVE CLUTCH	OVERDRIVE CLUTCH	2/4 CLUTCH	LOW/REVERSE CLUTCH	REVERSE CLUTCH
OD	1st gear	X			X	
	2nd gear	X		X		
	3rd gear	X	X			
	4th gear		X	X		
D	1st gear	X			X	
	2nd gear	X		X		
	3rd gear	X	X			
L	1st gear	X			X	
	2nd gear	X		X		
R	Reverse				X	X

Figure 8-10 Clutch and band application chart for late-model Chrysler transaxles.

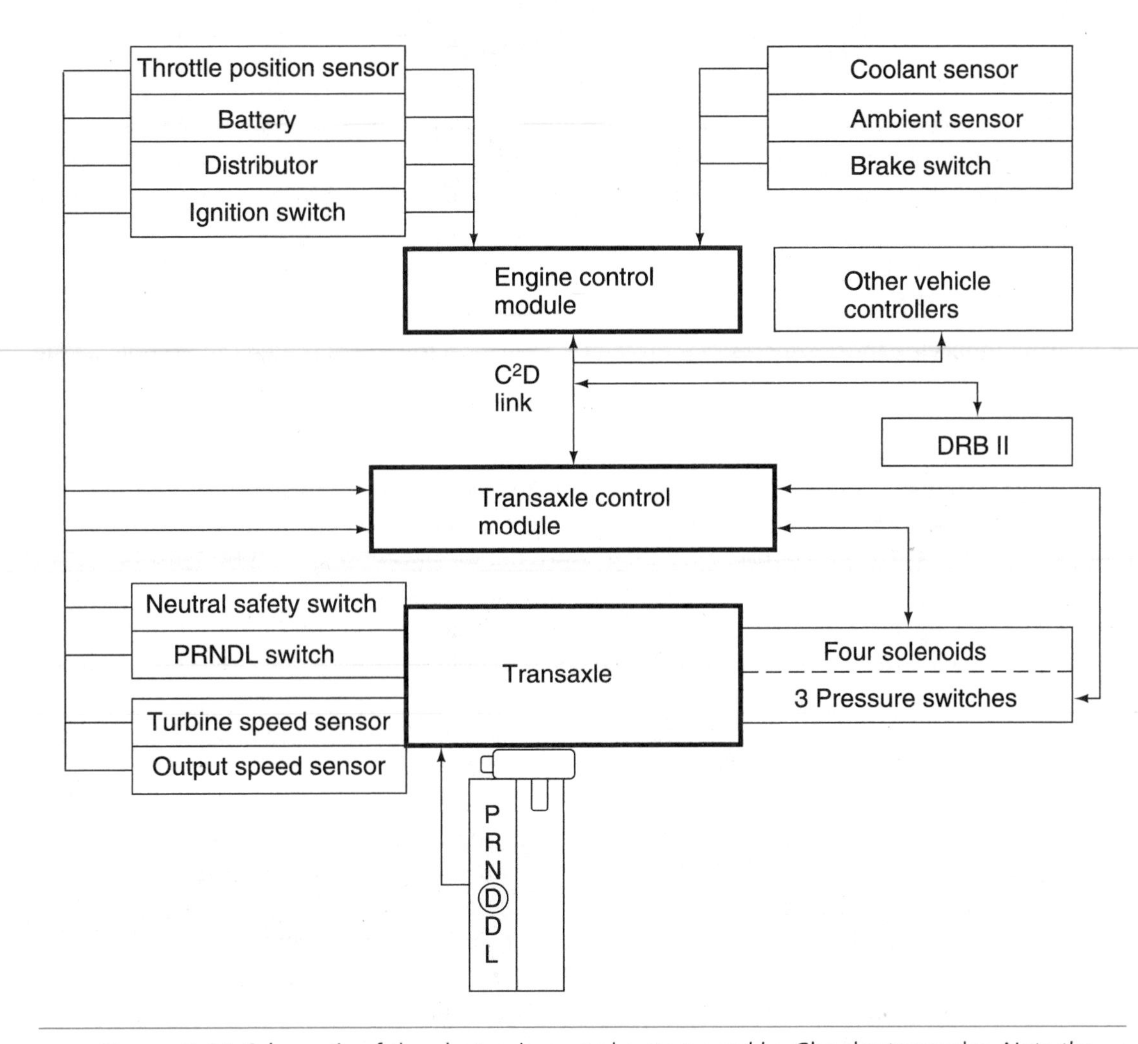

Figure 8-11 Schematic of the electronic control system used by Chrysler transaxles. Note the CCD (C^2D) bus link. (Courtesy of Chrysler Corp.)

The controller may receive information from two different sources: directly from a sensor or through a twisted-pair bus circuit, which connects all of the vehicle computer systems (Figure 8-11). This modulated bidirectional bus system is called **Chrysler Collision Detection (CCD) bus** and allows the various computers in the vehicle to share information. Bus wires are twisted to reduce the chance of the signals being disrupted by radio frequency interference. This interference can cause the sensitive voltage signals to be altered and to send false information to the PCM. The computers that share the bus have unique frequencies that serve as identification. The frequency may also be altered by radio frequency interferences. Radio frequency interference is a form of electromagnetic interference or electrical noise caused by the secondary ignition, high current flow, and the operation of devices such as motors and solenoids.

When the controller receives an ignition run from the ignition switch, it performs a series of circuit and relay checks. If no problem is found, the controller will provide voltage to the EATX relay, which causes its contacts to close. This sends voltage to the solenoid assembly.

Direct battery voltage is supplied to the controller. If the controller loses source voltage, the transaxle will enter into limp-in mode. The transaxle will also enter into limp-in if the controller senses a transmission failure. At this point, a fault code will be stored in the memory of the controller and the transaxle will remain in limp-in until the transaxle is repaired. While in limp-in, the transaxle will operate only in PARK, NEUTRAL, REVERSE, and SECOND gears. The transaxle will not upshift or downshift. This allows the vehicle to be operated, although its efficiency and performance are hurt.

The solenoids are controlled by the controller, which sends voltage to the EATX relay, which, in turn, sends voltage to the solenoid assembly. The controller also completes the ground circuit of the solenoids when a particular solenoid should be activated. The controller also monitors the operation of the solenoids through inputs from low/reverse, 2–4 pressure, and overdrive switches. These switches are located in the solenoid assembly and inform the controller when a hydraulic circuit is open.

Input Devices

Shop Manual
Chapter 8, page 323

The direct inputs are those sensors that provide information to the controller and do not use the CCD bus circuit. The CCD bus inputs use the bus circuit to supply the controller and other computers with information.

Typical CCD bus inputs used by the transaxle controller are from an ambient or battery temperature sensor, brake switch, engine coolant temperature (ECT) sensor, and manifold absolute pressure (MAP) sensor. Other information, such as engine and body identification, the SBEC's target idle speed, and speed control operation are not the result of monitoring by sensors; rather, these have been calculated or determined by the SBEC and made available on the bus. The ambient or battery temperature sensor monitors intake air temperature. The SBEC uses the temperature of intake air and current flow to calculate the temperature of the battery. The SBEC uses this temperature calculation to estimate transaxle fluid temperature.

The brake switch is used to disengage the torque converter clutch when the brakes are applied. Its input has little to do with the upshifting and downshifting of gears. The inputs from the ECT sensor are critical to the operation of the transaxle. If the engine's coolant temperature is cold, the controller may delay upshifts to improve drivability. The controller may also lock the converter clutch in second or third gear if the coolant temperature rises. The MAP sensor keeps the controller informed of changes in engine load.

Although engine speed information is available at the bus, the controller receives this signal directly from the distributor pickup coil or crank angle sensor. With the direct feed, time delay at the bus circuit is avoided and the controller is aware of current engine speeds.

Other direct inputs to the controller include battery voltage, selector lever positions, throttle position, turbine shaft speed, output shaft speed, and information from the low/reverse, overdrive, and 2–4 pressure switches at the solenoid assembly. The output shaft and turbine shaft speed sensors are magnetic pickup-type sensors that generate an ac voltage.

The controller processes these inputs and selects the proper shift schedule for the transaxle. The controller will then control the power feed to the solenoid assembly through the EATX relay and the ground circuit of the solenoids to force upshifts and downshifts. The controller will also activate the back-up lamp relay based on these inputs (Figure 8-12).

42LE Controls

Shop Manual
Chapter 8, page 337

The 42LE four-speed transaxle (Figure 8-13), released in 1994 by Chrysler for use in its New Yorker, Intrepid, LHS, and Vision models, also uses fully adaptive controls. Like the 41TE, this transaxle uses hydraulically applied clutches to shift gear, but it controls the hydraulics electronically (Figure 8-14).

The hydraulics of the transaxle provide for the control of torque converter fluid flow and oil cooler flow, the regulation of mainline pressure, and the movement of the manual shift valve. Fluid flow to the various apply devices is directly controlled by the solenoids. The basic hydraulic system also includes a logic-controlled solenoid switch valve, which locks out first gear when second, direct, or overdrive gears are engaged. This solenoid switch also redirects fluid to the torque converter lockup clutch. To regain access to first gear, a special sequence of solenoid commands from the control unit must be used to unlock and move the solenoid switch. This prevents the engagement of first gear when other gears are engaged (Figure 8-15).

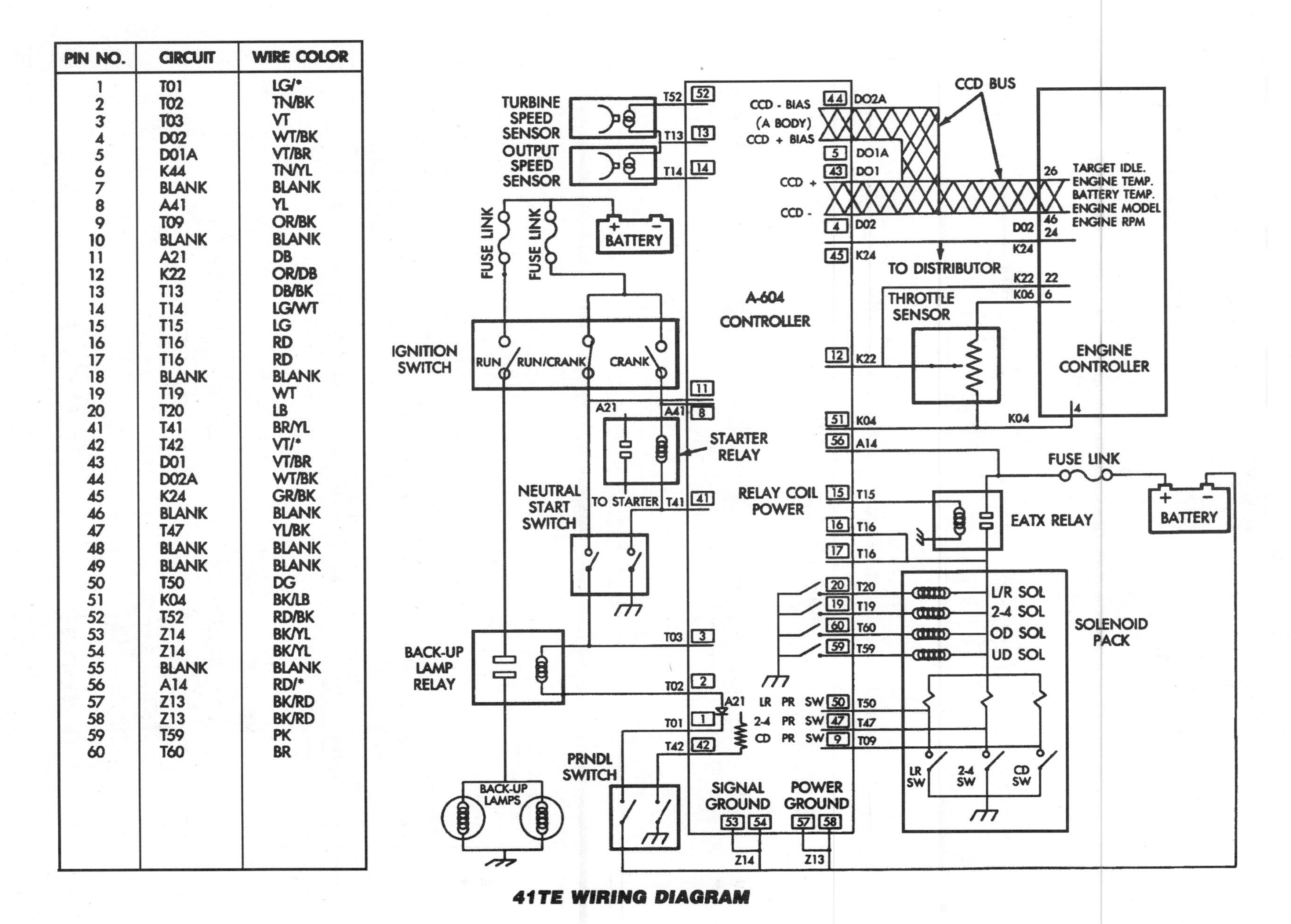

PIN NO.	CIRCUIT	WIRE COLOR
1	T01	LG/*
2	T02	TN/BK
3	T03	VT
4	D02	WT/BK
5	D01A	VT/BR
6	K44	TN/YL
7	BLANK	BLANK
8	A41	YL
9	T09	OR/BK
10	BLANK	BLANK
11	A21	DB
12	K22	OR/DB
13	T13	DB/BK
14	T14	LG/WT
15	T15	LG
16	T16	RD
17	T16	RD
18	BLANK	BLANK
19	T19	WT
20	T20	LB
41	T41	BR/YL
42	T42	VT/*
43	D01	VT/BR
44	D02A	WT/BK
45	K24	GR/BK
46	BLANK	BLANK
47	T47	YL/BK
48	BLANK	BLANK
49	BLANK	BLANK
50	T50	DG
51	K04	BK/LB
52	T52	RD/BK
53	Z14	BK/YL
54	Z14	BK/YL
55	BLANK	BLANK
56	A14	RD/*
57	Z13	BK/RD
58	Z13	BK/RD
59	T59	PK
60	T60	BR

Figure 8-12 Wiring diagram for the electronic transmission controls on a Chrysler 41TE transaxle. (Courtesy of Chrysler Corp.)

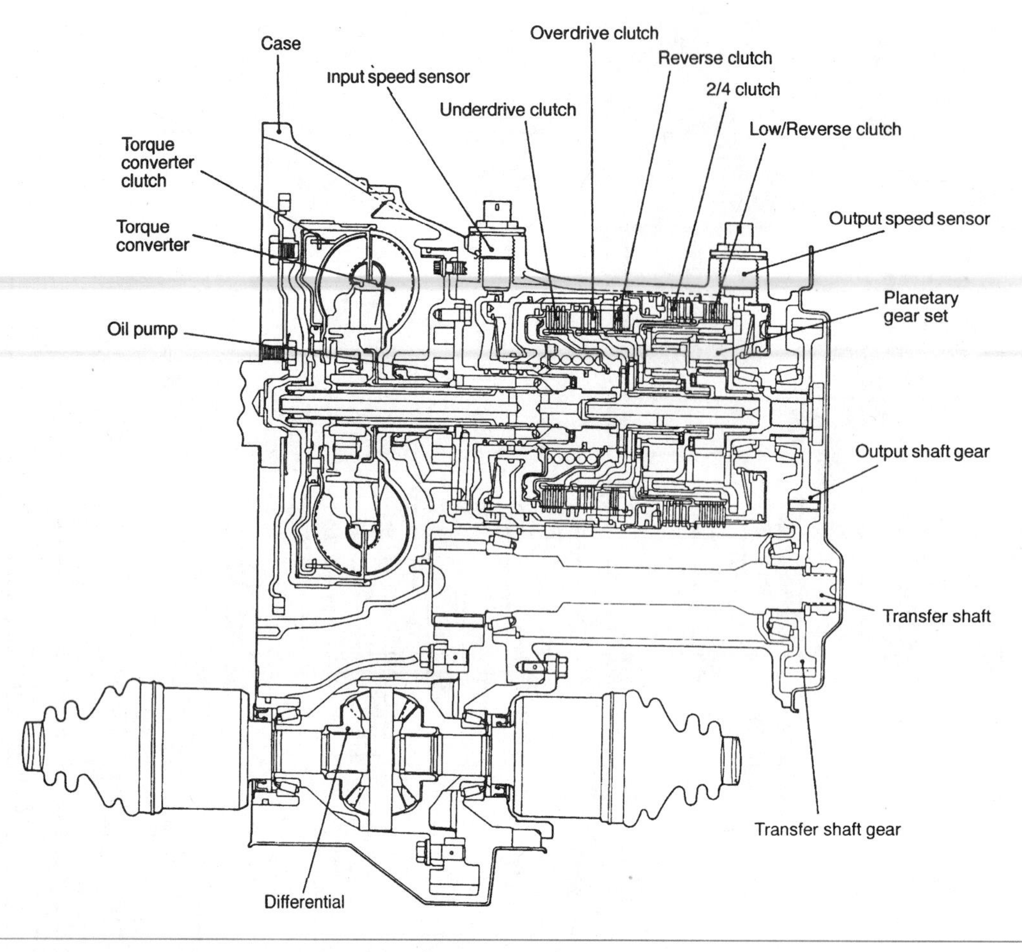

Figure 8-13 Chrysler 42LE transaxle. (Courtesy of Chrysler Corp.)

AUTOMATIC TRANSAXLE CONTROL COMPONENT LAYOUT

Name		Symbol	Name	Symbol
Data link connector		N	Output speed sensor	F
EATX-ECM	<SOHC>	D	Overdrive switch	L
	<DOHC>	O		
EATX relay		C	Pressure switches	I
Engine coolant temperature sensor		J	Solenoid assembly	H
Engine speed signal (Crankshaft position sensor)		K	Stoplight switch	M
Input speed sensor		G	Throttle position sensor	B
Manifold absolute pressure sensor		A	Transaxle range and park/neutral position switches	E

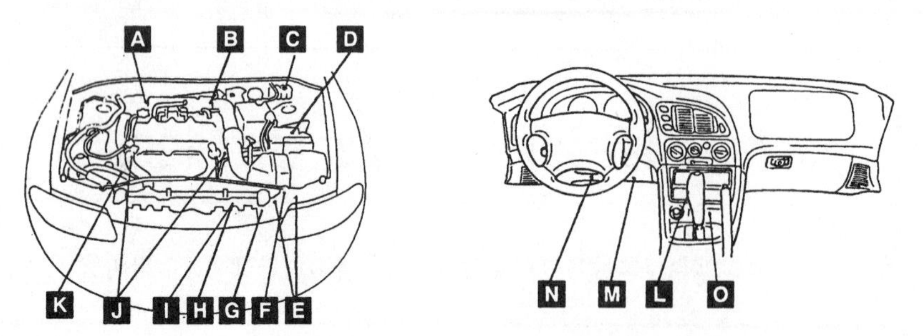

Figure 8-14 The layout of the electrical system for Chrysler's 42LE transaxle. (Courtesy of Chrysler Corp.)

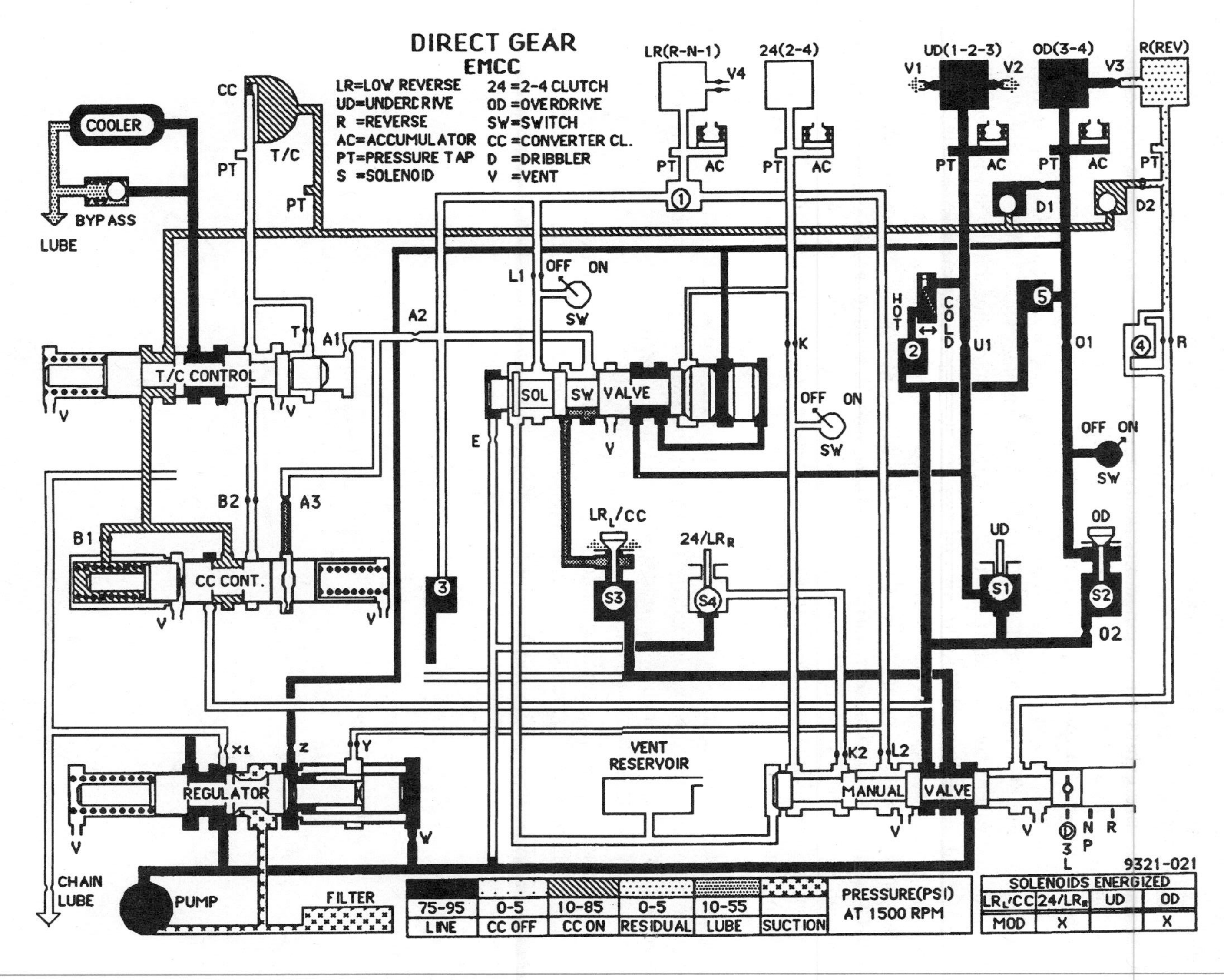

Figure 8-15 An oil circuit for direct drive in a 42LE transaxle. Notice how the fluid flow to low gear is blocked off. (Courtesy of Chrysler Corp.)

The solenoids act directly on steel poppet and ball valves. Two of the solenoid valves are normally venting and the other two are normally applying. This combination allows for a fault mode of operation. If electrical power to the transaxle is lost, the transaxle will provide only second gear in all forward drive ranges.

There are three pressure switches that give input to the transaxle controller. They are all located within the solenoid assembly. One speed sensor reads input speed at the turbine shaft and another speed sensor reads output speed (Figure 8-16). The third sensor monitors the position of the manual shift valve.

Shop Manual
Chapter 8, page 338

Engine speed, throttle position, temperature, engine load, and other typical engine-related inputs are also used by the controller to determine the best shift points. Many of these inputs are available through **multiplexing** and are inputted from the common bus.

The adaptive learning takes place as the controller reads input and output speeds over 140 times per second. The controller responds to each new reading. This learning process allows the transaxle controller to make adjustments to its program so that quality shifting always occurs.

The basic shift logic of the controller allows the releasing apply device to slip slightly during the engagement of the engaging apply device. Once the apply device has engaged and the next gear is driven, the releasing apply device is pulled totally away from its engaging member and the transmission is fully into its next gear. This allows for smooth shifting into all gears. The adaptive learning capability of the transaxle controller allows for this smooth shifting throughout the life of the transmission. The controller learns the characteristics of the transaxle and changes its programming accordingly.

On Chrysler cars equipped with a 42LE transaxle, an option is available that allows the driver to manually control shifting. This option is called AutoStick. The transaxle is not modified for this option, rather it is fitted with a special gear selector and switch assembly. The driver can either manually shift the gears or allow the transaxle to shift automatically.

When the shifter is moved into the AutoStick position, the transaxle remains in whatever gear it was in prior to activating AutoStick. Moving the gear selector to the left causes the transmission to downshift, moving it to the right causes an upshift. The selected gear is displayed on the instrument panel to keep the driver informed of the selected gear.

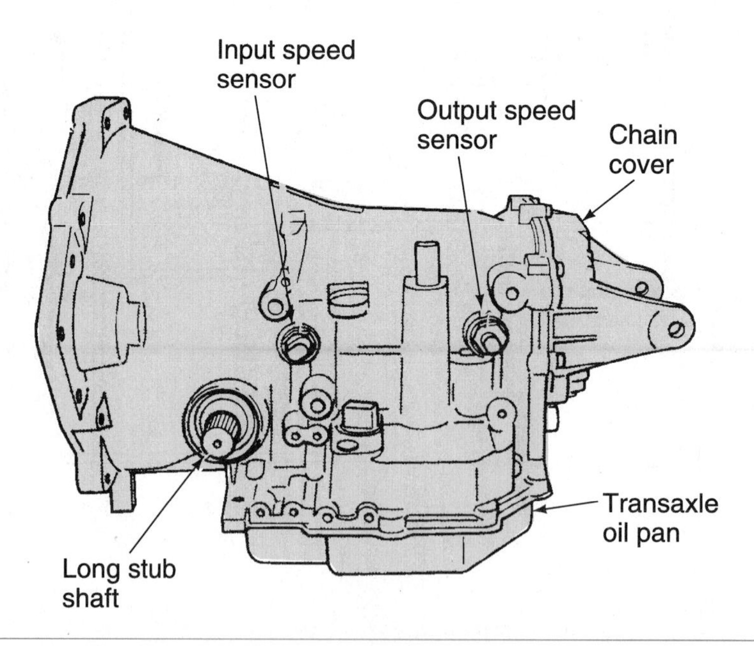

Figure 8-16 Location of the input (turbine) speed and output speed sensors. (Courtesy of Chrysler Corp.)

Although the driver has control of the shifting, the TCM will override the controls during some conditions. Regardless of the action taken by the driver, these automatic shifts will occur under these conditions:

4–3 coast downshift at 13 mph
3–2 coast downshift at 9 mph
2–1 coast downshift at 5 mph
1–2 upshift at 6300 engine rpm
2–3 upshift at 6300 engine rpm
4–3 kickdown shift at 13–31 mph and with sufficient throttle

Manual shifts are not permitted under the following conditions:

3–4 upshift when the vehicle is traveling at less than 15 mph
3–2 downshift when the vehicle is traveling above 74 mph with a closed throttle or 70 mph when the throttle is open
2–1 downshift when the vehicle is traveling above 41 mph with a closed throttle or 38 mph when the throttle is open

The AutoStick feature will also be deactivated if one of the following occurs:

DTC-28: This code relates to the TR sensor
DTC-70: This code relates to the AutoStick switch
DTC-71: This is a high engine and transmission temperature code

The AutoStick switch assembly is a nonserviceable unit. There are no adjustments nor are parts available to repair it. The switch unit is replaced as a unit whenever there is a fault. Diagnostics and service to the rest of the transaxle is the same for all 42LE transaxles.

Ford Motor Company Transmissions

The circuitry and components used by Ford Motor Company to control its transmissions vary with transmission models and applications. Although there are some similarities across models, the common systems should be looked at individually.

Ford AOD-E Controls

Shop Manual
Chapter 8, page 340

The AOD-E is a fully automatic four-speed transmission with electronic shifting, torque converter control, and line pressure controls. This transmission is identical to an AOD, except that it does not use a split-torque converter and its operation is controlled by the electronic engine control's (EEC) power train control module (PCM). Input sensors provide information to the PCM, which in turn controls actuators that control transmission operation.

The engine control system consists of the PCM, sensors, switches, and solenoids. It also has self-diagnostic capabilities. The engine sensors supply the PCM with the information necessary to determine engine operating conditions through voltage signals. The PCM then sends out electrical signals to control air/fuel ratio, idle speed, emission controls, ignition timing, and transmission solenoids.

The PCM uses logic to control shift scheduling, shift feel, and converter lockup. The PCM relies on information from the engine control system, through such sensors as the MAP and TPS (Figure 8-17), as well as information from the transmission to determine the optimal shift timing. Information about the operation of the transmission is received by the PCM through signals from an output shaft sensor (OSS), vehicle speed sensor (VSS), transmission oil temperature **(TOT)** sensor, and a transmission range sensor (TR sensor) (Figure 8-18).

The transmission control system uses four solenoids for control of operation. The MCCC solenoid is used for modulated converter clutch control. The EPC solenoid is used to control hydraulic pressures throughout the transmission. The remaining two solenoids are shift solenoids.

The EPC solenoid replaces the conventional T.V. cable setup to provide changes in pressure in response to engine load. This solenoid is a variable force solenoid **(VFS)** and contains a spool

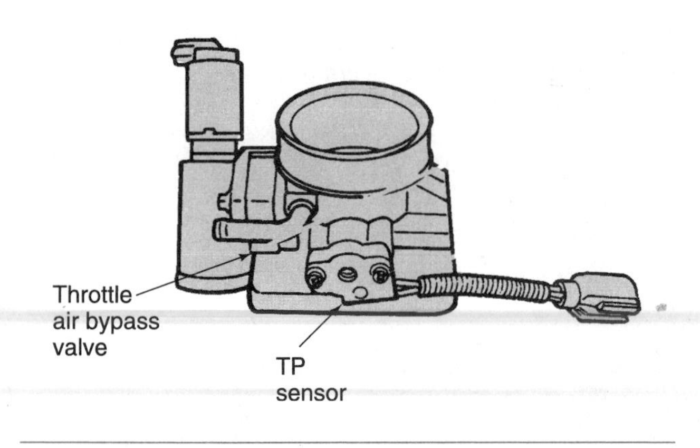

Figure 8-17 Typical TP sensor assembly. (Reprinted with the permission of Ford Motor Co.)

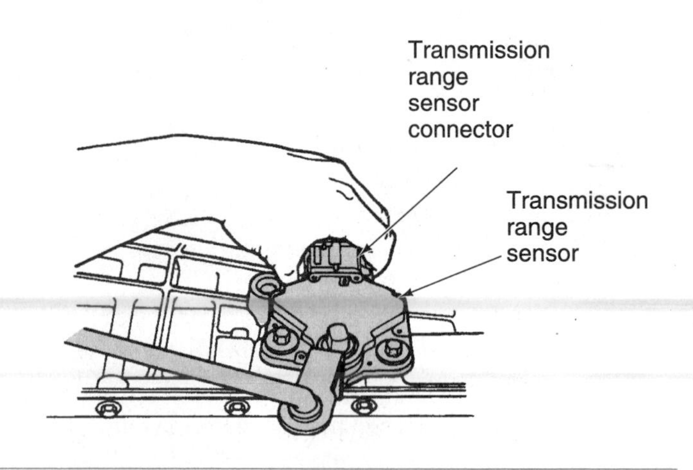

Figure 8-18 Typical transmission range (TR) sensor assembly and mounting. (Reprinted with the permission of Ford Motor Co.)

valve and spring (Figure 8-19). To control fluid pressure, the PCM sends a varying signal to the solenoid. This varies the amount the solenoid will cause the spool valve to move. When the solenoid is off, the spring tension keeps the valve in place to maintain maximum pressure. As more current is applied to the solenoid, the solenoid moves the spool valve more, which moves to uncover the exhaust port more, thereby causing a decrease in pressure (Figure 8-20). The shift solenoids offer four possible on/off combinations to control fluid to the various shift valves.

Ford 4R70W Controls

In 1993, Ford introduced a modified heavy-duty version of the AOD-E called the 4R70W. Initially, this transmission was used in passenger cars; now it is most commonly used in the redesigned F-150 pickup, Expedition, Mountaineer, and Explorer. The transmission system includes four solenoids: two shift solenoids, an EPC solenoid, and a modulated converter clutch solenoid.

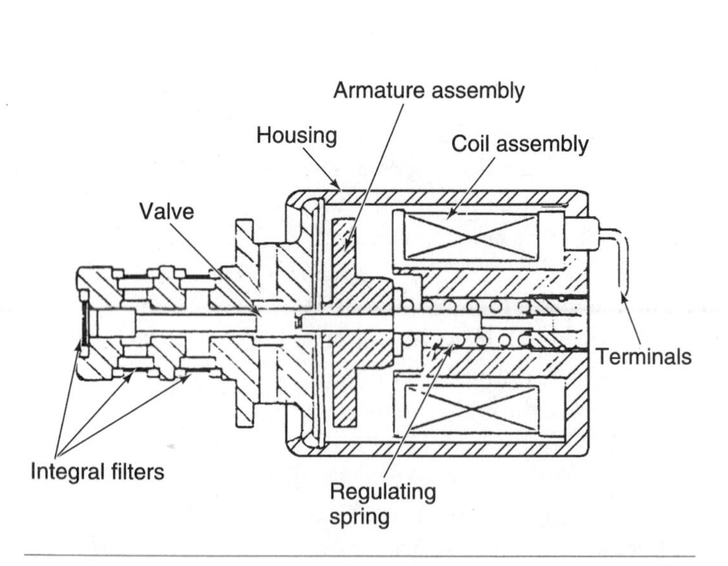

Figure 8-19 Typical variable force solenoid or motor. (Courtesy of the Hydra-Matic Division of General Motors Corp.)

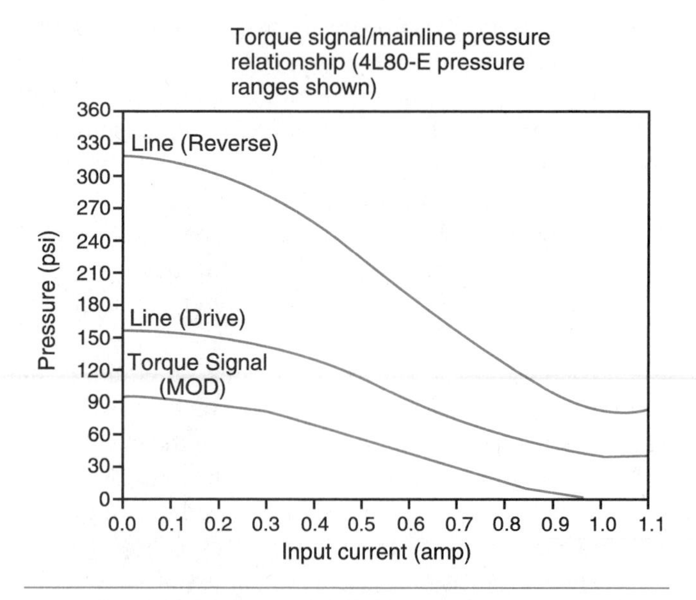

Figure 8-20 The EPC solenoid controls hydraulic pressures throughout the transmission by controlling the current going to the solenoid. (Courtesy of the Hydra-Matic Division of General Motors Corp.)

The EPC solenoid controls line pressure at all times, based on the programming of the system's computer. Likewise, the operation of the converter clutch is also totally controlled by computer. The only exception to this is during first gear and reverse gear operation when the clutch is hydraulically disabled to prevent lockup regardless of the commands by the computer.

Ford AX4S (AXODE) Controls

The AX4S is an electronically controlled four-speed transaxle. Shift control solenoids provide gear changes. These solenoids are controlled by the PCM, as is the converter lockup clutch solenoid (Figure 8-21). Five solenoids are mounted on the AX4S valve body to control transmission functions. One solenoid, the MCCC solenoid, is used for modulated converter clutch control. Another, the EPC solenoid, is used to control hydraulic pressures throughout the transmission. The remaining three solenoids are shift solenoids.

The EPC solenoid replaces the conventional T.V. cable setup to provide changes in pressure in response to engine load. This solenoid is a variable force solenoid (VFS) and contains a spool valve and spring. To control fluid pressure, the PCM sends a varying signal to the solenoid. This varies the amount the solenoid will cause the spool valve to move. When the solenoid is off, the spring tension keeps the valve in place to maintain maximum pressure. As more current is applied to the solenoid, the solenoid moves the spool valve more, which moves to uncover the exhaust port more, thereby causing a decrease in pressure.

The shift solenoids offer many possible on/off combinations to control fluid to the various shift valves. These solenoids are normally in the open position. Being open, the solenoid allows fluid flow to the bore of the shift valve. When the shift solenoids are energized, they block fluid flow to the valve.

A summary of the shift solenoid activity follows:

Selector Lever Position	Operating Gear	#1 Solenoid	#2 Solenoid	#3 Solenoid
OD	First gear	Off	On	Off
	Second gear	On	On	Off
	Third gear	Off	Off	On
	Fourth gear	On	Off	On
D (Drive)	First gear	Off	On	Off
	Second gear	On	On	Off
	Third gear	Off	Off	Off
1	First gear	Off	On	Off
	Second gear	Off	Off	Off
R	Reverse	Off	On	Off
P or N	Park/Neutral	Off	On	Off

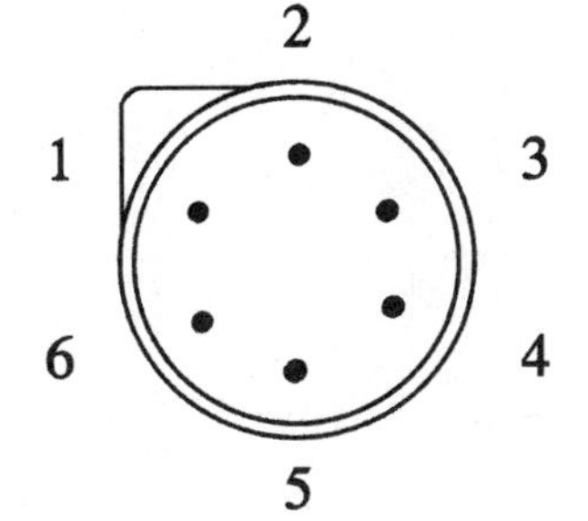

1. EPC solenoid ground
2. EPC solenoid power
3. TOT sensor
4. TOT sensor
5. LU solenoid power
6. LU solenoid ground

Figure 8-21 Identification of the pins in a typical electrical connector at the top of an AX4S housing. (Courtesy of ATRA)

Ford A4LD Controls

The A4LD is a four-speed automatic transmission with a lockup torque converter. The engine control system controls the operation of the lockup converter clutch and the operation of a shift solenoid for third to fourth gearshifting. Prior to 1987, third to fourth gearshifting was done hydraulically. Then in 1987, an electronic third to fourth shift solenoid was added.

This solenoid is normally closed, which inhibits the third to fourth gearshift. When the PCM receives information that the time is right to allow the third to fourth gearshift, the solenoid is energized and opens. This allows the shift into fourth gear.

Ford 4R44E, 4R55E, and 5R55E Controls

The 4R44E and 4R55E were introduced in 1995 for use in light-duty compact trucks. These transmissions are based on the previous models of the A4LD. The 4R55E is a heavy-duty version of the 4R44E. The 5R55E is a five-speed version of the 4R55E. This five-speed version is similar to the other designs, except for the controls necessary to provide the additional gear. A planetary gear set was not added; rather, new combinations of apply and holding devices were programmed into the design. The electronic controls for these transmissions are based on the A4LD and other Ford systems. These transmissions are equipped with five or six solenoids mounted to the valve body. These solenoids control shift quality and timing and torque converter clutch operation.

Ford AX4N Controls

The AX4N transaxle (Figure 8-22) is very similar in design to the AX4S (AXOD-E). It is a four-speed unit found in 1994 and newer Taurus, Sable, and Continental models. The unit uses two bands, five multiple-disc clutches, and three one-way clutches. The design of this transaxle allows it to be nonsynchronous. Nonsynchronous transmissions tend to have smoother shifts. In the AX4N, to eliminate the need for the synchronized release of the intermediate band with the application of the direct clutch, the intermediate band was eliminated. In its place, Ford uses a low/intermediate clutch and an intermediate one-way clutch. Actually, the intermediate band was not eliminated; it changed names and is now called the coast brake. With the new name came a new job; to allow for engine braking in first and second gears when the gear selector is in the Manual 1 position.

The electronic control system (Figure 8-23) is similar to that used on the AX4S, except the solenoids are mounted on the valve body. The shift solenoids are used for electronic shift scheduling (Figure 8-24). The EPC solenoid controls line pressure to ensure good shift feel.

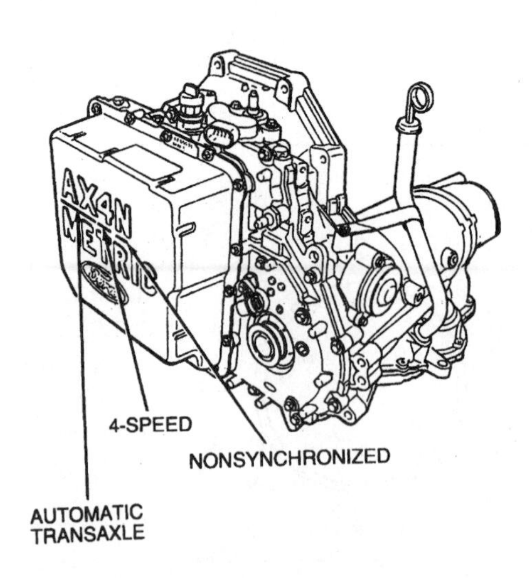

Figure 8-22 A Ford AX4N transaxle. (Reprinted with the permission of Ford Motor Co.)

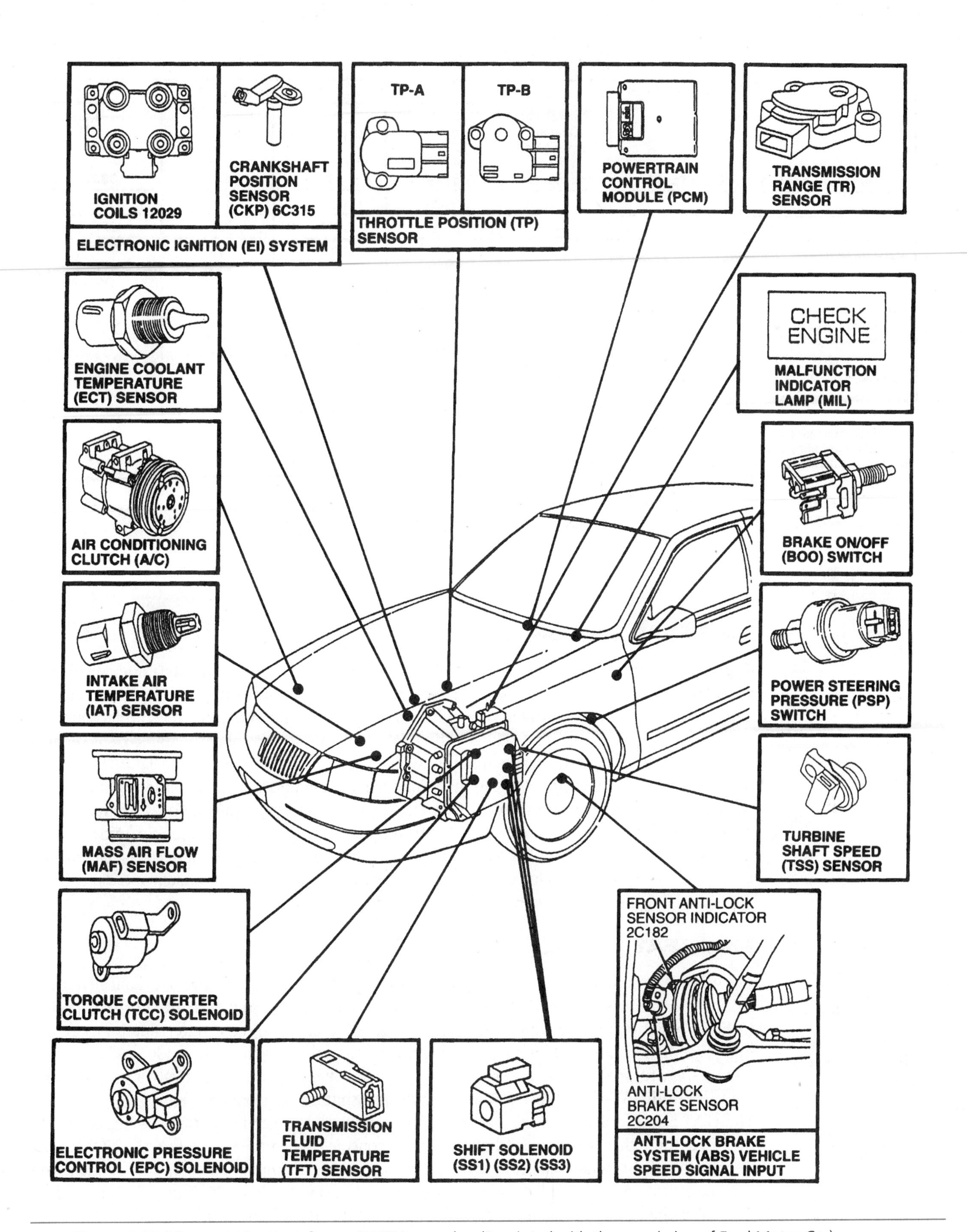

Figure 8-23 The control system for an AX4N transaxle. (Reprinted with the permission of Ford Motor Co.)

SHIFT SOLENOID OPERATION CHART

Transaxle Range Selector Lever Position	Powertrain Control Module (PCM) Gear Commanded	AX4N Solenoids			
		Eng Brake	SS1	SS2	SS3
P/N[c]	P/N	NO	OFF[a]	ON[a]	OFF
R	R	YES	OFF	ON	OFF
Ⓓ(OVERDRIVE)	1	NO	OFF	ON	OFF
	2	NO	OFF	OFF	OFF
	3	NO	ON	OFF	ON
	4	YES	ON	ON	ON
D (DRIVE)	1	NO	OFF	ON	OFF
	2	NO	OFF	OFF	OFF
	3	YES	ON	OFF	OFF
MANUAL 1	2[b]	YES	OFF	ON	OFF
	3[b]	YES	OFF	OFF	OFF
		YES	ON	OFF	OFF

c When transmission fluid temperature is below 50° then SS1=OFF, SS2=ON, SS3=ON to prevent cold creep.
a Not contributing to powerflow.
b When a manual pull-in occurs above calibrated speed the transaxle will downshift from the higher gear until the vehicle speed drops below this calibrated speed.

Figure 8-24 Shift solenoid operation for an AX4N transaxle. (Reprinted with the permission of Ford Motor Co.)

Ford CD4E Controls

The CD4E transaxle was introduced in 1994 and is used in Probe, Contour, Mystique, Mazda 626, and Mazda MX-6 models. The transaxle uses one band, five multiple-disc clutches, two one-way clutches, and five solenoids to provide four forward gears. Shift timing, shift feel, and torque converter clutch operation are controlled by the PCM. In the PCM, control of the transaxle is separate from the control of the engine systems, although some of the inputs are shared (Figure 8-25).

Using these inputs, the PCM can determine when the time and conditions are right for a shift or application of the converter clutch. The PCM can also determine the line pressure required to provide for a correct shift feel. The PCM controls the operation of the EPC solenoid, two shift solenoids, one pulse width modulated converter clutch solenoid, and a 3–2 timing/coast clutch solenoid. The 3–2 timing/coast clutch solenoid controls the release of the coast clutch and direct clutch while applying the low and intermediate band during a 3–2 downshift.

Ford 4EAT Controls

This transaxle is produced by Mazda but shared by Ford and Mazda, primarily in the Probe and MX-6.

The 4EAT is a four-speed transaxle that is controlled both electronically and hydraulically. Input signals from sensors are sent to the 4EAT control module, the **Mazda electronic control system (MECS),** which controls the appropriate gear range, shift timing, and converter clutch lockup timing. The 4EAT control module has self-diagnostic capabilities, a fail-safe mode, and a warning code display for the main input sensors and solenoid valves. Four solenoid valves are located in the valve body. These redirect fluid flow and cause shifting and converter lockup (Figure 8-26). The 4EAT can be operated in manual or automatic shift mode. Manual shifting can be selected by the driver through a switch located on the center console. When the switch is in the NORM position, the transmission automatically shifts gears to achieve maximum fuel economy. Some model cars have a switch that allows the driver to select in and out of a POWER range. Selection of the POWER range causes a delay in the shift timing to provide for maximum acceleration.

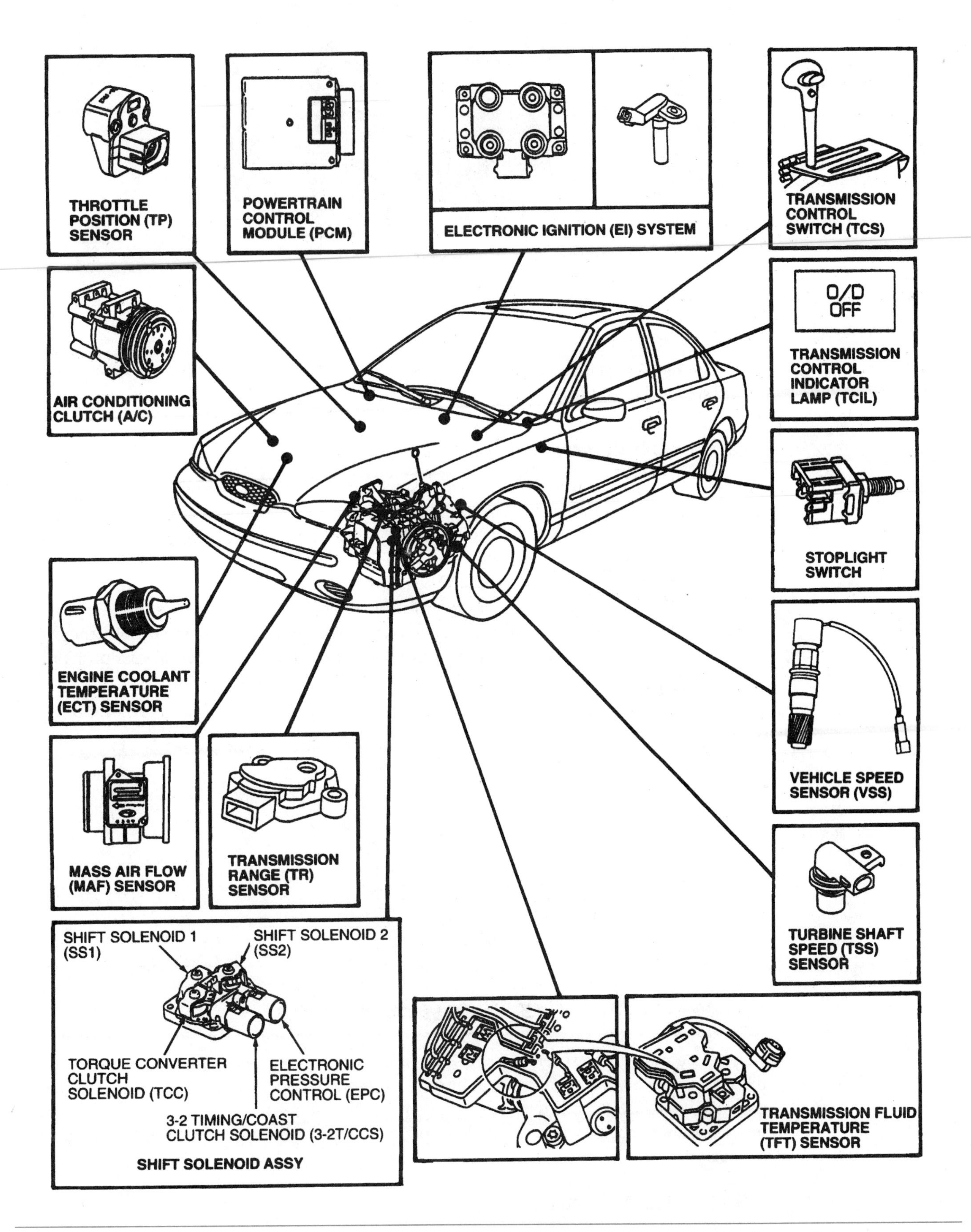

Figure 8-25 The control system for a CD4E transaxle. (Reprinted with the permission of Ford Motor Co.)

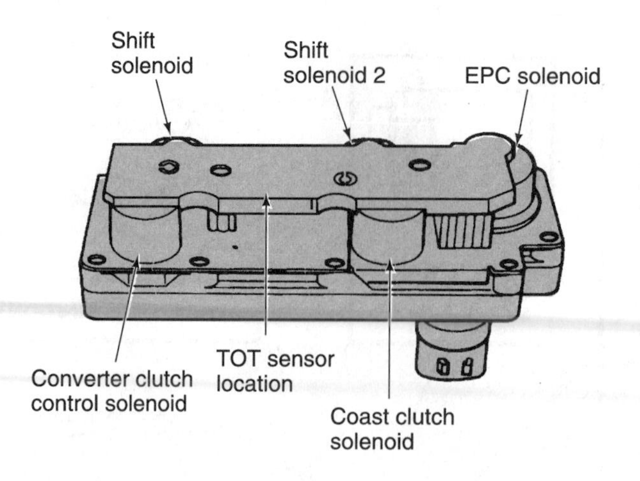

Figure 8-26 Typical transmission solenoid assembly. (Reprinted with the permission of Ford Motor Co.)

Ford E4OD Controls

This transmission was the first electronically controlled transmission for trucks and was introduced in 1989. The E4OD replaced the C-6 and is a modification of the C-6. To provide for a fourth gear and smoother operation, an overdrive planetary gear set, an overdrive band, an intermediate one-way clutch, and an intermediate clutch were added.

The action of the E4OD is controlled by the PCM, as is the operation of the converter clutch. Five solenoids are used. One solenoid controls converter clutch operation, and an EPC solenoid is used to control line pressure. There are also two shift solenoids and a coast clutch solenoid. The following chart lists the operation of these three solenoids according to gear selector position.

Selector Lever Position	Operating Gear	#1 Solenoid	#2 Solenoid	#3 Solenoid
OD	First gear	On	Off	Off
	Second gear	On	On	Off
	Third gear	Off	On	Off
	Fourth gear	Off	Off	Off
1	First gear	On	Off	On
	Second gear	Off	Off	On
2	Second gear	Off	Off	On
R	Reverse	On	Off	Off
P or N	Park/Neutral	On	Off	Off

General Motors Transmissions

General Motors is using several transmission models in its current line of automobiles. Although these transmissions are significantly different in design and operation, most of the electronically controlled units are controlled in the same way. A look at some of the more commonly used transmissions should provide an accurate description of all GM electronically controlled transmission systems.

Shop Manual
Chapter 8, page 343

THM 4L80-E Controls

The THM 4L80-E was introduced in 1991 and is based on the same design as the THM 400 with the THM 200-4R overdrive assembly and with an electronic control system. All automatic upshifts and downshifts are electronically controlled by the power train control module (PCM) for gasoline engines or the transmission control module (TCM) for diesel engines (Figure 8-27).

Park, neutral, and drive ranges are hydraulically controlled according to the position of the manual shift valve, as are selector-operated, forced downshifts. The PCM/TCM also has a self-diagnostic mode that is capable of storing transmission fault codes. The PCM is programmed to adjust its operating parameters in response to changes within the system, such as component wear. As component wear and shift overlap times increase, the PCM adjusts line pressure by controlling the VFM to maintain proper shift timing calibrations.

A **variable force motor (VFM)** controlled by the PCM is used to change line pressure in response to engine speed and vehicle load. By responding to current operating conditions, the VFM is able to match shift timing and feel with the current needs of the vehicle. The lockup converter is hydraulically applied and electrically controlled through a **pulse width modulated (PWM)** solenoid (Figure 8-28) and the PCM/TCM.

According to J1930, a VFM should be called a pressure control solenoid (PCS). Most literature still refers to it as a VFM.

CUSTOMER CARE: Vehicles with a 4L80-E transmission should have their transmission fluid changed on a regular basis as stated in the owner's manual. These transmissions use aluminum valves in the valve body instead of steel valves. This makes the valves subject to damage if the ATF contains metallic particles. The periodic fluid changes can prevent serious damage.

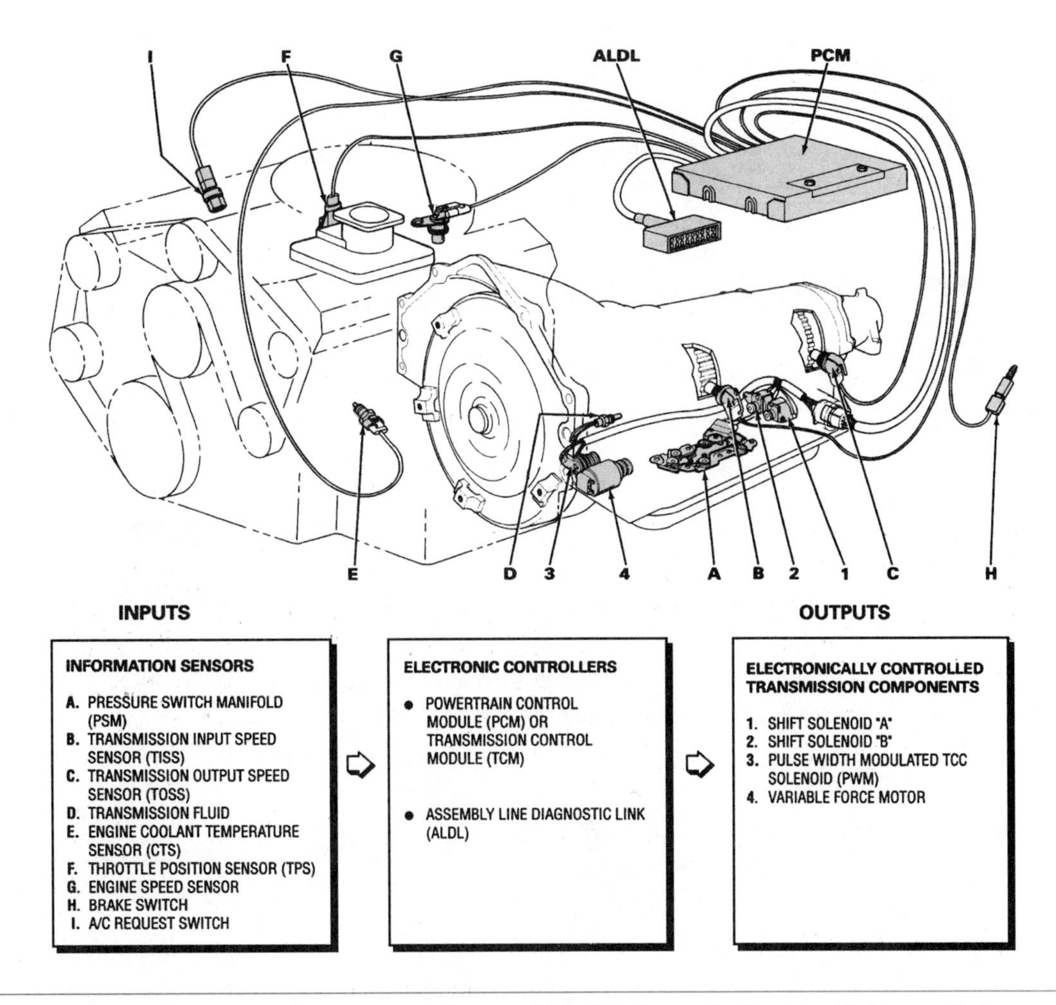

Figure 8-27 Layout of components for a typical GM electronically controlled transmission. (Courtesy of the Hydra-Matic Division of General Motors Corp.)

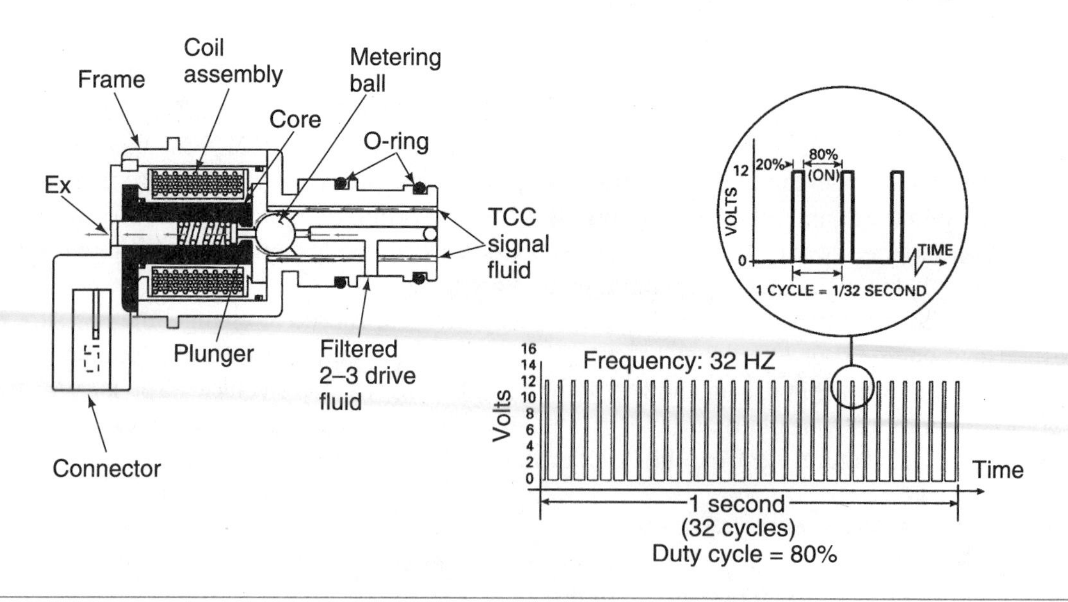

Figure 8-28 A typical PWM solenoid (left). The signal representing the control or ordered duty cycle from the computer (right). (Courtesy of the Hydra-Matic Division of General Motors Corp.)

There may be differences in valve shape or location according to vehicle application. For this reason, always refer to the correct service manual for the exact location of the valves.

The following valves and solenoids regulate system pressure and direct fluid to the proper apply devices. The TCC shift, TCC enable, pressure regulator, converter limit, and reverse boost valves are located in the oil pump cover. The torque signal compensator valve is in the accumulator housing. All other valves are in the valve body.

- Pressure regulator valve—regulates line pressure in the system and directs it to the converter limit valve and pump suction circuit.
- Converter limit valve—routes converter feed pressure to the regulated converter feed circuit and limits converter feed pressure to 93–107 psi (641–738 kPa).
- TCC regulator apply valve—regulates line pressure to the converter apply circuit when the PWM solenoid is energized.
- Manual control valve—mechanically linked to the shift selector lever. Directs line pressure to the appropriate passages for clutch and servo application by opening and closing feed passages.
- Reverse boost valve—boosts line pressure in reverse to accommodate increased torque requirements. Applies against the pressure regulator valve to boost line pressure relative to engine torque.
- Actuator feed limit valve—limits line pressure to a maximum of 105–125 psi (724–862 kPa).
- Variable force motor (VFM)—provides controlled pressure to the reverse boost valve, accumulator valve, and torque signal compensator valve.
- Torque signal compensator valve—dampens pressure irregularities in each gear range caused by VFM operation.
- TCC enable valve—routes regulated converter feed fluid to apply and release the TCC.
- TCC shift valve—controls TCC apply and release according to the position of the TCC regulator valve and PWM solenoid.
- Solenoids A and B—simple on/off solenoids that control upshifts and downshifts in all forward gear ranges. They either pressurize or exhaust fluid passages to the 1–2, 2–3, and 3–4 shift valves. They do not regulate fluid flow, but only turn the flow on or off.

- 1–2 shift valve—directs fluid flow according to shift solenoid operation, actuator feed pressure, and return spring force to control 1–2 and 2–1 shifts.
- 2–3 shift valve—directs fluid flow according to shift solenoid operation, actuator feed pressure, and return spring force to control 2–3 and 3–2 shifts.
- 3–4 shift valve—directs fluid flow according to shift solenoid operation, actuator and fourth clutch feed pressure, and return spring force to control 3–4 and 4–3 shifts.
- Accumulator valve—controls fluid flow and exhaust in the accumulator circuit according to VFM operation.
- Third clutch accumulator—works with the accumulator valve to cushion direct clutch application by absorbing some of the apply pressure.
- Fourth clutch accumulator—works with the accumulator valve to cushion fourth clutch application by absorbing some of the apply pressure.
- Rear accumulator—works with the accumulator valve to cushion intermediate clutch application during 1–2 upshifts by absorbing some of the apply pressure.

Electronic Control System

The PCM is a multifunction computer that controls all engine and transmission operations. It receives information on engine operating conditions from different sensors and switches. The throttle position (TP) sensor provides a variable voltage signal to the PCM. This signal is used to calculate throttle position angle. It helps the PCM determine the appropriate shift patterns and TCC apply and release timing. The engine coolant temperature (ECT) sensor also sends a variable signal to the PCM. This signal varies with the changes in engine coolant temperature. If the engine temperature is less than approximately 130°F (54°C), the PCM prevents TCC apply. The ECT is not used on vehicles equipped with a diesel engine.

The PCM also receives a variable signal from the ignition module, which indicates current engine speed. This input is used to determine shift timing and TCC apply and release. The ignition module converts the analog signal from the distributor pickup coil to a digital square-wave signal and sends it to the PCM, which compares it to a fixed-frequency clock signal to determine engine rpm.

The PCM also receives input from two switches. The brake switch simply tells the PCM when the brake pedal is depressed so that it will release the TCC. An A/C request switch informs the PCM that the A/C has been turned on. The PCM then changes line pressure and shift timing to accommodate the extra engine load created by the compressor operation.

In addition to these engine-related inputs, the PCM also receives information from various transmission sensors:

Shop Manual
Chapter 8, page 335

- **Transmission output speed sensor (TOSS)**—This variable reluctance magnetic pickup is installed in the transmission case. It is positioned opposite a 40-toothed rotor, which is pressed onto the output carrier. Rotor tooth rotation past the TOSS induces an alternating current. This ac voltage signal is proportional to vehicle speed and is sent either to the PCM or to a **digital radio adapter controller (DRAC),** depending on the transmission application. The PCM or DRAC converts the analog ac signal to a digital 5-volt square wave. This digital signal is compared to a fixed-frequency clock signal inside the PCM to determine transmission output speed.
- **Transmission input speed sensor (TISS)**—This sensor and its operation are identical to the TOSS except that the 31 serrations machined on the forward clutch housing serve as the rotor. The signal is converted in the same way and is used by the PCM to calculate converter turbine speed. Input and output speeds provided by the TOSS and TISS are used by the PCM to help determine line pressure, shift patterns, and TCC apply pressure and timing.

- Vehicle speed sensor (VSS)—Four-wheel-drive vehicles use a third speed sensor installed in the transfer case. The DRAC determines vehicle speed from this sensor, rather than from the TOSS. TOSS output is sent unbuffered to the PCM, where it is compared to TISS output. The VSS analog signal is converted to a digital signal by the DRAC.
- Pressure switch manifold—This multiple switch assembly consists of five normally open pressure switches installed on the bottom of the valve body. The manual switch feeds fluid to the switches according to selector lever position. Fluid pressures determine the digital logic at electrical pins A, B, and C. The PCM uses the logic to determine the transmission gear range.
- Temperature sensor—A temperature-sensitive resistor is installed in the valve body. Its electrical resistance varies with ATF temperature. The PCM integrates this input with others to control TCC operation through the PWM solenoid and the VFM. The PCM prevents TCC operation until fluid temperatures reach about 68°F (20°C). If fluid temperature reaches about 250°F (122°C), the PCM applies the TCC in second, third, or fourth gears. If mechanically coupling the converter to the transmission does not reduce fluid temperature and it reaches 300°F (150°C), the PCM releases the TCC to prevent damage to the converter clutch from excessive temperatures. If the fluid reaches about 310°F (154°C), the PCM sets a fluid temperature code. When this happens, the PCM uses a fixed value of 266°F (130°C) as the fluid temperature input signal and applies the TCC in second, third, and fourth gears.
- **Barometric pressure sensor (BARO)**—The signals from this sensor are used by the PCM to adjust line pressures according to changes in altitude. This sensor input may not be used. Its use depends on the type of intake air monitoring system the vehicle is equipped with. On those vehicles using the BARO sensor as an input, the sensor may be integrated on the PCM circuit board or mounted externally.

Operation

Once the PCM has processed the input signals, it controls transmission operation through two on/off solenoids, a pulse width modulated (PWM) solenoid, and a variable force motor (VFM) located in the valve body. The two shift solenoids are attached to the valve body and are normally open. There are four possible on/off combinations of the solenoids that determine fluid pressure flow to the shift valves in the valve body.

The shift solenoids receive voltage through the ignition switch and are grounded through the PCM. When a solenoid is energized by the PCM, a check ball held in place by the solenoid plunger blocks the fluid pressure feed. This closes the exhaust passage and causes signal fluid pressure to increase. When the solenoid is deenergized, fluid pressure moves the check ball and plunger off the check ball seat. This allows fluid to flow past the check ball and exhaust through the solenoid, decreasing signal pressure.

The variation in length of time the solenoid is energized per cycle is called the solenoid's duty cycle.

The pulse width modulated (PWM) solenoid is a normally closed valve installed in the valve body. It controls the position of the TCC apply valve. When the solenoid is off, TCC signal fluid exhausts and the converter clutch remains released. Once the solenoid is energized, the plunger moves the metering ball to allow TCC signal fluid to pass to the TCC regulator valve. The PCM cycles the PWM solenoid on and off 32 times per second, but it varies the length of time the PWM solenoid is energized in each cycle, or 1/32 second.

The VFM is an electrohydraulic actuator made of a variable force solenoid and a regulating valve. The VFM is installed in the valve body and controls mainline pressure by moving a pressure regulator valve against spring pressure. The VFM operates by a constantly changing current from the PCM. The PCM turns the solenoid on and off 292.5 times per second, but varies the solenoid's duty cycle. When the duty cycle is zero, line pressure is at maximum. Under normal operating conditions, the PCM pulses the VFM about every 10 seconds to either 100 percent duty cycle or 0 percent duty cycle. This prevents the VFM valve from sticking in any given position due to contamination. The torque signal compensator valve spring prevents major fluctuations in line pressure by absorbing the pulsations caused by the temporary full duty of the solenoid.

THM 4T60-E Controls

The 4T60-E transaxle is a four-speed automatic transmission with electronically controlled shifting and converter lockup. The torque converter clutch is controlled by two electronic solenoids, one for apply and release and the other to control the feel of the apply and release of the TCC. Some 4T60-E transaxles use a PWM solenoid to control converter lockup. The operation of this transaxle is very similar to the 4L80-E transmission (Figure 8-29). The transaxle relies on the energizing and deenergizing of two solenoids to cause a change of gears. A summary of the solenoid activity and power flow is shown in Figure 8-30.

The primary inputs for transmission operation are from the throttle position sensor, vehicle speed sensor, engine coolant temperature sensor, manifold air pressure sensor, and from the cruise control, brake, low gear pressure, and fourth gear pressure switches. With this information, the PCM is able to calculate the optimal time for gear changes.

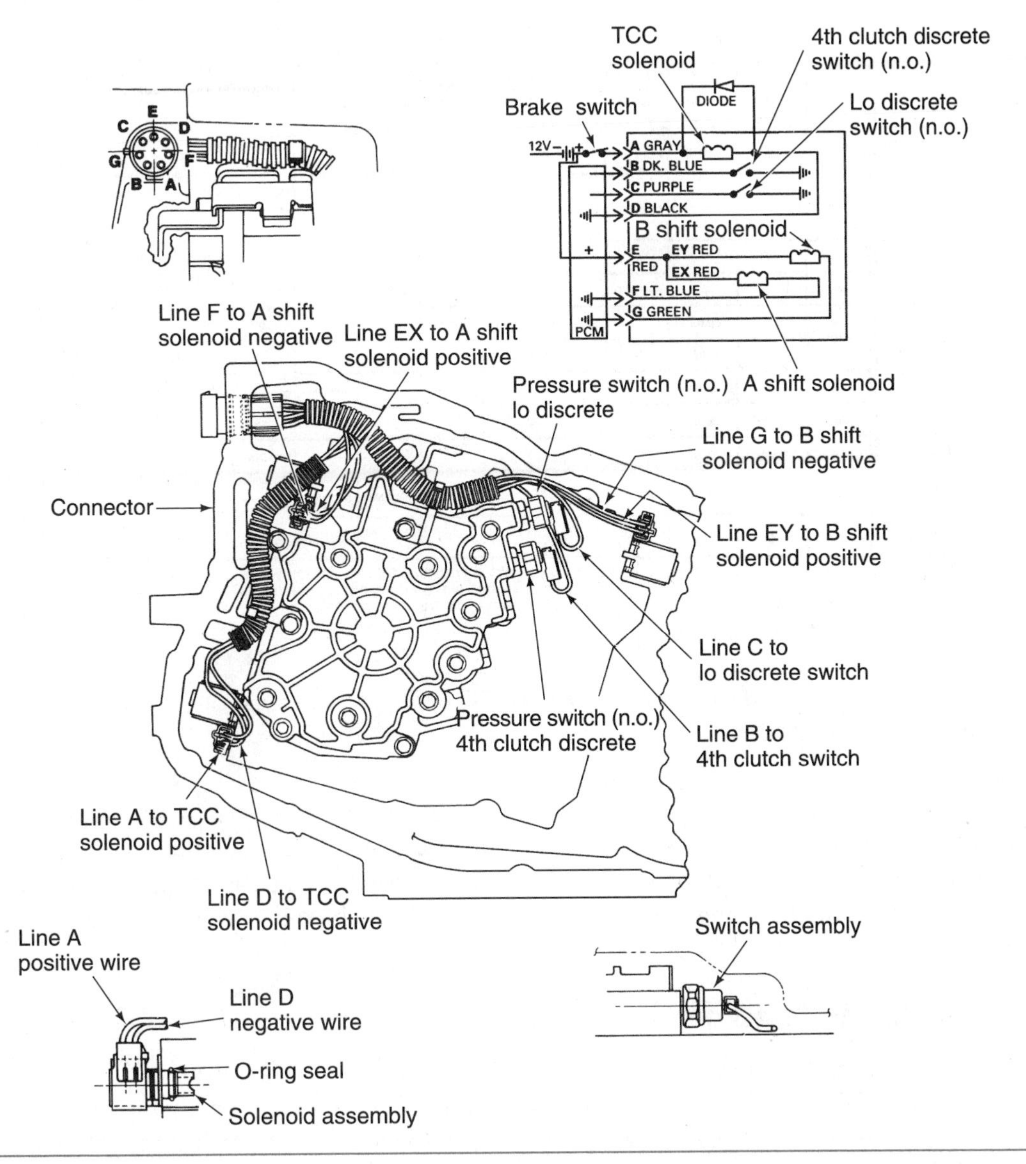

Figure 8-29 Wiring diagram for a 4T60-E transaxle. (Courtesy of the Chevrolet Division of General Motors Corp.)

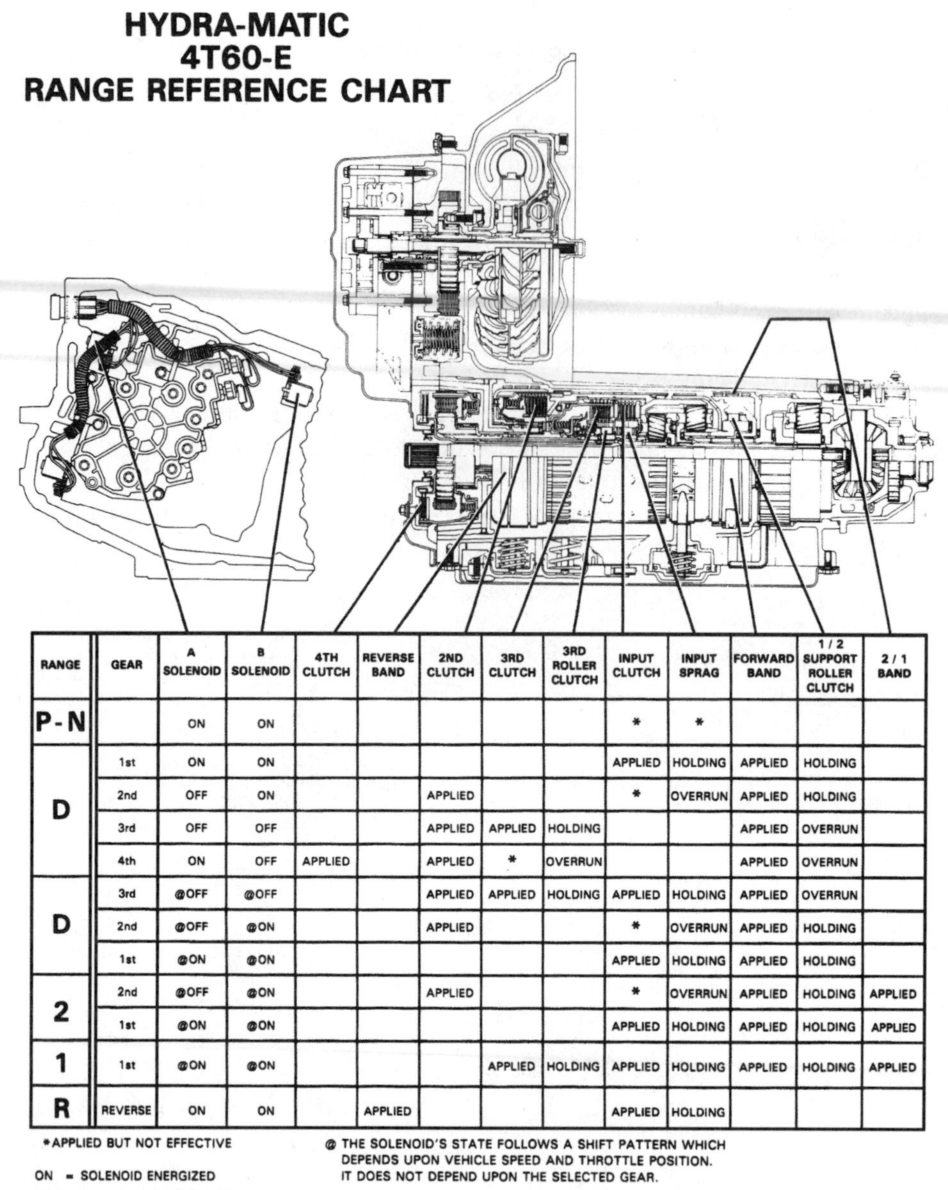

HYDRA-MATIC
4T60-E
RANGE REFERENCE CHART

RANGE	GEAR	A SOLENOID	B SOLENOID	4TH CLUTCH	REVERSE BAND	2ND CLUTCH	3RD CLUTCH	3RD ROLLER CLUTCH	INPUT CLUTCH	INPUT SPRAG	FORWARD BAND	1/2 SUPPORT ROLLER CLUTCH	2/1 BAND
P-N		ON	ON						*	*			
D	1st	ON	ON						APPLIED	HOLDING	APPLIED	HOLDING	
	2nd	OFF	ON			APPLIED			*	OVERRUN	APPLIED	HOLDING	
	3rd	OFF	OFF			APPLIED	APPLIED	HOLDING			APPLIED	OVERRUN	
	4th	ON	OFF	APPLIED		APPLIED	*	OVERRUN			APPLIED	OVERRUN	
D	3rd	@OFF	@OFF			APPLIED	APPLIED	HOLDING	APPLIED	HOLDING	APPLIED	OVERRUN	
	2nd	@OFF	@ON			APPLIED			*	OVERRUN	APPLIED	HOLDING	
	1st	@ON	@ON						APPLIED	HOLDING	APPLIED	HOLDING	
2	2nd	@OFF	@ON			APPLIED			*	OVERRUN	APPLIED	HOLDING	APPLIED
	1st	@ON	@ON						APPLIED	HOLDING	APPLIED	HOLDING	APPLIED
1	1st	@ON	@ON				APPLIED	HOLDING	APPLIED	HOLDING	APPLIED	HOLDING	APPLIED
R	REVERSE	ON	ON		APPLIED				APPLIED	HOLDING			

*APPLIED BUT NOT EFFECTIVE

ON - SOLENOID ENERGIZED

OFF - SOLENOID DE-ENERGIZED

@ THE SOLENOID'S STATE FOLLOWS A SHIFT PATTERN WHICH DEPENDS UPON VEHICLE SPEED AND THROTTLE POSITION. IT DOES NOT DEPEND UPON THE SELECTED GEAR.

Figure 8-30 Range reference chart for a 4T60-E transaxle. (Courtesy of the Chevrolet Division of General Motors Corp.)

On some models, the PCM is also connected to one or more other computers that operate the climate control system, antilock brake system, driver information center, and supplemental restraint system. Through multiplexing, these computers are able to share information and use common input sensors.

THM 4T65-E Controls

The 4T65-E is a modified version of the 4T60-E transaxle and was introduced in 1997. This transaxle provides smoother shifting and is more durable. Besides strengthening the case and clutch assemblies, the electronic control system was also modified. The vacuum modulator used on the 4T60 was replaced with a pressure control solenoid (PCS), and a pressure switch was added as an input for the PCM.

THM 4T40-E Controls

The 4T40-E was introduced in 1995 with the Chevrolet Cavalier and Pontiac Sunfire. The shifting of this four-speed transaxle is controlled by two shift solenoids that are controlled by the PCM. The transaxle also has a PCS to control line pressure. The basic electrical components of this transaxle are shown in Figure 8-31.

A feature of this transaxle is its ability to change line pressure in response to normal clutch, seal, and spring wear. The PCM controls the PCS to allow for this adapting to wear and to ensure proper shift quality during all operating conditions. The PCS is controlled by varying the duty cycle of the solenoid, which varies the current flow through its windings. As current flow increases, the magnetic field around the windings also increases. This increased strength moves the solenoid's plunger farther away from the fluid exhaust port. Allowing less fluid to exhaust will cause the pressure of the fluid to increase.

The shift solenoids operate and control the operating gears in the typical manner, as shown in the following chart.

Operating Gear	#1 Solenoid	#2 Solenoid
First gear	On	Off
Second gear	Off	Off
Third gear	Off	On
Fourth gear	On	On
Reverse	On	Off
Park/Neutral	On	Off

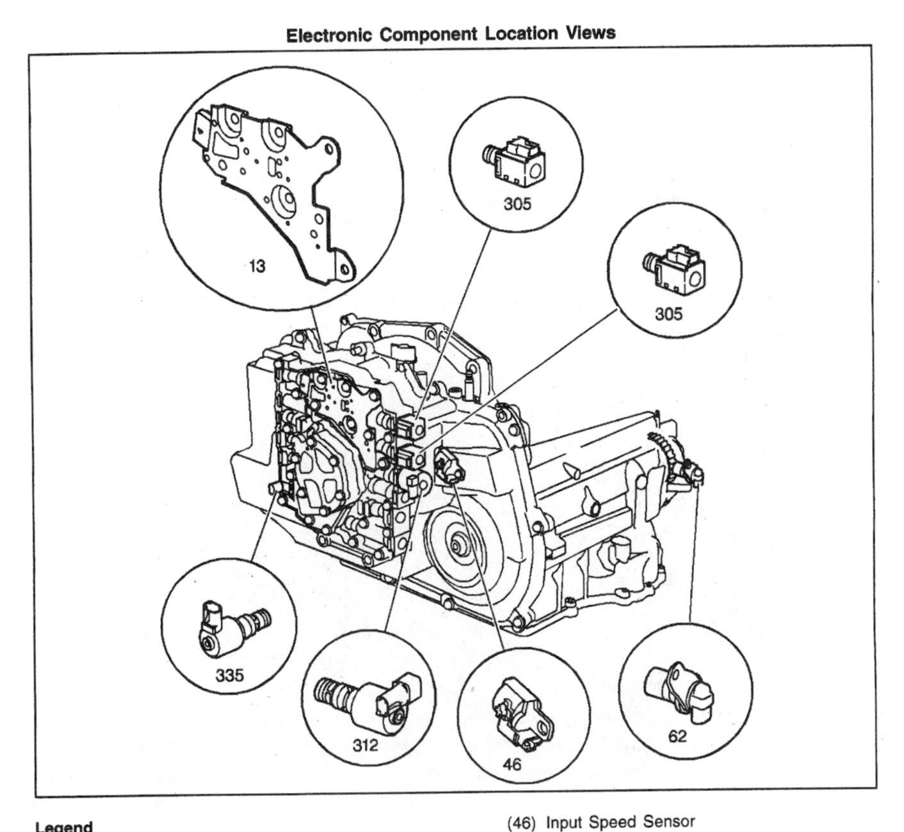

Figure 8-31 Electrical components on a 4T40-E transaxle. (Courtesy of General Motors Corp.)

Honda Transmission Controls

Shop Manual
Chapter 8, page 345

Honda automatic transmissions are unique in the industry because they use constant mesh gears (Figure 8-32), similar to a manual transmission, to provide for automatic gear changes. When certain combinations of these gears are engaged by clutches, power is transmitted from the mainshaft to the countershaft to provide the different gear ranges.

The electronic control system consists of the transmission control module (TCM), sensors, and four solenoid valves (Figure 8-33). Shifting and converter lockup are electronically controlled. Activating a shift solenoid valve changes modulator pressure, causing a shift valve to move. This allows pressurized fluid to flow through a line to engage the appropriate clutch and its corresponding gear set. Converter lockup is controlled by the action of the other two solenoid valves.

By receiving information from its input sensors, the TCM calculates the required gear changes and energizes or deenergizes the shift solenoids (Figure 8-34). A summary of the solenoid activity follows:

Selector Lever Position	Operating Gear	"A" Solenoid	"B" Solenoid
D4	First gear	Off	On
	Second gear	On	On
	Third gear	On	Off
	Fourth gear	Off	Off
D3	First gear	Off	On
	Second gear	On	On
	Third gear	On	Off
2	Second gear	On	On
1	First gear	Off	On
R	Reverse	On	Off

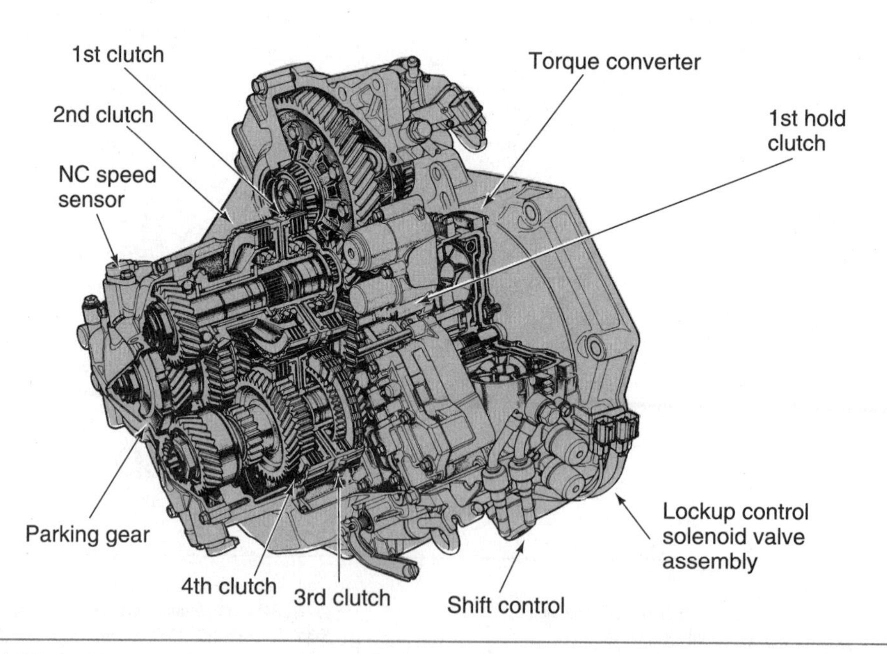

Figure 8-32 A Honda constant-mesh automatic transmission. (Courtesy of Honda Motor Company, Ltd.)

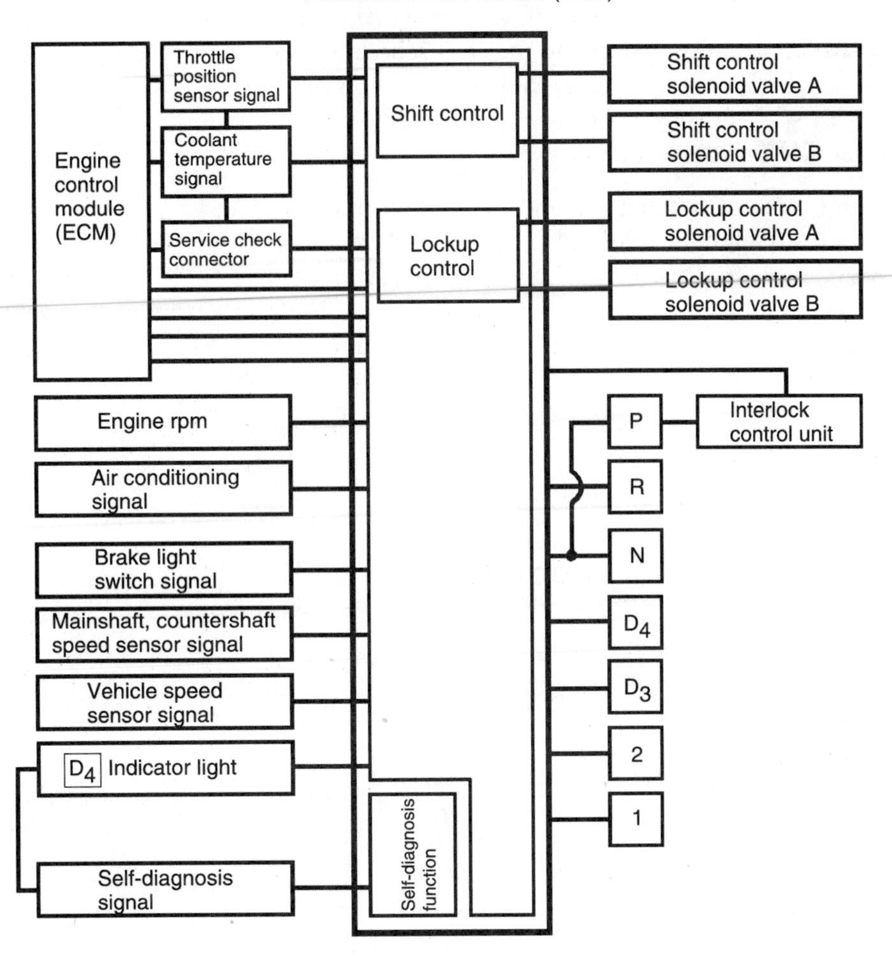

Figure 8-33 Schematic of Honda's electronically controlled transaxle. (Courtesy of Honda Motor Company, Ltd.)

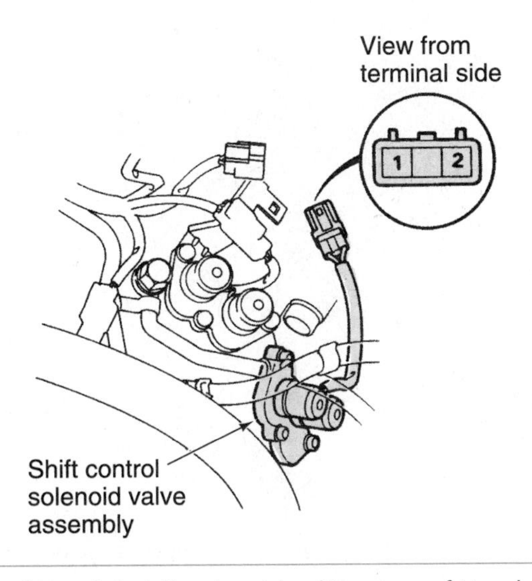

Figure 8-34 Location of Honda's shift solenoids. (Courtesy of Honda Motor Company, Ltd.)

The TCM receives information about the throttle position, engine coolant temperature, engine speed, A/C on/off, brake on/off, vehicle speed, and the speed of the transaxle's countershaft. Based on this information and the shift schedule and logic in the TCM, the transaxle is shifted to meet the current operating requirements of the vehicle.

Honda's CVT Controls

The electronic control system for Honda's CVT consists of a TCM, various sensors, three linear solenoids, and an inhibitor solenoid. Pulley ratios are always controlled by the control system (Figure 8-35). Input from the various sensors determines which linear solenoid the TCM will activate (Figure 8-36). Activating the shift control solenoid changes the shift control valve pressure, causing the shift valve to move. This changes the pressure applied to the driven and drive pulleys,

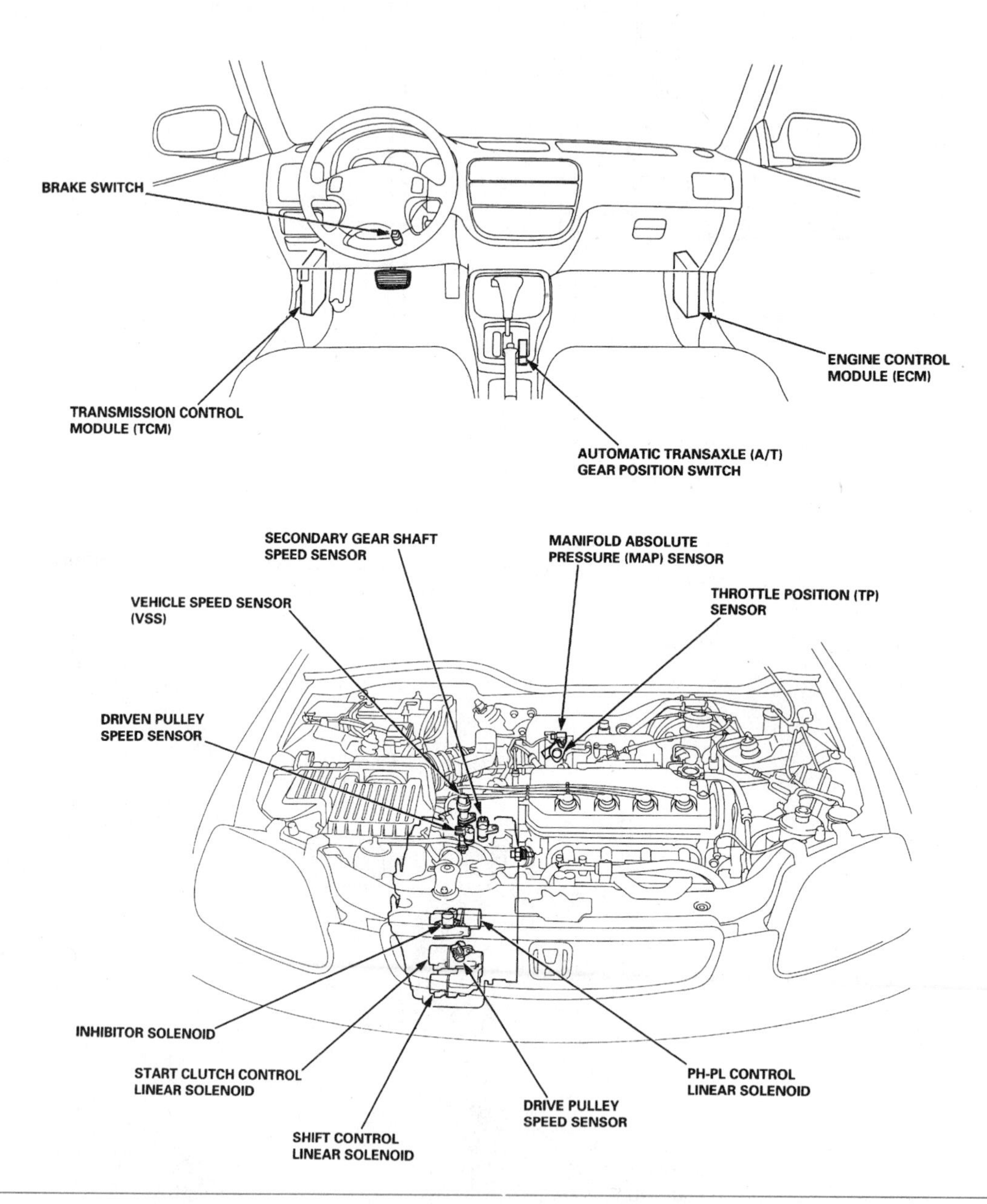

Figure 8-35 Component layout for the electronic control system for Honda's CVT. (Courtesy of Honda Motor Co., Ltd.)

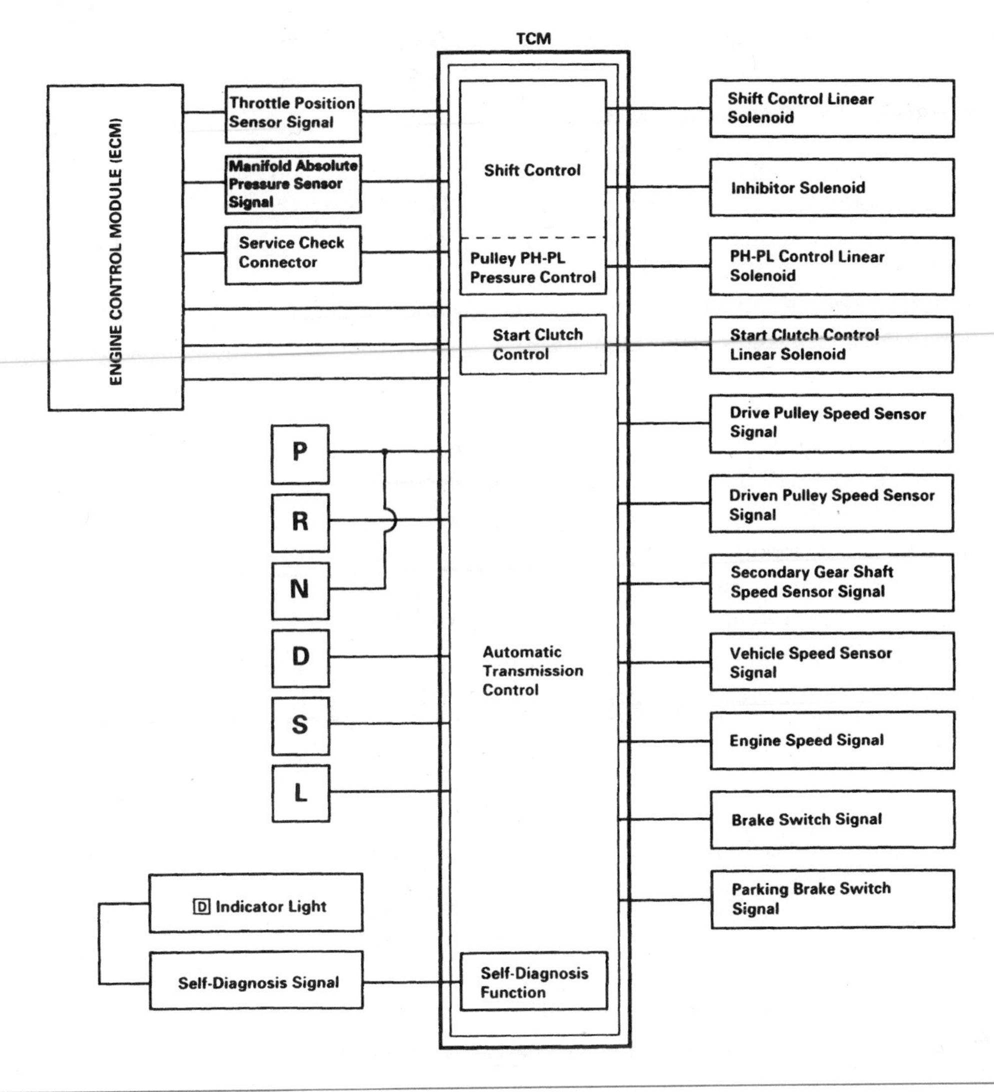

Figure 8-36 Schematic for the electronic control system for Honda's CVT. (Courtesy of Honda Motor Co., Ltd.)

which changes the effective pulley ratio. Activating the start clutch control solenoid moves the start clutch valve. This valve allows or disallows pressure to the start clutch assembly. When pressure is applied to the clutch, power is transmitted from the pulleys to the final drive gear set.

The start clutch allows for smooth starting. Since this transaxle does not have a torque converter, the start clutch is designed to slip just enough to get the car moving without stalling or straining the engine. The slippage is controlled by the hydraulic pressure applied to the start clutch. To compensate for engine loads, the TCM monitors the engine's vacuum and compares it to the measured vacuum of the engine while the transaxle was in PARK or NEUTRAL.

The TCM controls the pulley ratios to reduce engine speed and maintain ideal engine temperatures during acceleration. If the car is continuously driven at full throttle acceleration, the TCM causes an increase in pulley ratio. This reduces engine speed and maintains normal engine temperature while not adversely affecting acceleration. After the car has been driven at a lower speed or not accelerated for a while, the TCM lowers the pulley ratio. When the gear selector is placed in reverse, the TCM sends a signal to the PCM. The PCM then turns off the car's air conditioning and causes a slight increase in engine speed.

Aisin Warner Transmissions

The Aisin Warner 4 is sometimes referred to as the AW-4.

The Aisin Warner 4 is a four-speed overdrive electronically controlled automatic transmission used in some late-model Jeep Cherokee, Comanche, and Grand Cherokee vehicles. The transmission consists of a lockup torque converter, three planetary gear sets, apply devices, and a valve body fitted with three solenoids.

The solenoids are controlled by a transmission control unit (TCU) and are used to control the operation of the transmission. The solenoids #1 and #2 are used to control shifting and solenoid #3 is used for torque converter clutch lockup.

The TCU determines shift points and torque converter lockup timing based on input signals received from a TP sensor, neutral safety switch, speed sensor, and brake pedal switch.

The three solenoids are output devices controlled by signals received from the TCU. When the #1 and #2 solenoids are activated, the plungers of the solenoids are moved from their seats. This opens the exhaust port to release line pressure. When either of these solenoids is turned off, the exhaust port is closed. The result of the opening and closing of the exhaust ports is the change of fluid flow to the various apply devices. A summary of the solenoid activity follows:

Selector Lever Position	Operating Gear	#1 Solenoid	#2 Solenoid
D (Drive)	First gear	On	Off
	Second gear	On	On
	Third gear	Off	On
	Fourth gear	Off	Off
3	Third gear	Off	On
1–2	First gear	On	Off
	Second gear	On	On
R	Reverse	On	Off
P or N	Park/Neutral	On	Off

The #3 solenoid works in the opposite manner. When it is off, the plunger moves away from its seat, thereby opening the exhaust port to release line pressure. Likewise, when the solenoid is on, the exhaust port is closed.

The throttle position sensor, mounted on the throttle body, sends a voltage signal to the TCU in response to the angle of throttle opening. The TCU uses this input to control transmission shifts and for converter clutch application.

The neutral safety switch is an input device mounted on the shaft of the transmission's manual shift valve. This switch sends a signal to the TCU, which informs the computer of the location of the manual shift valve.

The speed sensor is normally mounted on the transmission's output shaft at the transmission's extension housing. It consists of a speed sensor rotor and a speed sensor. An input signal is sent to the TCU each time the output shaft makes one complete revolution. The TCU then compares the frequency of these signals to its program and determines actual vehicle speed.

The brake switch is mounted above the brake pedal and sends a brake signal to the computer when the brake pedal is depressed. The TCU uses this input for controlling the lockup converter through solenoid #3.

The TCU has an electronic self-diagnostic system that stores a fault code if certain transmission failures exist. If the fault goes away, the fault code will be erased after the ignition has been cycled approximately 75 times. On some models, a Power/Comfort switch is mounted on the instrument panel. This switch is an automatic transmission mode selection switch that allows the driver to select different modes to change transmission upshift characteristics.

Saturn Transmissions

Although Saturn is a division of General Motors Corporation, it does not use normal corporation transaxles. Saturn vehicles use an electronically controlled four-speed automatic transmission with a shift mode select function. This mode switch allows the driver to select Normal mode or Performance mode. When the switch is depressed, it provides a signal to the power train control module (PCM), which adjusts shift points and lockup timing for the selected mode.

Shop Manual
Chapter 8, page 351

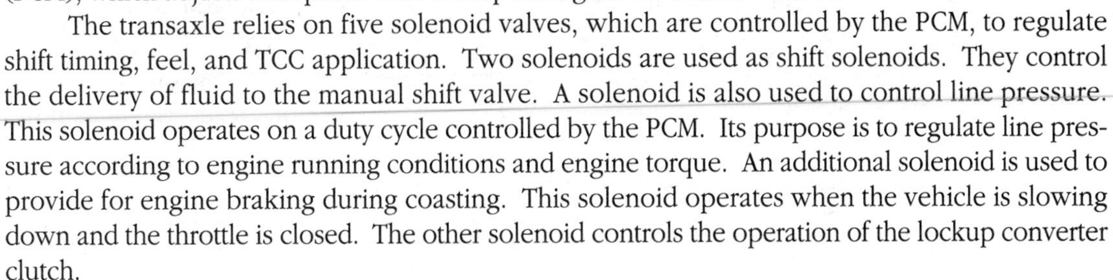

The transaxle relies on five solenoid valves, which are controlled by the PCM, to regulate shift timing, feel, and TCC application. Two solenoids are used as shift solenoids. They control the delivery of fluid to the manual shift valve. A solenoid is also used to control line pressure. This solenoid operates on a duty cycle controlled by the PCM. Its purpose is to regulate line pressure according to engine running conditions and engine torque. An additional solenoid is used to provide for engine braking during coasting. This solenoid operates when the vehicle is slowing down and the throttle is closed. The other solenoid controls the operation of the lockup converter clutch.

If an electrical circuit problem occurs, the Check Engine Soon or Shift to D2 light may come on or start flashing. The PCM will store fault codes or information flags in its memory. A stored fault code indicates that the PCM has detected a fault in the engine or transaxle circuits. Information flags help diagnose current or intermittent trouble codes.

The primary inputs to the PCM for transaxle operation are ATF temperature, brake on/off, an ignition signal, selected mode, selector lever position, torque converter turbine speed, and vehicle speed. Both the turbine speed and vehicle speed sensors are PM generators.

The ATF temperature sensor is a thermistor mounted in the line pressure port. The PCM supplies a 5-volt reference signal to the ATF temperature sensor through a resistor inside the PCM and uses a signal line to measure voltage. Signal voltage will be high when transaxle fluid is cold and it will be low when the ATF is warm.

Toyota Transmissions

Toyota led the way into electronically controlled automatic transmissions. Its transmission model, the A-140E, was the first fully electronic transmission (Figure 8-37). This transaxle is actually a three-speed unit with an add-on fourth, or overdrive, gear. The functions of the transaxle are controlled by a microprocessor located behind the glove box (Figure 8-38). The control module receives input signals from the water temperature-sensing switch, the throttle position

Shop Manual
Chapter 8, page 352

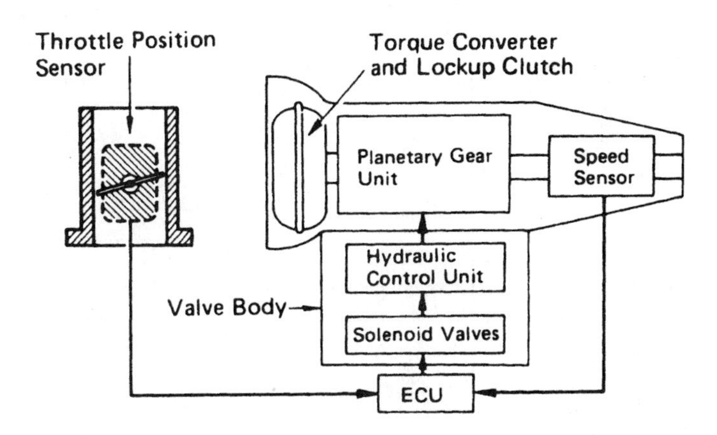

Figure 8-37 Basic layout of Toyota's A-140E transmission. (Reprinted with permission.)

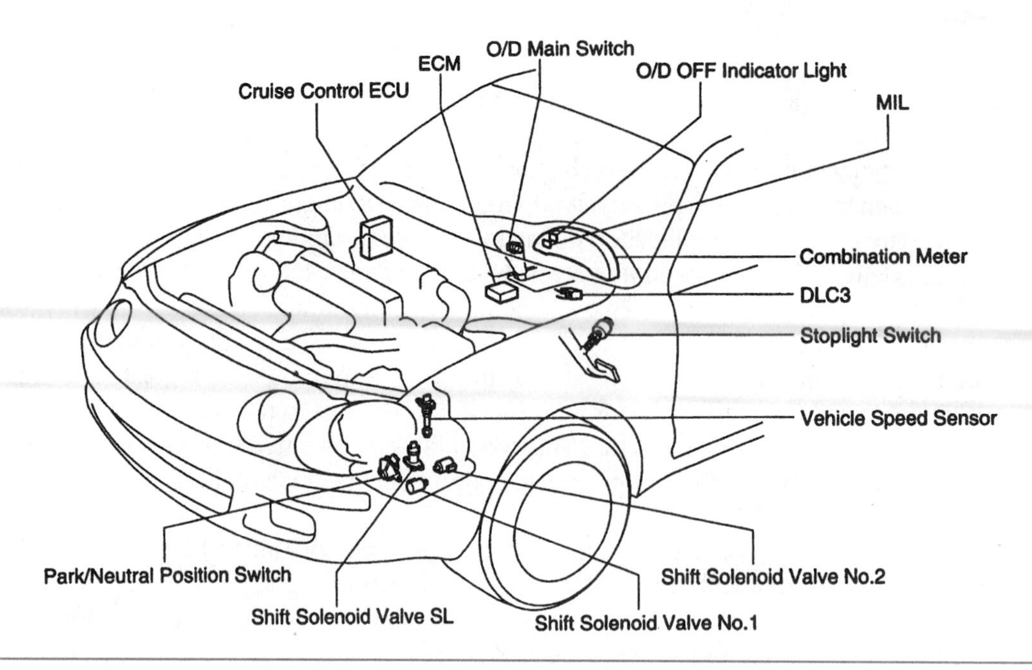

Figure 8-38 Location of the components for Toyota's A-140E electronic transmission. (Reprinted with permission.)

switch, and the shift pattern selector switch (Figure 8-39). The water temperature–sensing switch does not allow the transaxle to shift into overdrive until coolant has reached a minimum of 122°F. The throttle switch is located at the throttle body and the shift pattern switch is located at the instrument panel and is used for various driving conditions. The module receives signals from two speed sensors; one is located at the speedometer and the other at the transaxle. The back-up lamp/neutral safety switch signals the module for starting and back-up lamp circuits.

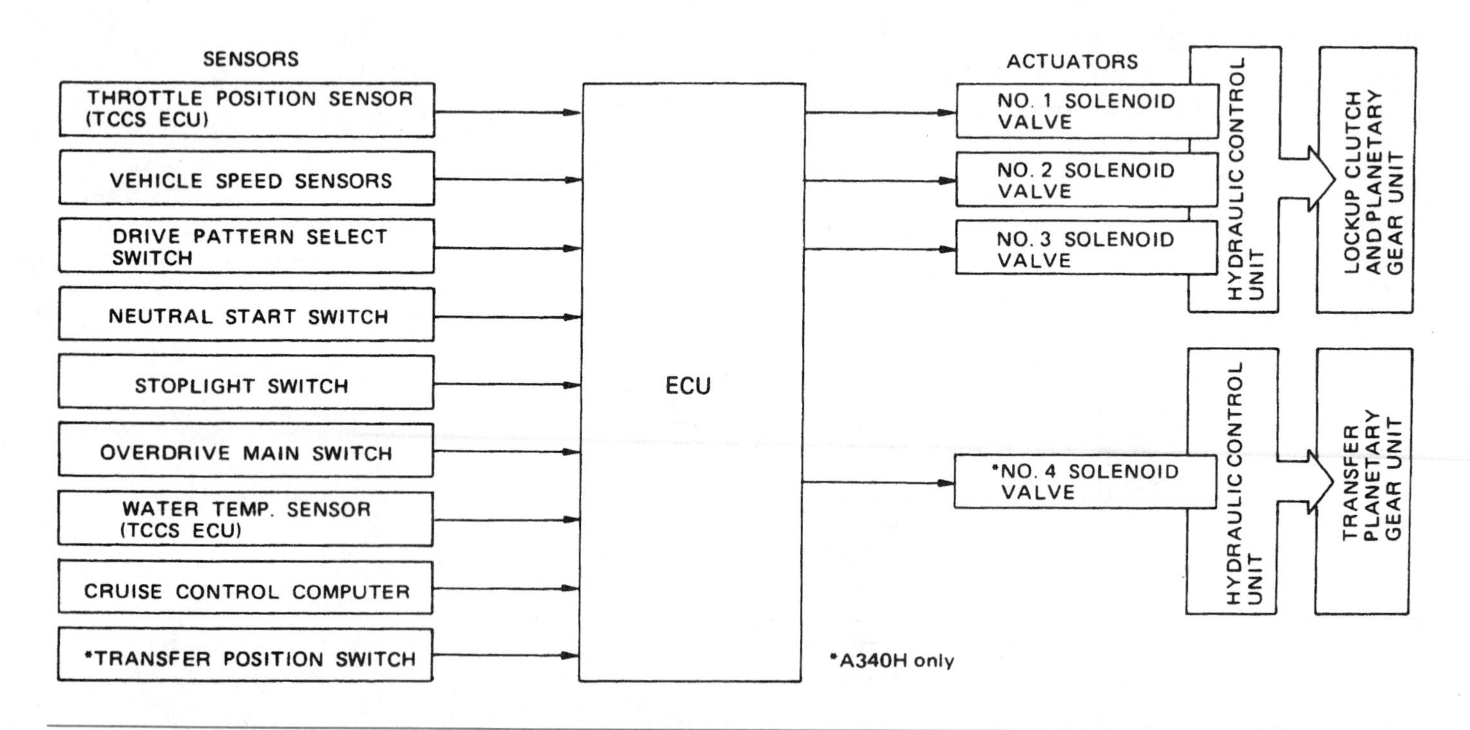

Figure 8-39 Schematic of Toyota's A-140E electronic transmission. (Reprinted with permission.)

The module controls and sends output signals to the stop and back-up lamps. The module also controls two shift solenoids, which are located within the transaxle. A summary of the solenoid activity follows:

Selector Lever Position	Operating Gear	#1 Solenoid	#2 Solenoid
Drive	First gear	On	Off
	Second gear	On	On
	Third gear	Off	On
	Overdrive	Off	Off
2—Second	First gear	On	Off
	Second gear	On	On
	Third gear	Off	On
L—Low	First gear	On	Off
	Second gear	On	On

Since the introduction of the A-140 with electronic controls, Toyota has added these controls to several other automatic transmissions. The A-240E/A-340E are four-speed transaxles with electronically controlled shifting. Gear changes are determined by an electronically controlled transmission computer (TCM), which activates solenoids in the valve body. One of the two different driving modes (Normal or Power) can be selected by the driver (Figure 8-40). The transaxle shift points are changed by the computer depending on the mode selected. The transmission is fitted with a lockup torque converter.

The A-340H is a modified version of the A-340E. It has been modified for four-wheel drive and has a hydraulically operated transfer case mounted to the rear of the transmission. The transmission is still a four-speed unit and gear changes are still determined by an ECT computer.

Variations of the A340 transmission are currently being used. These variations adapt the basic transmission to a specific application. The A340E is used in the Supra and two-wheel drive trucks. The A340F and A340H are used in four-wheel drive applications. The A340F (Figure 8-41) is a four-speed transmission with a mechanically controlled 4WD transfer case. The A340H (Figure 8-42) is also a four-speed transmission but it has an electronically controlled 4WD transfer case.

With the exception of the electronic controls added to the A340H, all A340 transmissions use the same electronic control system (Figure 8-43). This system includes three solenoid valves: two shift solenoids and a TCC solenoid.

The A540E and A541E transaxles are used in Toyota Camrys, Toyota Avalons, and the Lexus ES-300. This transaxle uses six multiple-disc clutches, three one-way clutches, and one brake band. The operation of the transaxle and the lockup converter is totally controlled by the PCM. However, a throttle pressure cable is used to mechanically modulate line pressure.

Taking electronics one step further, Toyota has recently introduced a Lexus high performance sedan. The car has many features designed to mix traditional Lexus luxury with performance. One

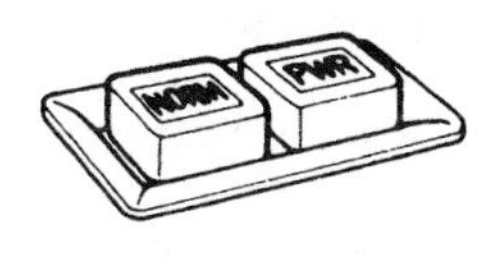

Figure 8-40 The transmission mode selector switch for Toyota's A-140E electronic transmission. (Reprinted with permission.)

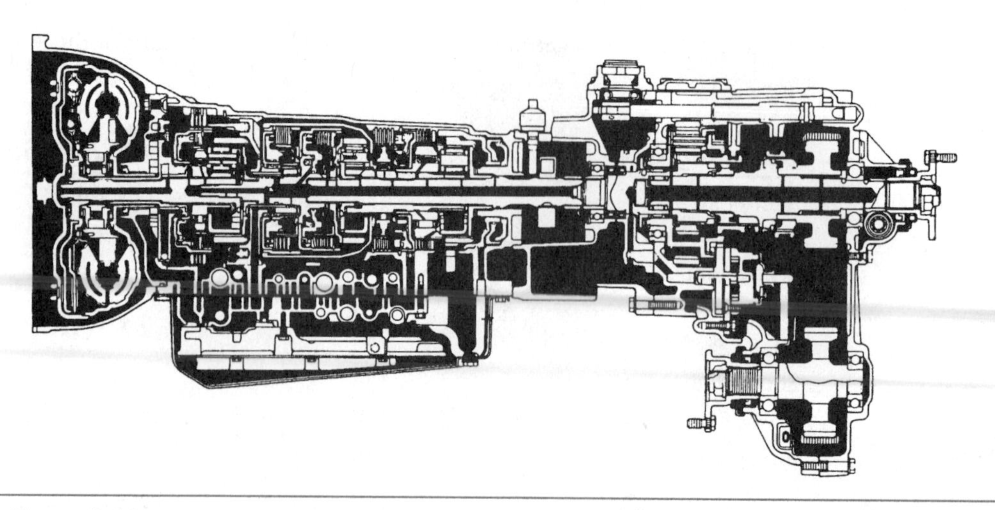

Figure 8-41 Toyota's A340F transmission with a mechanically controlled 4WD transfer case. (Reprinted with permission.)

of these is the five-speed transmission that can operate in either of two modes, providing fully automatic shifting or electronic manual control. Manual shifting is controlled by fingertip shifting buttons located on both horizontal spokes of the steering wheel (Figure 8-44). Downshifts are triggered by touching a button on the front of the steering wheel. Upshifts are controlled by contacting the buttons on the back side of the steering wheel. The buttons are located so that either thumb can be used to downshift and either index finger can be used to upshift.

The TCM is programmed to allow for rapid shifts in response to the driver's commands. It will also prevent shifting during conditions that may cause engine or transmission failure. The transmission may also be shifted by its gated console-mounted shift lever (Figure 8-45). The shift lever allows the driver to select individual gear ranges as well as the full-automatic mode.

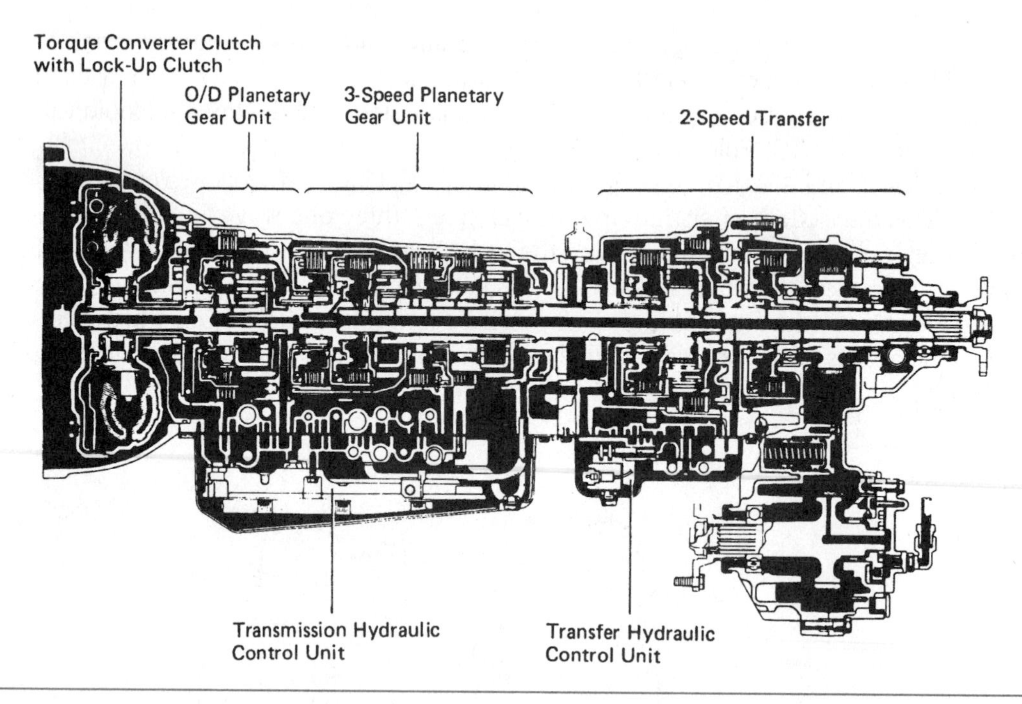

Figure 8-42 Toyota's A340H transmission with an electronically controlled 4WD transfer case. (Reprinted with permission.)

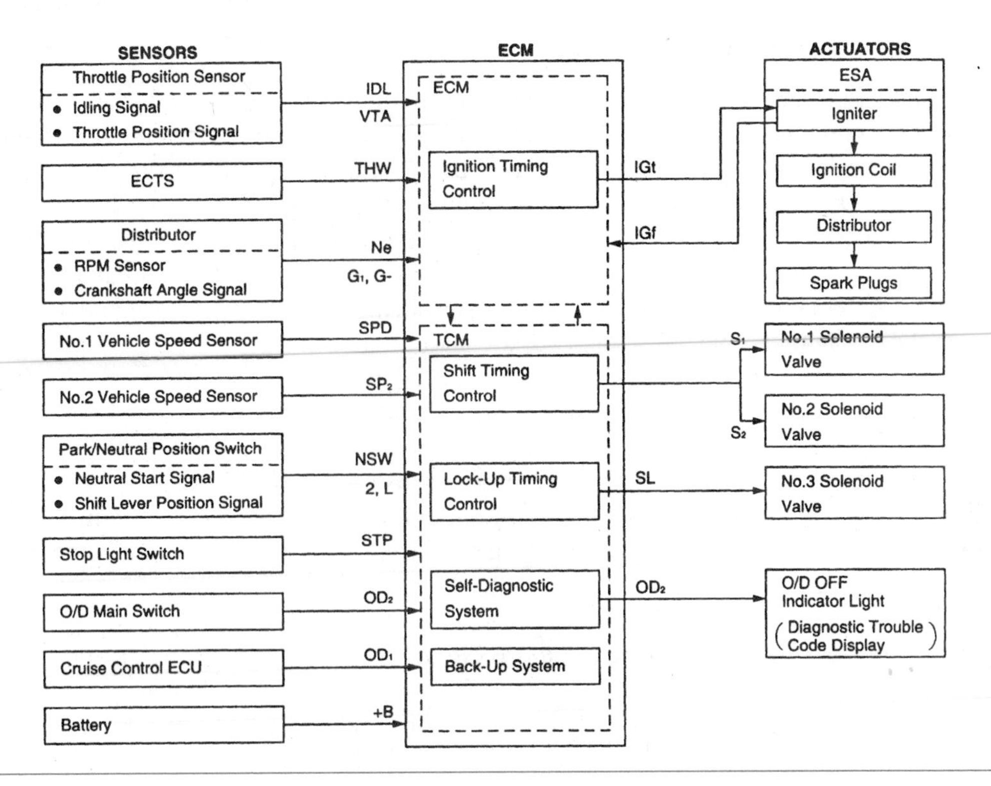

Figure 8-43 Schematic of the electronic controls for Toyota's A340F transmission. (Reprinted with permission.)

Figure 8-44 The gated shifter for Lexus' automatic/manual transmission. (Courtesy of Lexus Motor Corp.)

Figure 8-45 Fingertip controls for manually shifting Lexus' automatic/manual transmission. (Courtesy of Lexus Motor Corp.)

Shop Manual
Chapter 8, page 350

The first electronically controlled RE4F02A transaxle appeared in 1989 Maximas.

An inhibitor switch is a neutral/safety switch.

The revolution sensor monitors output shaft speed.

Nissan Transmissions

Nissan uses different transaxles and transmissions in their vehicles, but most have electronically controlled converter clutches. All RE4F02A-based transaxles also have electronic shift controls. Shifting is controlled by a transmission control unit (Figure 8-46). This microprocessor operates primarily in response to throttle position and vehicle speed. Based on inputs, the microprocessor provides the appropriate shift schedule for the current operating conditions. The control unit also has a fail-safe feature that allows the vehicle to be driven even if an important input fails.

This system relies on two shift solenoids controlled by the control unit (Figure 8-47). In addition to these solenoids, two other solenoids are incorporated into the system. The timing solenoid provides for smooth downshifting (Figure 8-48). The line pressure solenoid provides for smooth upshifting. Both of these solenoids control the engagement and disengagement of the transmission's apply units. The system has a fifth solenoid used to control converter lockup clutch activity.

The control unit relies on inputs from various sensors: a throttle position sensor, mode switch inhibitor switch, throttle valve switch, fluid temperature switch, revolution sensor, and vehicle speed sensor. Based on the throttle position and vehicle speed, the control unit is able to calculate the load on the engine. This information is then used to control the timing solenoid and the shift solenoids.

The driver can select among three driving modes: Power, Comfort, or Auto. The mode switch is located near the gearshift lever. When Power is selected, the timing for downshifting and upshifting is set for higher speeds. This mode is designed for rapid acceleration or operating under heavy loads. Normal shift patterns are followed when the mode selector is in the Comfort position. This mode is designed to achieve maximum fuel economy. The Auto mode is a mixture between the Comfort and the Power modes. Most of the time, the control unit will control the shifts in a normal way. However, if the throttle is quickly opened, the shift pattern will switch to the Power mode.

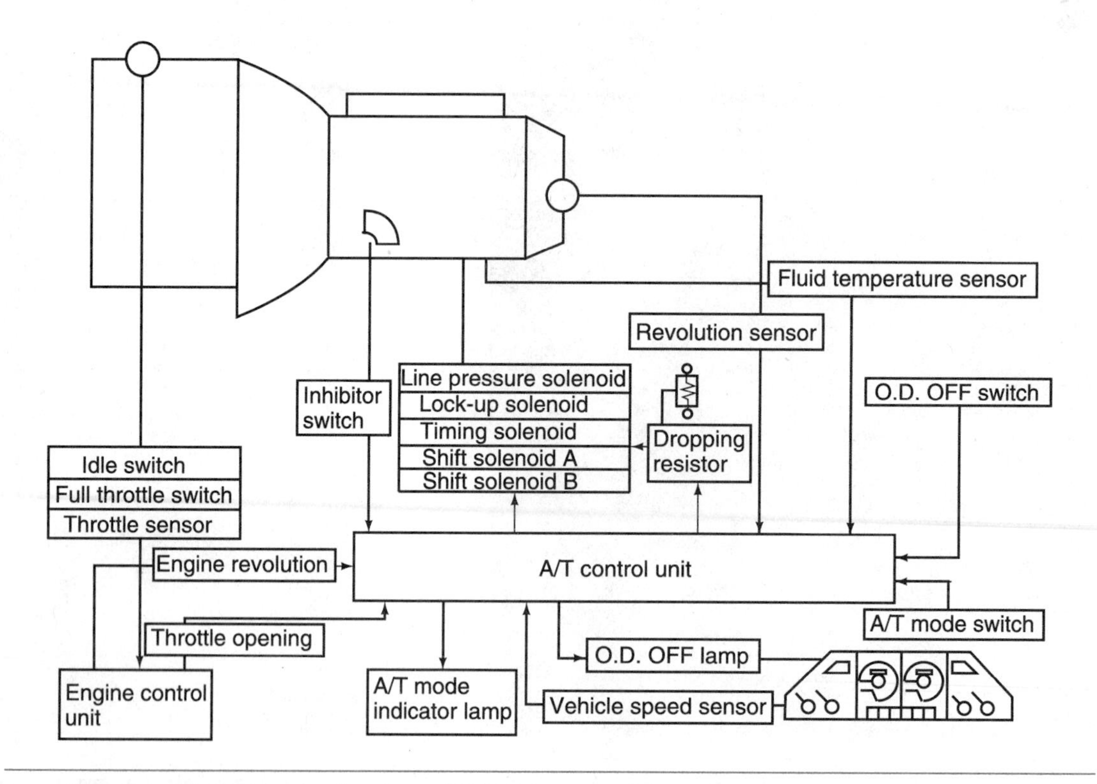

Figure 8-46 Schematic of a Nissan RE4F02A transaxle's electronic control system. (Courtesy of Nissan Motor Co., Ltd.)

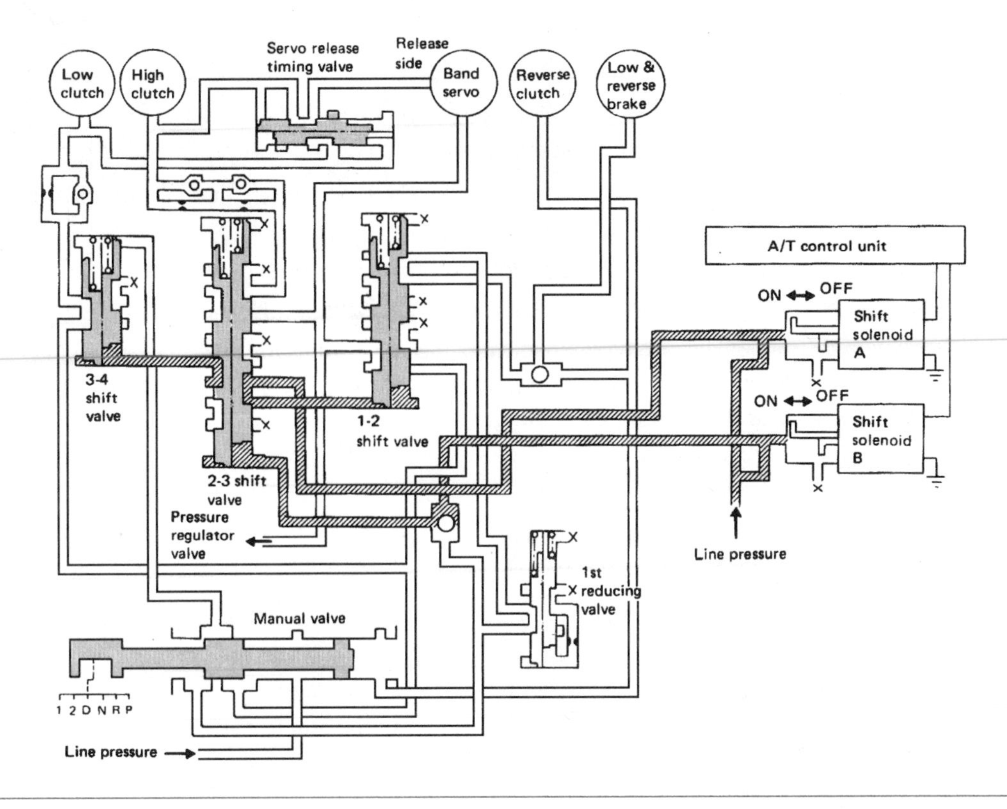

Figure 8-47 Two shift solenoids control all forward gear changes. (Courtesy of Nissan Motor Co., Ltd.)

The control unit activates the shift solenoids according to the shift schedule it selected in response to the signals received by its inputs. The shift solenoids are simple on/off solenoids. When the solenoid is energized, line pressure flows to the appropriate shift valve. One solenoid controls fluid flow to the 2–3 shift valve. The other solenoid activates the 1–2 or 3–4 shift valve through the 2–3 valve. In other words, the position of the first shift valve determines where fluid flow will be directed when the second solenoid is activated. When only the first solenoid is energized, the transaxle will operate in second gear. When both are energized, the transaxle will operate in first gear.

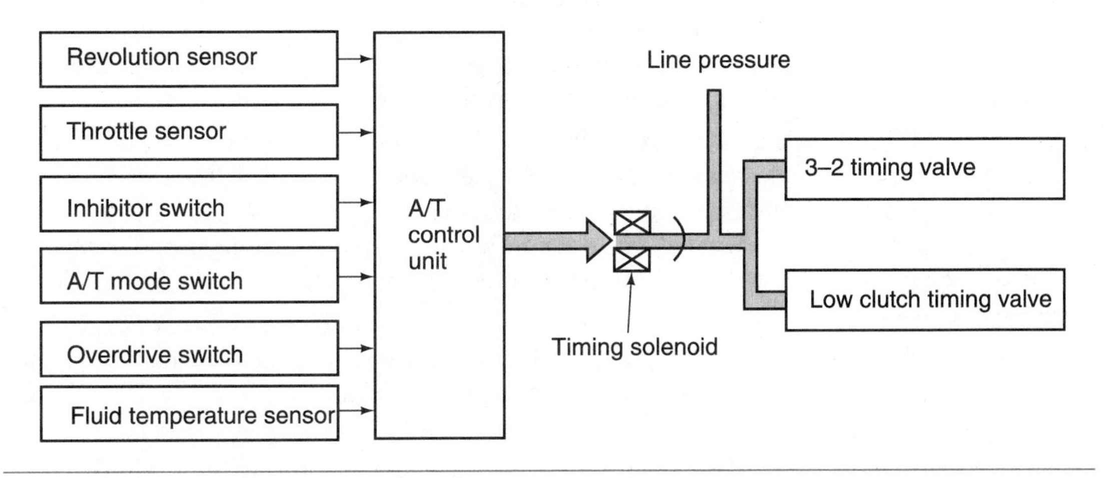

Figure 8-48 The timing solenoid controls the pressure to the 3–2 and low valves in order to control the quality of downshifts. (Courtesy of Nissan Motor Co., Ltd.)

Summary

Terms to Know

Adaptive learning
BARO
CCD bus
Computer
DRAC
EAT
EATX
MECS
Multiplexing
OSS
PCM
Potentiometers
Pressure sensors
PWM
Reference voltage sensors
SBEC
Shift schedules
TCM
Thermistors
TISS
TOSS
TOT
VFM
VFS
Voltage generator
VSS

- By using electronic controls for transmission operation, the amount of wasted power can be reduced and the overall operation of the transmission can be more responsive with better reliability.
- A computer is an electronic device that receives, stores, processes, and communicates information.
- The input devices used in electronic control systems vary with each system; however, they can be grouped into distinct categories: reference voltage sensors and voltage generators.
- Voltage generation devices are typically used to monitor rotational speeds. The most common of these is the PM generator used as a vehicle speed sensor. A speed sensor is a magnetic pickup that senses and transmits low-voltage pulses.
- Common voltage reference sensors include on/off switches, potentiometers, thermistors, and pressure sensors.
- Potentiometers, thermistors, and pressure sensors are designed to change their electrical resistance in response to something else changing.
- Thermistors also change their resistance values in response to conditions; however, they respond to changes in heat.
- Pressure sensors respond to pressure applied to a movable diaphragm in the switch. As pressure increases, so does the movement of the diaphragm and the amount of resistance in the sensor.
- If a switch is cycled on/off very rapidly, the return signal is a rapid on/off one. This describes a digital signal, a series of on/off pulses.
- The first processing task of a computer is to translate the analog signals into a digital signal.
- The typical output devices are solenoids and motors, which cause something mechanical or hydraulic to change.
- The decision to shift or not to shift is based on shift schedules and logic programmed into the memory of the computer.
- A shift schedule contains the actual shift points to be used by the computer according to the input data it receives from the sensors. Shift schedule logic chooses the proper shift schedule for the current conditions of the transmission.
- Late-model Chrysler A-500/A-518 transmissions and the transaxles used in most Chrysler/Mitsubishi vehicles rely on electronics to control the shifting into overdrive gear.
- Fourth gear operation is controlled by a manually operated overdrive switch on the instrument panel, center console, or shift lever. The overdrive switch is wired into a circuit with an overdrive solenoid, which is located on the valve body, and the single board engine controller. The torque converter clutch lockup solenoid is also wired into this circuit. The overdrive switch prevents a shift into fourth gear when the switch is off.
- The A-604/41TE transaxle is an electronically controlled four-speed transaxle. The transaxle uses hydraulically operated clutches controlled by the transaxle controller. The controller receives information from various inputs and controls a solenoid assembly through the electronic automatic transaxle relay. The solenoid assembly consists of four solenoids that control hydraulic pressure to four of the five clutches in the transaxle and to the lockup clutch of the torque converter.
- Adaptive learning allows the controller to compensate for wear and other events that might occur and cause the normal shift programming to be inefficient.
- The controller may receive information from two different sources: directly from a sensor, or through a twisted-pair bus circuit, which connects all the vehicle computer systems. This modulated bidirectional bus system is called Chrysler Collision Detection bus and allows the various computers in the vehicle to share information.

- If the controller loses source voltage, the transaxle will enter into limp-in mode. The transaxle will also enter into limp-in mode if the controller senses a transmission failure. While in limp-in mode, the transaxle will operate only in PARK, NEUTRAL, REVERSE, and SECOND gears. The transaxle will not upshift or downshift. This allows the vehicle to be operated, although its efficiency and performance are hurt.
- Typical CCD bus inputs used by the transaxle controller are from an ambient or battery temperature sensor, a brake switch, a coolant temperature sensor, and a manifold absolute pressure sensor.
- The Chrysler 42LE four-speed transaxle also uses fully adaptive controls. Like the 41TE, this transaxle uses hydraulically applied clutches to shift gears but controls the hydraulics electronically.
- Fluid flow to the various apply devices is directly controlled by the solenoids.
- There are three pressure switches that give input to the transaxle controller. They are all located within the solenoid assembly.
- Engine speed, throttle position, temperature, engine load, and other typical engine-related inputs are used by the controller to determine the best shift points. Many of these inputs are available through multiplexing and are inputted from the common bus.
- Adaptive learning takes place as the controller reads input and output speeds over 140 times per second.
- The basic shift logic of the controller allows the releasing apply device to slip slightly during the engagement of the engaging apply device.
- The AOD-E is identical to an AOD except that it no longer uses a split-torque converter and its operation is controlled by the electronic engine control's electronic control assembly.
- The ECA relies on information from the engine control system, through such sensors as the MAP and TPS, as well as information from the transmission to determine the optimal shift timing.
- The transmission control system uses four solenoids for control of operation. One solenoid, the MCCC solenoid, is used for modulated converter clutch control. Another, the EPC solenoid, is used to control hydraulic pressures throughout the transmission. The remaining two solenoids are shift solenoids.
- The EPC solenoid replaces the conventional T.V. cable setup to provide changes in pressure in response to engine load.
- The AX4S is an electronically controlled four-speed transaxle. Shift control solenoids provide gear changes and these solenoids are controlled by the PCM, as is the converter lockup clutch solenoid. Five solenoids are mounted on the AX4S valve body to control transmission functions. One solenoid, the MCCC solenoid, is used for modulated converter clutch control. Another, the EPC solenoid, is used to control hydraulic pressures throughout the transmission. The remaining three solenoids are shift solenoids.
- The A4LD is a four-speed automatic transmission with a lockup torque converter. The engine control system controls the operation of the lockup converter clutch and the operation of a shift solenoid for third-to-fourth gear shifting.
- The 4EAT is controlled both electronically and hydraulically.
- Four solenoid valves are located in the valve body. These redirect fluid flow and cause shifting and converter lockup.
- The 4EAT can be operated in manual or automatic shift mode.
- The THM 4L80-E is based on the same design as a THM 400 with a THM 200-4R overdrive assembly and an electronic control system. All automatic upshifts and downshifts are electronically controlled by the power train control module for gasoline engines or the transmission control module for diesel engines.

- The PCM is programmed to adjust its operating parameters in response to changes within the system, such as component wear. As component wear and shift overlap times increase, the PCM adjusts line pressure by controlling the VFM to maintain proper shift timing calibrations.
- A variable force motor controlled by the PCM is used to change line pressure in response to engine speed and vehicle load.
- The PCM is a multifunction computer that controls all engine and transmission operations. It receives information on engine operating conditions from different sensors and switches.
- Once the PCM has processed the input signals, it controls transmission operation through two on/off solenoids, a pulse width modulated solenoid, and a variable force motor located in the valve body.
- The shift solenoids receive voltage through the ignition switch and are grounded through the PCM.
- The pulse width modulated solenoid is a normally closed valve installed in the valve body. It controls the position of the TCC apply valve.
- The VFM is an electrohydraulic actuator made of a variable force solenoid and a regulating valve. The VFM is installed in the valve body and controls mainline pressure by moving a pressure regulator valve against spring pressure.
- The 4T60-E transaxle is an electronically controlled shifting transmission with converter lockup. The torque converter clutch is controlled by two electronic solenoids, one for apply and release and the other to control the feel of the apply and release of the TCC.
- On some models, the PCM is also connected to one or more other computers that operate the climate control system, antilock brake system, driver information center, and supplemental restraint system. Through multiplexing, these computers are able to share information and use common input sensors.
- Honda automatic transmissions are unique in the industry because they use constant mesh gears to provide for automatic gear changes.
- The electronic control system consists of the transmission control module, sensors, and four solenoid valves.
- Activating a shift solenoid valve changes modulator pressure, causing a shift valve to move. This allows pressurized fluid to flow through a line to engage the appropriate clutch and its corresponding gear set.
- The Aisin Warner 4 is an electronically controlled automatic transmission with a valve body fitted with three solenoids.
- The solenoids are controlled by a transmission control unit and are used to control the operation of the transmission. The #1 and #2 solenoids are used to control shifting, and solenoid #3 is used for torque converter clutch lockup.
- On some models, a Power/Comfort switch is mounted on the instrument panel. This switch is an automatic transmission mode selection switch that allows the driver to select different modes to change transmission upshift characteristics.
- Saturn vehicles use an electronically controlled four-speed automatic transmission with a shift mode select function. This mode switch allows the driver to select Normal mode or Performance mode.
- The transaxle relies on five solenoid valves controlled by the PCM to regulate shift timing, feel, and TCC application.
- Two solenoids are used as shift solenoids. They control the delivery of fluid to the manual shift valve. A solenoid is also used to control line pressure. This solenoid operates on a duty cycle controlled by the PCM. Its purpose is to regulate line pressure according to engine running conditions and engine torque. An additional solenoid is used to provide for engine braking during coasting. This solenoid operates when the vehicle is slowing down and the throttle is closed. The other solenoid controls the operation of the lockup converter clutch.

- ❑ Toyota's A-140E was the first fully electronic transmission.
- ❑ A control module controls two shift solenoids, which are located within the transaxle.
- ❑ Gear changes are determined by an electronically controlled transmission computer, which activates solenoids in the valve body. One of the two different driving modes (Normal or Power) can be selected by the driver.
- ❑ Nissan's RE4F02A-based transaxles also have electronic shift controls. Shifting is controlled by a transmission control unit. This microprocessor operates primarily in response to throttle position and vehicle speed.
- ❑ This system relies on two shift solenoids controlled by the control unit. In addition to these solenoids, two other solenoids are incorporated into the system. The timing solenoid provides for smooth downshifting. The line pressure solenoid provides for smooth upshifting. Both of these solenoids control the engagement and disengagement of the transmission's apply units. The system has a fifth solenoid that is used to control converter lockup clutch activity.
- ❑ The driver can select among three driving modes: Power, Comfort, or Auto.
- ❑ One solenoid controls fluid flow to the 2–3 shift valve. The other solenoid activates the 1–2 or 3–4 shift valve through the 2–3 valve. In other words, the position of the first shift valve determines where fluid flow will be directed when the second solenoid is activated. When only the first solenoid is energized, the transaxle will operate in second gear. When both are energized, the transaxle will operate in first gear.

Review Questions

Short Answer Essays

1. A computer relies on many different reference voltage sensors. These can be divided into types. Name the types and give a brief description of their operation.
2. Although computers receive different information from a variety of sensors, the decisions for shifting are actually based on more than the inputs. What are they based on?
3. Saturn vehicles rely on five different solenoids to control the operation of their transaxle. What do these solenoids do?
4. Some Chrysler transaxles receive information through multiplexing. How does this work?
5. Nissan transaxles use four solenoids to control shifting. What is the purpose of each of these solenoids?
6. Most late-model transmission control systems have adaptive learning. What does this mean?
7. What inputs are of prime importance to a computer in deciding when to shift gears?
8. What are the advantages of using electronic controls rather than relying on conventional hydraulic controls in a transmission?
9. What is unique about a Honda transaxle?
10. The Ford 4EAT transmission, like many other designs, has an additional switch that allows the driver to select a mode of operation. How does the transmission function differently in the different mode positions?

Fill in the Blanks

1. The ______________________ , ______________________ , ______________________ , and ______________________ are common electronically controlled transmissions used in Ford vehicles.

2. Most General Motors electronically controlled transmissions are similar to the ______________________ transmission or a ______________________ transaxle.

3. Chrysler has two families of automatic transaxles fitted with adaptive learning. They are the ______________________ and the ______________________ .

4. In some Ford transmissions, an EPC solenoid replaces the conventional ______________________ setup to provide changes in pressure in response to ______________________ .

5. A computer is an electronic device that ______________________ information, ______________________ information, ______________________ information, and ______________________ information.

6. The input devices used in electronic control systems vary with each system; however, they can be grouped into distinct categories: ______________________ ______________________ ______________________ and ______________________ ______________________ .

7. Common voltage reference sensors include ______________________ switches, ______________________ , ______________________ , and ______________________ sensors.

8. In an electronic control system, the typical output devices are ______________________ and ______________________ , which cause something ______________________ or ______________________ to change.

9. Most electronically controlled transmissions rely on ______________________ ______________________ solenoids to control all forward gears.

10. Voltage generation devices are typically used to monitor ______________________ ______________________ .

ASE-Style Review Questions

1. *Technician A* says voltage generation devices are typically used to monitor rotational speeds.
   Technician B says the most common voltage generation device used in electronic transmission control systems is the vehicle speed sensor.
   Who is correct?
   A. A only
   B. B only
   C. Both A and B
   D. Neither A nor B

2. While discussing Chrysler transaxles:
   Technician A says that if the controller loses source voltage, the transaxle will enter into limp-in mode.
   Technician B says the transaxle will enter into limp-in mode if the controller senses a transmission failure.
   Who is correct?
   A. A only
   B. B only
   C. Both A and B
   D. Neither A nor B

3. *Technician A* says throttle position is an important input in most electronic shift control systems. *Technician B* says vehicle speed is an important input for most electronic shift control systems. Who is correct?
   A. A only **C.** Both A and B
   B. B only **D.** Neither A nor B

4. *Technician A* says some systems use a special modulated shift control solenoid. *Technician B* says some systems use a special modulated converter clutch control solenoid. Who is correct?
   A. A only **C.** Both A and B
   B. B only **D.** Neither A nor B

5. While discussing various sensors used with an electronic transmission control system: *Technician A* says potentiometers are typically used to measure temperature changes. *Technician B* says vacuum modulators are used to measure engine load. Who is correct?
   A. A only
   B. B only
   C. Both A and B
   D. Neither A nor B

6. *Technician A* says shift solenoids direct fluid flow to and away from the various apply devices in the transmission. *Technician B* says shift solenoids are used to mechanically apply a friction band or multiple-disc clutch assembly. Who is correct?
   A. A only
   B. B only
   C. Both A and B
   D. Neither A nor B

7. While discussing shift logic: *Technician A* says the logic must be based on the current engine and transmission combination. *Technician B* says the logic must be based on the driver's driving habits. Who is correct?
   A. A only **C.** Both A and B
   B. B only **D.** Neither A nor B

8. While discussing line pressure control in a GM 4L80-E: *Technician A* says mainline pressure varies with the output of the governor. *Technician B* says the duty cycle of the VFM controls the amount of line pressure in the transmission. Who is correct?
   A. A only **C.** Both A and B
   B. B only **D.** Neither A nor B

9. *Technician A* says multiplexing allows information to be shared with many computers. *Technician B* says multiplexing reduces the number of wires and components needed in current vehicles. Who is correct?
   A. A only **C.** Both A and B
   B. B only **D.** Neither A nor B

10. While discussing valve body assemblies in late-model transmissions: *Technician A* says the valve body is no longer needed in some electronically controlled transmissions. *Technician B* says most shift solenoid assemblies are mounted directly to the valve body. Who is correct?
   A. A only **C.** Both A and B
   B. B only **D.** Neither A nor B

APPENDIX A

Abbreviations

The following abbreviations are some of the more common ones used today in the automotive industry.

TABLE 1—CROSS REFERENCE AND LOOK UP

Existing Usage	Acceptable Usage	Acceptable Acronized Usage
A/C (Air Conditioning)	Air Conditioning	A/C
A/C Cycling Switch	Air Conditioning Cycling Switch	A/C Cycling Switch
A/T (Automatic Transaxle)	Automatic Transaxle[1]	A/T[1]
A/T (Automatic Transmission)	Automatic Transmission[1]	A/T[1]
AAT (Ambient Air Temperature)	**Ambient Air Temperature**	**AAT**
AC (Air Conditioning)	Air Conditioning	A/C
ACC (Air Conditioning Clutch)	Air Conditioning Clutch	A/C Clutch
Accelerator	Accelerator Pedal	AP
Accelerator Pedal Position	**Accelerator Pedal Position**[1]	**APP**[1]
ACCS (Air Conditioning Cyclic Switch)	Air Conditioning Cycling Switch	A/C Cycling Switch
ACH (Air Cleaner Housing)	Air Cleaner Housing[1]	ACL Housing1
ACL (Air Cleaner)	Air Cleaner[1]	ACL[1]
ACL (Air Cleaner) Element	Air Cleaner Element[1]	ACL Element[1]
ACL (Air Cleaner) Housing	Air Cleaner Housing[1]	ACL Housing[1]
ACL (Air Cleaner) Housing Cover	Air Cleaner Housing Cover[1]	ACL Housing Cover[1]
ACS (Air Conditioning System)	Air Conditioning System	A/C System
ACT (Air Charge Temperature)	Intake Air Temperature[1]	IAT[1]
Adaptive Fuel Strategy	Fuel Trim[1]	FT[1]
AFC (Air Flow Control)	Mass Air Flow	MAF
AFC (Air Flow Control(	Volume Air Flow	VAF
AFS (Air Flow Sensor)	Mass Air Flow Sensor	MAF Sensor
AFS (Air Flow Sensor)	Volume Air Flow Sensor	VAF Sensor
After Cooler	Charge Air Cooler[1]	CAC[1]
AI (Air Injection)	Secondary Air Injection[1]	AIR[1]
AIP (Air Injection Pump)	Secondary Air Injection Pump[1]	AIR Pump[1]
AIR (Air Injection Reactor)	Pulsed Secondary Air Injection[1]	PAIR[1]
AIR (Air Injection Reactor)	Secondary Air Injection[1]	AIR[1]
AIRB (Secondary Air Injection Bypass)	Secondary Air Injection Bypass[1]	AIR Bypass[1]
AIRD (Secondary Air Injection Diverter)	Secondary Air Injection Diverter[1]	AIR Diverter[1]
Air Cleaner	Air Cleaner[1]	ACL[1]
Air Cleaner Element	Air Cleaner Element[1]	ACL Element[1]
Air Cleaner Housing	Air Cleaner Housing[1]	ACL Housing[1]
Air Cleaner Housing Cover	Air Cleaner Housing Cover[1]	ACL Housing Cover[1]
Air Conditioning	Air Conditioning	A/C
Air Conditioning Sensor	Air Conditioning Sensor	A/C Sensor
Air Control Valve	Secondary Air Injection Control Valve[1]	AIR Control Valve[1]
Air Flow Meter	Mass Air Flow Sensor[1]	MAF Sensor[1]
Air Flow Meter	Volume Air Flow Sensor[1]	VAF Sensor[1]
Air Intake System	Intake Air System[1]	IA System[1]
Air Flow Sensor	Mass Air Flow Sensor[1]	MAF Sensor[1]
Air Management 1	Secondary Air Injection Bypass[1]	AIR Bypass[1]
Air Management 2	Secondary Air Injection Diverter[1]	AIR Diverter[1]
Air Temperature Sensor	Intake Air Temperature Sensor[1]	IAT Sensor[1]
Air Valve	Idle Air Control Valve[1]	IAC Valve[1]
AIV (Air Injection Valve)	Pulsed Secondary Air Injection[1]	PAIR[1]
ALCL (Assembly Line Communication Link)	Data Link Connector[1]	DLC[1]
Alcohol Concentration Sensor	Flexible Fuel Sensor[1]	FF Sensor[1]
ALDL (Assembly Line Diagnostic Link)	Data Link Connector[1]	DLC[1]

TABLE 1—CROSS REFERENCE AND LOOK UP (CONTINUED)

Existing Usage	Acceptable Usage	Acceptable Acronized Usage
ALT (Alternator)	Generator	GEN
Alternator	Generator	GEN
Ambient Air Temperature	**Ambient Air Temperature**	**AAT**
AM1 (Air Management 1)	Secondary Air Injection Bypass[1]	AIR Bypass[1]
AM2 (Air Management 2)	Secondary Air Injection Diverter[1]	AIR Diverter[1]
APP (Accelerator Pedal Position)	**Accelerator Pedal Position[1]**	**APP[1]**
APS (Absolute Pressure Sensor)	Barometric Pressure Sensor[1]	BARO Sensor[1]
ATS (Air Temperature Sensor)	Intake Air Temperature Sensor[1]	IAT Sensor[1]
Automatic Transaxle	Automatic Transaxle[1]	A/T[1]
Automatic Transmission	Automatic Transmission[1]	A/T[1]
B+ (Battery Positive Voltage)	Battery Positive Voltage	B+
Backpressure Transducer	Exhaust Gas Recirculation Backpressure Transducer[1]	EGR Backpressure Transducer[1]
BARO (Barometric Pressure)	Barometric Pressure[1]	BARO[1]
Barometric Pressure Sensor	Barometric Pressure Sensor[1]	BARO Sensor[1]
Battery Positive Voltage	Battery Positive Voltage	B+
BLM (Block Learn Memory)	Long Term Fuel Trim[1]	Long Term FT[1]
BLM (Block Learn Multiplier)	Long Term Fuel Trim[1]	Long Term FT[1]
BLM (Block Learn Matrix)	Long Term Fuel Trim[1]	Long Term FT[1]
Block Learn Integrator	**Long Term Fuel Trim[1]**	**Long Term FT[1]**
Block Learn Matrix	Long Term Fuel Trim[1]	Long Term FT[1]
Block Learn Memory	Long Term Fuel Trim[1]	Long Term FT[1]
Block Learn Multiplier	Long Term Fuel Trim[1]	Long Term FT
BP (Barometric Pressure) Sensor	Barometric Pressure Sensor[1]	BARO Sensor[1]
BPP (Brake Pedal Position)	**Brake Pedal Position[1]**	**BPP[1]**
Brake Pressure	**Brake Pressure**	**Brake Pressure**
Brake Pedal Position	**Brake Pedal Position[1]**	**BPP[1]**
C3I (Computer Controlled Coil Ignition)	Electronic Ignition[1]	EI[1]
CAC (Charge Air Cooler)	Charge Air Cooler[1]	CAC[1]
Calculated Load Value	**Calculated Load Value**	**LOAD**
Camshaft Position	Camshaft Position[1]	CMP[1]
Camshaft Position Actuator	**Camshaft Position Actuator[1]**	**CMP Actuator[1]**
Camshaft Position Controller	**Camshaft Position Actuator[1]**	**CMP Actuator[1]**
Camshaft Position Sensor	Camshaft Position Sensor[1]	CMP Sensor[1]
Camshaft Sensor	Camshaft Position Sensor[1]	CMP Sensor[1]
Camshaft Timing Actuator	**Camshaft Position Actuator[1]**	**CMP Actuator[1]**
Canister	Canister[1]	Canister[1]
Canister	Evaporative Emission Canister[1]	EVAP Canister[1]
Canister Purge	**Evaporative Emission Canister Purge[1]**	**EVAP Canister Purge[1]**
Canister Purge Vacuum Switching Valve	Evaporative Emission Canister Purge Valve[1]	EVAP Canister Purge Valve[1]
Canister Purge Valve	Evaporative Emission Canister Purge Valve[1]	EVAP Canister Purge Valve[1]
Canister Purge VSV (Vacuum Switching Valve)	Evaporative Emission Canister Purge Valve[1]	EVAP Canister Purge Valve[1]
CANP (Canister Purge)	Evaporative Emission Canister Purge[1]	EVAP Canister Purge[1]
CARB (Carburetor)	Carburetor[1]	CARB[1]
Carburetor	Carburetor[1]	CARB[1]
Catalytic Converter Heater	**Catalytic Converter Heater**	**Catalytic Converter Heater**
CCC (Converter Clutch Control)	Torque Converter Clutch[1]	TCC[1]
CCO (Converter Clutch Override)	Torque Converter Clutch[1]	TCC[1]
CCS (Coast Clutch Solenoid)	**Coast Clutch Solenoid**	**CCS**

TABLE 1—CROSS REFERENCE AND LOOK UP (CONTINUED)

Existing Usage	Acceptable Usage	Acceptable Acronized Usage
CCS (Coast Clutch Solenoid) Valve	**Coast Clutch Solenoid Valve**	**CCS Valve**
CCRM (Constant Control Relay Module)	**Constant Control RM**	**Constant Control RM**
CDI (Capacitive Discharge Ignition)	Distributor Ignition[1]	DI[1]
CDROM (Compact Disc Read Only Memory)	Compact Disc Read Only Memory[1]	CDROM[1]
CES (Clutch Engage Switch)	Clutch Pedal Position Switch	CPP Switch[1]
Central Multiport Fuel Injection	Central Multiport Fuel Injection[1]	Central MFI[1]
Central Sequential Multiport Fuel Injection	**Central Sequential Multiport Fuel Injection**	**Central SFI**
CFI (Continuous Fuel Injection)	Continuous Fuel Injection[1]	CFI[1]
CFI (Central Fuel Injection)	Throttle Body Fuel Injection[1]	TBI[1]
CFV	**Critical Flow Venturi**	**CFV**
Charcoal Canister	Evaporative Emission Canister	EVAP Canister[1]
Charge Air Cooler	Charge Air Cooler	CAC[1]
Check Engine	Service Reminder Indicator[1]	SRI[1]
Check Engine	Malfunction Indicator Lamp[1]	MIL[1]
CID (Cylinder Identification) Sensor	Camshaft Position Sensor	CMP Sensor[1]
CIS (Continuous Injection System)	Continuous Fuel Injection[1]	CFI[1]
CIS-E (Continuous Injection System Electronic)	Continuous Fuel Injection[1]	CFI[1]
CKP (Crankshaft Position)	Crankshaft Position[1]	CKP[1]
CKP (Crankshaft Position) Sensor	Crankshaft Position Sensor[1]	CKP Sensor[1]
CL (Closed Loop)	Closed Loop[1]	CL[1]
Closed Bowl Distributor	Distributor Ignition[1]	DI[1]
Closed Throttle Position	Closed Throttle Position[1]	CTP[1]
Closed Throttle Switch	Closed Throttle Position Switch[1]	CTP Switch[1]
CLS (Closed Loop System)	Closed Loop[1]	CL[1]
CLV	**Calculated Load Value**	**LOAD**
Clutch Engage Switch	Clutch Pedal Position Switch[1]	CPP Switch[1]
Clutch Pedal Position Switch	Clutch Pedal Position Switch[1]	CPP Switch[1]
Clutch Start Switch	Clutch Pedal Position Switch[1]	CPP Switch[1]
Clutch Switch	Clutch Pedal Position Switch[1]	CPP Switch[1]
CMFI (Central Multiport Fuel Injection)	Central Multiport Fuel Injection[1]	Central MFI[1]
CMP (Camshaft Position)	Camshaft Position[1]	CMP[1]
CMP (Camshaft Position) Sensor	Camshaft Position Sensor[1]	CMP Sensor[1]
COC (Continuous Oxidation Catalyst)	Oxidation Catalytic Converter[1]	OC[1]
Coast Clutch Solenoid	**Coast Clutch Solenoid**	**CCS**
Coast Clutch Solenoid Valve	**Coast Clutch Solenoid Valve**	**CCS Valve**
Condenser	Distributor Ignition Capacitor[1]	DI Capacitor[1]
Constant Control Relay Module	**Relay Module**	**RM**
Constant Volume Sampler	**Constant Volume Sampler**	**CVS**
Continuous Fuel Injection	Continuous Fuel Injection[1]	CFI[1]
Continuous Injection System	Continuous Fuel Injection System[1]	CFI System[1]
Continuous Injection System-E	Electronic Continuous Fuel Injection System[1]	Electronic CFI System[1]
Continuous Trap Oxidizer	Continuous Trap Oxidizer[1]	CTOX[1]
Coolant Temperature Sensor	Engine Coolant Temperature Sensor[1]	ECT Sensor[1]
CP (Crankshaft Position)	Crankshaft Position[1]	CKP[1]
CPP (Clutch Pedal Position)	Clutch Pedal Position[1]	CPP[1]
CPP (Clutch Pedal Position) Switch	Clutch Pedal Position Switch	CPP Switch[1]
CPS (Camshaft Position Sensor)	Camshaft Position Sensor[1]	CMP Sensor[1]
CPS (Crankshaft Position Sensor)	Crankshaft Position Sensor[1]	CKP Sensor[1]
Crank Angle Sensor	Crankshaft Position Sensor[1]	CKP Sensor[1]

TABLE 1—CROSS REFERENCE AND LOOK UP (CONTINUED)

Existing Usage	Acceptable Usage	Acceptable Acronized Usage
Crankshaft Position	Crankshaft Position[1]	CKP[1]
Crankshaft Position Sensor	Crankshaft Position Sensor[1]	CKP Sensor[1]
Crankshaft Speed	Engine Speed[1]	RPM[1]
Crankshaft Speed Sensor	Engine Speed Sensor[1]	RPM Sensor[1]
Critical Flow Venturi	**Critical Flow Venturi**	**CFV**
CTO (Continuous Trap Oxidizer)	Continuous Trap Oxidizer[1]	CTOX[1]
CTOX (Continuous Trap Oxidizer)	Continuous Trap Oxidizer[1]	CTOX[1]
CTP (Closed Throttle Position)	Closed Throttle Position[1]	CTP[1]
CTS (Coolant Temperature Sensor)	Engine Coolant Temperature Sensor[1]	ECT Sensor[1]
CTS (Coolant Temperature Switch)	Engine Coolant Temperature Switch[1]	ECT Switch[1]
CVS	**Constant Volume Sampler**	**CVS**
Cylinder ID (Identification) Sensor	Camshaft Position Sensor[1]	CMP Sensor[1]
D-Jetronic	Multiport Fuel Injection[1]	MFI[1]
Data Link Connector	Data Link Connector[1]	DLC[1]
Detonation Sensor	Knock Sensor[1]	KS[1]
DFI (Direct Fuel Injection)	Direct Fuel Injection[1]	DFI[1]
DFI (Digital Fuel Injection)	Multiport Fuel Injection[1]	MFI[1]
DI (Direct Injection)	Direct Fuel Injection[1]	DFI[1]
DI (Distributor Ignition)	Distributor Ignition[1]	DI[1]
DI (Distributor Ignition) Capacitor	Distributor Ignition Capacitor[1]	DI Capacitor[1]
Diagnostic Test Mode	Diagnostic Test Mode[1]	DTM[1]
Diagnostic Trouble Code	Diagnostic Trouble Code[1]	DTC[1]
DID (Direct Injection - Diesel)	Direct Fuel Injection[1]	DFI[1]
Differential Pressure Feedback EGR (Exhaust Gas Recirculation) System	Differential Pressure Feedback Exhaust Gas Recirculation System[1]	Differential Pressure Feedback EGR System[1]
Digital EGR (Exhaust Gas Recirculation)	Exhaust Gas Recirculation[1]	EGR[1]
Direct Fuel Injection	Direct Fuel Injection[1]	DFI[1]
Direct Ignition System	Electronic Ignition System[1]	EI System[1]
DIS (Distributorless Ignition System)	Electronic Ignition System[1]	EI System[1]
DIS (Distributorless Ignition System) Module	Ignition Control Module[1]	ICM[1]
Distance Sensor	Vehicle Speed Sensor[1]	VSS[1]
Distributor Ignition	Distributor Ignition[1]	DI[1]
Distributorless Ignition	Electronic Ignition[1]	EI[1]
DLC (Data Link Connector)	Data Link Connector[1]	DLC[1]
DLI (Distributorless Ignition)	Electronic Ignition[1]	EI[1]
Driver	**Driver**	**Driver**
DS (Detonation Sensor)	Knock Sensor[1]	KS[1]
DTC (Diagnostic Trouble Code)	Diagnostic Trouble Code[1]	DTC[1]
DTM (Diagnostic Test Mode)	Diagnostic Test Mode[1]	DTM[1]
Dual Bed	Three Way + Oxidation Catalytic Converter[1]	TWC+OC[1]
Duty Solenoid for Purge Valve	Evaporative Emission Canister Purge Valve	EVAP Canister Purge Valve[1]
Dynamic Pressure Control	**Dynamic Pressure Control**	**Dynamic PC**
Dynamic Pressure Control Solenoid	**Dynamic Pressure Control Solenoid**[1]	**Dynamic PC Solenoid**[1]
Dynamic Pressure Control Solenoid Valve	**Dynamic Pressure Control Solenoid Valve**[1]	**Dynamic PC Solenoid Valve**[1]
E2PROM (Electrically Erasable Programmable Read Only Memory)	Electrically Erasable Programmable Read Only Memory[1]	EEPROM[1]
Early Fuel Evaporation	Early Fuel Evaporation[1]	EFE[1]
EATX (Electronic Automatic Transmission/ Transaxle)	Automatic Transmission[1]	A/T[1]
EC (Engine Control)	Engine Control[1]	EC[1]

TABLE 1—CROSS REFERENCE AND LOOK UP (CONTINUED)

Existing Usage	Acceptable Usage	Acceptable Acronized Usage
ECA (Electronic Control Assembly)	Powertrain Control Module[1]	PCM[1]
ECL (Engine Coolant Level)	Engine Coolant Level	ECL
ECM (Engine Control Module)	Engine Control Module[1]	ECM[1]
ECT (Engine Coolant Temperature)	Engine Coolant Temperature[1]	ECT[1]
ECT (Engine Coolant Temperature) Sender	Engine Coolant Temperature Sensor[1]	ECT Sensor[1]
ECT (Engine Coolant Temperature) Sensor	Engine Coolant Temperature Sensor[1]	ECT Sensor[1]
ECT (Engine Coolant Temperature) Switch	Engine Coolant Temperature Switch[1]	ECT Switch[1]
ECU4 (Electronic Control Unit 4)	Powertrain Control Module[1]	PCM[1]
EDF (Electro-Drive Fan) Control	Fan Control	FC
EDIS (Electronic Distributor Ignition System)	Distributor Ignition System[1]	DI System[1]
EDIS (Electronic Distributorless Ignition System)	Electronic Ignition System[1]	EI System[1]
EDIS (Electronic Distributor Ignition System) Module	Distributor Ignition Control Module[1]	Distributor ICM[1]
EEC (Electronic Engine Control)	Engine Control[1]	EC[1]
EEC (Electronic Engine Control) Processor	Powertrain Control Module[1]	PCM[1]
EECS (Evaporative Emission Control System)	Evaporative Emission System[1]	EVAP System[1]
EEPROM (Electrically Erasable Programmable Read Only Memory)	Electrically Erasable Programmable Read Only Memory[1]	EEPROM[1]
EFE (Early Fuel Evaporation)	Early Fuel Evaporation[1]	EFE[1]
EFI (Electronic Fuel Injection)	Multiport Fuel Injection[1]	MFI[1]
EFI (Electronic Fuel Injection)	Throttle Body Fuel Injection[1]	TBI[1]
EGO (Exhaust Gas Oxygen) Sensor	Oxygen Sensor[1]	O2S[1]
EGOS (Exhaust Gas Oxygen Sensor)	Oxygen Sensor[1]	O2S[1]
EGR (Exhaust Gas Recirculation)	Exhaust Gas Recirculation[1]	EGR[1]
EGR (Exhaust Gas Recirculation) Diagnostic Valve	Exhaust Gas Recirculation Diagnostic Valve[1]	EGR Diagnostic Valve[1]
EGR (Exhaust Gas Recirculation) System	Exhaust Gas Recirculation System[1]	EGR System[1]
EGR (Exhaust Gas Recirculation) Thermal Vacuum Valve	Exhaust Gas Recirculation Thermal Vacuum Valve[1]	EGR TVV[1]
EGR (Exhaust Gas Recirculation) Valve	Exhaust Gas Recirculation Valve[1]	EGR Valve[1]
EGR TVV (Exhaust Gas Recirculation Thermal Vacuum Valve)	Exhaust Gas Recirculation Thermal Vacuum Valve[1]	EGR TVV[1]
EGRT (Exhaust Gas Recirculation Temperature)	Exhaust Gas Recirculation Temperature	EGRT[1]
EGRT (Exhaust Gas Recirculation Temperature) Sensor	Exhaust Gas Recirculation Temperature Sensor[1]	EGRT Sensor[1]
EGRV (Exhaust Gas Recirculation Valve)	Exhaust Gas Recirculation Valve[1]	EGR Valve[1]
EGRVC (Exhaust Gas Recirculation Valve Control)	Exhaust Gas Recirculation Valve Control[1]	EGR Valve Control[1]
EGS (Exhaust Gas Sensor)	Oxygen Sensor[1]	O2S[1]
EI (Electronic Ignition) (With Distributor)	Distributor Ignition[1]	DI[1]
EI (Electronic Ignition) (Without Distributor)	Electronic Ignition[1]	EI[1]
Electrically Erasable Programmable Read Only Memory	Electrically Erasable Programmable Read Only Memory[1]	EEPROM[1]
Electronic Engine Control	Electronic Engine Control[1]	Electronic EC[1]
Electronic Ignition	Electronic Ignition[1]	EI[1]
Electronic Spark Advance	Ignition Control[1]	IC[1]
Electronic Spark Timing	Ignition Control[1]	IC[1]
EM (Engine Modification)	Engine Modification[1]	EM[1]
EMR (Engine Maintenance Reminder)	Service Reminder Indicator[1]	SRI[1]
Engine Control	Engine Control[1]	EC[1]
Engine Coolant Fan Control	Fan Control	FC
Engine Coolant Level	Engine Coolant Level	ECL
Engine Coolant Level Indicator	Engine Coolant Level Indicator	ECL Indicator

TABLE 1—CROSS REFERENCE AND LOOK UP (CONTINUED)

Existing Usage	Acceptable Usage	Acceptable Acronized Usage
Engine Coolant Temperature	Engine Coolant Temperature[1]	ECT[1]
Engine Coolant Temperature Sender	Engine Coolant Temperature Sensor[1]	ECT Sensor[1]
Engine Coolant Temperature Sensor	Engine Coolant Temperature Sensor[1]	ECT Sensor[1]
Engine Coolant Temperature Switch	Engine Coolant Temperature Switch[1]	ECT Switch[1]
Engine Modification	Engine Modification[1]	EM[1]
Engine Oil Pressure Sender	Engine Oil Pressure Sensor	EOP Sensor
Engine Oil Pressure Sensor	Engine Oil Pressure Sensor	EOP Sensor
Engine Oil Pressure Switch	Engine Oil Pressure Switch	EOP Switch
Engine Oil Temperature	Engine Oil Temperature	EOT
Engine Speed	Engine Speed[1]	RPM[1]
EOS (Exhaust Oxygen Sensor)	Oxygen Sensor[1]	O2S[1]
EOT (Engine Oil Temperature)	Engine Oil Temperature	EOT
EP (Exhaust Pressure)	Exhaust Pressure	EP
EPROM (Erasable Programmable Read Only Memory)	Erasable Programmable Read Only Memory[1]	EPROM[1]
Erasable Programmable Read Only Memory	Erasable Programmable Read Only Memory[1]	EPROM[1]
ESA (Electronic Spark Advance)	Ignition Control[1]	IC[1]
ESAC (Electronic Spark Advance Control)	Distributor Ignition[1]	DI[1]
EST (Electronic Spark Timing)	Ignition Control[1]	IC[1]
EVAP (Evaporate Emission) CANP (Canister Purge)	Evaporative Emission Canister Purge[1]	EVAP Canister Purge[1]
EVAP (Evaporative Emission)	Evaporative Emission[1]	EVAP[1]
EVAP (Evaporative Emission) Canister	Evaporative Emission Canister[1]	EVAP Canister[1]
EVAP (Evaporative Emission) Purge Valve	Evaporative Emission Canister Purge Valve[1]	EVAP Canister Purge Valve[1]
Evaporative Emission	Evaporative Emission[1]	EVAP[1]
Evaporative Emission Canister	Evaporative Emission Canister[1]	EVAP Canister[1]
EVP (Exhaust Gas Recirculation Valve Position) Sensor	Exhaust Gas Recirculation Valve Position Sensor[1]	EGR Valve Position Sensor[1]
EVR (Exhaust Gas Recirculation Vacuum Regulator) Solenoid	Exhaust Gas Recirculation Vacuum Regulator Solenoid[1]	EGR Vacuum Regulator Solenoid[1]
EVRV (Exhaust Gas Recirculation Vacuum Regulator Valve)	Exhaust Gas Recirculation Vacuum Regulator Valve[1]	EGR Vacuum Regulator Valve[1]
Exhaust Gas Recirculation	Exhaust Gas Recirculation[1]	EGR[1]
Exhaust Gas Recirculation Temperature	Exhaust Gas Recirculation Temperature[1]	EGRT[1]
Exhaust Gas Recirculation Temperature Sensor	Exhaust Gas Recirculation Temperature Sensor[1]	EGRT Sensor[1]
Exhaust Gas Recirculation Vacuum Solenoid Valve Regulator	Exhaust Gas Recirculation Vacuum Regulator Solenoid Valve[1]	EGR Vacuum Regulator Solenoid Valve[1]
Exhaust Gas Recirculation Vacuum Regulator Valve	Exhaust Gas Recirculation Vacuum Regulator Valve[1]	EGR Vacuum Regulator Valve[1]
Exhaust Gas Recirculation Valve	Exhaust Gas Recirculation Valve[1]	EGR Valve[1]
Exhaust Pressure	Exhaust Pressure	EP
4GR (Fourth Gear)	Fourth Gear	4GR
4WD (Four Wheel Drive)	Full Time Four Wheel Drive	F4WD
4WD (Four Wheel Drive)	Selectable Four Wheel Drive	S4WD
F4WD	Full Time Four Wheel Drive	F4WD
Fan Control	Fan Control	FC
Fan Control Module	Fan Control Module	FC Module
Fan Control Relay	Fan Control Relay	FC Relay
Fan Motor Control Relay	Fan Control Relay	FC Relay
Fast Idle Thermo Valve	Idle Air Control Thermal Valve[1]	IAC Thermal Valve[1]
FBC (Feed Back Carburetor)	Carburetor[1]	CARB[1]
FBC (Feed Back Control)	Mixture Control[1]	MC[1]

TABLE 1—CROSS REFERENCE AND LOOK UP (CONTINUED)

Existing Usage	Acceptable Usage	Acceptable Acronized Usage
FC (Fan Control)	Fan Control	FC
FC (Fan Control) Relay	Fan Control Relay	FC Relay
FEEPROM (Flash Electrically Erasable Programmable Read Only Memory)	Flash Electrically Erasable Programmable Read Only Memory[1]	FEEPROM[1]
FEPROM (Flash Erasable Programmable Read Only Memory)	Flash Erasable Programmable Read Only Memory[1]	FEPROM[1]
FF (Flexible Fuel)	Flexible Fuel[1]	FF[1]
FI (Fuel Injection)	Central Multiport Fuel Injection[1]	Central MFI[1]
FI (Fuel Injection)	Continuous Fuel Injection[1]	CFI[1]
FI (Fuel Injection)	Direct Fuel Injection[1]	DFI[1]
FI (Fuel Injection)	Indirect Fuel Injection[1]	IFI[1]
FI (Fuel Injection)	Multiport Fuel Injection[1]	MFI[1]
FI (Fuel Injection)	Sequential Multiport Fuel Injection[1]	SFI[1]
FI (Fuel Injection)	Throttle Body Fuel Injection[1]	TBI[1]
Flame Ionization Detector	**Flame Ionization Detector**	**FID**
Flash EEPROM (Electrically Erasable Programmable Read Only Memory)	Flash Electrically Erasable Programmable Read Only Memory[1]	FEEPROM[1]
Flash EPROM (Erasable Programmable Read Only Memory)	Flash Erasable Programmable Read Only Memory[1]	FEPROM[1]
Flexible Fuel	Flexible Fuel[1]	FF[1]
Flexible Fuel Sensor	Flexible Fuel Sensor[1]	FF Sensor
Fourth Gear	Fourth Gear	4GR
FP (Fuel Pump)	Fuel Pump	FP
FP (Fuel Pump) Module	Fuel Pump Module	FP Module
Freeze Frame	**Freeze Frame**	**See Table 4**
Front Wheel Drive	**Front Wheel Drive**	**FWD**
FRZF (Freeze Frame)	**Freeze Frame**	**See Table 4**
FT (Fuel Trim)	Fuel Trim[1]	FT[1]
Fuel Charging Station	Throttle Body[1]	TB[1]
Fuel Concentration Sensor	Flexible Fuel Sensor[1]	FF Sensor[1]
Fuel Injection	Central Multiport Fuel Injection[1]	Central MFI[1]
Fuel Injection	Continuous Fuel Injection[1]	CFI[1]
Fuel Injection	Direct Fuel Injection[1]	DFI[1]
Fuel Injection	Indirect Fuel Injection[1]	IFI[1]
Fuel Injection	Multiport Fuel Injection[1]	MFI[1]
Fuel Injection	Sequential Multiport Fuel Injection[1]	SFI[1]
Fuel Injection	Throttle Body Fuel Injection[1]	TBI[1]
Fuel Level Sensor	Fuel Level Sensor	Fuel Level Sensor
Fuel Module	Fuel Pump Module	FP Module
Fuel Pressure	Fuel Pressure[1]	Fuel Pressure[1]
Fuel Pressure	**Fuel Pressure**	**See Table 4**
Fuel Pressure Regulator	Fuel Pressure Regulator[1]	Fuel Pressure Regulator[1]
Fuel Pump	Fuel Pump	FP
Fuel Pump Relay	Fuel Pump Relay	FP Relay
Fuel Quality Sensor	Flexible Fuel Sensor[1]	FF Sensor[1]
Fuel Regulator	Fuel Pressure Regulator[1]	Fuel Pressure Regulator[1]
Fuel Sender	Fuel Pump Module	FP Module
Fuel Sensor	Fuel Level Sensor	Fuel Level Sensor
Fuel System Status	**Fuel System Status**	**See Table 4**
FUEL SYS	**Fuel System Status**	**See Table 4**
Fuel Tank Unit	Fuel Pump Module	FP Module

TABLE 1—CROSS REFERENCE AND LOOK UP (CONTINUED)

Existing Usage	Acceptable Usage	Acceptable Acronized Usage
Fuel Trim	Fuel Trim[1]	FT[1]
Full Time Four Wheel Drive	**Full Time Four Wheel Drive**	**F4WD**
Full Throttle	Wide Open Throttle[1]	WOT[1]
FWD	**Front Wheel Drive**	**FWD**
GCM (Governor Control Module)	Governor Control Module	GCM
GEM (Governor Electronic Module)	Governor Control Module	GCM
GEN (Generator)	Generator	GEN
Generator	Generator	GEN
Glow Plug	**Glow Plug**[1]	**Glow Plug**[1]
GND (Ground)	Ground	GND
Governor	Governor	Governor
Governor Control Module	Governor Control Module	GCM
Governor Electronic Module	Governor Control Module	GCM
Gram Per Mile	**Gram Per Mile**	**GPM**
GRD (Ground)	Ground	GND
Ground	Ground	GND
Heated Oxygen Sensor	Heated Oxygen Sensor[1]	HO2S[1]
HEDF (High Electro-Drive Fan) Control	Fan Control	FC
HEGO (Heated Exhaust Gas Oxygen) Sensor	Heated Oxygen Sensor[1]	HO2S[1]
HEI (High Energy Ignition)	Distributor Ignition[1]	DI[1]
High Speed FC (Fan Control) Switch	High Speed Fan Control Switch	High Speed FC Switch
HO2S (Heated Oxygen Sensor)	Heated Oxygen Sensor[1]	HO2S[1]
HOS (Heated Oxygen Sensor)	Heated Oxygen Sensor[1]	HO2S[1]
Hot Wire Anemometer	Mass Air Flow Sensor[1]	MAF Sensor[1]
IA (Intake Air)	Intake Air	IA
IA (Intake Air) Duct	Intake Air Duct	IA Duct
IAC (Idle Air Control)	Idle Air Control[1]	IAC[1]
IAC (Idle Air Control) Thermal Valve	Idle Air Control Thermal Valve[1]	IAC Thermal Valve[1]
IAC (Idle Air Control) Valve	Idle Air Control Valve[1]	IAC Valve[1]
IACV (Idle Air Control Valve)	Idle Air Control Valve[1]	IAC Valve[1]
IAT (Intake Air Temperature)	Intake Air Temperature[1]	IAT[1]
IAT (Intake Air Temperature) Sensor	Intake Air Temperature Sensor[1]	IAT Sensor[1]
IATS (Intake Air Temperature Sensor)	Intake Air Temperature Sensor[1]	IAT Sensor[1]
IC (Ignition Control)	Ignition Control[1]	IC[1]
ICM (Ignition Control Module)	Ignition Control Module[1]	ICM[1]
ICP (Injection Control Pressure)	**Injection Control Pressure**[1]	**ICP**[1]
IDFI (Indirect Fuel Injection)	Indirect Fuel Injection[1]	IFI[1]
IDI (Integrated Direct Ignition)	Electronic Ignition[1]	EI[1]
IDI (Indirect Diesel Injection)	Indirect Fuel Injection[1]	IFI[1]
Idle Air Bypass Control	Idle Air Control[1]	IAC[1]
Idle Air Control	Idle Air Control[1]	IAC[1]
Idle Air Control Valve	Idle Air Control Valve[1]	IAC Valve[1]
Idle Speed Control	Idle Air Control[1]	IAC[1]
Idle Speed Control	Idle Speed Control[1]	ISC[1]
Idle Speed Control Actuator	Idle Speed Control Actuator[1]	ISC Actuator[1]
IFI (Indirect Fuel Injection)	Indirect Fuel Injection[1]	IFI[1]
IFS (Inertia Fuel Shutoff)	Inertia Fuel Shutoff	IFS
Ignition Control	Ignition Control[1]	IC[1]
Ignition Control Module	Ignition Control Module[1]	ICM[1]
I/M (Inspection and Maintenance)	**Inspection and Maintenance**	**I/M**

TABLE 1—CROSS REFERENCE AND LOOK UP (CONTINUED)

Existing Usage	Acceptable Usage	Acceptable Acronized Usage
IMRC (Intake Manifold Runner Control)	**Intake Manifold Runner Control**	**IMRC**
In Tank Module	Fuel Pump Module	FP Module
Indirect Fuel Injection	Indirect Fuel Injection[1]	IFI[1]
Inertia Fuel Shutoff	Inertia Fuel Shutoff	IFS
Inertia Fuel - Shutoff Switch	Inertia Fuel Shutoff Switch	IFS Switch
Inertia Switch	Inertia Fuel Shutoff Switch	IFS Switch
Injection Control Pressure	**Injection Control Pressure**[1]	**ICP**[1]
Input Shaft Speed	**Input Shaft Speed**	**ISS**
INT (Integrator)	Short Term Fuel Trim[1]	Short Term FT[1]
Inspection and Maintenance	**Inspection and Maintenance**	**I/M**
Intake Air	Intake Air	IA
Intake Air Duct	Intake Air Duct	IA Duct
Intake Air Temperature	Intake Air Temperature[1]	IAT[1]
Intake Air Temperature Sensor	Intake Air Temperature Sensor[1]	IAT Sensor[1]
Intake Manifold Absolute Pressure Sensor	Manifold Absolute Pressure Sensor[1]	MAP Sensor[1]
Intake Manifold Runner Control	**Intake Manifold Runner Control**	**IMRC**
Integrated Relay Module	Relay Module	RM
Integrator	Short Term Fuel Trim[1]	Short Term FT[1]
Inter Cooler	Charge Air Cooler[1]	CAC[1]
ISC (Idle Speed Control)	Idle Air Control[1]	IAC[1]
ISC (Idle Speed Control)	Idle Speed Control[1]	ISC[1]
ISC (Idle Speed Control) Actuator	Idle Speed Control Actuator[1]	ISC Actuator[1]
ISC BPA (Idle Speed Control By Pass Air)	Idle Air Control[1]	IAC
ISC (Idle Speed Control) Solenoid Vacuum Valve	Idle Speed Control Solenoid Vacuum Valve[1]	ISC Solenoid Vacuum Valve[1]
ISS (Input Shaft Speed)	**Input Shaft Speed**	**ISS**
K-Jetronic	Continuous Fuel Injection[1]	CFI[1]
KAM (Keep Alive Memory)	Non Volatile Random Access Memory[1]	NVRAM[1]
KAM (Keep Alive Memory)	Keep Alive Random Access Memory[1]	Keep Alive RAM[1]
KE-Jetronic	Continuous Fuel Injection[1]	CFI[1]
KE-Motronic	Continuous Fuel Injection[1]	CFI[1]
Knock Sensor	Knock Sensor[1]	KS[1]
KS (Knock Sensor)	Knock Sensor[1]	KS[1]
L-Jetronic	Multiport Fuel Injection[1]	MFI[1]
Lambda	Oxygen Sensor[1]	O2S[1]
LH-Jetronic	Multiport Fuel Injection[1]	MFI[1]
Light Off Catalyst	Warm Up Three Way Catalytic Converter[1]	WU-TWC[1]
Light Off Catalyst	Warm Up Oxidation Catalytic Converter[1]	WU-OC[1]
Line Pressure Control Solenoid Valve	**Line Pressure Control Solenoid Valve**	**Line PC Solenoid Valve**
LOAD (Calculated Load Value)	**Calculated Load Value**	**LOAD**
Lock Up Relay	Torque converter Clutch Relay[1]	TCC Relay[1]
Long Term FT (Fuel Trim)	Long Term Fuel Trim[1]	Long Term FT[1]
Long Term Fuel Trim	**Long Term FT**	**Long Term FT**
LONG FT	**Long Term Fuel Trim**	**See Table 4**
Low Speed FC (Fan Control) Switch	Low Speed Fan Control Switch	Low Speed FC Switch
LUS (Lock Up Solenoid) Valve	Torque Converter Clutch Solenoid Valve[1]	TCC Solenoid Valve[1]
M/C (Mixture Control)	Mixture Control[1]	MC[1]
MAF (Mass Air Flow)	Mass Air Flow[1]	MAF[1]
MAF (Mass Air Flow) Sensor	Mass Air Flow Sensor[1]	MAF Sensor[1]

TABLE 1—CROSS REFERENCE AND LOOK UP (CONTINUED)

Existing Usage	Acceptable Usage	Acceptable Acronized Usage
Malfunction Indicator Lamp	Malfunction Indicator Lamp[1]	MIL[1]
Manifold Absolute Pressure	Manifold Absolute Pressure[1]	MAP[1]
Manifold Absolute Pressure Sensor	Manifold Absolute Pressure Sensor	MAP Sensor[1]
Manifold Differential Pressure	Manifold Differential Pressure[1]	MDP[1]
Manifold Surface Temperature	Manifold Surface Temperature[1]	MST[1]
Manifold Vacuum Zone	Manifold Vacuum Zone[1]	MVZ[1]
Manual Lever Position Sensor	Transmission Range Sensor[1]	TR Sensor[1]
MAP (Manifold Absolute Pressure)	Manifold Absolute Pressure[1]	MAP[1]
MAP (Manifold Absolute Pressure) Sensor	Manifold Absolute Pressure Sensor[1]	MAP Sensor[1]
MAPS (Manifold Absolute Pressure Sensor)	Manifold Absolute Pressure Sensor[1]	MAP Sensor[1]
Mass Air Flow	Mass Air Flow[1]	MAF[1]
Mass Air Flow Sensor	Mass Air Flow Sensor[1]	MAF Sensor[1]
MAT (Manifold Air Temperature)	Intake Air Temperature[1]	IAT[1]
MATS (Manifold Air Temperature Sensor)	Intake Air Temperature Sensor[1]	IAT Sensor[1]
MC (Mixture Control)	Mixture Control[1]	MC[1]
MCS (Mixture Control Solenoid)	Mixture Control Solenoid[1]	MC Solenoid[1]
MCU (Microprocessor Control Unit)	Powertrain Control Module[1]	PCM[1]
MDP (Manifold Differential Pressure)	Manifold Differential Pressure[1]	MDP[1]
MFI (Multiport Fuel Injection)	Multiport Fuel Injection[1]	MFI[1]
MIL (Malfunction Indicator Lamp)	Malfunction Indicator Lamp[1]	MIL[1]
Mixture Control	Mixture Control[1]	MC[1]
MLPS (Manual Lever Position Sensor)	**Transmission Range Sensor[1]**	**TR Sensor[1]**
Modes	Diagnostic Test Mode[1]	DTM[1]
Mono-Jetronic	**Throttle Body Injection[1]**	**TBI[1]**
Mono-Motronic	**Throttle Body Injection[1]**	**TBI[1]**
Monotronic	Throttle Body Fuel Injection[1]	TBI[1]
Motronic-Pressure	**Multiport Fuel Injection[1]**	**MFI[1]**
Motronic	Multiport Fuel Injection[1]	MFI[1]
MPI (Multipoint Injection)	Multiport Fuel Injection[1]	MFI[1]
MPI (Multiport Injection)	Multiport Fuel Injection[1]	MFI[1]
MRPS (Manual Range Position Switch)	Transmission Range Switch	TR Switch
MST (Manifold Surface Temperature)	Manifold Surface Temperature[1]	MST[1]
Multiport Fuel Injection	Multiport Fuel Injection[1]	MFI[1]
MVZ (Manifold Vacuum Zone)	Manifold Vacuum Zone[1]	MVZ[1]
NDS (Neutral Drive Switch)	Park/Neutral Position Switch[1]	PNP Switch[1]
Neutral Safety Switch	Park/Neutral Position Switch[1]	PNP Switch[1]
NGS (Neutral Gear Switch)	Park/Neutral Position Switch[1]	PNP Switch[1]
Non Dispersive Infrared	**Non Dispersive Infrared**	**NDIR**
Non Volatile Random Access Memory	Non Volatile Random Access Memory[1]	NVRAM[1]
NPS (Neutral Position Switch)	Park/Neutral Position Switch[1]	PNP Switch[1]
NVM (Non Volatile Memory)	Non Volatile Random Access Memory[1]	NVRAM[1]
NVRAM (Non Volatile Random Access Memory)	Non Volatile Random Access Memory[1]	NVRAM[1]
O2 (Oxygen) Sensor	Oxygen Sensor[1]	O2S[1]
O2S (Oxygen Sensor)	Oxygen Sensor[1]	O2S[1]
Oxygen Sensor Location	**Oxygen Sensor Location**	**See Table 4**
OBD (On Board Diagnostic)	On Board Diagnostic[1]	OBD[1]
OBD Status	**OBD Status**	**see Table 4**
OBD STAT	**OBD Status**	**see Table 4**
OC (Oxidation Catalyst)	Oxidation Catalytic Converter[1]	OC[1]

TABLE 1—CROSS REFERENCE AND LOOK UP (CONTINUED)

Existing Usage	Acceptable Usage	Acceptable Acronized Usage
Oil Pressure Sender	Engine Oil Pressure Sensor	EOP Sensor
Oil Pressure Sensor	Engine Oil Pressure Sensor	EOP Sensor
Oil Pressure Switch	Engine Oil Pressure Switch	EOP Switch
OL (Open Loop)	Open Loop[1]	OL[1]
On Board Diagnostic	On Board Diagnostic[1]	OBD[1]
Open Loop	Open Loop[1]	OL[1]
OS (Oxygen Sensor)	Oxygen Sensor[1]	O2S[1]
OSS (Output Shaft Speed) Sensor	**Output Shaft Speed Sensor**[1]	**OSS Sensor**[1]
Output Driver	**Driver**	**Driver**
Output Shaft Speed Sensor	**Output Shaft Speed Sensor**[1]	**OSS Sensor**[1]
Oxidation Catalytic Converter	Oxidation Catalytic Converter[1]	OC[1]
OXS (Oxygen Sensor) Indicator	Service Reminder Indicator[1]	SRI[1]
Oxygen Sensor	Oxygen Sensor[1]	O2S[1]
P/N (Park/Neutral)	Park/Neutral Position[1]	PNP[1]
P/S (Power Steering) Pressure Switch	Power Steering Pressure Switch	PSP Switch
P- (Pressure) Sensor	Manifold Absolute Pressure Sensor[1]	MAP Sensor[1]
PAIR (Pulsed Secondary Air Injection)	Pulsed Secondary Air Injection[1]	PAIR[1]
Parameter Identification	**Parameter Identification**	**PID**
Parameter Identification Supported	**Parameter Identification Supported**	**See Table 4**
Park/Neutral Position	Park/Neutral Position[1]	PNP[1]
PC (Pressure Control) Solenoid Valve	**Pressure Control Solenoid Valve**[1]	**PC Solenoid Valve**[1]
PCM (Powertrain Control Module)	Powertrain Control Module[1]	PCM[1]
PCV (Positive Crankcase Ventilation)	Positive Crankcase Ventilation[1]	PCV[1]
PCV (Positive Crankcase Ventilation) Valve	Positive Crankcase Ventilation Valve[1]	PCV Valve[1]
Percent Alcohol Sensor	Flexible Fuel Sensor[1]	FF Sensor[1]
Periodic Trap Oxidizer	Periodic Trap Oxidizer[1]	PTOX[1]
PFE (Pressure Feedback Exhaust Gas Recirculation Sensor	**Feedback Pressure** Exhaust Gas Recirculation Sensor[1]	**Feedback Pressure** EGR Sensor[1]
PFI (Port Fuel Injection)	Multiport Fuel Injection[1]	MFI[1]
PG (Pulse Generator)	Vehicle Speed Sensor[1]	VSS[1]
PGM-FI (Programmed Fuel Injection)	Multiport Fuel Injection[1]	MFI[1]
PID (Parameter Identification)	**Parameter Identification**	**PID**
PID SUP	**Parameter Identification Supported**	**See Table 4**
PIP (Position Indicator Pulse)	Crankshaft Position[1]	CKP[1]
PNP (Park/Neutral Position)	Park/Neutral Position[1]	PNP[1]
Positive Crankcase Ventilation	Positive Crankcase Ventilation[1]	PCV[1]
Positive Crankcase Ventilation Valve	Positive Crankcase Ventilation Valve[1]	PCV Valve[1]
Power Steering Pressure	Power Steering Pressure	PSP
Power Steering Pressure Switch	Power Steering Pressure Switch	PSP Switch
Powertrain Control Module	Powertrain Control Module[1]	PCM[1]
Pressure Control Solenoid Valve	**Pressure Control Solenoid Valve**[1]	**PC Solenoid Valve**[1]
Pressure Feedback EGR (Exhaust Gas Recirculation)	Feedback Pressure Exhaust Gas Recirculation[1]	Feedback Pressure EGR[1]
Pressure Sensor	Manifold Absolute Pressure Sensor[1]	MAP Sensor[1]
Pressure Feedback EGR (Exhaust Gas **Recirculation) System**	**Feedback Pressure** Exhaust Gas Recirculation System[1]	**Feedback Pressure** EGR System[1]
Pressure Transducer EGR (Exhaust Gas Recirculation) System	Pressure Transducer Exhaust Gas Recirculation System[1]	Pressure Transducer EGR System[1]
PRNDL (Park- Reverse- Neutral- Drive- Low)	Transmission Range	TR
Programmable Read Only Memory	Programmable Read Only Memory[1]	PROM[1]

TABLE 1—CROSS REFERENCE AND LOOK UP (CONTINUED)

Existing Usage	Acceptable Usage	Acceptable Acronized Usage
PROM (Programmable Read Only Memory)	Programmable Read Only Memory[1]	PROM[1]
PSP (Power Steering Pressure)	Power Steering Pressure	PSP
PSP (Power Steering Pressure) Switch	Power Steering Pressure Switch	PSP Switch
PSPS (Power Steering Pressure Switch)	Power Steering Pressure Switch	PSP Switch
PTOX (Periodic Trap Oxidizer)	Periodic Trap Oxidizer[1]	PTOX[1]
Pulsair	Pulsed Secondary Air Injection[1]	PAIR[1]
Pulsed Secondary Air Injection	Pulsed Secondary Air Injection	PAIR[1]
Pulse Width Modulation	**Pulse Width Modulation**	**PWM**
PWM	**Pulse Width Modulation**	**PWM**
QDM (Quad Driver Module)	**Driver**	**Driver**
Quad Driver Module	**Driver**	**Driver**
Radiator Fan Control	Fan Control	FC
Radiator Fan Relay	Fan Control Relay	FC Relay
RAM (Random Access Memory)	Random Access Memory[1]	RAM[1]
Random Access Memory	Random Access Memory[1]	RAM[1]
Read Only Memory	Read Only Memory[1]	ROM[1]
Rear Wheel Drive	**Rear Wheel Drive**	**RWD**
Recirculated Exhaust Gas Temperature Sensor	Exhaust Gas Recirculation Temperature Sensor	EGRT Sensor[1]
Reed Valve	Pulsed Secondary Air Injection Valve[1]	PAIR Valve[1]
REGTS (Recirculated Exhaust Gas Temperature Sensor)	Exhaust Gas Recirculation Temperature Sensor[1]	EGRT Sensor[1]
Relay Module	Relay Module	RM
Remote Mount TFI (Thick Film Ignition)	Distributor Ignition[1]	DI[1]
Revolutions per Minute	Engine Speed[1]	RPM[1]
RM (Relay Module)	Relay Module	RM
ROM (Read Only Memory)	Read Only Memory[1]	ROM[1]
RPM (Revolutions per Minute)	Engine Speed[1]	RPM[1]
RWD	**Rear Wheel Drive**	**RWD**
S4WD	**Selectable Four Wheel Drive**	**S4WD**
SABV (Secondary Air Bypass Valve)	Secondary Air Injection Bypass Valve[1]	AIR Bypass Valve[1]
SACV (Secondary Air Check Valve)	Secondary Air Injection Control Valve[1]	AIR Control Valve[1]
SASV (Secondary Air Switching Valve)	Secondary Air Injection Switching Valve[1]	AIR Switching Valve[1]
SBEC (Single Board Engine Control)	Powertrain Control Module[1]	PCM[1]
SBS (Supercharger Bypass Solenoid)	Supercharger Bypass Solenoid[1]	SCB Solenoid[1]
SC (Supercharger)	Supercharger[1]	SC[1]
Scan Tool	Scan Tool[1]	ST[1]
SCB (Supercharger Bypass)	Supercharger Bypass[1]	SCB[1]
Secondary Air Bypass Valve	Secondary Air Injection Bypass Valve[1]	AIR Bypass Valve[1]
Secondary Air Check Valve	Secondary Air Injection Check Valve[1]	AIR Check Valve[1]
Secondary Air Injection	Secondary Air Injection[1]	AIR[1]
Secondary Air Injection Bypass	Secondary Air Injection Bypass[1]	AIR Bypass[1]
Secondary Air Injection Diverter	Secondary Air Injection Diverter[1]	AIR Diverter[1]
Secondary Air Switching Valve	Secondary Air Injection Switching Valve[1]	AIR Switching Valve[1]
Selectable Four Wheel Drive	**Selectable Four Wheel Drive**	**S4WD**
SEFI (Sequential Electronic Fuel Injection)	Sequential Multiport Fuel Injection[1]	SFI[1]
Self Test	On Board Diagnostic[1]	OBD[1]
Self Test Codes	Diagnostic Trouble Code[1]	DTC[1]
Self Test Connector	Data Link Connector[1]	DLC[1]
Sequential Multiport Fuel Injection	Sequential Multiport Fuel Injection[1]	SFI[1]

TABLE 1—CROSS REFERENCE AND LOOK UP (CONTINUED)

Existing Usage	Acceptable Usage	Acceptable Acronized Usage
Service Engine Soon	Service Reminder Indicator[1]	SRI[1]
Service Engine Soon	Malfunction Indicator Lamp[1]	MIL[1]
Service Reminder Indicator	Service Reminder Indicator[1]	SRI[1]
SFI (Sequential Fuel Injection)	Sequential Multiport Fuel Injection[1]	SFI[1]
Shift Solenoid	**Shift Solenoid**[1]	**SS**[1]
Shift Solenoid Valve	**Shift Solenoid Valve**[1]	**SS Valve**[1]
Short Term FT (Fuel Trim)	Short Term Fuel Trim[1]	Short Term FT[1]
Short Term Fuel Trim	**Short Term Fuel Trim**[1]	**Short Term FT**[1]
SHRT FT	**Short Term Fuel Trim**[1]	**See Table 4**
SLP (Selection Lever Position)	Transmission Range	TR
SMEC (Single Module Engine Control)	Powertrain Control Module[1]	PCM[1]
Smoke Puff Limiter	Smoke Puff Limiter[1]	SPL[1]
SPARK ADV	**Spark Advance**	**See Table 4**
Spark Advance	**Spark Advance**	**See Table 4**
Spark Plug	**Spark Plug**[1]	**Spark Plug**[1]
SPI (Single Point Injection)	Throttle Body Fuel Injection[1]	TBI[1]
SPL (Smoke Puff Limiter)	Smoke Puff Limiter[1]	SPL[1]
SS (Shift Solenoid)	**Shift Solenoid**[1]	**SS**[1]
SRI (Service Reminder Indicator)	Service Reminder Indicator[1]	SRI[1]
SRT (System Readiness Test)	System Readiness Test[1]	SRT[1]
ST (Scan Tool)	Scan Tool[1]	ST[1]
Supercharger	Supercharger[1]	SC[1]
Supercharger Bypass	Supercharger Bypass[1]	SCB[1]
Sync Pickup	Camshaft Position[1]	CMP[1]
System Readiness Test	System Readiness Test[1]	SRT[1]
3-2TS (3-2 Timing Solenoid)	**3-2 Timing Solenoid**	**3-2TS**
3-2TS Valve (3-2 Timing Solenoid)Valve	**3-2 Timing Solenoid Valve**	**3-2TS Valve**
3-2 Timing Solenoid	**3-2 Timing Solenoid**	**3-2TS**
3-2 Timing Solenoid Valve	**3-2 Timing Solenoid Valve**	**3-2TS Valve**
3GR (Third Gear)	Third Gear	3GR
TAB (Thermactor Air Bypass)	Secondary Air Injection Bypass[1]	AIR Bypass[1]
TAC (Throttle Actuator Control)	**Throttle Actuator Control**	**TAC**
TAC (Throttle Actuator Control) Module	**Throttle Actuator Control Module**[1]	**TAC Module**[1]
TAD (Thermactor Air Diverter)	Secondary Air Injection Diverter[1]	AIR Diverter[1]
TB (Throttle Body)	Throttle Body[1]	TB[1]
TBI (Throttle Body Fuel Injection)	Throttle Body Fuel Injection[1]	TBI[1]
TBT (Throttle Body Temperature)	Intake Air Temperature[1]	IAT[1]
TC (Turbocharger)	Turbocharger[1]	TC[1]
TC (Turbocharger) Wastegate	**Turbocharger Wastegate**[1]	**TC Wastegate**[1]
TC (Turbocharger) Wastegate Regulating Valve	**Turbocharger Wastegate Regulating Valve**[1]	**TC Wastegate Regulating Valve**[1]
TCC (Torque Converter Clutch)	Torque Converter Clutch[1]	TCC[1]
TCC (Torque Converter Clutch) Relay	Torque Converter Clutch Relay[1]	TCC Relay[1]
TCC (Torque Converter Clutch) Solenoid	**Torque Converter Clutch Solenoid**[1]	**TCC Solenoid**[1]
TCC (Torque Converter Clutch) Solenoid Valve	**Torque Converter Clutch Solenoid Valve**[1]	**TCC Solenoid Valve**[1]
TCM (Transmission Control Module)	Transmission Control Module	TCM
TCCP (Torque Converter Clutch Pressure)	**Torque Converter Clutch Pressure**	**TCCP**
TFI (Thick Film Ignition)	Distributor Ignition[1]	DI[1]
TFI (Thick Film Ignition) Module	Ignition Control Module[1]	ICM[1]

TABLE 1—CROSS REFERENCE AND LOOK UP (CONTINUED)

Existing Usage	Acceptable Usage	Acceptable Acronized Usage
TFP (Transmission Fluid Pressure)	**Transmission Fluid Pressure**	**TFP**
TFT (Transmission Fluid Temperature) Sensor	**Transmission Fluid Temperature Sensor**	**TFT Sensor**
Thermac	Secondary Air Injection[1]	AIR[1]
Thermac Air Cleaner	Air Cleaner[1]	ACL[1]
Thermactor	Secondary Air Injection[1]	AIR[1]
Thermactor Air Bypass	Secondary Air Injection Bypass[1]	AIR Bypass[1]
Thermactor Air Diverter	Secondary Air Injection Diverter[1]	AIR Diverter[1]
Thermactor II	Pulsed Secondary Air Injection[1]	PAIR[1]
Thermal Vacuum Switch	Thermal Vacuum Valve[1]	TVV[1]
Thermal Vacuum Valve	Thermal Vacuum Valve[1]	TVV[1]
Third Gear	Third Gear	3GR
Three Way + Oxidation Catalytic Converter	Three Way + Oxidation Catalytic Converter[1]	TWC+OC[1]
Three Way Catalytic Converter	Three Way Catalytic Converter[1]	TWC[1]
Throttle Actuator Control	**Throttle Actuator Control**	**TAC**
Throttle Actuator Control Module	**Throttle Actuator Control Module**	**TAC Module**
Throttle Body	Throttle Body[1]	TB[1]
Throttle Body Fuel Injection	Throttle Body Fuel Injection[1]	TBI[1]
Throttle Opener	Idle Speed Control[1]	ISC[1]
Throttle Opener Vacuum Switching Valve	Idle Speed Control Solenoid Vacuum Valve[1]	ISC Solenoid Vacuum Valve[1]
Throttle Opener VSV (Vacuum Switching Valve)	Idle Speed Control Solenoid Vacuum Valve[1]	ISC Solenoid Vacuum Valve[1]
Throttle Position	Throttle Position[1]	TP
Throttle Position Sensor	Throttle Position Sensor[1]	TP Sensor[1]
Throttle Position Switch	Throttle Position Switch[1]	TP Switch[1]
Throttle Potentiometer	Throttle Position Sensor[1]	TP Sensor[1]
TOC (Trap Oxidizer - Continuous)	Continuous Trap Oxidizer[1]	CTOX[1]
TOP (Trap Oxidizer - Periodic)	Periodic Trap Oxidizer[1]	PTOX[1]
Torque Converter Clutch	Torque Converter Clutch[1]	TCC[1]
Torque Converter Clutch Pressure	**Torque Converter Clutch Pressure**	**TCCP**
Torque Converter Clutch Relay	Torque Converter Clutch Relay[1]	TCC Relay[1]
Torque Converter Clutch Solenoid	**Torque Converter Clutch Solenoid**[1]	**TCC Solenoid**[1]
Torque Converter Clutch Solenoid Valve	**Torque Converter Clutch Solenoid Valve**[1]	**TCC Solenoid Valve**[1]
TP (Throttle Position)	Throttle Position[1]	TP[1]
TP (Throttle Position) Sensor	Throttle Position Sensor[1]	TP Sensor[1]
TP (Throttle Position) Switch	Throttle Position Switch[1]	TP Switch[1]
TPI (Tuned Port Injection)	Multiport Fuel Injection[1]	MFI[1]
TPNP (Transmission Park Neutral Position)	**Park/Neutral Position**[1]	**PNP**[1]
TPS (Throttle Position Sensor)	Throttle Position Sensor[1]	TP Sensor[1]
TPS (Throttle Position Switch)	Throttle Position Switch[1]	TP Switch[1]
TR (Transmission Range)	Transmission Range	TR
Track Road Load Horsepower	**Track Road Load Horsepower**	**TRLHP**
Transmission Control Module	Transmission Control Module	TCM
Transmission Fluid Pressure	**Transmission Fluid Pressure**	**TFP**
Transmission Fluid Temperature Sensor	**Transmission Fluid Temperature Sensor**	**TFT Sensor**
Transmission Park Neutral Position	**Park/Neutral Position**[1]	**PNP**[1]
Transmission Position Switch	Transmission Range Switch	TR Switch
Transmission Range Selection	Transmission Range	TR
Transmission Range Sensor	**Transmission Range Sensor**	**TR Sensor**
TRS (Transmission Range Selection)	Transmission Range	TR

TABLE 1—CROSS REFERENCE AND LOOK UP (CONTINUED)

Existing Usage	Acceptable Usage	Acceptable Acronized Usage
TRSS (Transmission Range Selection Switch)	Transmission Range Switch	TR Switch
TSS (Turbine Shaft Speed) Sensor	**Turbine Shaft Speed Sensor**[1]	**TSS Sensor**[1]
Tuned Port Injection	Multiport Fuel Injection[1]	MFI[1]
Turbine Shaft Speed Sensor	**Turbine Shaft Speed Sensor**[1]	**TSS Sensor**[1]
Turbo (Turbocharger)	Turbocharger[1]	TC[1]
Turbocharger	Turbocharger[1]	TC[1]
Turbocharger Wastegate	**Turbocharger Wastegate**[1]	**TC Wastegate**[1]
Turbocharger Wastegate Regulating Valve	**Turbocharger Wastegate Regulating Valve**[1]	**TC Wastegate Regulating Valve**[1]
TVS (Thermal Vacuum Switch)	Thermal Vacuum Valve[1]	TVV[1]
TVV (Thermal Vacuum Valve)	Thermal Vacuum Valve[1]	TVV[1]
TWC (Three Way Catalytic Converter)	Three Way Catalytic Converter[1]	TWC[1]
TWC + OC (Three Way + Oxidation Catalytic Converter)	Three Way + Oxidation Catalytic Converter[1]	TWC+OC[1]
VAC (Vacuum) Sensor	Manifold Differential Pressure Sensor[1]	MDP Sensor[1]
Vacuum Switches	Manifold Vacuum Zone Switch	MVZ Switch[1]
VAF (Volume Air Flow)	Volume Air Flow[1]	VAF[1]
Valve Position EGR (Exhaust Gas Recirculation) System	**Valve Position Exhaust Gas Recirculation System**[1]	**Valve Position EGR System**[1]
Vane Air Flow	Volume Air Flow[1]	VAF[1]
Variable Control Relay Module	**Variable Control Relay Module**	**VCRM**
Variable Fuel Sensor	Flexible Fuel Sensor	FF Sensor[1]
VAT (Vane Air Temperature)	Intake Air Temperature[1]	IAT[1]
VCC (Viscous Converter Clutch)	Torque Converter Clutch[1]	TCC[1]
VCM	**Vehicle Control Module**	**VCM**
VCRM	**Variable Control Relay Module**	**VCRM**
Vehicle Control Module	**Vehicle Control Module**	**VCM**
Vehicle Identification Number	**Vehicle Identification Number**	**VIN**
Vehicle Speed Sensor	Vehicle Speed Sensor[1]	VSS[1]
VIN (Vehicle Identification Number)	**Vehicle Identification Number**	**VIN**
VIP (Vehicle In Process) Connector	Data Link Connector[1]	DLC[1]
Viscous Converter Clutch	Torque Converter Clutch[1]	TCC[1]
Voltage Regulator	Voltage Regulator	VR
Volume Air Flow	Volume Air Flow[1]	VAF[1]
VR (Voltage Regulator)	Voltage Regulator	VR
VSS (Vehicle Speed Sensor)	Vehicle Speed Sensor[1]	VSS[1]
VSV (Vacuum Solenoid Valve) (Canister)	Evaporative Emission Canister Purge Valve[1]	EVAP Canister Purge Valve[1]
VSV (Vacuum Solenoid Valve) (EVAP)	Evaporative Emission Canister Purge Valve[1]	EVAP Canister Purge Valve[1]
VSV (Vacuum Solenoid Valve) (Throttle)	Idle Speed Control Solenoid Vacuum Valve[1]	ISC Solenoid Vacuum Valve[1]
Warm Up Oxidation Catalytic Converter	Warm Up Oxidation Catalytic Converter[1]	WU-OC[1]
Warm Up Three Way Catalytic Converter	Warm Up Three Way Catalytic Converter[1]	WU-OC[1]
Wide Open Throttle	Wide Open Throttle[1]	WOT[1]
WOT (Wide Open Throttle)	Wide Open Throttle[1]	WOT[1]
WOTS (Wide Open Throttle Switch)	Wide Open Throttle Switch[1]	WOT Switch[1]
WU-OC (Warm Up Oxidation Catalytic Converter)	Warm Up Oxidation Catalytic Converter[1]	WU-OC[1]
WU-TWC (Warm Up Three Way Catalytic Converter)	Warm Up Three Way Catalytic Converter[1]	WU-TWC[1]

Recommended Terms and Recommended Acronyms See Table 2

[1] Emission-Related Term

Bold indicates new/revised entry

GLOSSARY

Note: Terms are highlighted in color, followed by Spanish translation in bold.

Abrasion Wearing or rubbing away of a part.
Abrasión El desgaste o consumo por rozamiento de una parte.

Acceleration An increase in velocity or speed.
Aceleración Un incremento en la velocidad.

Accumulator A device used in automatic transmissions to cushion the shock of shifting between gears, providing a smoother feel inside the vehicle.
Acumulador Un dispositivo que se usa en las transmisiones automáticas para suavizar el choque de cambios entre las velocidades, así proporcionando una sensación más uniforme en el interior del vehículo.

Adhesives Chemicals used to hold gaskets in place during the assembly of an engine. They also aid the gasket in maintaining a tight seal by filling in the small irregularities on the surfaces and by preventing the gasket from shifting due to engine vibration.
Adhesivo Los productos químicos que se usan para sujetar a los empaques en una posición correcta mientras que se efectua la asamblea de un motor. También ayuden para que los empaques mantengan un sello impermeable, rellenando a las irregularidades pequeñas en las superficies y previniendo que se mueva el empaque debido a las vibraciones del motor.

Antifriction bearing A bearing designed to reduce friction. This type of bearing normally uses ball or roller inserts to reduce the friction.
Cojinetes de antifricción Un cojinete diseñado con el fin de disminuir la fricción. Este tipo de cojinete suele incorporar una pieza inserta esférica o de rodillos para disminuir la fricción.

Antiseize Thread compound designed to keep threaded connections from damage due to rust or corrosion.
Antiagarrotamiento Un compuesto para filetes diseñado para protejer a las conecciones fileteados de los daños de la oxidación o la corrosión.

Apply devices Devices that hold or drive members of a planetary gear set. They may be hydraulically or mechanically applied.
Dispositivos de aplicación Los dispositivos que sujeten o manejan los miembros de un engranaje planetario. Se pueden aplicar mecánicamente o hidráulicamente.

Asbestos A material that was commonly used as a gasket material in places where temperatures are extreme. This material is being used less frequently today because of health hazards that are inherent to the material.
Amianto Una materia que se usaba frecuentemente como materia de empaques en sitios en los cuales las temperaturas son extremas. Esta materia se usa menos actualmente debido a los peligros al salud que se atribuyan a esta materia.

ATF Automatic transmission fluid.
ATF Fluido de transmisión automática.

Automatic transmission A transmission in which gear or ratio changes are self-activated, eliminating the necessity of hand-shifting gears.
Transmisión automática Una transmisión en la cual un cambio deengranajes o los cambios en relación son por mando automático, así eliminando la necesidad de cambios de velocidades manual.

Axial Parallel to a shaft or bearing bore.
Axial Paralelo a una flecha o al taladro del cojinete.

Axis The centerline of a rotating part, a symmetrical part, or a circular bore.
Eje La linea de quilla de una parte giratoria, una parte simétrica, o un taladro circular.

Axle The shaft or shafts of a machine upon which the wheels are mounted.
Semieje El eje o los ejes de una máquina sobre los cuales se montan las ruedas.

Axle carrier assembly A cast-iron framework that can be removed from the rear axle housing for service and adjustment of the parts.
Asamblea del portador del eje Un armazón de acero vaciado que se puede remover del cárter de los ejes traseros para afectuar el mantenimiento o los ajustes de las partes.

Axle housing Designed in the removable carrier or integral carrier types to house the drive pinion, ring gear, differential, and axle shaft assemblies.
Cárter del eje Diseñado en los tipos de portador removible o de portador integral para encajar el piñon de ataque, la corona, la diferencial y las asambleas de las flechas de los ejes.

Axle ratio The ratio between the rotational speed (rpm) of the driveshaft and that of the driven wheel; gear reduction through the differential, determined by dividing the number of teeth on the ring gear by the number of teeth on the drive pinion.
Relación del eje La relación entre la velocidad giratorio (rpm) del árbol propulsor y la de la rueda arrastrada; reducción de los engranajes por medio del diferencial, que se determina por dividir el número de dientes de la corona por el número de los dientes en el pinión de ataque.

Axle shaft A shaft on which the road wheels are mounted.
Flecha del semieje Una flecha en la cual se monta las ruedas.

Axle shaft end thrust A force exerted on the end of an axle shaft that is most pronounced when the vehicle turns corners and curves.
Golpe en la flecha del semieje Una fuerza que se aplica en el extremo de la flecha del semieje que se pronuncia más cuando un vehículo da la vuelta.

Axle shaft tubes These tubes are attached to the axle housing center section to surround the axle shaft and bearings.
Tubos de los semiejes Estos tubos se conectan a la sección central del cárter del eje para rodear la flecha del eje y los cojinetes.

Backlash The amount of clearance or play between two meshed gears.
Juego La cantidad de holgura o juego entre dos engranajes endentados.

Balance Having equal weight distribution. The term is usually used to describe the weight distribution around the circumference and between the front and back sides of a wheel.
Equilibrio Lo que tiene una distribución igual de peso. El término suele usarse para describir la distribución del peso alrededor de la circunferencia y entre los lados delanteros y traseros de una rueda.

Balance valve A regulating valve that controls a pressure of just the right value to balance other forces acting on the valve.
Válvula niveladora Una válvula de reglaje que controla a la presión del valor correcto para mantener el equilibrio contra las otras fuerzas que afectan a la válvula.

Ball bearing An antifriction bearing consisting of a hardened inner and outer race with hardened steel balls that roll between the two races, and supports the load of the shaft.

Rodamiento de bolas Un cojinete de antifricción que consiste de una pista endurecida interior e exterior y contiene bolas de acero endurecidos que ruedan entre las dos pistas, y sostiene la carga de la flecha.

Ball joint A suspension component that attaches the control arm to the steering knuckle and serves as the lower pivot point for the steering knuckle. The ball joint gets its name from its ball-and-socket design. It allows both up-and-down motion as well as rotation. In a MacPherson strut FWD suspension system, the two lower ball joints are nonload carrying.

Articulación esférica Un componente de la suspensión que une el brazo de mando a la articulación de la dirección y sirve como un punto pivote inferior de la articulación de la dirección. La articulación esférica derive su nombre de su diseño de bola y casquillo. Permite no sólo el movimiento de arriba y abajo sino también el de rotación. En un sistema de suspensión tipo FWD con poste de MacPherson, las articulaciones esféricas inferiores no soportan el peso.

Band A steel band with an inner lining of frictional material. Device used to hold a clutch drum at certain times during transmission operation.

Banda Una banda de acero que tiene un forro interior de una materia de fricción. Un dispositivo que retiene al tambor del embrague en algunos momentos durante la operación de la transmisión.

Bearing The supporting part that reduces friction between a stationary and rotating part or between two moving parts.

Cojinete La parte portadora que reduce la fricción entre una parte fija y una parte giratoria o entre dos partes que muevan.

Bearing cage A spacer that keeps the balls or rollers in a bearing in proper position between the inner and outer races.

Jaula del cojinete Un espaciador que mantiene a las bolas o a los rodillos del cojinete en la posición correcta entre las pistas interiores e exteriores.

Bearing caps In the differential, caps held in place by bolts or nuts which, in turn, hold bearings in place.

Tapones del cojinete En un diferencial, las tapas que se sujeten en su lugar por pernos o tuercas, los cuales en su turno, retienen y posicionan a los cojinetes.

Bearing cone The inner race, rollers, and cage assembly of a tapered roller bearing. Cones and cups must always be replaced in matched sets.

Cono del cojinete La asamblea de la pista interior, los rodillos, y el jaula de un cojinete de rodillos cónico. Se debe siempre reemplazar a ambos partes de un par de conos del cojinete y los anillos exteriores a la vez.

Bearing cup The outer race of a tapered roller bearing or ball bearing.

Anillo exterior La pista exterior de un cojinete cónico de rodillas o de bolas.

Bearing race The surface upon which the rollers or balls of a bearing rotate. The outer race is the same thing as the cup, and the inner race is the one closest to the axle shaft.

Pista del cojinete La superficie sobre la cual rueden los rodillos o las bolas de un cojinete. La pista exterior es lo mismo que un anillo exterior, y la pista interior es la más cercana a la flecha del eje.

Belleville spring A tempered spring steel cone-shaped plate used to aid the mechanical force in a pressure plate assembly.

Resorte de tensión Belleville Un plato de resorte del acero revenido en forma cónica que aumenta a la fuerza mecánica de una asamblea del plato opresor.

Bell housing A housing that fits over the clutch components and connects the engine and the transmission.

Concha del embrague Un cárter que encaja a los componentes del embrague y conecta al motor con la transmisión.

Bolt torque The turning effort required to offset resistance as the bolt is being tightened.

Torsión del perno El esfuerzo de torsión que se requiere para compensar la resistencia del perno mientras que esté siendo apretado.

Brake horsepower (bhp) Power delivered by the engine and available for driving the vehicle; bhp = torque x rpm/5252.

Caballo indicado al freno (bhp) Potencia que provee el motor y que es disponible para el uso del vehículo; bhp = de par motor x rpm/5252.

Bronze An alloy of copper and tin.

Bronce Una aleación de cobre y hojalata.

Burnish To smooth or polish by the use of a sliding tool under pressure.

Bruñir Pulir o suavizar por medio de una herramienta deslizando bajo presión.

Burr A feather edge of metal left on a part being cut with a file or other cutting tool.

Rebaba Una lima espada de metal que permanece en una parte que ha sido cortado con una lima u otro herramienta de cortar.

Bushing A cylindrical lining used as a bearing assembly made of steel, brass, bronze, nylon, or plastic.

Buje Un forro cilíndrico que se usa como una asamblea de cojinete que puede ser hecho del acero, del latón, del bronze, del nylon, o del plástico.

Butt-end locking ring A locking ring whose ends are cut to butt up or contact each other once in place. There is no gap between the ends of a butt-end ring.

Anillo de enclavamiento a tope Un anillo de enclavamiento cuyos extremidades son cortadas para toparse o ajustarse una contra la otra en lugar. No hay holgura entre las extremidades de un anillo de retén a tope.

C-clip A C-shaped clip used to retain the drive axles in some rear axle assemblies.

Grapa de C Una grapa en forma de C que retiene a las flechas motrices en algunas asambleas de ejes traseras.

Cage A spacer used to keep the balls or rollers in proper relation to one another. In a constant-velocity joint, the cage is an open metal framework that surrounds the balls to hold them in position.

Jaula Una espaciador que mantiene una relación correcta entre los rodillos o las bolas. En una junta de velocidad constante, la jaula es un armazón abierto de metal que rodea a las bolas para mantenerlas en posición.

Cardan Universal Joint A nonconstant velocity universal joint consisting of two yokes with their forked ends joined by a cross. The driven yoke changes speed twice in 360 degrees of rotation.

Junta Universal Cardan Una junta universal de velocidad no constante que consiste de dos yugos cuyos extremidades ahorquilladas se unen en cruz. El yugo de arrastre cambia su velocidad dos veces en 360 grados de rotación.

Case-harden To harden the surface of steel. The carburizing method used on low-carbon steel or other alloys to make the case or outer layer of the metal harder than its core.

Cementar Endurecer la superficie del acero. El método de carburación que se emplea en el acero de bajo carbono o en otros aleaciones para que el cárter o capa exterior queda más dura que lo que esta al interior.

Castellate Formed to resemble a castle battlement, as in a castellated nut.
Acanalado De una forma que parece a las almenas de un castillo (véa la palabra en inglés), tal como una tuerca con entallas.

Castellated nut A nut with six raised portions or notches through which a cotter pin can be inserted to secure the nut.
Tuerca con entallas Una tuerca que tiene seis porciones elevadas o muescas por los cuales se puede insertar un pasador de chaveta para retener a la tuerca.

Centrifugal clutch A clutch that uses centrifugal force to apply a higher force against the friction disc as the clutch spins faster.
Embrague centrífugo Un embrague que emplea a la fuerza centrífuga para aplicar una fuerza mayor contra el disco de fricción mientras que el embrague gira más rapidamente.

Centrifugal force The force acting on a rotating body that tends to move it outward and away from the center of rotation. The force increases as rotational speed increases.
Fuerza centrífuga La fuerza que afecta a un cuerpo en rotación moviendolo hacia afuera y alejándolo del centro de rotación. La fuerza aumenta al aumentar la velocidad de rotación.

Chamfer A bevel or taper at the edge of a hole or a gear tooth.
Chaflán Un bisél o cono en el borde de un hoyo o un diente del engranaje.

Chamfer face A beveled surface on a shaft or part that allows for easier assembly. The ends of FWD drive shafts are often chamfered to make installation of the CV joints easier.
Cara achaflanada Una superficie biselada en una flecha o una parte que facilita la asamblea. Los extremos de los árboles de mando de FWD suelen ser achaflandos para facilitar la instalación de las juntas CV.

Chassis The vehicle frame, suspension, and running gear. On FWD cars, it includes the control arms, struts, springs, trailing arms, sway bars, shocks, steering knuckles, and frame. The drive shafts, constant-velocity joints, and transaxle are not part of the chassis or suspension.
Chasis El armazón de un vehículo, la suspensión, y el engranaje de marcha. En los coches de FWD, incluye los brazos de mando, los postes, los resortes (chapas), los brazos traseros, las estabilizadoras, las articulaciones de la dirección y el armazón. Los árboles de mando, las juntas de velocidad constante, y la flecha impulsora no son partes del chasis ni de la suspensión.

Circlip A split steel snapring that fits into a groove to hold various parts in place. Circlips are often used on the ends of FWD drive shafts to retain the constant-velocity joints.
Grapa circular Un seguro partido circular de acero que se coloca en una ranura para posicionar a varias partes. Las grapas circulares se suelen usar en las extremidades de los árboles de mando en FWD para retener las juntas de velocidad constante.

Clearance The space allowed between two parts, such as between a journal and a bearing.
Holgura El espacio permitido entre dos partes, tal como entre un muñon y un cojinete.

Clutch A device for connecting and disconnecting the engine from the transmission or for a similar purpose in other units.
Embrague Un dispositivo para conectar y desconectar el motor de la transmisión o para tal propósito en otros conjuntos.

Clutch packs A series of clutch discs and plates installed alternately in a housing to act as a driving or driven unit.
Conjuntos de embrague Una seria de discos y platos de embrague que se han instalado alternativamente en un cárter para funcionar como una unedad de propulsión o arrastre.

Clutch slippage Engine speed increases but increased torque is not transferred through to the driving wheels because of clutch slippage.
Resbalado del embrague La velocidad del motor aumenta pero la torsión aumentada del motor no se transfere a las ruedas de marcha por el resbalado del embrague.

Coefficient of friction The ratio of the force resisting motion between two surfaces in contact to the force holding the two surfaces in contact.
Coeficiente de la fricción La relación entre la fuerza que resiste al movimiento entre dos superficies que tocan y la fuerza que mantiene en contacto a éstas dos superficies.

Coil preload springs Coil springs are made of tempered steel rods formed into a spiral that resist compression; located in the pressure plate assembly.
Muelles de embrague Los muelles espirales son fabricadas de varillas de acero revenido y resisten la compresión; se ubican en el conjunto del plato opresor.

Coil spring A heavy wire-like steel coil used to support the vehicle weight while allowing for suspension motions. On FWD cars, the front coil springs are mounted around the MacPherson struts. On the rear suspension, they may be mounted to the rear axle, to trailing arms, or around rear struts.
Muelles de embrague Un resorte espiral hecho de acero en forma de alambre grueso que soporte el peso del vehículo mientras que permite a los movimientos de la suspensión. En los coches de FWD, los muelles de embrague delanteros se montan alrededor de los postes Macpherson. En la suspensión trasera, pueden montarse en el eje trasero, en los brasos traseros, o alrededor de los postes traseros.

Compound A mixture of two or more ingredients.
Compuesto Una combinación de dos ingredientes o más.

Concentric Two or more circles having a common center.
Concéntrico Dos círculos o más que comparten un centro común.

Constant-velocity joint A flexible coupling between two shafts that permits each shaft to maintain the same driving or driven speed regardless of operating angle, allowing for a smooth transfer of power. The constant-velocity joint (also called CV joint) consists of an inner and outer housing with balls in between, or a tripod and yoke assembly.
Junta de velocidad constante Un acoplador flexible entre dos flechas que permite que cada flecha mantenga la velocidad de propulsión o arrastre sin importar el ángulo de operación, efectuando una transferencia lisa del poder. La junta de velociadad constante (también llamado junta CV) consiste de un cárter interior e exterior entre los cuales se encuentran bolas, o de un conjunto de trípode y yugo.

Contraction A reduction in mass or dimension; the opposite of expansion.
Contración Una reducción en la masa o en la dimensión; el opuesto de expansión.

Control arm A suspension component that links the vehicle frame to the steering knuckle or axle housing and acts as a hinge to allow up-and-down wheel motions. The front control arms are attached to the frame with bushings and bolts and are connected to the steering knuckles with ball joints. The rear control arms attach to the frame with bushings and bolts and are welded or bolted to the rear axle or wheel hubs.
Brazo de mando Un componente de la suspención que une el armazón del vehículo al articulación de dirección o al cárter del eje y que se porta como una bisagra para permitir a los movimientos verticales de las ruedas. Los brazos de mando delanteros se conectan al armazón por medio de pernos y bujes y se conectan al articulación de dirección por medio de los articulaciones esféricos. Los brazos de mando traseros se conectan al armazón por medio de pernos y bujes y son soldados o empernados al eje trasero o a los cubos de la rueda.

Corrode To eat away gradually as if by gnawing, especially by chemical action.
Corroer Roído poco a poco, primariamente por acción químico.

Corrosion Chemical action, usually by an acid, that eats away (decomposes) a metal.
Corrosión Un acción químico, por lo regular un ácido, que corroe (descompone) un metal.

Cotter pin A type of fastener, made from soft steel in the form of a split pin, that can be inserted in a drilled hole. The split ends are spread to lock the pin in position.
Pasador de chaveta Un tipo de fijación, hecho de acero blando en forma de una chaveta que se puede insertar en un hueco tallado. Las extremidades partidas se despliegen para asegurar la posición de la chaveta.

Counterclockwise rotation Rotating in the opposite direction of the hands on a clock.
Rotación en sentido inverso Girando en el sentido opuesto de las agujas de un reloj.

Coupling A connecting means for transferring movement from one part to another; may be mechanical, hydraulic, or electrical.
Acoplador Un método de conección que transfere el movimiento de una parte a otra; puede ser mecánico, hidráulico, o eléctrico.

Coupling phase Point in torque converter operation where the turbine speed is 90% of impeller speed and there is no longer any torque multiplication.
Fase del acoplador El punto de la operación del convertidor de la torsión en el cual la velocidad de la turbina es el 90% de la velocidad del impulsor y no queda ningún multiplicación de la torsión.

Cover plate A stamped steel cover bolted over the service access to the manual transmission.
Cubrejuntas Un cubierto de acero estampado que se emperna en la apertura de servicio de la transmisión manual.

Deflection Bending or movement away from normal due to loading.
Desviación Curvación o movimiento fuera de lo normal debido a la carga.

Degree A unit of measurement equal to 1/360th of a circle.
Grado Una uneda de medida que iguala al 1/360 parte de un círculo.

Density Compactness; relative mass of matter in a given volume.
Densidad La firmeza; una cantidad relativa de la materia que ocupa a un volumen dado.

Detent A small depression in a shaft, rail, or rod into which a pawl or ball drops when the shaft, rail, or rod is moved. This provides a locking effect.
Detención Un pequeño hueco en una flecha, una barra o una varilla en el cual cae una bola o un linguete al moverse la flecha, la barra o la varilla. Esto provee un efecto de enclavamiento.

Detent mechanism A shifting control designed to hold the manual transmission in the gear range selected.
Aparato de detención Un control de desplazamiento diseñado a sujetar a la transmisión manual en la velocidad selecionada.

Diagnosis A systematic study of a machine or machine parts to determine the cause of improper performance or failure.
Diagnóstico Un estudio sistemático de una máquina o las partes de una máquina con el fín de determinar la causa de una falla o de un operación irregular.

Differential A mechanism between drive axles that permits one wheel to run at a different speed than the other while turning.
Diferencial Un mecanismo entre dos semiejes que permite que una rueda gira a una velocidad distincta que la otra en una curva.

Differential action An operational situation where one driving wheel rotates at a slower speed than the opposite driving wheel.
Acción del diferencial Una situación durante la operación en la cual una rueda propulsora gira con una velocidad más lenta que la rueda propulsora opuesta.

Differential case The metal unit that encases the differential side gears and pinion gears, and to which the ring gear is attached.
Caja de satélites La unidad metálica que encaja a los engranajes planetarios (laterales) y a los satélites del diferencial, y a la cual se conecta la corona.

Differential drive gear A large circular helical gear that is driven by the transaxle pinion gear and shaft and drives the differential assembly.
Corona Un engranaje helicoidal grande circular que es arrastrado por el piñon de la flecha de transmisión y la flecha y propela al conjunto del diferencial.

Differential housing Cast-iron assembly that houses the differential unit and the drive axles. Also called the rear axle housing.
Cárter del diferencial Una asamblea de acero vaciado que encaja a la unedad del diferencial y los semiejes. También se llama el cárter del eje trasero.

Differential pinion gears Small beveled gears located on the differential pinion shaft.
Satélites Engranajes pequeños biselados que se ubican en la flecha del piñon del diferencial.

Differential pinion shaft A short shaft locked to the differential case. This shaft supports the differential pinion gears.
Flecha del piñon del diferencial Una flecha corta clavada en la caja de satélites. Esta flecha sostiene a los satélites.

Differential ring gear A large circular hypoid-type gear enmeshed with the hypoid drive pinion gear.
Corona Un engranaje helicoidal grande circular endentado con el piñon de ataque hipoide.

Differential side gears The gears inside the differential case that are internally splined to the axle shafts, and which are driven by the differential pinion gears.
Planetarios (laterales) Los engranajes adentro de la caja de satélites que son acanalados a los semiejes desde el interior, y que se arrastran por los satélites.

Dipstick A metal rod used to measure the fluid in an engine or transmission.
Varilla de medida Una varilla de metal que se usa para medir el nivel de flúido en un motor o en una transmisión.

Direct drive One turn of the input driving member compared to one complete turn of the driven member, such as when there is direct engagement between the engine and driveshaft where the engine crankshaft and the driveshaft turn at the same rpm.
Mando directo Una vuelta del miembro de ataque o propulsión que se compara a una vuelta completa del miembro de arrastre, tal como cuando hay un enganchamiento directo entre el motor y el árbol de transmisión cuando el cigueñal y el árbol de transmisión giran al mismo rpm.

Disengage When the operator moves the clutch pedal toward the floor to disconnect the driven clutch disc from the driving flywheel and pressure plate assembly.
Desembragar Cuando el operador mueva el pedal de embrague hacia el piso para desconectar el disco de embrague del volante impulsor y del conjunto del plato opresor.

DMM The acronym for a digital multimeter.
DMM La sigla en inglés por un multímeter digital.

Dowel A metal pin attached to one object which, when inserted into a hole in another object, ensures proper alignment.

Espiga Una clavija de metal que se fija a un objeto, que al insertarla en el hoyo de otro objeto, asegura una alineación correcta.

Dowel pin A pin inserted in matching holes in two parts to maintain those parts in fixed relation one to another.

Clavija de espiga Una clavija que se inserte en los hoyos alineados en dos partes para mantener ésos dos partes en una relación fijada el uno al otro.

Downshift To shift a transmission into a lower gear.

Cambio descendente Cambiar la velocidad de una transmision a una velocidad más baja.

Drive line The universal joints, drive shaft, and other parts connecting the transmission with the driving axles.

Flecha motríz Las juntas universales, el árbol de mando, u otras partes que conectan a la transmisión con los ejes impulsores.

Drive line torque Relates to rear-wheel drive line and is the transfer of torque between the transmission and the driving axle assembly.

Potencia de la flecha motríz Se relaciona a la flecha motríz de las ruedas traseras y transfere la potencia de la torsión entre la transmisión y el conjunto del eje trasero.

Driven gear The gear meshed directly with the driving gear to provide torque multiplication, reduction, or a change of direction.

Engranaje de arrastre El engranaje endentado directamente al engranaje de ataque para proporcionar la multiplicación, la reducción, o los cambios de dirección de la potencia.

Drive pinion gear One of the two main driving gears located within the transaxle or rear driving axle housing. Together the two gears multiply engine torque.

Engranaje de piñon de ataque Uno de dos engranajes de ataque principales que se ubican adentro de la flecha de transmisión o en el cárter del eje de propulsión. Los dos engranajes trabajan juntos para multiplicar la potencia.

Drive shaft An assembly of one or two universal joints connected to a shaft or tube; used to transmit power from the transmission to the differential. Also called the propeller shaft.

Árbol de mando Una asamblea de uno o dos uniones universales que se conectan a un árbol o un tubo; se usa para transferir la potencia desde la transmisión al diferencial. También se le refiere como el árbol de propulsión.

Drop forging A piece of steel shaped between dies while hot.

Estampado Un pedazo de acero que se forma entre bloques mientras que esté caliente.

Dry friction The friction between two dry solids.

Fricción seca Fricción entre dos sólidos secos.

DTC The acronym for diagnostic trouble code.

DTC La sigla en inglés por un código diagnóstico de averías.

Dynamic In motion.

Dinámico En movimiento.

Dynamic balance The balance of an object when it is in motion; for example, the dynamic balance of a rotating drive shaft.

Balance dinámico El balance de un objeto mientras que esté en movimiento: por ejemplo el balance dinámico de un árbol de mando giratorio.

Eccentric One circle within another circle wherein both circles do not have the same center or a circle mounted off center. On FWD cars, front-end camber adjustments are accomplished by turning an eccentric cam bolt that mounts the strut to the steering knuckle.

Excéntrico Se dice de dos círculos, el uno dentro del otro, que no comparten el mismo centro o de un círculo ubicado descentrado. En los coches FWD, los ajustes de la inclinación se efectuan por medio de un perno excéntrico que fija el poste sobre el articulación de dirección.

Efficiency The ratio between the power of an effect and the power expended to produce the effect; the ratio between an actual result and the theoretically possible result.

Eficiencia La relación entre la potencia de un efecto y la potencia que se gasta para producir el efecto; la relación entre un resultado actual y el resultado que es una posibilidad teórica.

Elastomer Any rubber-like plastic or synthetic material used to make bellows, bushings, and seals.

Elastómero Cualquiera materia plástic parecida al hule o una materia sintética que se utiliza para fabricar a los fuelles, los bujes y las juntas.

Engage When the vehicle operator moves the clutch pedal up from the floor, this engages the driving flywheel and pressure plate to rotate and drive the driven disc.

Accionar Cuando el operador del vehículo deja subir el pedal del embrague del piso, ésto acciona la volante de ataque y el plato opresor para impulsar al disco de arrastre.

Engagement chatter A shaking, shuddering action that takes place as the driven disc makes contact with the driving members. Chatter is caused by a rapid grip and slip action.

Chasquido de enganchamiento Un movimiento de sacudo o temblor que resulta cuando el disco de ataque viene en contacto con los miembros de propulsión. El chasquido se causa por una acción rápida de agarrar y deslizar.

Engine The source of a power for most vehicles. It converts burned fuel energy into mechanical force.

Motor El orígen de la potencia para la mayoría de los vehículos. Convierte la energía del combustible consumido a la fuerza mecánica.

Engine torque A turning or twisting action developed by the engine, measured in foot-pounds or kilogram meters.

Torsión del motor Una acción de girar o torcer que crea el motor, ésta se mide en libras-pie o kilos-metros.

Extension housing An aluminum or iron casting of various lengths that encloses the transmission output shaft and supporting bearings.

Cubierta de extensión Una pieza moldeada de aluminio o acero que puede ser de varias longitudes que encierre a la flecha de salida de la transmisión y a los cojinetes de soporte.

External gear A gear with teeth across the outside surface.

Engranaje exterior Un engranaje cuyos dientes estan en la superficie exterior.

Externally tabbed clutch plates Clutch plates that are designed with tabs around the outside periphery to fit into grooves in a housing or drum.

Placas de embrague de orejas externas Las placas de embrague que se diseñan de un modo para que las orejas periféricas de la superficie se acomoden en una ranura alrededor de un cárter o un tambor.

Extreme-pressure lubricant A special lubricant for use in hypoid-gear differentials; needed because of the heavy wiping loads imposed on the gear teeth.

Lubricante de presión extrema Un lubricante especial que se usa en las diferenciales de tipo engranaje hipóide; se requiere por la carga de transmisión de materia pesada que se imponen en los dientes del engranaje.

Face The front surface of an object.

Cara La superficie delantera de un objeto.

Final drive ratio The ratio between the drive pinion and ring gear.

Relación del mando final La relación entre el piñon de ataque y la corona.

Fit The contact between two machined surfaces.

Ajuste El contacto entre dos superficies maquinadas.

Fixed-type constant-velocity joint A joint that cannot telescope or plunge to compensate for suspension travel. Fixed joints are always found on the outer ends of the drive shafts of FWD cars. A fixed joint may be of either Rzeppa or tripod type.

Junta tipo fijo de velocidad constante Una junta que no tiene la capacidad de los movimientos telescópicos o repentinos que sirven para compensar en los viajes de suspensión. Las juntas fijas siempre se ubican en las extremidades exteriores de los árboles de mando en los coches de FWD. Una junta tipo fijo puede ser de un tipo Rzeppa o de trípode.

Flange A projecting rim or collar on an object for keeping it in place.

Reborde Una orilla o un collar sobresaliente de un objeto cuyo función es de mantenerlo en lugar.

Flexplate A lightweight flywheel used only on engines equipped with an automatic transmission. The flexplate is equipped with a starter ring gear around its outside diameter and also serves as the attachment point for the torque converter.

Placa articulada Un volante ligera que se usa solamente en los motores que se equipan con una transmisión automática. El diámetro exterior de la placa articulada viene equipado con un anillo de engranajes para arrancar y también sirve como punto de conección del convertidor de la torsión.

Fluid coupling A device in the power train consisting of two rotating members; transmits power from the engine, through a fluid, to the transmission.

Acoplamiento de fluido Un dispositivo en el tren de potencia que consiste de dos miembros rotativos; transmite la potencia del motor, por medio de un fluido, a la transmisión.

Fluid drive A drive in which there is no mechanical connection between the input and output shafts, and power is transmitted by moving oil.

Dirección fluido Una dirección en la cual no hay conecciones mecánicas entre las flechas de entrada o salida, y la potencia se transmite por medio del aceite en movimiento.

Flywheel A heavy metal wheel that is attached to the crankshaft and rotates with it; helps smooth out the power surges from the engine power strokes; also serves as part of the clutch and engine-cranking system.

Volante Una rueda pesada de metal que se fija al cigueñal y gira con ésta; nivela a los sacudos que provienen de la carrera de fuerza del motor; también sirve como parte del embrague y del sistema de arranque.

Flywheel ring gear A gear fitted around the flywheel that is engaged by teeth on the starting-motor drive to crank the engine.

Engranaje anular del volante Un engranaje, colocado alrededor del volante que se acciona por los dientes en el propulsor del motor de arranque y arranca al motor.

Foot-pound (ft.-lb.) A measure of the amount of energy or work required to lift 1 pound a distance of 1 foot.

Pie libra Una medida de la cantidad de energía o fuerza que requiere mover una libra a una distancia de un pie.

Force Any push or pull exerted on an object; measured in pounds and ounces, or in newtons (N) in the metric system.

Fuerza Cualquier acción empujado o jalado que se efectua en un objeto; se mide en pies y onzas, o en newtones (N) en el sistema métrico.

Four-wheel drive On a vehicle, driving axles at both front and rear, so that all four wheels can be driven.

Tracción a cuatro ruedas En un vehículo, se trata de los ejes de dirección fronteras y traseras, para que cada uno de las ruedas puede impulsar.

Frame The main understructure of the vehicle to which everything else is attached. Most FWD cars have only a subframe for the front suspension and drive train. The body serves as the frame for the rear suspension.

Armazón La estructura principal del vehículo al cual todo se conecta. La mayoría de los coches FWD sólo tiene un bastidor auxiliar para la suspensión delantera y el tren de propulsión. El carrocería del coche sirve de chassis par la suspensión trasera.

Free-wheel To turn freely and not transmit power.

Volante libre Da vueltas libremente sin transferir la potencia.

Freewheeling clutch A mechanical device that will engage the driving member to impart motion to a driven member in one direction but not the other. Also known as an "overrunning clutch."

Embrague de volante libre Un dispositivo mecánico que acciona el miembro de tracción y da movimiento al miembro de tracción en una dirección pero no en la otra. También se conoce bajo el nombre de un "embrague de sobremarcha."

Friction The resistance to motion between two bodies in contact with each other.

Fricción La resistencia al movimiento entre dos cuerpos que estan en contacto.

Friction bearing A bearing in which there is sliding contact between the moving surfaces. Sleeve bearings, such as those used in connecting rods, are friction bearings.

Rodamientos de fricción Un cojinete en el cual hay un contacto deslizante entre las superficies en movimiento. Los rodamientos de manguitos, como los que se usan en las bielas, son rodamientos de fricción.

Friction disc In the clutch, a flat disc, faced on both sides with frictional material and splined to the clutch shaft. It is positioned between the clutch pressure plate and the engine flywheel. Also called the clutch disc or driven disc.

Disco de fricción En el embrague, un disco plano al cual se ha cubierto ambos lados con una materia de fricción y que ha sido estriado a la flecha del embrague. Se posiciona entre el plato opresor del embrague y el volante del motor. También se llama el disco del embrague o el disco de arrastre.

Friction facings A hard-molded or woven asbestos or paper material that is riveted or bonded to the clutch driven disc.

Superficie de fricción Un recubrimiento remachado o aglomerado al disco de arrastre del embrague que puede ser hecho del amianto moldeado o tejido o de una materia de papel.

Front pump Pump located at the front of the transmission. It is driven by the engine through two dogs on the torque converter housing. It supplies fluid whenever the engine is running.

Bomba delantera Una bomba ubicado en la parte delantera de la transmisión. Se arrastre por el motor al través de dos álabes en el cárter del convertidor de la torsión. Provee el fluido mientras que funciona el motor.

Front-wheel drive (FWD) The vehicle has all drive train components located at the front.

Tracción de las ruedas delanteras (FWD) El vehículo tiene todos los componentes del tren de propulsión en la parte delantera.

FWD Abbreviation for front-wheel drive.

FWD Abreviación de tracción de las ruedas delanteras.

Gasket A layer of material, usually made of cork, paper, plastic, composition, or metal, or a combination of these, placed between two parts to make a tight seal.

Empaque Una capa de una materia, normalmente hecho del corcho, del papel, del plástico, de la materia compuesta o del metal, o de cualquier combinación de éstos, que se coloca entre dos partes para formar un sello impermeable.

Gasket cement A liquid adhesive material or sealer used to install gaskets.

Mastique para empaques Una substancia líquida adhesiva, o una substancia impermeable, que se usa para instalar a los empaques.

Gear A wheel with external or internal teeth that serves to transmit or change motion.

Engranaje Una rueda que tiene dientes interiores o exteriores que sirve para transferir o cambiar el movimiento.

Gear lubricant A type of grease or oil blended especially to lubricate gears.

Lubricante para engranaje Un tipo de grasa o aceite que ha sido mezclado específicamente para la lubricación de los engranajes.

Gear ratio The number of revolutions of a driving gear required to turn a driven gear through one complete revolution. For a pair of gears, the ratio is found by dividing the number of teeth on the driven gear by the number of teeth on the driving gear.

Relación de los engranajes El número de las revoluciones requeridas del engranaje de propulsión para dar una vuelta completa al engranaje arrastrado. En una pareja de engranajes, la relación se calcula al dividir el número de los dientes en el engranaje de arrastre por el númber de los dientes en el engranaje de propulsión.

Gear reduction When a small gear drives a large gear, there is an output speed reduction and a torque increase that result in a gear reduction.

Velocidad descendente Cuando un engranaje pequeño impulsa a un engranaje grande, hay una reducción en la velocidad de salida y un incremento en la torsión que resultan en una cambio descendente de los velocidades.

Gearshift A linkage-type mechanism by which the gears in an automobile transmission are engaged and disengaged.

Varillaje de cambios Un mecanismo tipo eslabón que acciona y desembraga a los engranajes de la transmisión.

Gear whine A high-pitched sound developed by some types of meshing gears.

Ruido del engranaje Un sonido agudo que proviene de algunos tipos de engranajes endentados.

Governor pressure The transmission's hydraulic pressure that is directly related to output shaft speed. It is used to control shift points.

Regulador de presión La presión hidráulica de una transmisión se relaciona directamente a la velocidad de la flecha de salida. Se usa para controllar los puntos de cambios de velocidad.

Governor valve A device used to sense vehicle speed. The governor valve is attached to the output shaft.

Válvula reguladora Un dispositivo que se usa para determinar la velocidad de un vehículo. La válvula reguladora se monta en la flecha de salida.

Graphite Very fine carbon dust with a slippery texture used as a lubricant.

Grafito Un polvo de carbón muy fino con una calidad grasosa que se usa para lubricar.

Heat treatment Heating, followed by fast cooling, to harden metal.

Tratamiento térmico Calentamiento, seguido por un enfriamiento rápido, para endurecer a un metal.

Horsepower A measure of mechanical power, or the rate at which work is done. One horsepower equals 33,000 ft.-lbs. (foot-pounds) of work per minute. It is the power necessary to raise 33,000 pounds a distance of 1 foot in 1 minute.

Caballo de fuerza Una medida de fuerza mecánica, o el régimen en el cual se efectua el trabajo. Un caballo de fuerza iguala a 33,000 lb.p. (libras pie) de trabajo por minuto. Es la fuerza requerida para transportar a 33,000 libras una distancia de 1 pie en 1 minuto.

Hub The center part of a wheel, to which the wheel is attached.

Cubo La parte central de una rueda, a la cual se monta la rueda.

Hydraulic pressure Pressure exerted through the medium of a liquid.

Presión hidráulica La presión esforzada por medio de un líquido.

ID Inside diameter.

DI Diámetro interior.

Idle Engine speed when the accelerator pedal is fully released and there is no load on the engine.

Marcha lenta La velocidad del motor cuando el pedal accelerador esta completamente desembragada y no hay carga en el motor.

Impeller The pump or driving member in a torque converter.

Impulsor La bomba o el miembro impulsor en un convertidor de torsión.

Increments Series of regular additions from small to large.

Incrementos Una serie de incrementos regulares que va de pequeño a grande.

Index To orient two parts by marking them. During reassembly the parts are arranged so the index marks are next to each other. Used to preserve the orientation between balanced parts.

Índice Orientar a dos partes marcándolas. Al montarlas, las partes se colocan para que las marcas de índice estén alinieadas. Se usan los índices para preservar la orientación de las partes balanceadas.

Input shaft The shaft carrying the driving gear by which the power is applied, as to the transmission.

Flecha de entrada La flecha que porta el engranaje propulsor por el cual se aplica la potencia, como a la transmisión.

Inspection cover A removable cover that permits entrance for inspection and service work.

Cubierta de inspección Una cubierta desmontable que permite a la entrada para inspeccionar y mantenimiento.

Integral Built into, as part of the whole.

Integral Incorporado, una parte de la totalidad.

Internal gear A gear with teeth pointing inward, toward the hollow center of the gear.

Engranaje internal Un engranaje cuyos dientes apuntan hacia el interior, al hueco central del engranaje.

Jam nut A second nut tightened against a primary nut to prevent it from working loose. Used on inner and outer tie-rod adjustment nuts and on many pinion-bearing adjustment nuts.

Contra tuerca Una tuerca secundaria que se aprieta contra una tuerca primaria para prevenir que ésta se afloja. Se emplean en las tuercas de ajustes interiores e exteriores para las barras de acoplamiento y también en muchas de las tuercas de ajuste de portapiñones.

Journal A bearing with a hole in it for a shaft.

Manga de flecha Un cojinete que tiene un hoyo para una flecha.

Key A small block inserted between the shaft and hub to prevent circumferential movement.

Chaveta Un tope pequeño que se meta entre la flecha y el cubo para prevenir un movimiento circunferencial.

Keyway A groove or slot cut to permit the insertion of a key.

Ranura de chaveta Un corte de ranura o mortaja que permite insertar una chaveta.

Knock A heavy metallic sound usually caused by a loose or worn bearing.

Golpe Un sonido metálico fuerte que suele ser causado por un cojinete suelto o gastado.

Knurl To indent or roughen a finished surface.

Moletear Indentar or desbastar a una superficie acabada.

LED The common acronym for a light-emitting diode.

LED La sigla común en inglés por un diódo emisor de luz.

Linkage Any series of rods, yokes, levers, and so on used to transmit motion from one unit to another.

Biela Cualquiera serie de barras, yugos, palancas, y todo lo demás, que se usa para transferir los movimientos de una unedad a otra.

Locknut A second nut turned down on a holding nut to prevent loosening.

Contra tuerca Una tuerca segundaria apretada contra una tuerca de sostén para prevenir que ésta se afloja.

Lock pin Used in some ball sockets (inner tie-rod end) to keep the connecting nuts from working loose. Also used on some lower ball joints to hold the tapered stud in the steering knuckle.

Clavija de cerrojo Se usan en algunas rótulas (las extremidades interiores de la barra de acoplamiento) para prevenir que se aflojan las tuercas de conexión. También se emplean en algunas juntas esféricas inferiores para retener al perno cónico en la articulación de dirección.

Locking ring A type of sealing ring that has ends that meet or lock together during installation. There is no gap between the ends when the ring is installed.

Anillo de enclavamiento Un tipo de anillo obturador que tiene las extremidades que se tocan o se enclavan durante la instalación. No hay holgura entre las extremidades cuando se ha instalado el anillo.

Lockplates Metal tabs bent around nuts or bolt heads.

Placa de cerrojo Chavetas de metal que se doblan alrededor de las tuercas o las cabezas de los pernos.

Lockwasher A type of washer which, when placed under the head of a bolt or nut, prevents the bolt or nut from working loose.

Arandela de freno Un tipo de arandela que, al colocarse bajo la cabeza de un perno, previene que el perno o la tuerca se aflojan.

Low speed The gearing that produces the highest torque and lowest speed of the wheels.

Velocidad baja La velocidad que produce la torsión más alta y la velocidad más baja a las ruedas.

Lubricant Any material, usually a petroleum product such as grease or oil, that is placed between two moving parts to reduce friction.

Lubricante Cualquier substancia, normalmente un producto de petróleo como la grasa o el aciete, que se coloca entre dos partes en movimiento para reducir la fricción.

Mainline pressure The hydraulic pressure that operates apply devices and is the source of all other pressures in an automatic transmission. It is developed by pump pressure and regulated by the pressure regulator.

Línea de presión La presión hidráulica que opera a los dispositivos de applicación y es el orígen de todas las presiones en la transmisión automática. Proviene de la bomba de presión y es regulada por el regulador de presión.

Main oil pressure regulator valve Regulates the line pressure in a transmission.

Válvula reguladora de la linea de presión Regula la presión en la linea de una transmisión.

Manual control valve A valve used to manually select the operating mode of the transmission. It is moved by the gearshift linkage.

Válvula de control manual Una válvula que se usa para escojer a una velocidad de la transmisión por mano. Se mueva por la biela de velocidades.

Meshing The mating or engaging of the teeth of two gears.

Engrane Embragar o endentar a los dientes de dos engranajes.

Micrometer A precision measuring device used to measure small bores, diameters, and thicknesses. Also called a mike.

Micrómetro Un dispositivo de medida precisa que se emplea a medir a los taladros pequeños y a los espesores. También se llama un mike (mayk).

MIL The malfunction indicator lamp for a computer control system. Prior to J1930, the MIL was commonly called a Check Engine or Service Engine Soon lamp.

MIL La lámpara de indicación de averías para un sistema de control de computadora. Antes del J1930, la MIL se llamaba la lámpara de Revise Motor o Servicia el Motor Pronto.

Misalignment When bearings are not on the same centerline.

Desalineamineto Cuando los cojinetes no comparten la misma linea central.

Modulator A vacuum diaphragm device connected to a source of engine vacuum. It provides an engine load signal to the transmission.

Modulador Un dispositivo de diafragma de vacío que se conecta a un orígen de vacío en el motor. Provee un señal de carga del motor a la transmisión.

Mounts Made of rubber to insulate vibrations and noise while they support a power train part, such as engine or transmission mounts.

Monturas Hecho de hule para insular a las vibraciones y a los ruidos mientras que sujetan una parte del tren de propulsión, tal como las monturas del motor o las monturas de la transmisión.

Multiple disc A clutch with a number of driving and driven discs as compared to a single plate clutch.

Discos múltiples Un embrague que tiene varios discos de propulsión o de arraste al contraste con un embrague de un sólo plato.

Needle bearing An antifriction bearing using a great number of long, small-diameter rollers. Also known as a quill bearing.

Rodamiento de agujas Un rodamiento (cojinete) antifricativo que emplea un gran cantidad de rodillos largos y de diámetro muy pequeños.

Neoprene A synthetic rubber that is not affected by the various chemicals that are harmful to natural rubber.

Neoprene Un hule sintético que no se afecta por los varios productos químicos que pueden dañar al hule natural.

Neutral In a transmission, the setting in which all gears are disengaged and the output shaft is disconnected from the drive wheels.

Neutral En una transmisión, la velocidad en la cual todos los engranajes estan desembragados y el árbol de salida esta desconectada de las ruedas de propulsión.

Neutral-start switch A switch wired into the ignition switch to prevent engine cranking unless the transmission shift lever is in neutral or the clutch pedal is depressed.

Interruptor de arranque en neutral Un interruptor eléctrico instalado en el interruptor de encendido que previene el arranque del motor al menos de que la palanca de cambio de velocidad esté en una posición neutral o que se pisa en el embrague.

Newton-meter (Nm) Metric measurement of torque or twisting force.

Metro newton (Nm) Una medida métrica de la fuerza de torsión.

Nominal shim A shim with a designated thickness.
Laminilla fina Una cuña de un espesor especificado.

Nonhardening A gasket sealer that never hardens.
Sinfragua Un cemento de empaque que no endurece.

Nut A removable fastener used with a bolt to lock pieces together; made by threading a hole through the center of a piece of metal that has been shaped to a standard size.
Tuerca Un retén removable que se usa con un perno o tuerca para unir a dos piezas; se fabrica al filetear un hoyo taladrado en un pedazo de metal que se ha formado a un tamaño especificado.

OD Outside diameter.
DE Diámetro exterior.

Oil seal A seal placed around a rotating shaft or other moving part to prevent leakage of oil.
Empaque de aciete Un empaque que se coloca alrededor de una flecha giratoria para prevenir el goteo de aceite.

One-way clutch See Sprag clutch.
Embrague de una via Vea Sprag clutch.

O-ring A type of sealing ring, usually made of rubber or a rubber-like material. In use, the O-ring is compressed into a groove to provide the sealing action.
Anillo en O Un tipo de sello anular, suele ser hecho de hule o de una materia parecida al hule. Al usarse, el anillo en O se comprime en una ranura para proveer un sello.

Oscillate To swing back and forth like a pendulum.
Oscilar Moverse alternativamente en dos sentidos contrarios como un péndulo.

Outer bearing race The outer part of a bearing assembly on which the balls or rollers rotate.
Pista exterior de un cojinete La parte exterior de una asamblea de cojinetes en la cual ruedan las bolas o los rodillos.

Out-of-round Wear of a round hole or shaft which, when viewed from an end, will appear egg-shaped.
Defecto de circularidad Desgaste de un taladro o de una flecha circular, que al verse de una extremidad, tendrá una forma asimétrica, como la de un huevo.

Output shaft The shaft or gear that delivers the power from a device, such as a transmission.
Flecha de salida La flecha o la velocidad que transmite la potencia de un dispositivo, tal como una transmisión.

Overall ratio The product of the transmission gear ratio multiplied by the final drive or rear axle ratio.
Relación global El producto de multiplicar la relación de los engranajes de la transmisión por la relación del impulso final o por la relación del eje trasero.

Overdrive Any arrangement of gearing that produces more revolutions of the driven shaft than of the driving shaft.
Sobremultiplicación Un arreglo de los engranajes que produce más revoluciones de la flecha de arrastre que los de la flecha de propulsión.

Overdrive ratio Identified by the decimal point indicating less than one driving input revolution compared to one output revolution of a shaft.
Relación del sobremultiplicación Se identifica por el punto decimal que indica menos de una revolución del motor comparado a una revolución de una flecha de salida.

Overrun coupling A freewheeling device to permit rotation in one direction but not in the other.
Acoplamiento de sobremarcha Un dispositivo de marcha de rueda libre que permite las giraciones en una dirección, pero no en la otra dirección.

Overrunning clutch A device consisting of a shaft or housing linked together by rollers or sprags operating between movable and fixed races. As the shaft rotates, the rollers or sprags jam between the movable and fixed races. This jamming action locks together the shaft and housing. If the fixed race should be driven at a speed greater than the movable race, the rollers or sprags will disconnect the shaft.
Embrague de sobremarcha Un dispositivo que consiste de una flecha o un cárter eslabonados por medio de rodillos o palancas de detención que operan entre pistas fijas y movibles. Al girar la flecha, los rodillos o palancas de detención se aprietan entre las pistas fijas y movibles. Este acción de apretarse enclava el cárter con la flecha. Si la pista fija se arrastra en una velocidad más alta que la pista movible, los rodillos o palancas de detención desconectarán a la flecha.

Oxidation Burning or combustion; the combining of a material with oxygen. Rusting is slow oxidation, and combustion is rapid oxidation.
Oxidación Quemando o la combustión; la combinación de una materia con el oxígeno. El orín es una oxidación lenta, la combustión es la oxidación rápida.

Pascal's Law The law of fluid motion.
Ley de pascal La ley del movimiento del fluido.

Parallel The quality of two items being the same distance from each other at all points; usually applied to lines and, in automotive work, to machined surfaces.
Paralelo La calidad de dos artículos que mantienen la misma distancia el uno del otro en cada punto; suele aplicarse a las líneas y, en el trabajo automotívo, a las superficies acabadas a máquina.

Pawl A lever that pivots on a shaft. When lifted, it swings freely and when lowered, it locates in a detent or notch to hold a mechanism stationary.
Trinquete Una palanca que gira en una flecha. Levantado, mueve sín restricción, bajado, se coloca en una endentación o una muesca para mantener sín movimiento a un mecanismo.

PCM The powertrain control module of a computer control system. Prior to J1930, the PCM was commonly called a ECA, ECM, or one of many acronyms used by the various manufacturers.
PCM El módulo de control del sistema de transmisión de fuerza de un sistema de control de una computadora. Antes d3l J1930, el PCM se llamaba un ECA, un ECM, o una de varias siglas usadas por los varios fabricantes.

Peen To stretch or clinch over by pounding with the rounded end of a hammer.
Martillazo Estirar o remachar con la extremidad redondeado de un martillo de bola.

Pitch The number of threads per inch on any threaded part.
Paso El número de filetes por pulgada de cualquier parte fileteada.

Pivot A pin or shaft upon which another part rests or turns.
Pivote Una chaveta o una flecha que soporta a otra parte o sirve como un punto para girar.

Planetary gear set A system of gearing that is modeled after the solar system. A pinion is surrounded by an internal ring gear and planet gears are in mesh between the ring gear and pinion around which all revolve.
Conjunto de engranajes planetarios Un sistema de engranaje cuyo patrón es el sistema solar. Un engranaje propulsor (la corona interior) rodea al piñon de ataque y los engranajes satélites y planetas se endentan entre la corona y el piñon alrededor del cual todo gira.

Planet carrier The carrier or bracket in a planetary gear system that contains the shafts upon which the pinions or planet gears turn.
Perno de arrastre planetario El soporte o la abrazadera que contiene las flechas en las cuales giran los engranajes planetarios o los piñones.

Planet gears The gears in a planetary gear set that connect the sun gear to the ring gear.

Engranages planetarios Los engranajes en un conjunto de engranajes planetario que connectan al engranaje propulsor interior (el engranaje sol) con la corona.

Planet pinions In a planetary gear system, the gears that mesh with, and revolve about, the sun gear; they also mesh with the ring gear.

Piñones planetarios En un sistema de engranajes planetarios, los engranajes que se endentan con, y giran alrededor, el engranaje propulsor (sol); también se endentan con la corona.

Porosity A statement of how porous or permeable to liquids a material is.

Porosidad Una expresión de lo poroso o permeable a los líquidos es una materia.

Power train The mechanisms that carry the power from the engine crankshaft to the drive wheels; these include the clutch, transmission, drive line, differential, and axles.

Tren impulsor Los mecanismos que transferen la potencia desde el cigueñal del motor a las ruedas de propulsión; éstos incluyen el embrague, la transmisión, la flecha motríz, el diferencial y los semiejes.

Preload A load applied to a part during assembly so as to maintain critical tolerances when the operating load is applied later.

Carga previa Una carga aplicada a una parte durante la asamblea para asegurar sus tolerancias críticas antes de que se le aplica la carga de la operación.

Press-fit Forcing a part into an opening that is slightly smaller than the part itself to make a solid fit.

Ajustamiento a presión Forzar a una parte en una apertura que es de un tamaño más pequeño de la parte para asegurar un ajustamiento sólido.

Pressure Force per unit area, or force divided by area. Usually measured in pounds per square inch (psi) or in kilopascals (kPa) in the metric system.

Presión La fuerza por unidad de una area, o la fuerza divida por la area. Suele medirse en libras por pulgada cuadrada (lb/pulg2) o en kilopascales (kPa) en el sistema métrico.

Pressure plate That part of the clutch that exerts force against the friction disc; it is mounted on and rotates with the flywheel.

Plato opresor Una parte del embraque que aplica la fuerza en el disco de fricción; se monta sobre el volante, y gira con éste.

Propeler shaft See Driveshaft.

Flecha de Propulsion Vea Flecha motríz.

Prussian blue A blue pigment; in solution, useful in determining the area of contact between two surfaces.

Azul de Prusia Un pigmento azul; en forma líquida, ayuda en determinar la area de contacto entre dos superficies.

PSI Abbreviation for pounds per square inch, a measurement of pressure.

Lb/pulg2 Una abreviación de libras por pulgada cuadrada, una medida de la presión.

Pulsation To move or beat with rhythmic impulses.

Pulsación Moverse o batir con impulsos rítmicos.

Race A channel in the inner or outer ring of an antifriction bearing in which the balls or rollers roll.

Pista Un canal en el anillo interior o exterior de un cojinete antifricción en el cual ruedan las bolas o los rodillos.

Radial The direction moving straight out from the center of a circle. Perpendicular to the shaft or bearing bore.

Radial La dirección al moverse directamente del centro de un círculo. Perpendicular a la flecha o al taladro del cojinete.

Radial clearance Clearance within the bearing and between balls and races perpendicular to the shaft. Also called radial displacement.

Holgura radial La holgura en un cojinete entre las bolas y las pistas que son perpendiculares a la flecha. También se llama un desplazamiento radial.

Radial load A force perpendicular to the axis of rotation.

Carga radial Una fuerza perpendicular al centro de rotación.

Ratio The relation or proportion that one number bears to another.

Relación La correlación o proporción de un número con respeto a otro.

Rear-wheel drive A term associated with a vehicle where the engine is mounted at the front and the driving axle and driving wheels are at the rear of the vehicle.

Tracción trasera Un término que se asocia con un vehículo en el cual el motor se ubica en la parte delantera y el eje propulsor y las ruedas propulsores se encuentran en la parte trasera del vehículo.

Relief valve A valve used to protect against excessive pressure in the case of a malfunctioning pressure regulator.

Válvula de seguridad Una válvula que se usa para guardar contra una presión excesiva en caso de que malfulciona el regulador de presión.

Retaining ring A removable fastener used as a shoulder to retain and position a round bearing in a hole.

Anillo de retén Un seguro removible que sirve de collarín para sujetar y posicionar a un cojinete en un agujero.

RFI Radio frequency interference. This acronym is used to describe a type of electrical interference that may affect voltage signals in a computerized system.

RFI Interferencia de frecuencias de radio. Esta sigla se usa en describir un tipo de interferencia eléctrica que puede afectar los señales de voltaje en un sistema computerizado.

Rivet A headed pin used for uniting two or more pieces by passing the shank through a hole in each piece and securing it by forming a head on the opposite end.

Remache Una clavija con cabeza que sirve para unir a dos piezas o más al pasar el vástago por un hoyo en cada pieza y asegurarlo por formar una cabeza en el extremo opuesto.

Roller bearing An inner and outer race upon which hardened steel rollers operate.

Cojinete de rodillos Una pista interior y exterior en la cual operan los rodillos hecho de acero endurecido.

Rollers Round steel bearings that can be used as the locking element in an overrunning clutch or as the rolling element in an antifriction bearing.

Rodillos Articulaciones redondos de acero que pueden servir como un elemento de enclavamiento en un embrague de sobremarcha o como el elemento que rueda en un cojinete antifricción.

Rotary flow A fluid force generated in the torque converter that is related to vortex flow. The vortex flow leaving the impeller is not only flowing out of the impeller at high speed but is also rotating faster than the turbine. The rotating fluid striking the slower turning turbine exerts a force against the turbine that is defined as rotary flow.

Flujo rotativo Una fuerza fluida producida en el convertidor de torsión que se relaciona al flujo torbellino. El flujo torbellino saliendo del rotor no sólo viaja en una alta velocidad sino también gira más rápidamente que el turbino. El fluido rotativo chocando contra el turbino que gira más lentamente, impone una fuerza contra el turbino que se define como flujo rotativo.

RPM Abbreviation for revolutions per minute, a measure of rotational speed.

RPM Abreviación de revoluciones por minuto, una medida de la velocidad rotativa.

RWD Abbreviation for rear-wheel drive.
RWD Abreviación de tracción trasera.

SAE Society of Automotive Engineers.
SAE La Sociedad de Ingenieros Automotrices.

Seal A material shaped around a shaft, used to close off the operating compartment of the shaft, preventing oil leakage.
Sello Una materia, formado alrededor de una flecha, que sella el compartimiento operativo de la flecha, previniendo el goteo de aceite.

Sealer A thick, tacky compound, usually spread with a brush, that may be used as a gasket or sealant to seal small openings or surface irregularities.
Sellador Un compuesto pegajoso y espeso, comúnmente aplicado con una brocha, que puede usarse como un empaque o un obturador para sellar a las aperturas pequeñas o a las irregularidades de la superficie.

Seat A surface, usually machined, upon which another part rests or seats; for example, the surface upon which a valve face rests.
Asiento Una superficie, comúnmente maquinada, sobre la cual yace o se asienta otra parte; por ejemplo, la superficie sobre la cual yace la cara de la válvula.

Servo A device that converts hydraulic pressure into mechanical movement, often multiplying it. Used to apply the bands of a transmission.
Servo Un dispositivo que convierte la presión hidráulica al movimiento mecánico, frequentemente multiplicándola. Se usa en la aplicación de las bandas de una transmisión.

Shift lever The lever used to change gears in a transmission. Also the lever on the starting motor that moves the drive pinion into or out of mesh with the flywheel teeth.
Palanca del cambiador La palanca que sirve para cambiar a las velocidades de una transmisión. También es la palanca del motor de arranque que mueva al piñon de ataque para engranarse o desegranarse con los dientes del volante.

Shift valve A valve that controls the shifting of the gears in an automatic transmission.
Válvula de cambios Una válvula que controla a los cambios de las velocidades en una transmisión automática.

Shim Thin sheets used as spacers between two parts, such as the two halves of a journal bearing.
Laminilla de relleno Hojas delgadas que sirven de espaciadores entre dos partes, tal como las dos partes de un muñon.

Shim stock Sheets of metal of accurately known thickness that can be cut into strips and used to measure or correct clearances.
Materia de laminillas Las hojas de metal cuyo espesor se conoce precisamente que pueden cortarse en tiras y usarse para medir o correjir a las holguras.

Side clearance The clearance between the sides of moving parts when the sides do not serve as load-carrying surfaces.
Holgura lateral La holgura entre los lados de las partes en movimiento mientras que los lados no funcionan como las superficies de carga.

Sliding-fit Where sufficient clearance has been allowed between the shaft and journal to allow free running without overheating.
Ajuste corredera Donde se ha dejado una holgura suficiente entre la flecha y el muñon para permitir una marcha libre sin sobrecalentamiento.

Snapring Split spring-type ring located in an internal or external groove to retain a part.
Anillo de seguridad Un anillo partido tipo resorte que se coloca en una muesca interior o exterior para retener a una parte.

Spindle The shaft on which the wheels and wheel bearings mount.
Husillo La flecha en la cual se montan las ruedas y el conjunto del cojinete de las ruedas.

Spline Slot or groove cut in a shaft or bore; a splined shaft onto which a hub, wheel, gear, and so on, with matching splines in its bore is assembled so that the two must turn together.
Acanaladura (espárrago) Una muesca o ranura cortada en una flecha o en un taladro; una flecha acanalada en la cual se asamblea un cubo, una rueda, un engranaje, y todo lo demás que tiene un acanaladura pareja en el taladro de manera de que las dos deben girar juntos.

Split lip seal Typically, a rope seal sometimes used to denote any two-part oil seal.
Sello hendido Típicamente, un sello de cuerda que se usa a veces para demarcar cualquier sello de aceite de dos partes

Split pin A round split spring steel tubular pin used for locking purposes; for example, locking a gear to a shaft.
Chaveta hendida Una chaveta partida redonda y tubular hecho de acero para resorte que sirve para el enclavamiento; por ejemplo, para enclavar un engranaje a una flecha.

Spool valve A cylindrically shaped valve with two or more valleys between the lands. Spool valves are used to direct fluid flow.
Válvula de carrete Una válvula de forma cilíndrica que tiene dos acanaladuras de cañon o más entre las partes planas. Las válvulas de carrete sirven para dirigir el flujo del fluido.

Sprag clutch A member of the overrunning clutch family using a sprag to jam between the inner and outer races used for holding or driving action.
Embrague de puntal Un miembro de la familia de embragues de sobremarcha que usa a una palanca de detención trabada entre las pistas interiores e exteriores para realizar una acción de asir o marchar.

Spring A device that changes shape when it is stretched or compressed, but returns to its original shape when the force is removed; the component of the automotive suspension system that absorbs road shocks by flexing and twisting.
Resorte Un dispositivo que cambia de forma al ser estirado o comprimido, pero que recupera su forma original al levantarse la fuerza; es un componente del sistema de suspensión automotívo que absorba los choques del camino al doblarse y torcerse.

Spring retainer A steel plate designed to hold a coil or several coil springs in place.
Retén de resorte Una chapa de acero diseñado a sostener en su posición a un resorte helicoidal o más.

Squeak A high-pitched noise of short duration.
Chillido Un ruido agudo de poca duración.

Squeal A continuous high-pitched noise.
Alarido Un ruido agudo continuo.

Stall A condition where the engine is operating and the transmission is in gear, but the drive wheels are not turning because the turbine of the torque converter is not moving.
Paro Una condición en la cual opera el motor y la transmisión esta embragada pero las ruedas de impulso no giran porque no mueva el turbino del convertidor de la torsión.

Stall test A test of the one-way clutch in a torque converter.
Prueba de paro Una prueba del embrague de una vía en un convertidor de la torsión.

Stress The force to which a material, mechanism, or component is subjected.
Esfuerzo La fuerza a la cual se somete a una materia, un mecanísmo o un componente.

Sun gear The central gear in a planetary gear system around which the rest of the gears rotate. The innermost gear of the planetary gear set.
Engranaje principal (sol) El engranaje central en un sistema de engranajes planetarios alrededor del cual giran los otros engranajes. El engranaje más interno del conjunto de los engranajes planetarios.

Tap To cut threads in a hole with a tapered, fluted, threaded tool.
Roscar con macho Cortar las roscas en un agujero con una herramienta cónica, acanalada y fileteada.

Temper To change the physical characteristics of a metal by applying heat.
Templar Cambiar las características físicas de un metal mediante una aplicación del calor.

Tension Effort that elongates or "stretches" a material.
Tensión Un esfuerzo que alarga o "estira" a una materia.

Thickness gauge Strips of metal made to an exact thickness, used to measure clearances between parts.
Calibre de espesores Las tiras del metal que se han fabricado a un espesor exacto, sirven para medir las holguras entre las partes.

Thrust bearing A bearing designed to resist or contain side or end motion as well as reduce friction.
Cojinete de empuje Un cojinete diseñado a detener o reprimir a los movimientos laterales o de las extremidades y también reducir la fricción.

Thrust load A load that pushes or reacts through the bearing in a direction parallel to the shaft.
Carga de empuje Una carga que empuja o reacciona por el cojinete en una dirección paralelo a la flecha.

Thrust washer A washer designed to take up end thrust and prevent excessive endplay.
Arandela de empuje Una arandela diseñada para rellenar a la holgura de la extremidad y prevenir demasiado juego en la extremidad.

Tolerance A permissible variation between the two extremes of a specification or dimension.
Tolerancia Una variación permisible entre dos extremos de una especificación o de un dimensión.

Torque A twisting motion, usually measured in ft.-lb. (Nm).
Torsión Un movimiento giratorio, suele medirse en pies/libra (Nm).

Torque converter A turbine device utilizing a rotary pump, one or more reactors (stators), and a driven circular turbine or vane, whereby power is transmitted from a driving to a driven member by hydraulic action. It provides varying drive ratios; with a speed reduction, it increases torque.
Convertidor de la torsión Un dispositivo de turbino que utilisa a una bomba rotativa, a un reactor o más, y un molinete o turbino circular impulsado, por cual se transmite la energía de un miembro de impulso a otro arrastrado mediante la acción hidráulica. Provee varias relaciones de impulso; al descender la velocidad, aumenta la torsión.

Torque curve A line plotted on a chart to illustrate the torque personality of an engine. When the engine operates on its torque curve, it is producing the most torque for the quantity of fuel being burned.
Curva de la torsión Una linea delineada en una carta para ilustrar las características de la torsión del motor. Al operar un motor en su curva de la torsión, produce la torsión óptima para la cantidad del combustible que se consuma.

Torque multiplication The result of meshing a small driving gear and a large driven gear to reduce speed and increase output torque.
Multiplicación de la torsión El resultado de engranar a un engranaje pequeño de ataque con un engranaje más grande arrastrado para reducir la velocidad y incrementar la torsión de salida.

Torque steer An action felt in the steering wheel as the result of increased torque.
Dirección la torsión Una acción que se nota en el volante de dirección como resultado de un aumento de la torsión.

Traction The gripping action between the tire tread and the road's surface.
Tracción La acción de agarrar entre la cara de la rueda y la superficie del camino.

Transaxle Type of construction in which the transmission and differential are combined in one unit.
Flecha de transmisión Un tipo de construcción en el cual la transmisión y el diferencial se combinan en una unidad.

Transaxle assembly A compact housing most often used in front-wheel drive vehicles that houses the manual transmission, final drive gears, and differential assembly.
Asamblea de la flecha de transmisión Un cárter compacto que se usa normalmente en los vehículos de tracción delantera que contiene la transmisión manual, los engranajes de propulsión, y la asamblea del diferencial.

Transfer case An auxiliary transmission mounted behind the main transmission. Used to divide engine power and transfer it to both front and rear differentials, either full-time or part-time.
Cárter de la transferencia Una transmisión auxiliar montada detrás de la transmisión principal. Sirve para dividir la potencia del motor y transferirla a ambos diferenciales delanteras y traseras todo el tiempo o la mitad del tiempo.

Transmission The device in the power train that provides different gear ratios between the engine and drive wheels as well as reverse.
Transmisión El dispositivo en el trén de potencia que provee las relaciones diferentes de engranaje entre el motor y las ruedas de impulso y también la marcha de reversa.

Transverse Power train layout in a front-wheel drive automobile extending from side to side.
Transversal Una esquema del tren de potencia en un automóvil de tracción delantera que se extiende de un lado a otro.

U-joint A four-point cross connected to two U-shaped yokes that serves as a flexible coupling between shafts.
Junta de U Una cruceta de cuatro puntos que se conecta a dos yugos en forma de U que sirven de acoplamientos flexibles entre las flechas.

Universal joint A mechanical device that transmits rotary motion from one shaft to another shaft at varying angles.
Junta Universal Un dispositivo mecánico que transmite el movimiento giratorio desde una flecha a otra flecha en varios ángulos.

Upshift To shift a transmission into a higher gear.
Cambio ascendente Cambiar a la velocidad de una transmisión a una más alta.

Valve body Main hydraulic control assembly of a transmission containing the components necessary to control the distribution of pressurized transmission fluid throughout the transmission.
Cuerpo de la válvula Asamblea principal del control hidráulico de una transmisión que contiene los componentes necesarios para controlar a la distribución del fluido de la transmisión bajo presión por toda la transmisión.

Vehicle identification number (VIN) The number assigned to each vehicle by its manufacturer, primarily for registration and identification purposes.
Número de identificacíon del vehículo El número asignado a cada vehículo por su fabricante, primariamente con el propósito de la registración y la identificación.

Vibration A quivering, trembling motion felt in the vehicle at different speed ranges.

Vibración Un movimiento de estremecer o temblar que se siente en el vehículo en varios intervalos de velocidad.

Viscosity The resistance to flow exhibited by a liquid. A thick oil has greater viscosity than a thin oil.

Viscosidad La resistencia al flujo que manifiesta un líquido. Un aceite espeso tiene una viscosidad mayor que un aceite ligero.

Vortex Path of fluid flow in a torque converter. The vortex may be high, low, or zero depending on the relative speed between the pump and turbine.

Vórtice La vía del flujo de los fluidos en un convertido de torsión. El vórtice puede ser alto, bajo, o cero depende de la velocidad relativa entre la bomba y la turbina.

Vortex flow Recirculating flow between the converter impeller and turbine that causes torque multiplication.

Flujo del vórtice El fluyo recirculante entre el impulsor del convertidor y la turbina que causa la multiplicación de la torsión.

Wet-disc clutch A clutch in which the friction disc (or discs) is operated in a bath of oil.

Embrague de disco flotante Un embrague en el cual el disco (o los discos) de fricción opera en un baño de aceite.

Wheel A disc or spokes with a hub at the center that revolves around an axle, and a rim around the outside for mounting the tire on.

Rueda Un disco o rayo que tiene en su centro un cubo que gira alrededor de un eje, y tiene un rim alrededor de su exterior en la cual se monta el neumático.

Yoke In a universal joint, the drivable torque-and-motion input and output member attached to a shaft or tube.

Yugo En una junta universal, el miembro de la entrada y la salida que transfere a la torsión y al movimiento, que se conecta a una flecha o a un tubo.

INDEX

D

E

F

G

H

I

T

U

V

W